BIOLOGY

Nelson Science

Michael Roberts

Nelson

Thomas Nelson and Sons Ltd
Nelson House Mayfield Road
Walton-on-Thames Surrey
KT12 5PL UK

First published by Thomas Nelson and Sons Ltd 1995

I(T)P® Thomas Nelson is an International Thomson Publishing Company
I(T)P® is used under licence

ISBN 0-17-438677-X
NPN 9 8 7 6 5 4

Printed in China

Acknowledgements

The publishers are grateful to the following for permission to reproduce photographs. While every effort has been made to trace copyright holders, if any acknowledgement has been inadvertently omitted, the publishers will be pleased to make the necessary arrangements at the first opportunity.

A–Z Botanical, 249.5; ACE, 112.6; Aerofilms, 89.5; AFRC Rothamstead, 80.6; Heather Angel, 44.1, 48.1, 56.3, 56.4, 61.3, 62.7, 62.8, 89.4, 169.2, 172.2, 176.1, 178 *bottom right*, 218.1 *top left*, 219.3 *top right*, 241.5, 242.9, 259.4, 268.7, 286.2, 286.3, 295.6 *bottom*; Ardea, 69.3, 166.1, 186.2; Barnaby's, 76.7; BBC Hulton Picture Library, 271; Bensons International Press Service, 285.3; Biblioteca Ambrosiana, 114.2; Biofotos, 249.6; Biophoto Associates, 106.1, 116.5, 138.5, 196.6; Anthony Blake, 25 *bottom*, 93.4; Joe Blossom/Wildfowl and Wetlands Trust, 90.7; Brewers' Society, 127.5; Bridgeman, 273 *top right*, 290.1; Isobel Campbell, 296.7 bottom; J Allan Cash, 4.3 *lower right*, *upper right*, 8.6 *bottom*, *middle*, 10.1, 11.3, 46.6, 56.2, 60.2, 75.6, 76.8, 90.6, 173.5, 187.4 *left*, *right*, 198.1, 242.11, 242.7, 295.6 *middle*; Cellmark Diagnostics, 280 *right*; Chelsea and Westminister Hospital, 216.3; John Urling Clark, 72.3 *left*, *right*, 104.3, 130.1, 136.1, 184.1, 229.7 *upper*, *lower*, 237.2, 249.4, 262.1, 289.1, 289.2, 295.5 *bottom right*, 296.7 top; Bruce Coleman, 10.2, 56.5, 61.4, 73.4, 87, 240.3, 242.1, 256.9, 285.4 *left*, *right*, 291.3, 291.4, 294.1, 295.5 *bottom left*, *top left*, *top right*; Colorific, 282.1; Colorsport, 201.2; Gene Cox 19.1, Neil Croft, 25 *top* ; Cystic Fibrosis Research Trust, 270.1; Danish Dairy Board, 93.3; Ecoscene, 127.4, 177.4, 286.1; Nick Evans, 13.1; Farm Electric Centre, 172.1; Farmers Weekly, 297.1; Flour Advisory Bureau, 88.1; Format, 275.4; M Frazer, Chelsea College, 78.2; Garden Picture Library, 15.4 *left*, *right*, 250.1, 266.2, 18.2, 98.2, 98.3, 178.7; Yvonne Gay, 230.1; Glasshouse Crops Research Centre, 172.3; Sally & Richard Greenhill, 110.1, 255.4, 284.1; GSF, 276 *centre*, 288.6; Robert Harding,4.3 *bottom*, 89.3, 122.1, 146.1, 212.1; Harry Smith Horticulture, 173.4; BJW Heath, 8.7, 64.1, 110.2, 168.1 *upper*, *lower*, 242.12; BJW Heath/MBV Roberts, 24.4; Holt, 53.5, 85.6 *bottom right*, *top right*, 108.6; Hulton Deutsch, 258.1; Image Bank, 162.2; Impact, 239.2; Dr MW Jenison, Syracuse University NY courtesy of Society of American Bacteriologists, 96.1; Frank Lane, 294.3; Levington Horticulture, 79.3, 84.3; London Scientific Fotos, 126.1; GA Maclean, 84.4; Antony Miles, 4.3 *middle*; National Medical Slide Bank, 107.4, 114.1, 131.4, 134.1 *upper*, *lower*, 142.1 *upper*, *lower*, 190.4, 193.1, 203.4 *upper*, *lower*; Natural History Museum, 287.5 *top*; Thomas Nelson & Sons, 12.3 *left and right*, 13.4, 18.1, 25 middle, 54.6, 93.5, 127.3, 127.6, 139.7, 150.9, 155.2, 176.3, 204.7, 232.1, 250.7, 255.6, 255.7, 287.5 bottom; NHPA, 240.1; Oxford Scientific Films, 3.2, 4.3 *top*, 8.6 *top*, 28.2, 28.3, 37.3, 49.3, 51.1, 52.1, 56.1, 59.1, 61.5, 67.1, 67.2 top, *bottom*, 74.1, 81.8, 84.1, 86.7, 88.2, 104.2, 166.2, 173.6, 186.3, 195.5, 202.2, 242.8, 245.4, 295.6 *top*; Photo Deutsches Museum, 27.1 *top right*; Dr R Porter, Harvard University, 13.2; Rex Features, 43.1, 68.2, 119.3 *left*, *right*, 222.1; Chris Ridgers, 12.1, 111.3, 111.4, 208.1, 244.1; RNIB, 207.1; MBV Roberts, 111.5, 115.4, 120.5, 125.1, 146.2, 170.5, 178.8; Royal Botanic Gardens, Kew, 180.1; Royal College of Surgeons, 12.2, 141.1, 215.1, 217.4 ; Smith Kline Beecham, 6.1, 7.4, 102.6; Science Photo Library, 2.1 *bottom right*, *upper right*, *lower left*, *top left*, 5, 7.5, 40.1, 60.1, 72.1, 78.1, 79.4, 83.1, 84.2, 85.6 *bottom left*, *bottom middle*, 92.1, 98.1, 99.1 *bottom*, *lower middle*, *top*, *upper middle*, 101.4, 104.1, 109.1, 141.2 *left*, *right*, 144.7, 148.1, 151.11, 153.1, 156.1, 157.5, 162.1, 180.2, 191.6, 192.1, 198.2, 217.1 *left*, right, 224.4, 224.5, 226.1, 227.5, 228.5, 234.2, 239.3, 241.6, 245.5, 246.6, 248.2, 256.8, 261.2, 269.1, 272.8, 274.1, 274.2, 281.1, 281.2, 295.4; St Bartholomew's Hospital, 159.8; St George's Hospital Medical School, 101.3; St Mary's Hospital Medical School, 95.1, 95.2, 201.3 *left*, *right*, 238.1; Tunnel Refineries, 27.1 ; UNFAO, 80.5; Unilever, 252.1, 252.2; University of Utah Medical Centre, 100.1; Bob Watkins, 234.1; Wellcome Centre for Medical Science, 97.1; World Health Organization, 99.3; Zefa, 101.5

Certain material in this volume is taken from or based on already published sources as follows:

p 19, Picture 1: *The Sunday Times*, 19 May 1968.
p 27 *The Science of Life* , Gordon Rattray Taylor (Thames & Hudson).
p 51 and p289 *Vanishing Birds*, Tim Halliday (Sidgwick and Jackson).
p 65, Picture 3, *Urban Ecology*, David Gilman (MacDonald Educational).
p 72 , picture 3 and p73, questions, data after GF Gause;
p 72 , picture *Fundamentals of Ecology*, EP Odum (3rd edn 1971) (WB Saunders).
p 112 , Picture 2, *Nutrition for Developing Countries*, Maurice King et al. by permission of Oxford University Press.
p 118, Tables 1 and 2. data from the Ministry of Agriculture Fisheries and Food
p 120 , Picture 6, data from the UN Food and Agriculture Organization.
p 125, *Reminiscences and Reflections*, Hans Krebs (1981) by permission of Oxford University Press.
p 142, Pictutre 2 and p144 Picture 8, Professional Dental Services.
p 152, Picture 1, *The Cardiovascular System*, PP Turner (2nd edn 1985), (Churchill Livingstone).
p 154 Picture 1 *Sunday Times Book of Body Maintenance* (Michael Joseph).
p 160, questions, data from FG Hall.
p 163 *Introduction to Nutrition, Exercise and Health* FI and WD McArdle (4th edn, 1993) (Lea & Febiger).
p 212, Picture 2 *Principles of Human Anatomy*, GT Tortora (Canfield Press).
p 226, Picture 2 *Conception, Birth and Contraception*, RJ Demarest and JJ Sciarra (Hodder and Stoughton).
p 227, Picture 3, *Understanding the Human Body*, ER and EMcl Tudor (Pitman) by pernission of Longman Group Ltd.
p 227, Picture 4, *Sex and Fertility*, C Wood (Thames and Hudson).
p 254, Picture 3, Biology: *A modern introduction* BS Becket © Oxford University Press. reprinted by permission of Oxford University Press
p255, Picture 5, *Comparative Morphology of vascular plants*, AS Foster and EM Gifford © 1988 by W.H. Freeman and Company. Used with permission.
p 297, The Times 13 March 1989.

To the reader

This book is about biology. It is about the way living things work, particularly the human, and how our knowledge of living things can be used for the benefit of humankind. One of the most important problems in biology at the moment is how we can get the best out of our environment without damaging it. You will find a lot about that in this book.

You will also find a lot about how biology can be used in industry. This is an important aspect of the subject with an exciting future.

The book is split up into short topics. Each topic includes Activities (with a blue background) – things to do in the laboratory and at home – and Questions (mauve background). There are also case studies and extension exercises (with a green background) which take some of the ideas a bit further.

Michael Roberts

Many people helped to write this book, particularly John Holman, Ken Dobson, Stuart Elford and Dr Tim King. In addition I would like to thank the following for helping me with particular topics.

David Alford, formerly National Association of British and Irish Millers
Janet Alford, Estate Office, Longford Castle
John Barker, King's College, London
Philip Bunyan, ASE Laboratory Safeguards Committee
Bill Butler, Ministry of Agriculture, Fisheries and Food
Dr Jonathan Cooper, Queen Mary's Hospital for Children, Carshalton
Dr Arthur Cruickshank, Open University
Dr P G Debenham, ICI Cellmark
Dr Nicholas Denny, Queen Elizabeth Hospital, King's Lynn
Craig Egner, Sea Fish Industry Authority, Edinburgh
John Finagin, Rhyl High School
Malcolm Hardstaff, Marlborough College
Dr June Hassall, formerly Ministry of Education, Jamaica
Dr Neil Ingram, Clifton College
Roy Johnson, Alcohol Concern, London
Peng Tee Khaw, Moorfields Eye Hospital
Dr Patricia Kohn, University of Sheffield
Dr John Land, Marlborough College
B J R Mack, Unilever Plantation Group
Grace Monger, The Holt School, Wokingham
Pamela Parker, formerly Amersham General Hospital
Dr James Parkyn, formerly Marlborough College
Dr Alan Radford, University of Leeds
The Rev Dr Michael Reiss, University of Cambridge
Dr Steve Smith, St Helier's Hospital, Carshalton
Bruce Tulloh, formerly Marlborough College

Contents

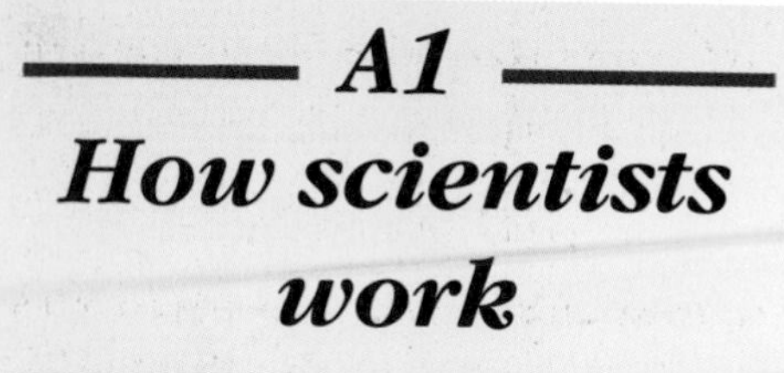

A1 How scientists work

Here we look at what science is and how scientists work.

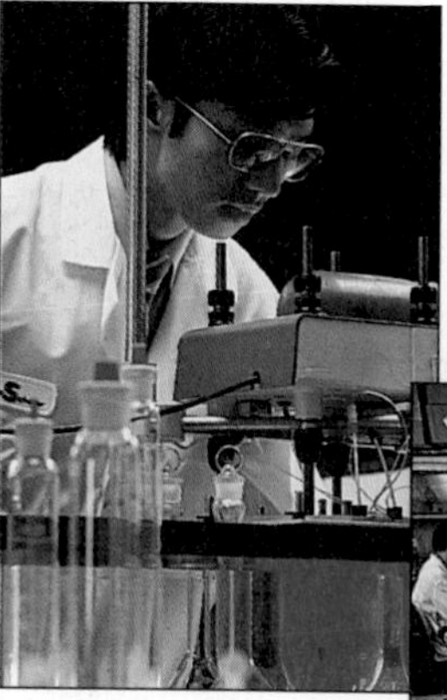

A chemist carrying out quality control tests on a new drug.

A geneticist doing research on human genes.

An astronaut carrying out investigations in space.

A geologist examining a sample of rock.

Picture 1 Scientists at work.

What's happened to my stereo?

Suppose you switch on your stereo and nothing happens. How do you go about finding what is wrong with it?

Well, first you think what the fault might be. Then you carry out a simple test to see if your idea is right. If the result of the test supports your idea, you then set about mending the machine in the right way. On the other hand, if the result of the test suggests that your idea is wrong, you give up that idea and think of another one.

This is exactly what scientists do when they tackle a problem. People call it the **scientific method.** The name makes it sound rather grand, but it is really what people usually do when wanting to find something out. For example, the same procedure is used by a doctor who wants to find what's wrong with you, a garage mechanic who has to mend your car, and a police detective who is trying to solve a crime.

So really the scientific method is just common sense. With that in mind, let's look at it in detail.

The scientific method

The scientific method starts off with an **observation**: you notice something interesting – it may be an object of some sort, or it may be something happening. You then ask questions about it: what is it, what is it doing, why is it there?

The next step is to think of a possible explanation for it. We call this an **hypothesis**. From the hypothesis, various **predictions** can be made. A prediction is simply a consequence which follows logically from the hypothesis.

You then test the predictions to find out if they are true. This is usually done by carrying out **experiments**. If your predictions turn out to be right, the hypothesis may well be true.

If your hypothesis appears to be true, you can make further predictions from it. These, too, can be tested by experiments. In this way, new discoveries are made, and new problems crop up.

Picture 2 describes an investigation in terms of the scientific method. Now let's look at the main steps in the scientific method more closely.

The main steps in the scientific method

Observing things

The natural world is full of interesting things. It may be the way clouds form, or the way substances react when mixed together, or the way animals move. Scientists are always on the look-out for interesting things to ask questions about.

To be useful, scientific observations must be accurate and carefully recorded. This is one reason why you will need to keep a notebook in your science course.

Making hypotheses

This involves thinking up explanations, and that means using your imagination. However, you must not let your imagination run away with you, otherwise your hypotheses may be just science fiction!

To be of any scientific use, a hypothesis must be testable. This means that you must be able to make predictions from it which can be tested by experiments.

Doing experiments

In picture 2 an experiment is done to find out if plants grow towards light. A potted plant is lit from one side, and you see if it grows towards the light.

However, there is something more that has to be done; we must obtain a second plant and light it from above. We need this second plant to provide a standard with which to compare the first plant. The second plant in the experiment is called the **control**.

It is essential that the two plants are kept in exactly the same conditions, except for the light they receive. To put it in a general way, *we must keep all the variables constant except the one whose effect we are investigating*.

An experiment of this kind, in which the experimenter controls the conditions, is called a **controlled experiment**.

A control is not needed for all experiments. For example, suppose you do an experiment to find out if heavy objects fall faster than light ones. You don't need a control for this experiment. (Can you see why?) However, you do need to keep all variables the same when making your comparison. These variables include the shape and density of the objects and the height from which you drop them. The two objects should differ only in weight.

Drawing conclusions

Ideally, an experiment should be so well planned that only one conclusion can be drawn from the results. However, this is not always possible. Often two or more different conclusions may be drawn, though usually one is more likely than the others.

When you look at the results of your own experiments, it is a great temptation to draw only the conclusion that you want to draw. This is usually the one that supports your hypothesis. Always think of other possible conclusions, however unlikely they may seem to be.

Look again at picture 2. The conclusion from the experiment is that plants grow towards light. But at least two other conclusions are possible. Can you think what they are?

Repeating experiments

When you do an experiment at school, you probably do it only once. There is not usually time to do it more than once, even if you want to. But real science is not like that. In a proper scientific investigation, you do the experiment over and over again until consistent results are given every time. Only then can you be sure that you are doing the experiment correctly, so that valid conclusions may be drawn.

Even that is not the end of it. In real science experiments are repeated by other scientists, and they too should get the same results.

What are variables?

In any experiment you are faced with **variables**, that is, things which can change or vary. Suppose you want to find the effect of temperature on the rate of a chemical reaction. In this case the relevant variables are temperature and reaction rate. The reaction rate is called the **dependent variable** because it depends on other things (including temperature). Temperature, however, is an **independent variable** – so are other conditions such as pressure and light intensity.

In an experiment we alter (i.e. *manipulate*) one of the independent variables (in this case the temperature) and see what effect this has on the dependent variable (in this case the reaction rate). All other independent variables must be kept constant.

1 Which are the dependent and independent variables in the investigation summarised in picture 2? Which variable is manipulated, which one is measured and which ones should be kept constant?

2 Can you think of examples where temperature is the dependent variable?

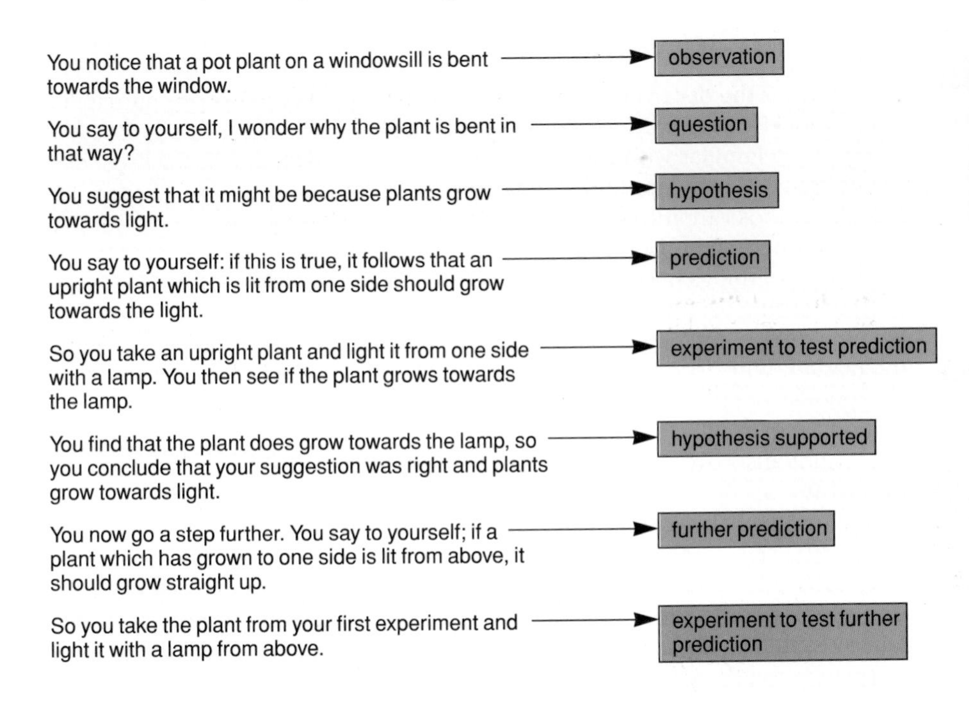

Picture 2 An investigation presented in terms of the scientific method.

Picture 3 Technology over the centuries. Top to bottom: Cheops pyramid and temple in Egypt; a Roman aqueduct in Spain; Salisbury Cathedral, the tallest spire in England; Clifton suspension bridge in Bristol; Mariner IV rocket.

Playing fair

We can sum up what we have said so far like this: when you test a hypothesis, you must carry out the experiment in such a way that it really does test the hypothesis. This means that you must:

- include any necessary controls,
- keep all independent variables constant except the one which you are investigating,
- repeat the experiment until you get similar results every time.

If you test your hypothesis in this way, then it is a **fair test.** If, however, your test falls short in any way, then it is an unfair test. Unfair tests are useless in science, and the results can be very misleading.

Proving things

In picture 2 the plant that was lit from one side bent over towards the light. Does this *prove* that plants grow towards light? Surely the answer is no. The results of this experiment support the hypothesis, but they do not prove it. They provide **evidence**, not proof. To prove the hypothesis you would have to do some other experiments. What should they be?

Rarely is a hypothesis proved beyond all doubt. However, more and more evidence may be obtained to support it. When enough evidence has been found, we can call the hypothesis a **theory**. A theory is like a hypothesis, but much more certain.

Generalisations

The experiment in picture 2 has not been done on every species of plant in the world. However, it has been done on enough species for us to feel certain that plants in general grow towards light. A statement of this sort is called a **generalisation**.

Generalisations are useful because they help us to cut through the detail of the subject and see patterns. They also enable us to make predictions about things we do not yet know. For example, think of the generalisation that plants grow towards light. From this we can predict that plants have some way of sensing where the light is coming from. Having made this prediction we can test it by doing appropriate experiments.

Making use of science

For centuries the discoveries of science have been used to improve human life. Using science in this way is called **technology**.

We are all familiar with the way physics and chemistry affect our lives: electricity, transport, computers, medicines, dyes and so on. Your physics and chemistry books are full of examples. Some of the ways biology affects our lives are mentioned in the next topic.

All sorts of things have to be done before a scientific discovery can be turned into something useful. Take penicillin, for example (see page 95). Sir Alexander Fleming discovered penicillin in 1929, but it was ten years before it was put on the market. Why do you think there was such a long delay? In Britain we have a very good reputation for making scientific discoveries, but not such a good reputation for marketing them.

Technologists are concerned with making things work. Many of them are **inventors**, applying the discoveries of science to making things that people want. How many inventions can you think of which directly affect your life?

As with science itself, an invention often starts with an observation. James Watt watched a kettle boiling, and this led to his inventing the steam engine. But it took people with other talents to turn Watt's invention into a full-scale railway system with stations, tunnels, bridges, signals and so on. This is the world of technology for which science is only the beginning.

Activities

A Observing things and asking questions

On your way home from school, or the next time you go for a walk, look for interesting scientific things. They may be objects such as animals, plants or rocks, or they may be events such as birds feeding or leaves falling off a tree.

Make a list of ten such things. In each case write down one question relating to the observation, and say how it might be investigated further.

B Testing a hypothesis

Your teacher will give you some Plasticine, a tall measuring cylinder and a stopwatch. Use them to test the following hypothesis: a streamlined object moves through water faster than a non-streamlined object.

Write an account of your experiment, explaining your method and describing your results.

What predictions can you make from your results about the shapes of animals that live in water? In what way might the results be useful in industry?

What design problems did you encounter in your experiment, and how did you overcome them?

Questions

1 Someone has suggested that cows yield more milk if you play music to them. How could you test this hypothesis?

2 There is a large tree growing in the middle of my lawn. I have noticed that there are lots of daisies in the lawn, but none under the tree.

a Suggest two hypotheses to explain why there are no daisies under the tree.
b Choose one of your hypotheses and make a prediction from it which can be tested by an experiment.
c Describe an experiment which could be done to find out if the prediction is correct.
d Assuming that your hypothesis turns out to be true, make a further prediction which could be tested by an experiment.

3 When carrying out an experiment you must keep all variables constant except the one you are investigating.

a Explain the meaning of the word variable. Give an example so as to make the meaning clear.
b Why is it necessary to keep all variables constant except the one you are investigating?

4 The experiment in picture 2 on page 3 is not, as described, a fair test of the hypothesis. In what way or ways is it not a fair test? Rewrite the experiment in your own words so that it is a fair test.

5 The picture shows a body louse which feeds by sucking human blood through the skin.

a Write down *two* observations from the picture.
b Ask a question about *one* of the observations and put forward an hypothesis to answer it.
c How would you test the hypothesis?

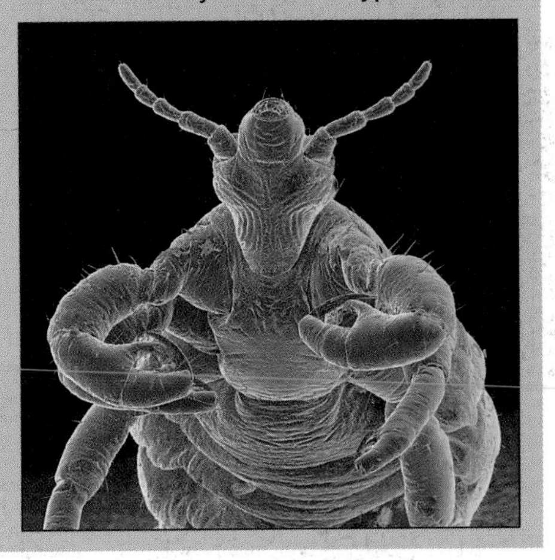

Thinking scientifically

In studying science one of the things you have to do is to learn to think scientifically. What does this mean? Well, it can best be understood by knowing the difference between a scientific statement and an unscientific statement.

A scientific statement is one which can be disproved by investigation. An example of a scientific statement is 'potatoes contain starch'. We can test this statement by carrying out a simple chemical test on a potato.

Here is an example of an unscientific statement: 'Britain is better under the Conservatives.' This is an unscientific statement because it cannot be tested by investigation and therefore cannot be disproved. It's a statement not of science but of opinion.

This exercise is about unscientific statements which manufacturers sometimes make about their products. Here is an example. It is written on the side of bottles containing a certain type of bleach which is used as a powerful disinfectant in the home:

"THE FRESH SMELL TELLS YOU IT KEEPS ON KILLING GERMS"

1 Explain why it would be impossible to test this statement scientifically.

Suggest a *scientific* statement which the manufacturer might make about this particular product.

2 Visit a supermarket and look at the labels on the various items. Find five statements which you think are scientific and five which are unscientific.

Choose one of the unscientific statements and write a letter to the manufacturer criticising this kind of advertising. (There's no need to post it!)

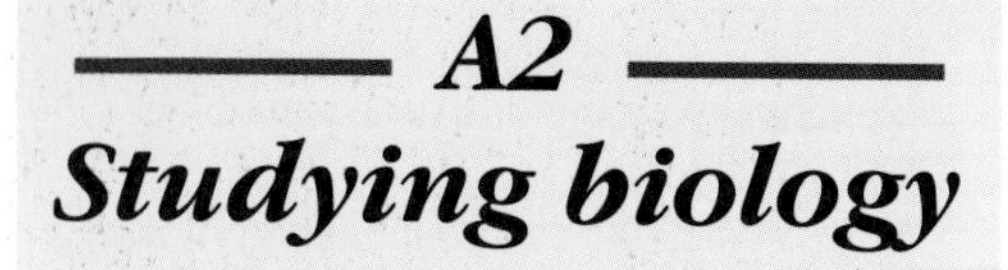

A2 Studying biology

In this topic you will find out what biology is and why we study it.

Picture 1 A biologist at work in the laboratory.

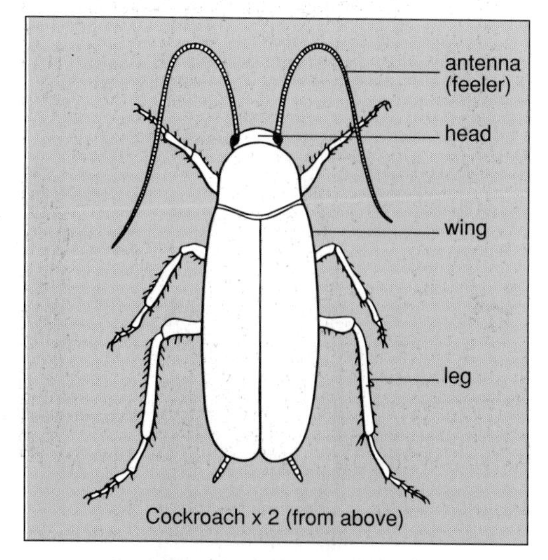

Picture 2 A biological drawing should be realistic but simple. There is no need to shade it.

What is biology?

Biology is the study of living things. Living things are called **organisms**. Organisms include animals and plants. Some organisms are so small that they cannot be seen with the unaided eye; we call them **microorganisms**, or **microbes** for short. Microbes are very important to us as we shall see later; studying them is therefore an important part of biology.

The different branches of biology

Zoology:	the study of animals
Botany:	the study of plants
Human biology:	the study of humans
Microbiology:	the study of microorganisms (microbes)
Anatomy:	the study of the structure of living things
Physiology:	the study of how the body works
Nutrition:	the study of food and how living things feed
Heredity (genetics):	the study of how characteristics are passed from parents to offspring
Ecology:	the study of where organisms live.

Biology as a science

Biologists investigate organisms and make discoveries about them. Biology is a scientific subject, just like chemistry and physics. We can therefore study organisms using the methods of science, as described in the last topic. This involves making careful observations and carrying out experiments.

Drawings

As you study biology you will find that you need to make a record of your observations. Often the best way of doing this is to make a **drawing**. The aim is to produce an accurate record of what you see, so the drawing must look like the real thing. Of course, if you are drawing a fly you need not put every hair in the right place. However, the general impression must be as realistic as possible.

A useful biological drawing is shown in picture 2. It was done by a pupil in a school. Notice that she has given her drawing a title and has labelled the various parts.

When you make a drawing always write down how many times larger, or smaller, your drawing is than the real thing. This is called the **scale**. If your drawing is twice as large as the specimen, the scale is ×2. If you draw it life-size, the scale is ×1. And if you draw it half the natural size, the scale is ×0.5.

Diagrams

Look at the map of the London Underground in picture 3. It doesn't look much like a railway does it? In reality the railway lines are not straight, as in the map, but wander about all over the place.

The designer of the map has straightened out the lines and simplified them. He has got rid of all unimportant detail and has concentrated on the thing that really matters, namely the relative positions of the stations. He has also used colour to distinguish between the different lines, so the overall pattern is easy to see. What he has done is to produce a **diagram**, rather than a drawing, of the London Underground.

We often make diagrams in biology. A biological diagram shows how things relate to each other, not what they actually look like. Look, for example, at the diagram of the human blood system on page 148. It shows the general plan of the circulation, but not the individual blood vessels and where each one goes. Note that arrows have been put in to show the direction in which the blood flows, and colour has been used to distinguish between the different types of blood. As with drawings, a diagram must always have a title and be fully

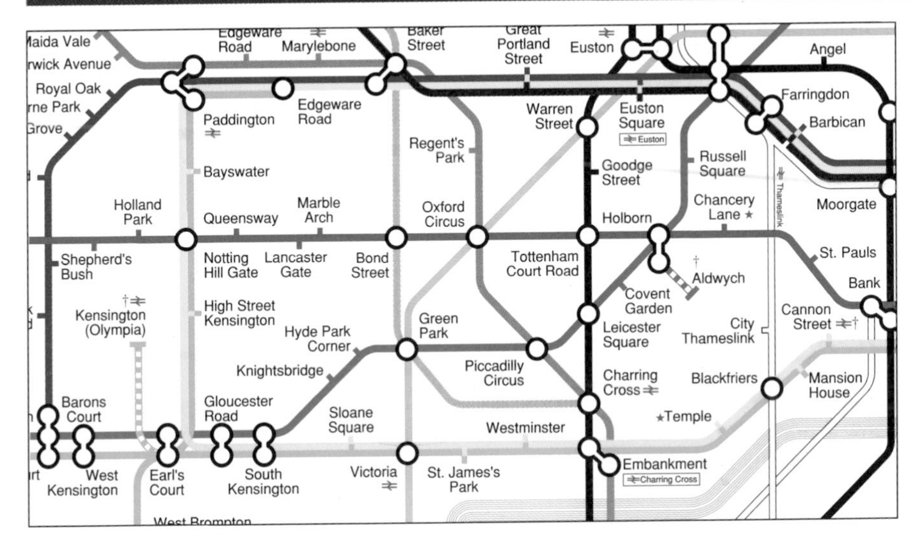

Picture 3 The central part of the London Underground.

labelled. Think how useless the London Underground map would be if the stations had no names!

Picture 4 This vet uses her knowledge of biology to care for the health of animals.

Jobs involving biology

Not all biologists spend their time doing experiments and making drawings! Some apply their knowledge of biology to other things. Doctors, nurses, physiotherapists, vets, farmers and foresters are all biologists in different ways. They all study biology in their training, and biology comes into everything they do (picture 4).

Can you think of any other jobs which involve biology?

Biology and industry

For thousands of years humans have used yeast, a tiny microorganism, for making bread and alcoholic drinks such as beer and wine (see page 126). These processes are now major industries. They are examples of how biology can be applied to making commercially useful products.

Today microorganisms are used in many other production processes. Bacteria and fungi are particularly useful in this respect. In some cases the organism's genes are first changed to make it produce the thing we want. This is called **genetic engineering**. Microorganisms reproduce very quickly, so vast numbers of them can be used for mass production in large-scale industrial plants (picture 5).

The application of the science of biology to the manufacturing industry is known as **biotechnology**. Large amounts of money have been invested in this new type of industry.

Picture 5 Inside a biotechnology plant. The technician is adjusting a fermentation unit in which millions of microorganisms produce a particular type of protein for human use.

Biology and the environment

Our environment is threatened by pollution – not just our own local surroundings but the environment of the whole world.

The most important part of the environment which is under threat is the atmosphere. Biology comes into this because the gases in our atmosphere are controlled by the natural activities of living organisms, notably respiration and photosynthesis.

Over the last two centuries humans have changed the composition of the atmosphere. This in turn has brought about other changes, for example an increase in the world's temperature. If nothing is done about this, life itself could be threatened.

Picture 6 Fishing, gardening and bird-watching are all hobbies involving biology.

Safeguarding our environment is one of the greatest challenges facing scientists today. We shall return to this later.

Biology and everyday life

Biology comes into our lives all the time. The food we eat, our health and our environment are just three ways in which biology affects us personally.

Many hobbies and outdoor pursuits also involve biology. Gardening, fishing and bird-watching are three examples (picture 6). Can you think of others?

Looking at small things

In biology we often have to look at small organisms or parts of organisms. Sometimes a **hand lens** can help us. A typical hand lens has a magnifying power of ×10. This means that we see the object ten times larger than it really is.

What do we do if the object is too small to be seen with a hand lens? The answer is that we use a **light microscope** (pictures 7 and 8). If you have not used a microscope before, you should do activity A very carefully. When looking at an object under the microscope it is important to appreciate how much it has been magnified. Activity B will help you with this.

If you were to measure the width of a coin you would probably express it in millimetres. However, for small objects seen down the microscope we use a smaller unit called the **micrometre** (μm). A micrometre is one thousandth of a millimetre.

What about objects that are too small to be seen even with a light microscope? To see these we must use an **electron microscope**. This is explained on page 13.

Sometimes scientists want to look at the microscopic structure of an organ such as the liver or a plant stem. This is done by cutting thin slices (called **sections**) of the organ and examining them under a microscope. Usually the sections are first stained to show up the different parts.

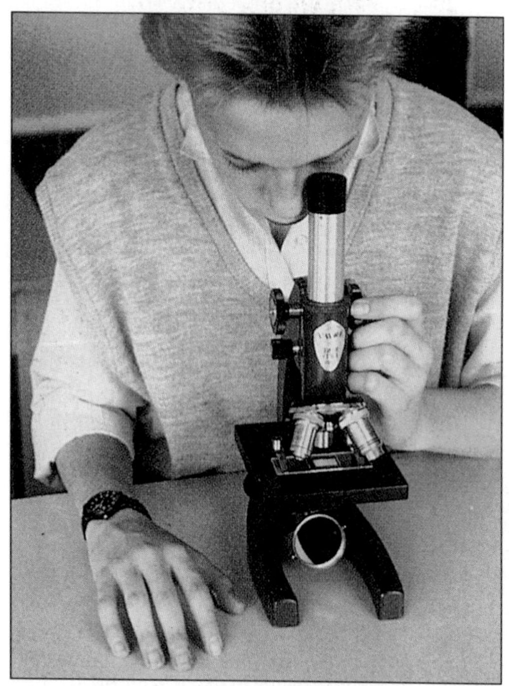

Picture 7 This kind of light microscope is used in many schools and colleges.

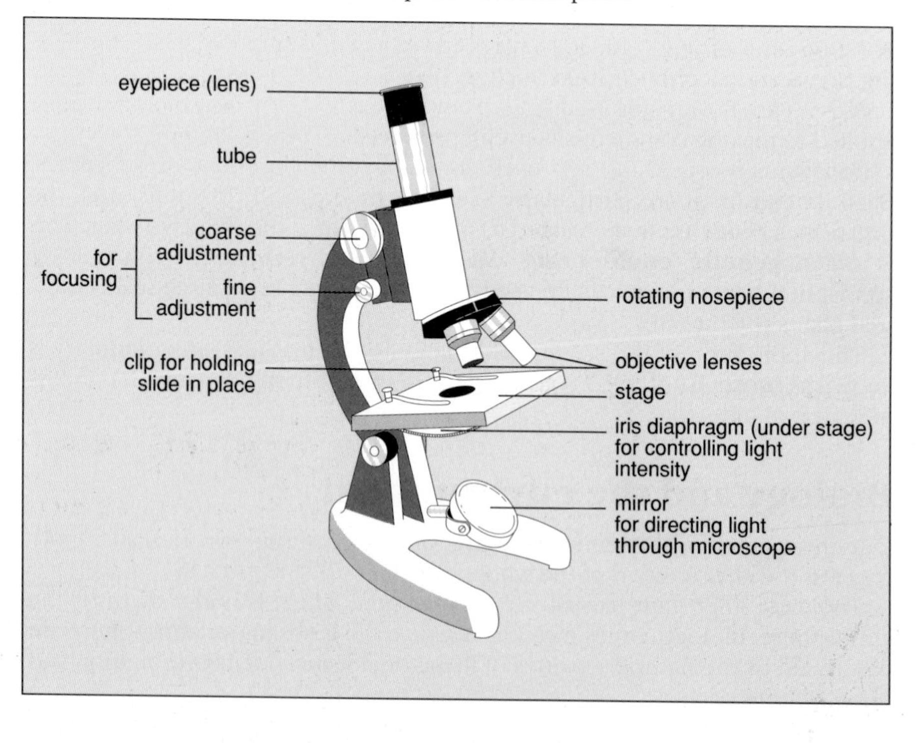

Picture 8 The main parts of a typical light microscope. It is known as a compound microscope because it consists of two lenses in series, the eyepiece lens at the top and the objective lens lower down.

Activities

A Learning to use the microscope

1 Study your microscope carefully and compare it with picture 8. Yours may be slightly different. Make sure you understand it before you use it.

2 Objects to be viewed under the microscope are first placed on a glass slide and covered with a thin piece of glass called a coverslip. Your teacher will give you a specimen which has been mounted in this way.

3 Place the slide on the stage of your microscope: arrange it so the specimen is in the centre of the hole in the stage.

4 Fix the slide in place with the two clips.

5 Rotate the nosepiece so the small objective lens is immediately above the specimen: the nosepiece should click into position.

6 Place a lamp in front of the microscope, and set the angle of the mirror so the light is directed up through the microscope.

7 Look down the microscope through the eyepiece. You will see a white or grey circular area. This is called the field of view. Adjust the iris diaphragm so the field of view is bright but not dazzling.

8 Look at the microscope from the side. Turn the coarse adjustment knob in the direction of the arrow in picture 8. This will make the tube move downwards.

9 Continue turning the knob until the tip of the objective lens is close to the slide.

10 Now look down the microscope again. Slowly turn the coarse adjustment knob in the other direction, so the tube gradually moves upwards. The specimen on the slide should eventually come into view.

11 Use the coarse and fine adjustment knobs to focus the object as sharply as possible.

12 If necessary readjust the iris diaphragm so the specimen is correctly illuminated. You will get a much better picture if you don't have too much light coming through the microscope.

You are now looking at the specimen under low power, i.e. at low magnification. To look at it under high power, i.e. at a greater magnification, proceed as follows:

13 Rotate the nosepiece so the large objective lens is immediately above the specimen. The nosepiece should click into position, as before.

14 If the specimen is not in focus, focus it with the fine adjustment knob. Be careful that the tip of the objective lens does not touch the slide.

15 Readjust the illumination if necessary.

You are now looking at the specimen under high power. Do you agree that it is now much more enlarged?

Remember: always treat the microscope with the greatest care: it is an expensive precision instrument. Always carry it with both hands, and keep it covered when you are not using it. Do not rest it on books or near the edge of the bench. Make sure the lenses never get scratched or damaged: if they need cleaning tell your teacher.

B Using a microscope to investigate a feather

Your teacher will give you a feather of a bird such as a pigeon.

1 With scissors, cut out a small piece of the feather approximately 5 mm square.

2 With a spatula, place the piece of feather on a microscope slide.

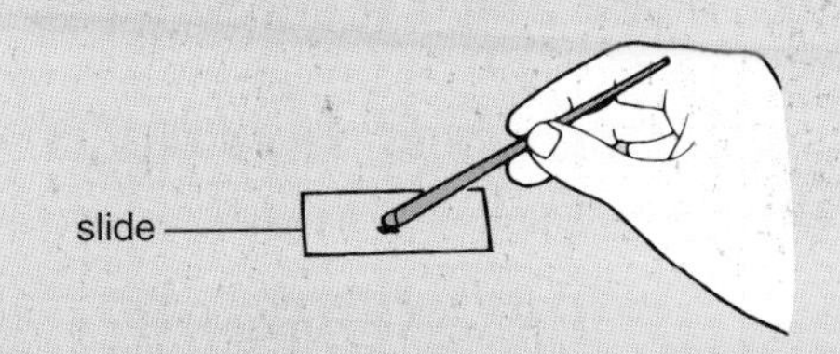

3 Add a drop of clove oil or olive oil from a pipette.

4 Cover the piece of feather with a coverslip. Lower it carefully onto the slide. The oil will spread out beneath it.

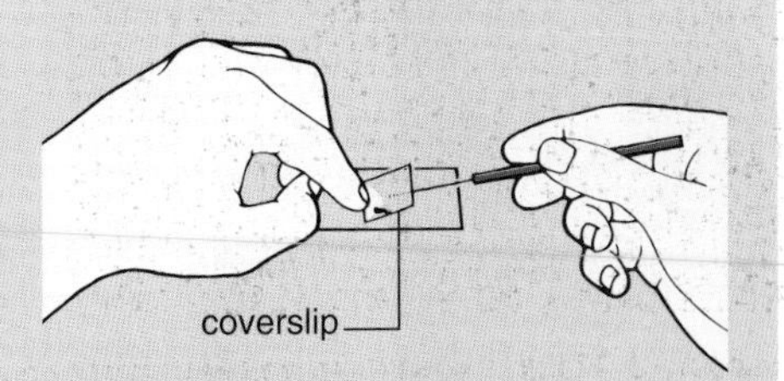

5 Examine the feather under the microscope. Make at least one drawing to illustrate its detailed structure.

6 Give your drawing a scale. To do this place a transparent ruler on the stage of your microscope and work out the diameter of the field of view in millimetres.

Questions

1 Which branch or branches of biology listed at the beginning of this topic must each of the following people know about:

a farmer, a gardener, a nurse, a family doctor, a game warden, a person who breeds dogs, a PE teacher, a forester, a surgeon, a keeper in a zoo?

2 Suppose you use a microscope with one eyepiece lens and two objective lenses. The magnifying power of the eyepiece lens is ×10, and the magnifying powers of the two objective lenses are ×10 and ×40.

What are the low and high power magnifications of your microscope?

3 A certain specimen is 0.5 mm long. What is its length in micrometres?

Suppose you drew the specimen and gave it a length of 2 cm. What would be the scale of your drawing?

Why is it necessary for biologists to give their drawings a scale?

4 Humans make use of other organisms in all sorts of ways. In some cases this involves the organism being exploited in some way.

a What does exploited mean?
b Give examples of ways in which organisms are exploited.
c Is it right to save a seal but swat a fly?

This question raises important moral issues which every biologist should think about. You might like to discuss it with friends, parents and grandparents.

Have attitudes towards the use of organisms changed over the years?

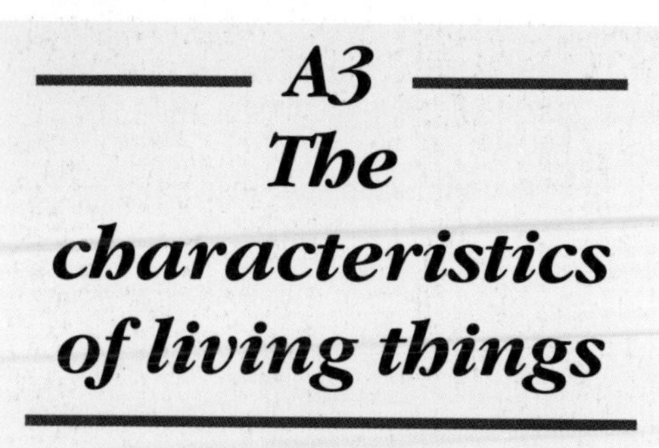

A3 The characteristics of living things

Certain features are common to all living things.

Living things move

Movement is obvious in the case of an animal like the human. We move our arms and legs by means of **muscles**. Most organisms can move by one means or another, at least at some period in their life cycle.

Movement is usually less obvious in plants. To see movement in a plant you must look inside it, under a microscope (activity A).

Picture 1 The gymnast is showing one of the basic features of life, movement.

Living things are sensitive

If you sit on a drawing pin, you jump up quickly. You are **sensitive** to the sharp point. The pricking of your bottom is called the **stimulus** (plural: **stimuli**). Your jumping is called the **response**.

Living things respond to different kinds of stimuli. The main ones are touch, chemicals, heat, light and sound.

At first sight you might think that plants are an exception to the rule that all organisms respond to stimuli. After all, if you hit a tree, it does not move away. However, plants *do* respond to certain stimuli, but much more slowly than animals. They do not have muscles. Instead they respond by *growing* in a particular direction. For example, most plants grow towards light.

An interesting case of a plant responding quickly to touch is investigated in activity B.

Living things grow

As an animal or plant develops, it gets larger and heavier. In other words, it **grows**. In this process its volume and mass increase.

Growth takes place by substances being taken into the organism from outside. These substances are then built up into the structures of the body: they become part of the organism. This is called **assimilation**.

Living things feed

We have just seen that, in order to grow, an organism must take substances into its body. This is achieved by **feeding (nutrition)**.

Animals and plants feed in quite different ways. Animals feed on complex organic substances (**heterotrophic nutrition**). These substances are often solid and have to be broken down into a soluble form: this process is called **digestion**.

In contrast to animals, plants make their own food (**autotrophic nutrition**). They take in simple things like carbon dioxide and water and build them up into complex organic substances. Energy is needed for this: it comes from sunlight. The green pigment **chlorophyll** enables the plant to use sunlight in this way: this is why plants are usually green. The process by which plants make food is called **photosynthesis**.

Picture 2 All organisms feed. This lizard, known as a bearded dragon, is eating a grasshopper.

Living things transfer energy

Living things need energy to move, grow, replace worn-out structures, and so on. They obtain this energy from their food. The food is broken down into carbon dioxide and water, and energy is transferred. This process is known as **respiration**.

Respiration normally requires oxygen. Organisms get this vital gas from the air or water around them.

Living things get rid of poisonous waste

In many ways an organism is like a chemical factory. Substances are constantly being broken down to release energy, or built up to make things.

Some of the by-products of these chemical reactions are poisonous. They must not be allowed to accumulate inside the organisms or they will kill it. So the body must get rid of them. This is called **excretion**.

Living things produce offspring

Organisms produce offspring. This is known as **reproduction**. Usually it involves the union of two individuals, a male and a female. This of course is **sexual reproduction**.

Some organisms can reproduce on their own without the help of another individual. This is called **asexual reproduction**. At its simplest, the organism merely splits in two. In good conditions asexual reproduction may take place very quickly and sometimes a very large number of offspring are produced.

When organisms reproduce, instructions in the form of **genes** get passed from the parents to the offspring. Genes are made of a substance called **DNA**. This stands for **deoxyribonucleic acid**. DNA belongs to a group of substances called **nucleic acids**.

DNA occurs in the cells of all organisms from bacteria to humans. It gives every individual organism its unique characteristics and has been aptly described as the molecule of life.

Picture 3 Reproduction is one of the basic features of living things. Here, a labrador bitch is suckling her young.

Activities

A Detecting movement in a plant

Movement is difficult to see in most plants, but here is an exception.

1 Obtain a sprig of the water plant Canadian pondweed (*Elodea*) which has been kept in the light for several hours.

2 Cut off one of the leaves and put it in a drop of water on a microscope slide.

3 Cover the leaf with a coverslip.

4 Look at the leaf under the low power of your microscope.
Can you see lots of small green objects inside the leaf? These are called chloroplasts and they contain chlorophyll. If they are moving, describe their movement as fully as you can. What use do you think these movements are to the plant?

B Observing a plant responding to touch

Few plants respond quickly when you touch them, but certain sensitive plants do, for example *Mimosa pudica*.

1 Obtain a potted specimen of a sensitive plant.

2 Gently touch the top side of a leaf with a needle. What happens?

3 Gently touch other parts of the plant, including the lower side of the leaves, and the stem.

Describe what happens in each case.

4 Pipette a drop of water onto one of the leaves. What happens?

What use do you think this response might be to the plant? How do you think the response might be brought about? How would a similar response be brought about in an animal?

Can you think of any other plants that respond to touch? Why is it useful to them to respond to touch?

C Recognising the characteristics of living organisms

1 Make a list of the characteristics of living things given in this topic, starting with 'living things move' and finishing with 'living things produce offspring'.

2 Examine various organisms or pictures of organisms, provided by your teacher.

They might include the following: an earthworm, a locust, a frog or toad, a green plant, a clam, a snail, a mould, a lichen, yeast, and yourself.

3 For each organism write down the particular characteristics which you can *see* it possesses. Do not write down the characteristic unless you can actually see it.

After doing this, you will realise that some of the characteristics of life are difficult to see in organisms.

How could you find out if an organism possesses a characteristic of life which you cannot actually see?

Questions

1 Of all the characteristics of living things mentioned in this topic, which ones are most important in each of the following?

The number of characteristics which you should mention in each case is given in brackets.

a A person watching television (1),
b a footballer kicking a ball (2),
c a lion stalking a zebra (2),
d germs spreading through your body when you are ill (1),
e a plant bending towards the light (1),
f a person panting after a race (1),
g a bean plant climbing up a bamboo cane (2).

2 If you blow up a balloon and then hold it in front of the fire, it increases in size.

Is the balloon growing in a biological sense? Give reasons for your answer.

3 A visitor to our planet from outer space thinks motor cars are alive.

In order to straighten the matter out for our extra-terrestrial friend, make a list of ways in which a motor car is similar to living organisms, and a list of ways in which it is different.

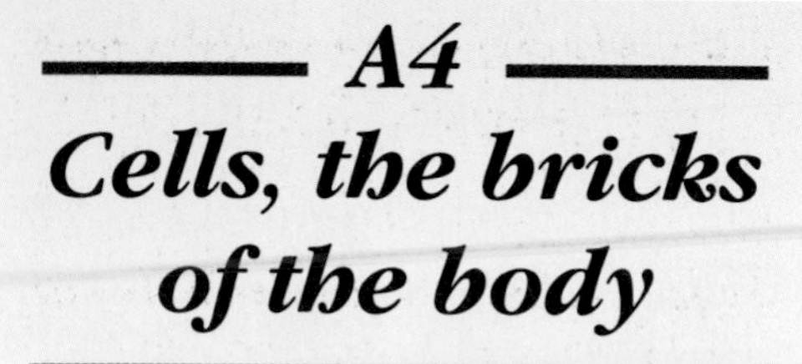

A4
Cells, the bricks of the body

An organism is made of cells in the same kind of way that a house is made of bricks.

Picture 1 Animals and plants are made of cells in the same kind of way that a house is made of bricks.

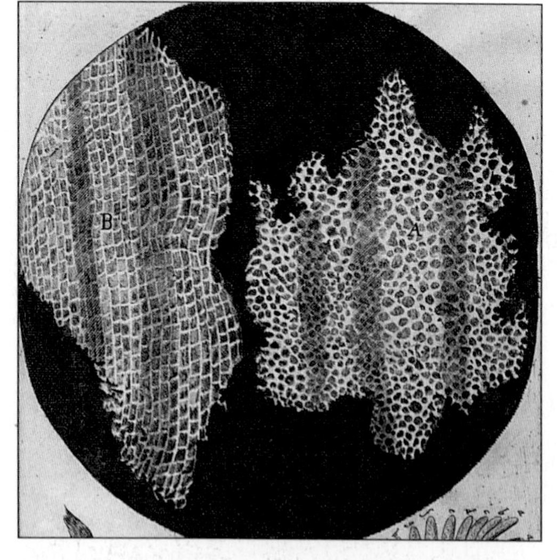

Picture 2 The first drawing of cells ever made. These cells were observed in a piece of cork by Robert Hooke. This drawing was published in Hooke's famous *Micrographia* in 1665.

How were cells discovered?

Cells were discovered in 1665 by the English inventor and scientist, Robert Hooke. Hooke examined a piece of bark which he stripped from a tree. Near the surface of bark is a layer of cork. Hooke cut a thin slice of the cork and placed it under a microscope which he had made himself. Hooke described the cork as being made up of hundreds of little boxes, giving a kind of honeycomb appearance (picture 2). He called these little boxes **cells**.

As more and more organisms were examined under the microscope, it became clear to scientists that virtually all living things are made of cells. And so cells came to be regarded as the basic unit of which organisms are made.

How can we see cells?

The human body consists of about one hundred million million cells, and each one is very small. Because they are so small, we usually give their size in micrometres (see page 8). A typical cell is about 20 μm wide.

Objects this size are too small to be seen with the naked eye, or even with a magnifying glass. To see them you must use a **light microscope** such as the one shown on page 8.

Even greater detail can be seen by looking at cells with the much more powerful **electron microscope**. This is explained on the opposite page. However, here we shall concentrate on what you can see with a good light microscope.

Inside a typical animal cell

Picture 3 shows the structure of a typical animal cell. The cell is bounded by a thin **cell surface membrane**. In the centre is a tiny ball, the **nucleus**. This is surrounded by a material called the **cytoplasm**.

The cell surface membrane

The cell surface membrane is very delicate. It holds the cell together and plays an important part in controlling what passes into and out of it. We shall come back to this in a moment.

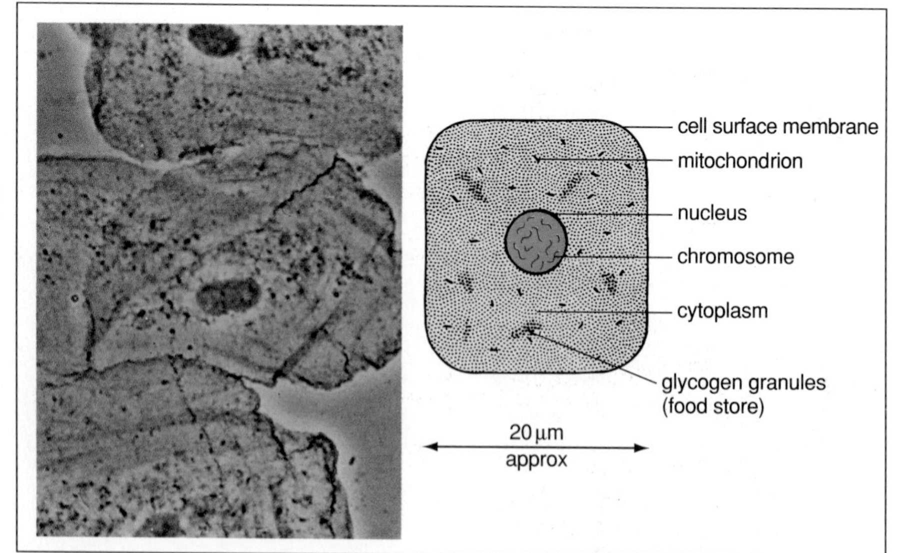

Picture 3 This diagram shows a typical animal cell. On the left are some cheek cells as they actually appear under the light microscope.

The nucleus

It is possible to remove the nucleus from certain cells by sucking it out with a very fine pipette. If this is done, the cell dies. From this experiment we conclude that the nucleus is essential for the life of the cell. It controls the various processes which go on inside it.

The nucleus contains a number of thread-like bodies called **chromosomes**. However, these can only be seen clearly when the cell is about to divide in two.

The chromosomes contain **genes** which tell the organism what sort of features to develop (see page 11).

The cytoplasm

The cytoplasm transfers energy, makes things, and stores food. Hundreds of chemical reactions take place inside it. Together, these reactions make up **metabolism** (see page 23). Scattered about in the cytoplasm are what look like little dots. The larger ones are **mitochondria** (singular: **mitochondrion**). The mitochondria have been described as the 'power-house of the cell': their job is to supply energy.

The smaller dots are tiny particles of stored food. Many of them consist of a substance called **glycogen**.

Inside a typical plant cell

A typical plant cell is shown in picture 4. It differs from animal cells in the following ways:

- In addition to the cell surface membrane, the plant cell has a **cell wall** . It is made of **cellulose**, a tough rubbery material.
- In the centre of the cell there is a large cavity called the **vacuole**, which is filled with a watery fluid called **cell sap**. This means that the cytoplasm is pushed towards the edge of the cell. The nucleus is usually found in this layer of cytoplasm. However, in some plant cells the nucleus is suspended in the middle of the vacuole by fine strands of cytoplasm.
- The cytoplasm contains **starch grains**. This is how plants store food. The starch grains are equivalent to the glycogen granules in animal cells.
- Many plant cells possess **chloroplasts**. These are located in the cytoplasm, and they contain the green pigment **chlorophyll** which is used in **photosynthesis**. Chloroplasts only occur in the green parts of the plant which are exposed to the light. Roots and other underground structures lack them.

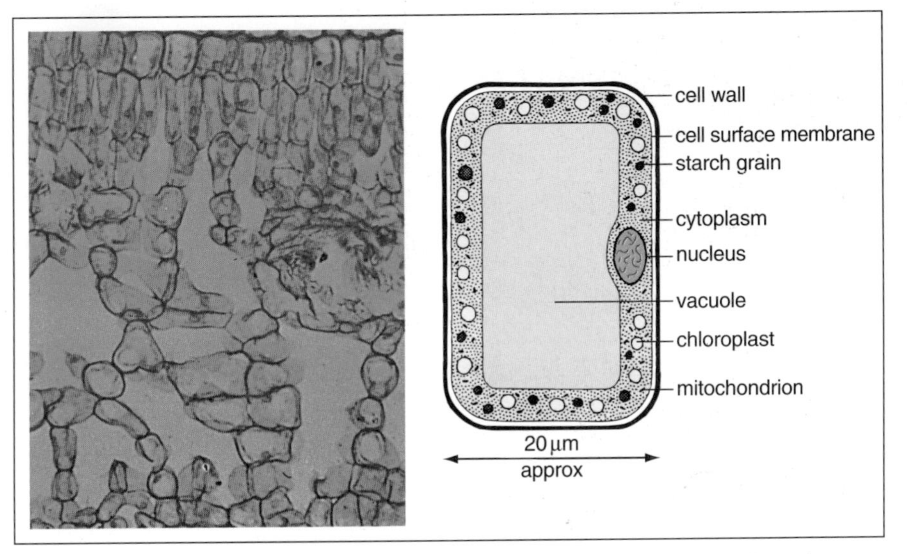

Picture 4 This diagram shows a typical plant cell. On the left are some leaf cells as they appear under the light microscope.

A new look at the cell

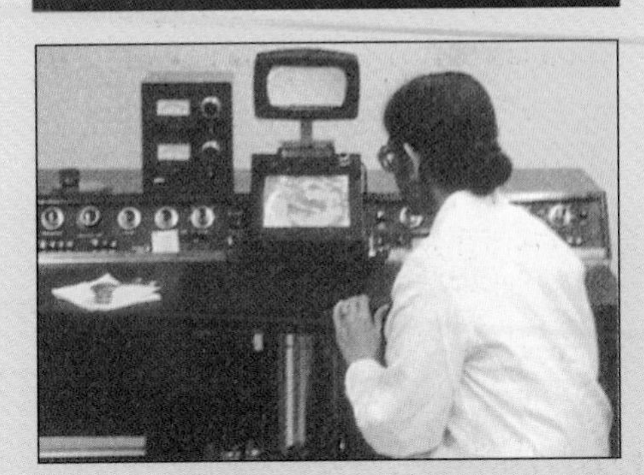

Picture 1 A scientist viewing a specimen in an electron microscope. The material is carefully prepared beforehand and placed in a vacuum. The image shows up on the fluorescent screen in the centre.

In the late 1930s a new kind of microscope was invented: the **electron microscope.** You can see one being used in the picture above. It uses a beam of electrons instead of light rays, and the image shows up on the fluorescent screen in the centre. It is much more powerful than the light microscope and can magnify things as much as *half a million times.* Enlarged to this extent, a pinhead would cover ten football pitches!

Below is part of an animal cell as it appears in the electron microscope. Notice the detail compared with picture 3. For example, you can see that the cytoplasm contains lots of membranes and channels. These help to transport materials inside the cell. The mitochondria are not just little dots. They are sausage-shaped objects with membranes inside. The energy-transferring reactions take place on the membranes. These are just two of the many special features of cells which have been discovered with this wonderful microscope.

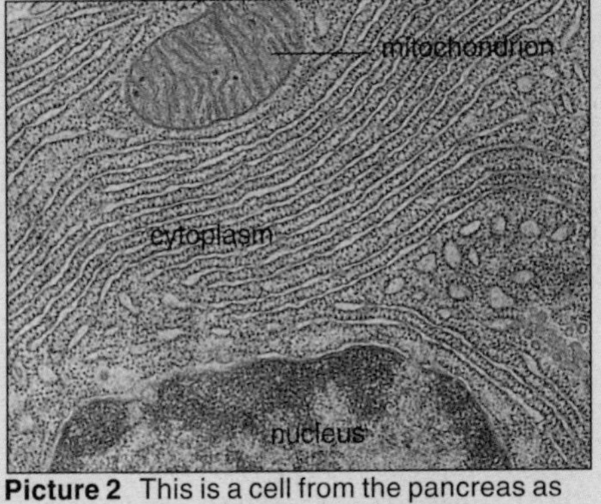

Picture 2 This is a cell from the pancreas as seen in the electron microscope. It is magnified 10 000 times.

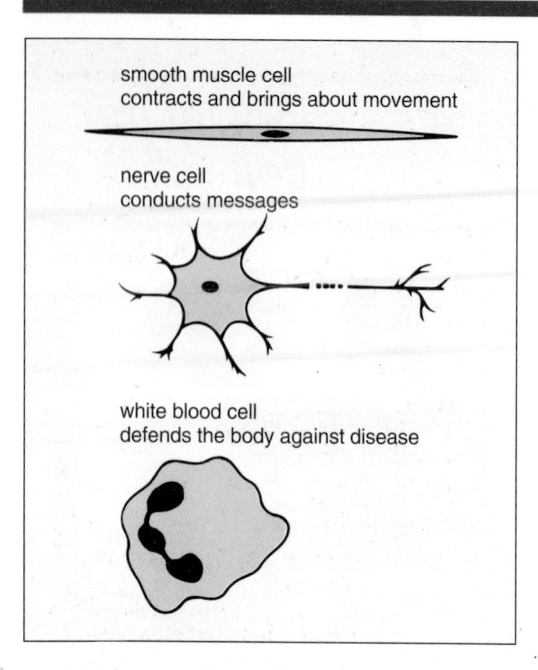

Picture 5 Three types of cell found in the human body.

Different cells for different jobs

Practically all cells contain a nucleus and cytoplasm. However, they vary tremendously in their shape and form. In the human body there are at least twenty different types of cell, each specialised to do a particular job. Three are shown in picture 5.

There is thus a **division of labour** between cells. It's rather like a factory or an office in which each person has his or her own job to do. This is more efficient than if each individual tried to do everything.

How substances get in and out of cells

Many substances pass in and out of cells by **diffusion**. To understand diffusion, you must first appreciate that a substance may be more concentrated in one region than another. In other words, there may be a **concentration gradient** between the two regions.

Diffusion is the net movement of particles from a region where they are at a higher concentration to a region where they are at a lower concentration, i.e. *down* a concentration gradient. This is a purely physical process; it does not require energy from respiration. Cells take up oxygen and get rid of carbon dioxide by diffusion (picture 6).

Water passes in and out of cells by a special type of diffusion called **osmosis**. This is explained on the opposite page.

Cells take up certain substances by transporting them across the cell surface membrane *against* a concentration gradient, that is from a region of lower concentration to a region of higher concentration. This is an active process and needs energy from respiration. It is called **active transport**. Roots take up certain substances from the soil this way.

Only very small particles, such as ions and small molecules, can get in and out of cells by diffusion and active transport. Certain cells can take in larger particles by a process called **phagocytosis**. This word comes from Greek and means 'cell eating'. Picture 7 shows how it happens. The single-celled organism *Amoeba* feeds this way, and it is how some of our white blood cells destroy germs (see page 157).

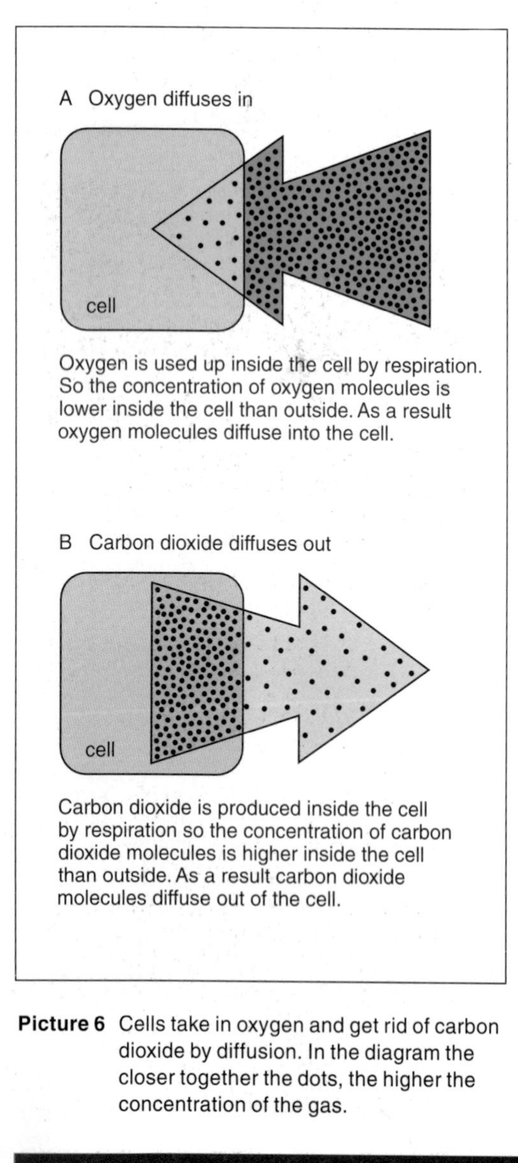

Picture 6 Cells take in oxygen and get rid of carbon dioxide by diffusion. In the diagram the closer together the dots, the higher the concentration of the gas.

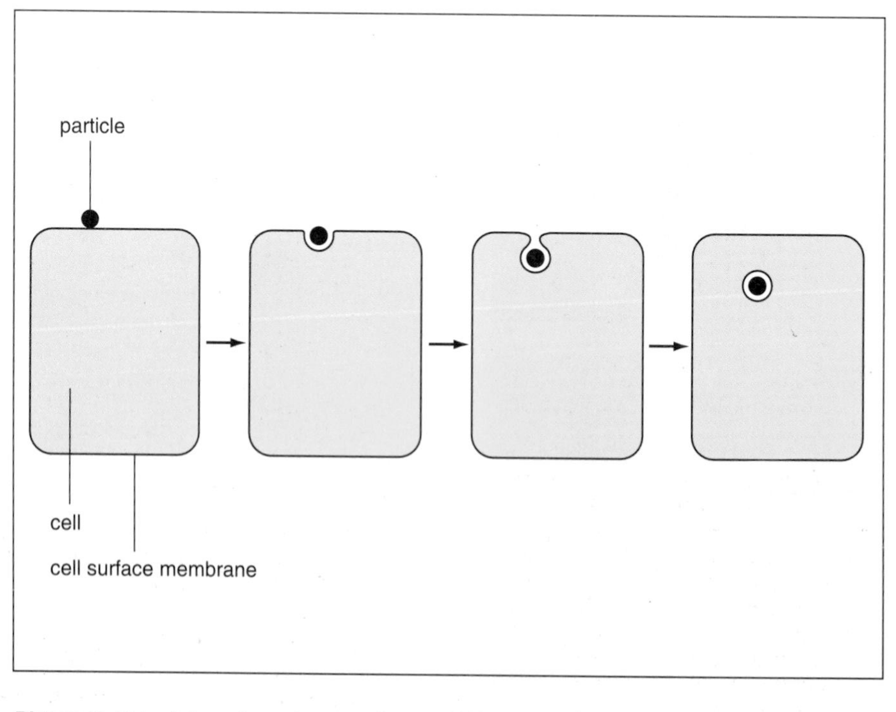

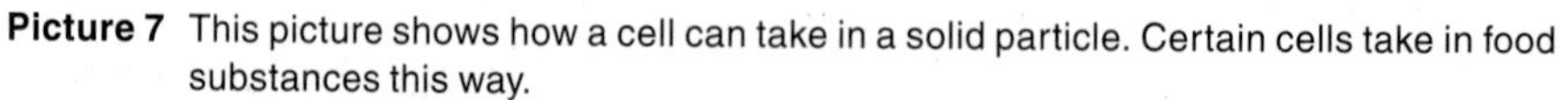

Picture 7 This picture shows how a cell can take in a solid particle. Certain cells take in food substances this way.

Osmosis

Carry out Activity C on page 16. A bag is made out of a thin membrane and filled with a sugar solution. The open end of the bag is tied to the end of a capillary tube. The bag is then suspended in a beaker of water.

After a short time, water passes from the beaker into the bag. As a result, the sugar solution moves up the capillary tube. To understand why this happens, look at picture 1. The sugar molecules are larger than the water molecules. The membrane of which the bag is made has tiny holes in it. These holes are large enough to let the water molecules through, but too small to let the sugar molecules through. We call this kind of membrane a **partially permeable membrane.**

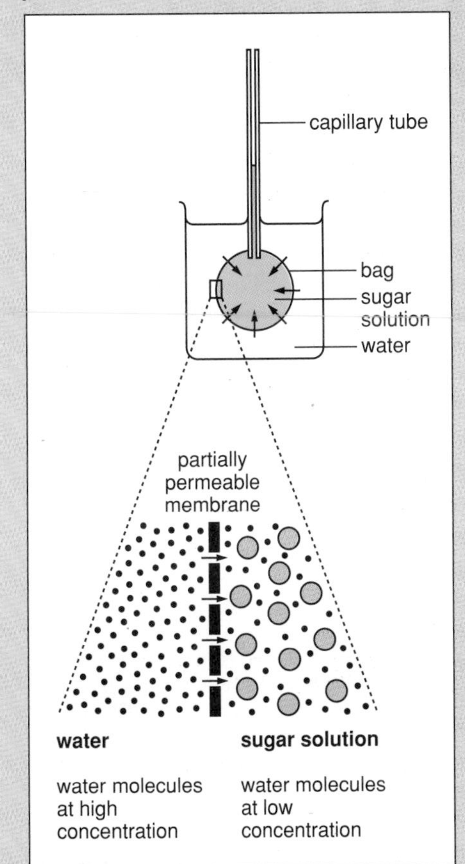

Picture 1 Osmosis is the one-way diffusion of water across a partially permeable membrane.

Now the presence of the sugar molecules in the bag means that there is less room for water molecules in the bag than in the beaker. So the water molecules in the bag are less concentrated than in the beaker. As a result, water molecules *diffuse* into the bag.

The diffusion of water across a partially permeable membrane is called **osmosis**. *Osmosis will take place wherever two solutions of different water concentrations are separated by a partially permeable membrane.*

The general term for any substance dissolved in a solution is solute. In the case we have been considering, the solute is sugar, but it could be some other substance such as salt.

What has all this got to do with biology? Well, cell surface membranes are partially permeable, so osmosis is important in organisms.

Osmosis in human cells

Human cells contain a solution of salts and other solutes. These are enclosed inside the partially permeable cell surface membrane.

Suppose you immerse a human red blood cell in water. What happens? Water enters the cell by osmosis, and the cell swells up and bursts (picture 2). The cell bursts because the cell surface membrane is too thin and delicate to withstand the pressure inside the cell.

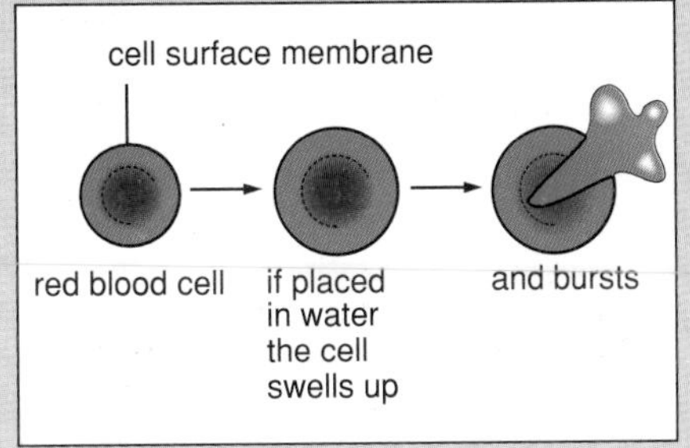

Picture 2 The result of immersing a red blood cell in water.

Obviously cells must not be allowed to burst inside our bodies. What stops this happening? The answer is that our blood and other body fluids have the same concentration as the cells, so water does not enter the cells by osmosis. In a later Topic we shall see how the concentration of the blood and body fluids is maintained at just the right value.

Osmosis in plant cells

Plant cells have a cell surface membrane, just like animal cells, but outside this there is the cellulose cell wall. Inside a plant cell there is a solution of salts and other solutes, many of which are located in the vacuole. The cell surface membrane is partially permeable, just as it is in animal cells. But the cellulose cell wall is *fully* permeable to solutes as well as to water.

What happens if you immerse a plant cell in water? Water flows through the cell wall and cell surface membrane into the vacuole from the outside. As a result, the cell swells up. But it does not burst. This is because the cell wall stops the cell expanding too much. The wall stretches but it does not break. It is like trying to blow up a football into which no more air can be forced. When this point is reached, we say the cell is **fully turgid** or at **full turgor** (picture 3).

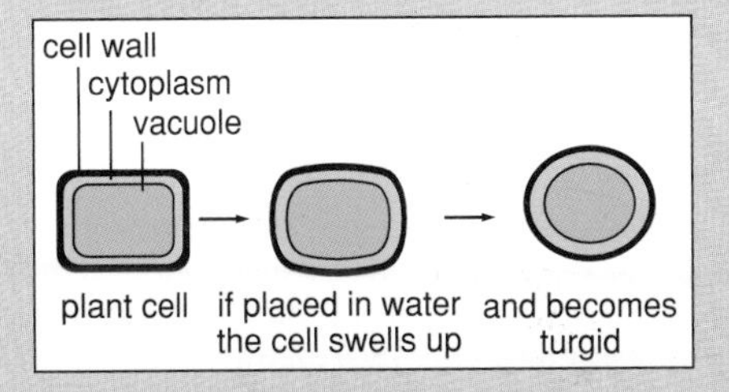

Picture 3 The result of immersing a plant cell in water.

Turgor is very important in land plants. It helps to make them firm. For example, when all the cells in a leaf are fully turgid, the cells press against each other and the leaf is held out in a open, expanded position. If the plant runs short of water, the cells may lose their turgor and the leaves droop. This is called wilting. Herbaceous plants which do not contain much wood depend on turgor to keep their stems erect. When such a plant wilts, the stem bends.

Picture 4 A melon plant when it is turgid (left) and when it has wilted (right).

1 Suggest an explanation for each of the following:
   a If a lettuce becomes floppy, you may be able to make it firm and crisp by putting it in water for a while.
   b If you sprinkle sugar on a bowl of strawberries, juice oozes out of them.

2 When a person swims in fresh water, you might think that water would enter the body through the skin by osmosis. But it does not. Suggest a reason.

3 What would happen to (a) a red blood cell and (b) a plant cell if they were immersed in a solution whose solute concentration was greater than that inside the cells?

Activities

A Looking at cheek cells

To avoid any possibility of infection follow these instructions carefully. They are based on recommendations by the Institute of Biology.

Saliva is a potential source of cross-infection.

1 Your teacher will give you a cotton bud from a newly opened pack. Gently rub the inside of your cheek with it.

2 Smear the cotton bud over a small area in the centre of a slide. Then put the cotton bud in a polythene bag provided by your teacher.

3 Pipette a drop of methylene blue onto the smear. This will stain the cells and help you to see them.

4 Cover with a coverslip.

5 Examine the slide under the microscope. First use low power to find a group of cells. Then look at one of the cells under high power.

Which of the structures in picture 3 on page 12 can you see?

6 Draw the cell and label it as fully as you can.

7 When you have finished your drawing, put the slide and coverslip in a bowl of disinfectant provided by your teacher.

Full instructions and background information for a safe way of examining human cheek cells are given in *The Biologist,* Volume 35, Number 4, September 1988.

B Looking at plant cells

1 Slice an onion in two lengthways.

2 Take out one of the thick 'leaves' from inside it.

3 With forceps pull away the thin lining from the inner surface of the 'leaf'.

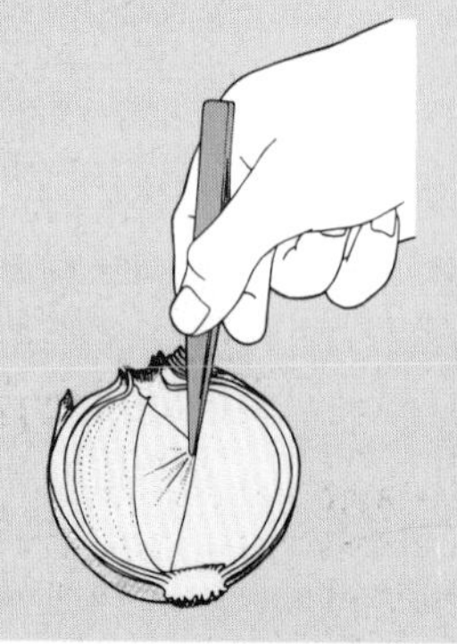

4 With scissors cut out a small piece of the lining, about 5 mm square.

5 Place the piece of lining on a slide and add a drop of dilute iodine solution.

Make sure the iodine solution goes under the lining as well as above it.

The iodine will stain the cells and make their nuclei easier to see.

6 Put on a coverslip.

7 Examine the slide under the microscope, first under low power, then high power. Choose an area of the lining where the cells are clear.

Which of the structures shown in picture 4 on page 13 can you see?

8 Draw one of the onion cells and label it as fully as you can.

Although an onion is part of a plant, you won't have seen any chloroplasts in the cells. Why not?

9 Obtain a moss plant and pull off one of its smaller leaves with forceps.

10 Mount the leaf in a drop of water on a slide and put on a coverslip.

11 Examine the slide under low power.

What structures, absent in onion cells, can you see in the moss cells? Explain their presence.

12 Select one of the clearest cells and examine it under high power.

13 Draw the cell and label it as fully as you can. You will have found it difficult or impossible to see a nucleus in the moss cells. Why?

What could you do to make the nuclei show up?

C Observing osmosis

1 Cut a length of Visking tubing, about 8 cm long, and wet it thoroughly with water.

2 Tie one end of the tube with strong thread, so that it forms a bag.

3 Fill the bag with a 20 per cent solution of sucrose.

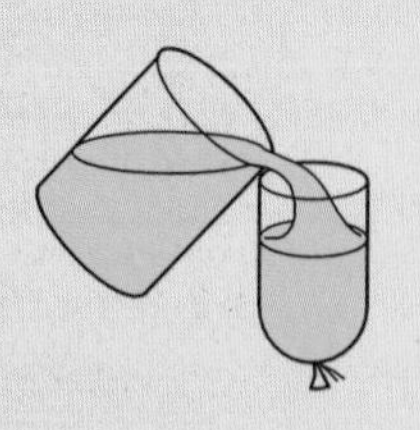

4 With another thread, tie the other end of the bag to the bottom of a capillary tube.

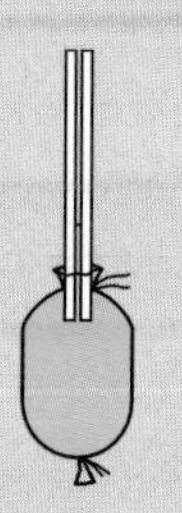

5 Clamp the capillary tube to a stand, and lower the bag into a beaker of distilled water. Mark the level of the sucrose solution on the capillary tube.

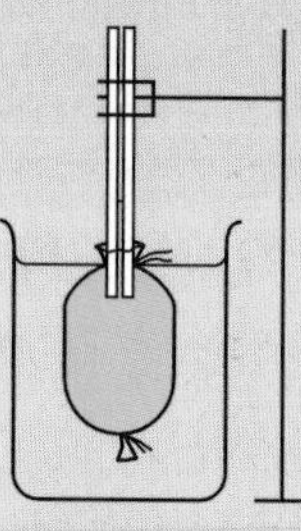

6 Observe any changes in the level of the sucrose solution in the capillary tube. Explain your observations.

7 Repeat the experiment, but this time make it *quantitative*, i.e. measure the *rate* at which changes occur in the level of the sucrose solution in the capillary tube. Graph your results.

D Osmosis in Visking tubing bags

As you will now realise from Activity C, you can make a partially permeable membrane out of Visking tubing. Plan an experiment to investigate the effect of varying the concentration of sucrose solutions in Visking tubing bags immersed in water. Consider carefully how you will observe and/or measure any effects obtained. Discuss your plan with your teacher before beginning the experiment.

Questions

1 Each word in the left-hand column below is related to one of the words in the right-hand column.

Write them down in the correct pairs.

glycogen	inheritance
chloroplast	energy
mitochondrion	sunlight
chromosomes	elastic
cellulose	storage

2 A typical cell is twenty micrometres wide.

→ Suppose that cells of this size were placed side by side. How many would there be in a row that was the same length as the third line of this question (arrowed).

3 Which of the structures listed below are found (a) in animals cells only, (b) in plant cells only, and (c) in both animal and plant cells?

cytoplasm	glycogen granules
chloroplasts	cell wall
starch grains	chromosomes
nucleus	cell surface membrane
vacuole	mitochondria

4 Cells can be likened to the bricks of a house. In what ways are cells more than just bricks?

5 A student did an experiment to find the effect of oxygen on the uptake of bromide ions by the roots of barley plants. The temperature was kept at 17°C. Here are her results:

	Uptake (relative units)
Oxygen present	100
Oxygen absent	10

a Why was the temperature kept at 17°C?

b What effect does oxygen have on the uptake of ions by the roots?

c What do the results suggest about the way ions are taken up by the roots?

Diffusion and surface area

Imagine that the box below is an organism. It is a cube whose sides are all one centimetre long:

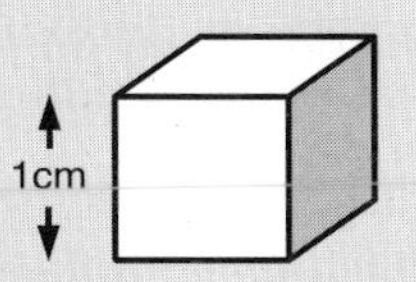

Its surface area is 6 cm^2, and its volume is 1 cm^3.

Now suppose we double the size of the box like this:

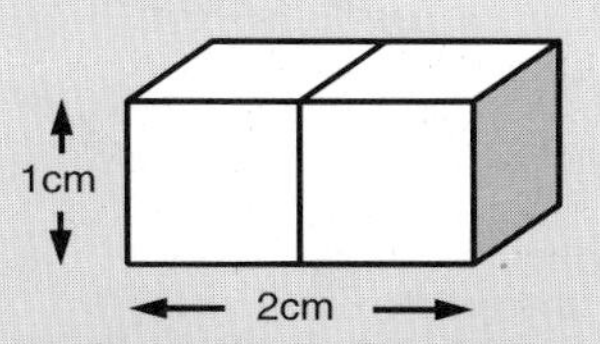

By how much have we increased its volume and its surface area? Well, its volume is now 2 cm^3, twice what it was. However, its surface area is 10 cm^2, which is less than twice what it was.

In other words, we have doubled its volume, but its surface area has less than doubled. This is because, in the process of doubling the volume, we have lost part of the original surface (the part shaded in the first diagram).

So we can make this general statement: *as an object increases in size, the amount of surface relative to volume (the surface–volume ratio) decreases.*

This is important to organisms which take in things by diffusion. Think of it this way. A small organism, like an amoeba, has a large surface–volume ratio, so it can take in all the oxygen it needs by diffusion across the body surface. However, a large organism, like the human, has a much smaller surface–volume ratio, so it could not get all the oxygen it needs this way. Organisms of this kind need a special **gaseous exchange surface** for taking in oxygen. Examples include the **lungs** of mammals and the **gills** of fishes. These organs consist of sheets of tissue which are folded many times, thus providing a large surface area across which oxygen can be absorbed.

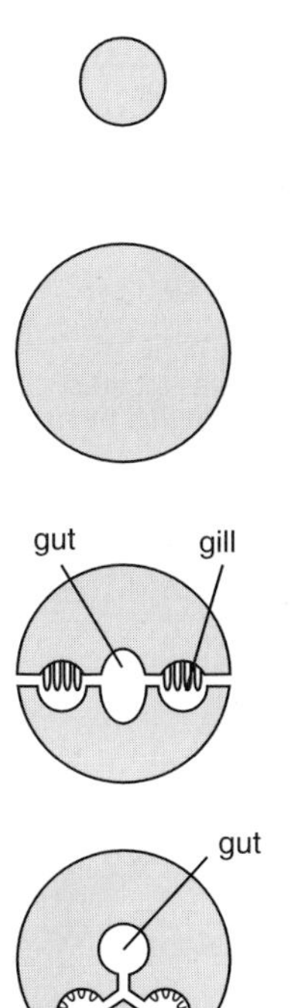

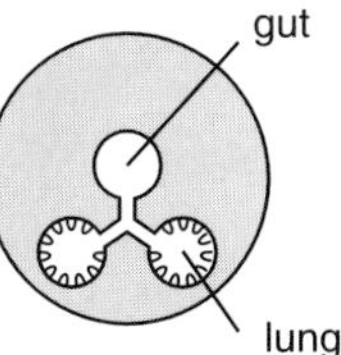

Picture 1 How large organisms may overcome the problem of having a small surface–volume ratio.

Larger organisms, including the human, also have a **circulatory system** which transports oxygen quickly from the gaseous exchange surface to all parts of the body. In many smaller organisms oxygen moves through the body by diffusion.

1 Having a gaseous exchange surface and circulatory system allows an organism to be more active than it would be otherwise. Why?

2 Do plants have special gaseous exchange surfaces? If so, where are they, and how are they suited for gaseous exchange?

3 Find out as much as you can about how oxygen gets from the surrounding air to the innermost parts of an amoeba, a flatworm, an earthworm, an insect, a fish and a mammal. Use books or (better still) observe the organisms themselves.

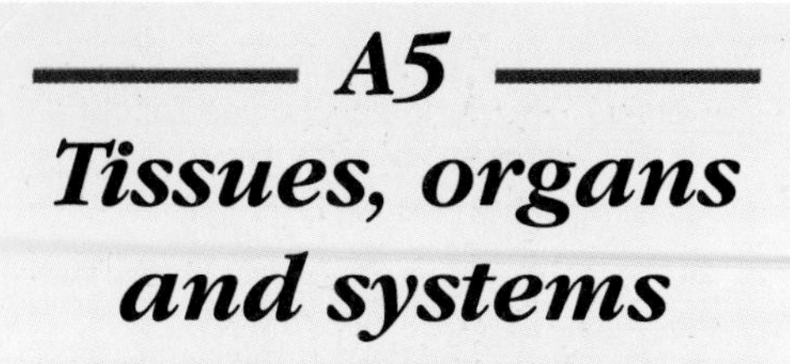

Cells are grouped into tissues

Cells do not normally exist on their own. Usually large numbers of them are massed together into a tissue.

One of the simplest tissues is shown in picture 1. It is called **epithelium**. It consists of a sheet of cells. The cells fit neatly together, like paving stones. This kind of tissue forms the lining of spaces and tubes inside the body and is also found on the surface of the skin. Plants have a similar surface tissue called **epidermis** (picture 2). These surface tissues protect the structures underneath.

The main tissues of animals and plants are summarised in tables 1 and 2. Some of them consist of just one type of cell. However, most of them contain two or three types of cells together.

Tissues are combined into organs

In most animals, including the human, tissues are combined together to form **organs** (picture 3). An organ is a complex structure which has a particular job to do. The main organs in the human body are shown in picture 4.

Some organs do just one job. For example, the only job the **heart** does is pump blood round the body. Other organs do more than one job. For example, the **kidneys** get rid of poisonous waste substances and control the amount of water in the body. The organ with the greatest number of jobs is the **liver**; it does at least 500 jobs.

Animals are not the only living things to have organs. Plants have them too. Leaves, flowers, bulbs and tubers are all organs. Each has a particular job to do.

Organs are grouped into systems

In the human body certain tasks are carried out by several different organs working together. These organs all belong to a **system**. An example is the **digestive system**. This consists of the gut, together with the liver, pancreas and gall bladder. Its job is to digest and absorb food.

The various systems found in the human body are summarised in table 3. Some organs belong to more than one system. The liver, for example, belongs to the digestive and excretory systems.

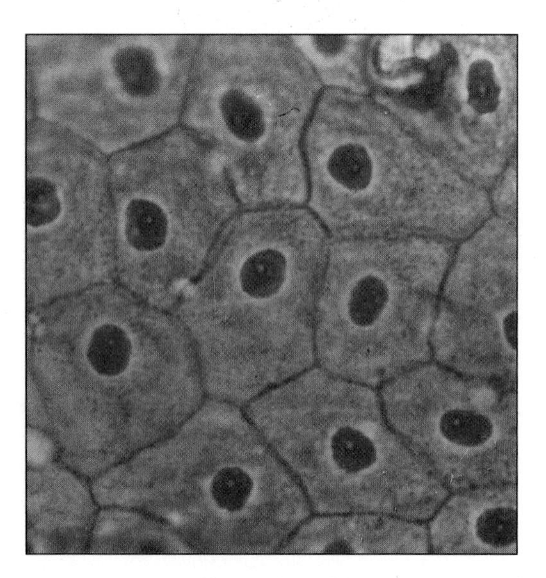

Picture 1 A simple type of epithelium seen under the microscope. It comes from one of the thin membranes inside the body.

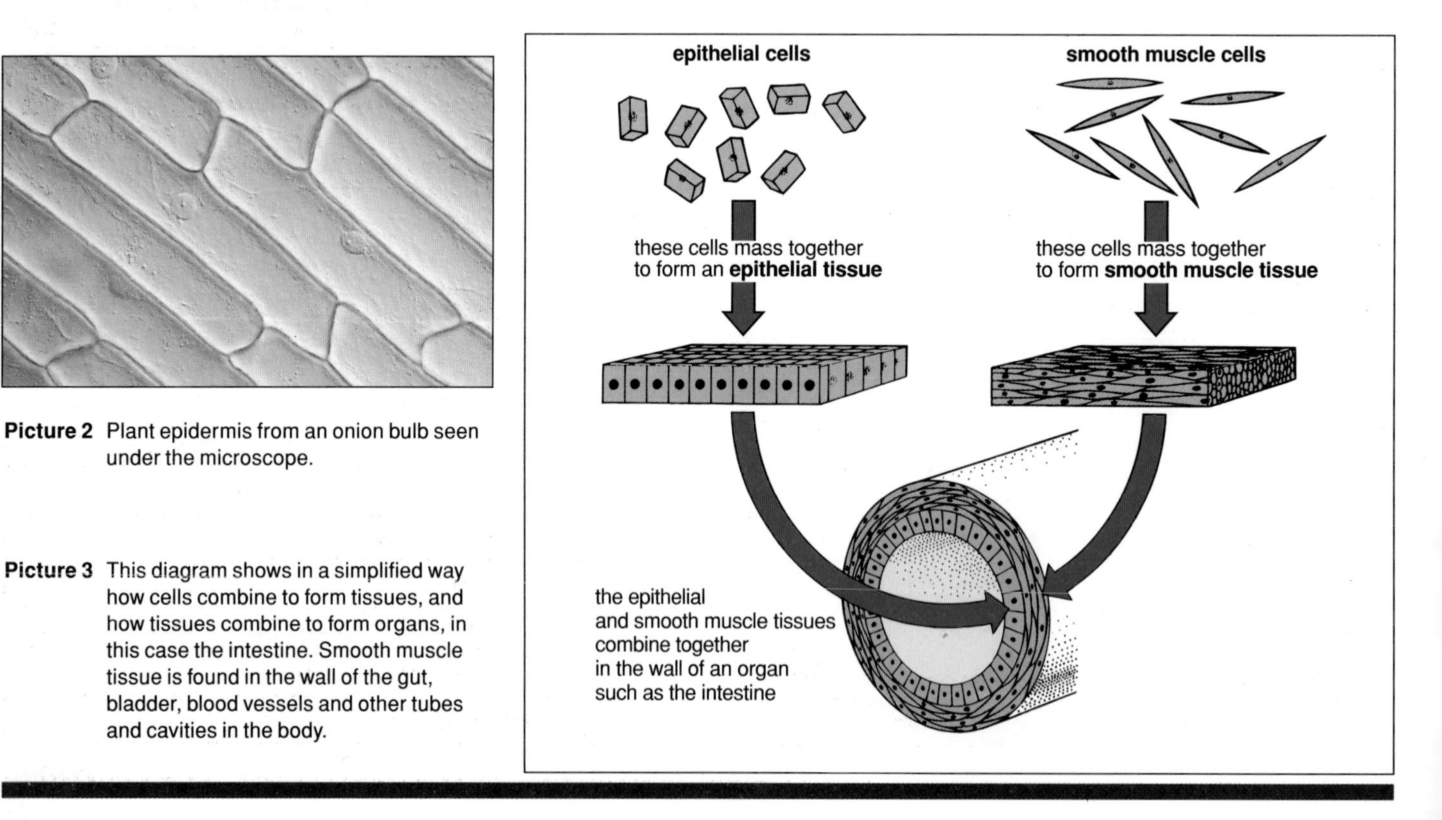

Picture 2 Plant epidermis from an onion bulb seen under the microscope.

Picture 3 This diagram shows in a simplified way how cells combine to form tissues, and how tissues combine to form organs, in this case the intestine. Smooth muscle tissue is found in the wall of the gut, bladder, blood vessels and other tubes and cavities in the body.

Specialisation

When a cell, tissue or organ does a particular job, we say that it has a specialised function. Specialised cells, tissues and organs are adapted in their structure and the way they work to carry out their functions.

The picture alongside shows a specialised epithelial tissue which contains two special types of cells:

- **Ciliated cells** have tiny hair-like **cilia** (singular: **cilium**) projecting from the surface. The cilia beat by moving backwards and forwards very quickly. An epithelium with ciliated cells is called **ciliated epithelium**.
- **Gland cells** secrete the slimy substance mucus onto the surface of the epithelium. An epithelium with mucus-secreting cells is called a **mucous membrane**.

Our breathing passages are lined with a ciliated mucous membrane; germs and dust get caught up in the mucus, which is then wafted away from the lungs by the beating cilia (see page 131). A similar surface tissue is found on the underside of snails and flatworms; the cilia enable the animal to glide on a trail of mucus.

Mucous membranes are not always ciliated: our gut, for example, is lined with a mucous membrane which has lots of gland cells but no cilia.

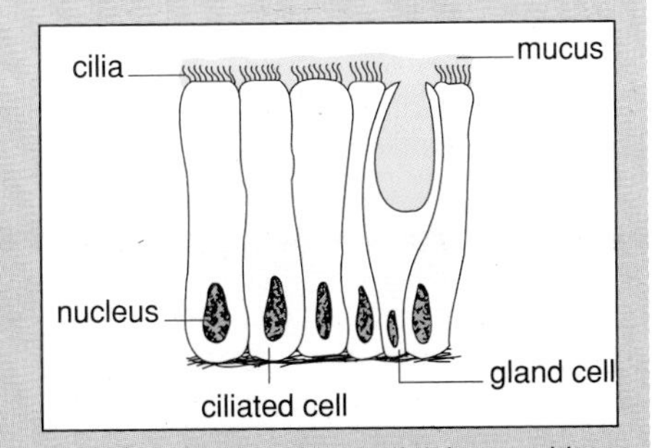

Picture 1 Ciliated epithelial tissue with mucous cell. Below: how this looks under a microscope.

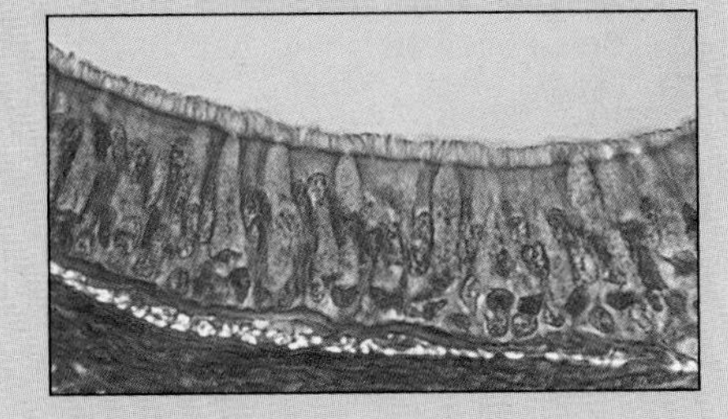

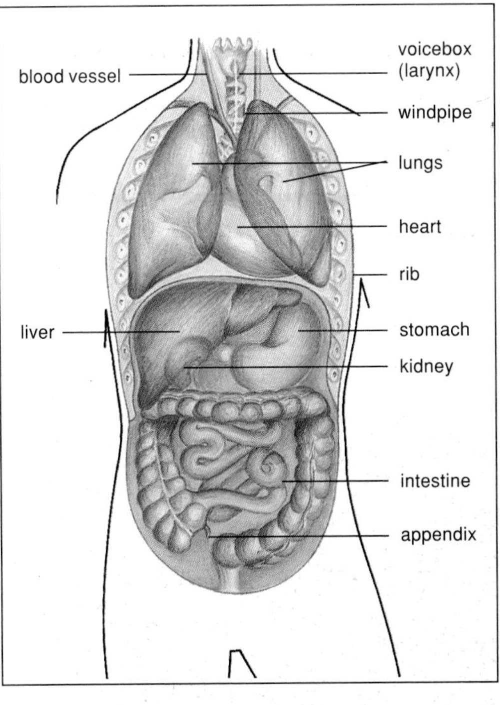

Picture 4 This diagram shows some of the main organs in the human body, as seen from the belly (ventral) side.

Table 1 (left) Tissues found in the human

Name of tissue	Main functions
Epithelial tissue	To line tubes and spaces and the skin
Connective tissue	To bind and connect other tissues together
Skeletal tissue	To support and protect the body and permit movement
Blood tissue	To carry oxygen and food substances round the body
Nerve tissue	To transmit and coordinate messages
Muscle tissue	To bring about movement

Table 2 (below) Tissues found in the flowering plant

Name of tissue	Main functions
Epidermal tissue	To line and protect the surface of the plant
Photosynthetic tissue	To feed the plant by photosynthesis
Packing tissue (parenchyma)	To fill in spaces inside the plant
Vascular tissue	To transport water and food substances
Strengthening tissue	To support and strengthen the plant

Table 3 Systems in the human body

Name of system	Main organs in the system	Main functions
Digestive system	Gut, liver and pancreas	To digest and absorb food
Breathing system	Windpipe and lungs	To take in oxygen and get rid of carbon dioxide
Circulatory system	Heart, blood vessels	To carry oxygen and food round the body
Excretory system	Kidneys, bladder, liver	To get rid of poisonous waste substances
Sensory system	Eyes, ears, nose	To detect stimuli
Nervous system	Brain and spinal cord	To conduct messages from one part of the body to another
Musculo-skeletal system	Muscles and skeleton	To support and move the body
Reproductive system	Testes and ovaries	To produce offspring

Activities

A Looking at epithelium

1 Your teacher will give you a prepared slide of epithelial tissue. It has been stained so as to show details of the cells that make up the tissue.

2 Examine your slide under the microscope: low power first, then high power.

How does the tissue compare with the one shown in picture 1?

Draw a small group of cells, showing how they fit together.

What job does this tissue do?

B Looking at plant packing tissue

1 Obtain a soft fruit such as a tomato.

2 Cut the fruit in half.

3 With a spatula remove a small piece of the soft pulpy material from inside. This is *packing tissue.*

4 Put the tissue on a slide and spread it out.

5 With a pipette add a drop of water to the tissue.

6 Cover it with a coverslip.

7 Examine your slide under the microscope: low power first, then high power.

How does the tissue compare in appearance with the epithelial tissue in activity A?

Draw a small group of the cells.

What job does this tissue do?

C Looking at tissues

Your teacher will give you a plant, or part of one.

Using whatever means you have available, investigate the tissues of which the plant is composed.

Describe your method and record your observations as fully as possible. Make drawings whenever you think they might be helpful. Be sure to label them, and give them a scale.

D Looking at organs

1 Examine some or all of the following organs obtained from a butcher: lungs, stomach, intestine, tongue, liver, pancreas, heart, kidney, muscle, brain and eye.

2 Try tearing, or cutting, each organ to see how tough it is.

3 Cut open each organ with a knife in order to see its inside.

Be careful not to cut your fingers.

DANGER

4 Describe what each organ looks and feels like.

5 Find out what each organ has to do in the body.

In what ways does the structure of each organ suit it to its job?

6 For each organ write down the main tissues which you would expect to find in it.

How would you relate the presence of the particular tissue to the job which the organ has to do?

What would you have to do to find out if the tissue really is present?

Questions

1 Explain each of these words: tissue, organ, muscle, epithelium, multicellular.

2 Each of the tissues listed in the left-hand column is related to one of the words in the right-hand column. Write them down in the correct pairs.

photosynthetic tissue	transport
epithelial tissue	protection
connective tissue	messages
blood tissue	feeding
nerve tissue	strength

3 What kind of tissue:

a fills spaces inside a plant stem?
b carries oxygen round the human body?
c supplies a plant with organic food?
d transports water through a plant?
e conducts messages from one part of your body to another?

4 Why is it an advantage to have a division of labour between different organs in the body?

5 Name two organs in the human body which occur in pairs and two which occur singly.

Why is it an advantage to have pairs of organs rather than single ones?

6 Make a list of all the functions you can think of which are performed by your head.

Why is it an advantage to an animal to have its head at the front end of its body?

7 It is possible to take small pieces of tissue from the human body and grow them on a suitable surface in the laboratory. This is called *tissue culture.* The cells which have been removed from the body multiply so more tissue is formed.

a What can you say about the 'surface' on which the tissue is placed? What special features should it have?
b What possible uses might be made of tissue culture?

New parts for old

If you cut yourself, the skin eventually knits together and the wound heals. Most tissues are quite good at mending themselves like this. But if an organ gets badly damaged, or stops working properly, a new one won't grow in its place.

Fortunately some of our organs occur in pairs, the lungs and kidneys for example, and we can manage with only one. People who have had a lung removed because of cancer can live for years. In fact it's amazing what we can do without. A person can survive with less than half the intestine and only fifteen per cent of the stomach.

You may have heard of people who have been given a new organ in a **transplant operation** – a kidney perhaps, or a heart.

Transplant operations are carried out on patients who have an organ that is working so badly that their life is in danger. The bad organ is removed and replaced with a healthy one taken from another person, the **donor**.

The donor is usually a healthy person who has died suddenly, often as a result of a car or motorcycle accident. The organ is removed from the donor's body and quickly taken to the hospital where the operation is to be performed. The organ is kept alive in a sealed bag containing a special fluid.

The trouble with transplant operations is that the transplanted organ may be destroyed after it has been put into the patient's body. Why this happens, and how it is overcome, is explained on page 193.

In the last thirty years or so, great progress has been made in replacing parts of the body, not only with transplants but also with artificial structures. The picture shows some of the 'spare parts' now available.

'Spare part surgery' is important because the human body is so bad at regenerating its own lost parts. However, certain other organisms are very good at regeneration. For example, an earthworm accidently cut in two by a gardener's spade will grow a new back end; and small pieces of certain plants can grow into complete new plants. We make use of this in propagating plants (page 250).

1 What features would you expect the 'special fluid' mentioned in this section to have?

2 Mr X is in hospital waiting to have a heart transplant. Make a list of all the difficulties facing the team of doctors who are hoping to carry out the operation.

3 Many people carry a card saying that if they die they are willing for their organs to be used in transplant operations. Suppose a card-carrying person has a fatal car accident. Should the relatives have any say in whether or not the person's organs are used?

4 Some people object to transplants on ethical grounds. Discuss possible ethical problems connected with transplant operations.

5 The picture showing spare parts is not complete. Other parts are available, and other materials can be used. Ask a doctor (who isn't too busy!) to bring the picture up to date for you.

Key

Transplants
1 Cornea
2 Heart
3 Lung
4 Heart valve
5 Liver
6 Kidney
7 Bone
8 Bone marrow
9 Hair
10 Brain tissue
11 Small intestine
12 Pancreas

Artificial structures
13 Skull plate (metal)
14 Ear flap (plastic or silicone rubber)
15 Ear ossicles (stainless steel)
16 Eye lens (plastic)
17 Nose cartilage (plastic)
18 Jaw bone (metal)
19 Teeth (ceramic with titanium jaw attachment)
20 Hearing aid
21 Blood pressure regulator (electronic)
22 Windpipe (plastic)
23 Shoulder joint (metal or plastic)
24 Elbow joint (metal or plastic)
25 Wrist joint (metal)
26 Knuckle (plastic or silicone rubber)
27 Breast (silicone rubber)
28 Heart valve (metal or plastic)
29 Heart pacemaker (electronic)
30 Arm (metal or plastic, powered)
31 Artery (Fluoro-ethene polymer or cultured tissue)
32 Bladder stimulator (electronic)
33 Hip joint (femur head: metal; cup: high density polythene)
34 Bone plate (plastic, dissolves as bone heals)
35 Blood vessel (plastic, dissolves as vessel knits)
36 Knee joint (metal or plastic)
37 Tendon (plastic)
38 Leg (metal or plastic, powered)
39 Calf stimulator (activated by foot)
40 Voice tone generator
41 Penis (silicone rubber erector)

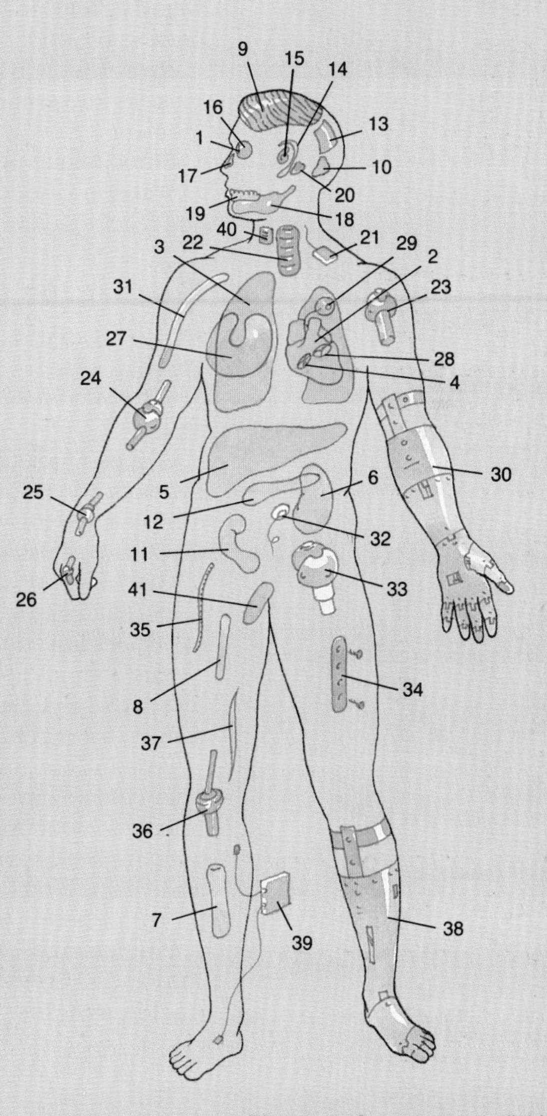

Picture 1 'Spare parts' available to humans.

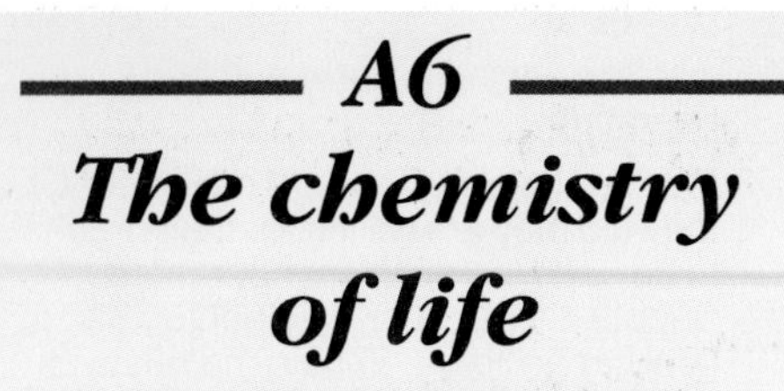

A6 The chemistry of life

Organisms are like chemical factories. What are the chemicals and what happens to them?

What are living things made of?

All living things, including humans, are made of chemical substances. Here is a summary of the main substances found in the body:

Organic substances:	Carbohydrates Fats and oils (lipids) Proteins
Inorganic substances:	Salts Water

Picture 1 shows the proportions of each of these substances in the human body. You may be surprised to see that our bodies contain far more water than any other substance. The fact is that we are really very wet.

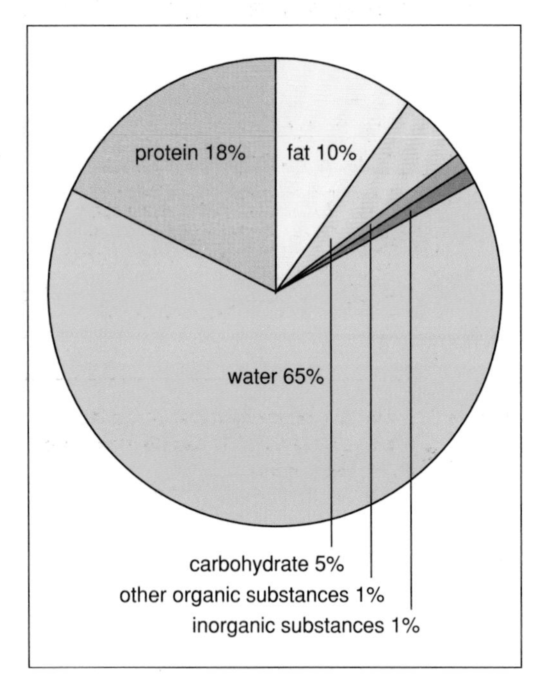

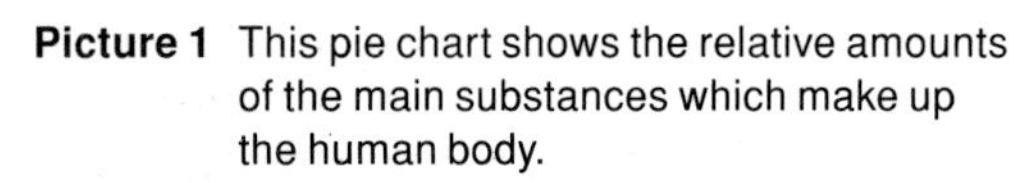

Picture 1 This pie chart shows the relative amounts of the main substances which make up the human body.

What are the functions of these substances?

Each substance has certain jobs to do. These are their main jobs:

Carbohydrates give us energy; one of the best known carbohydrates is sugar of which there are several different types.

Fats and **oils** (**lipids**) also give us energy; there is a lot of fat under the skin where it helps to keep us warm, and oils on the surface can help to make the body waterproof.

Proteins help to build up the body; they form important structures like muscles and tendons, and are also important as enzymes as we shall see in a moment.

Salts help to make the tissues and organs in the body work properly.

Water provides a fluid in which other substances can move about the body and react together within the cells.

How are the substances constructed?

All carbohydrates contain carbon, hydrogen and oxygen. One of the simplest is **glucose** which is a type of sugar. Its formula is $C_6H_{12}O_6$: this shows that a molecule of it contains six carbon atoms, twelve hydrogen atoms, and six oxygen atoms.

Glucose molecules can be linked together to form more complex carbohydrates, for example **starch** and **glycogen**. Their function is to store energy for use when it is needed (picture 2). The different types of carbohydrate are explained on page 26.

One of the most complex carbohydrates is **cellulose**, the material of which plant cell walls are made. Like other carbohydrates, it consists of lots of glucose molecules linked together.

Fats and oils are complex substances too. Their molecules are made up of two parts: **glycerol** and **fatty acids**. Like carbohydrates, fats and oils contain carbon, hydrogen and oxygen.

Proteins are amongst the most complex substances found in living things. They contain nitrogen as well as carbon, hydrogen and oxygen. Sulphur, too, is usually present. A protein molecule consists of one or more chains of chemical building blocks called **amino acids**. Sulphur helps to cross-link the amino acid chains. Some proteins are solid, others are in solution. Muscle is a solid protein, whereas egg white is a soluble protein.

Proteins may combine with other molecules to form even more complex substances. For example, they combine with nucleic acids such as DNA to form **nucleoproteins**.

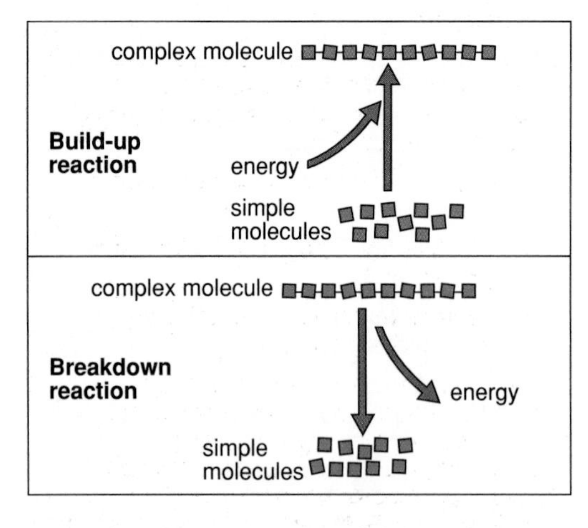

Picture 2 This diagram illustrates two important types of chemical reaction which take place in living organisms.

Chemical reactions in living things

Lots of chemical reactions take place inside living cells. Together, they comprise **metabolism**. Some metabolic reactions build things up (**anabolism**), others break things down (**catabolism**). Build-up reactions use up energy: they are **endothermic**. Breakdown reactions release energy: they are **exothermic** (picture 2).

An example of a build-up reaction is the linking of glucose molecules to form glycogen or starch for storage (picture 3). Another example is the linking of amino acid molecules to form proteins for body-building.

One of the most important examples of breakdown reactions is the oxidation of glucose to carbon dioxide and water in respiration (see page 122). This happens in almost all living cells.

Chemical reactions also take place in the gut. They are not part of metabolism but are concerned with **digestion** (see page 136). Starch, proteins and lipids are all broken down into simpler substances in the gut.

What enables all these reactions to occur and keeps them going? The answer is **enzymes**.

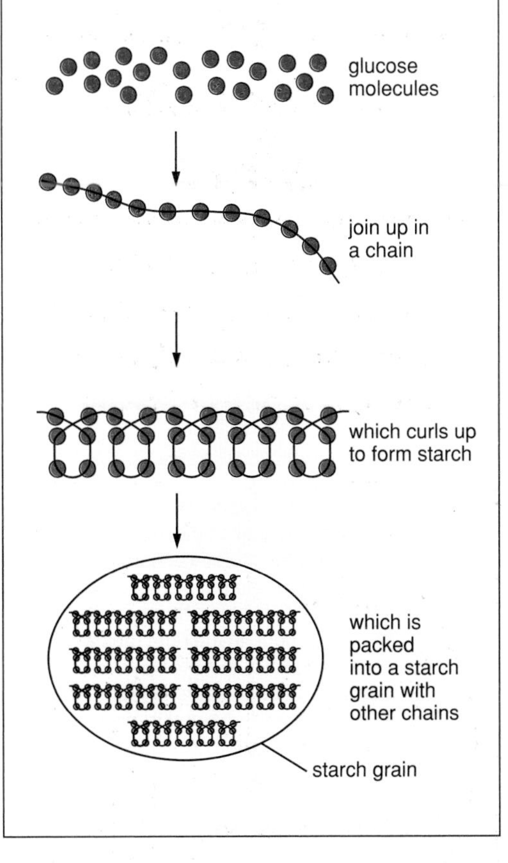

Picture 3 Starch is a convenient way of storing glucose molecules. These diagrams are not drawn to scale.

What are enzymes?

Enzymes are **biological catalysts**. They speed up the chemical reactions which go on inside living things. Without them the reactions would be so slow that life would grind to a halt!

Enzymes are extremely efficient at doing their job. Here is an example. Some of the chemical reactions which take place in our cells, for example in the liver, produce a by-product called hydrogen peroxide. Hydrogen peroxide is very poisonous so it must be got rid of quickly. Under the influence of an enzyme called **catalase**, the hydrogen peroxide is broken down into harmless water and oxygen. Catalase acts very quickly: one molecule of it can deal with six million molecules of hydrogen peroxide in one minute!

Types of enzymes

Enzymes are made inside cells. Once formed, the enzyme may leave the cell and do its job outside. Such enzymes are called **extracellular enzymes**. They include the digestive enzymes which break down food substances in our gut.

Other enzymes do their job inside the cell. They are called **intracellular enzymes**. Their job is to speed up the chemical reactions occurring in our cells. But they do more than just speed up the reactions; they also control them.

At this moment, thousands of chemical reactions are taking place in your body. Each reaction is controlled by a particular enzyme. Our enzymes make sure that the right reactions occur in the right place and at the right time.

An enzyme-controlled reaction

Here is an example of a reaction which is controlled by an enzyme:

$$\text{maltose (substrate)} \xrightarrow{\text{maltase (enzyme)}} \text{glucose (product)}$$

The substance which the enzyme acts on is called the **substrate** – in this case maltose. The new substance or substances formed as a result of the reaction are the **products**. In this case there is just one product: glucose. The enzyme catalysing this particular reaction is maltase.

This reaction will go in either direction. In other words the reaction is **reversible**: maltose can be turned into glucose, or glucose into maltose. The enzyme will work either way. If there is a lot of maltose present compared with glucose, the reaction will go from left to right; if there is a lot of glucose present compared with maltose, it will go from right to left. Most metabolic reactions are reversible.

Naming enzymes

Enzymes are usually named by putting 'ase' on the end of the name of the substance, or type of substance, on which the enzyme acts. So

- enzymes which act on carbohydrates are called **carbohydrases**,
- enzymes which act on lipids are called **lipases**, and
- enzymes which act on proteins are called **proteases**.

Each of these major groups of enzymes includes specific enzymes which act on particular substances. For example, carbohydrases include **amylase** which catalyses the breakdown of starch to maltose, and **maltase** which catalyses the breakdown of maltose to glucose.

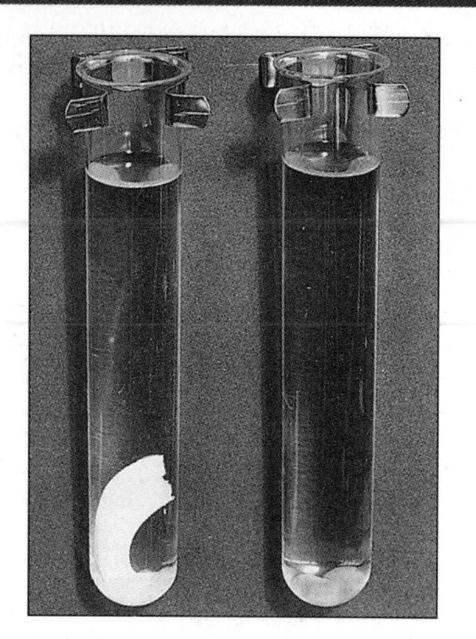

Picture 4 The effect of heating trypsin, a protein-digesting enzyme in the small intestine. In the left tube a piece of hard egg-white was covered with trypsin which had been heated to 50°C beforehand and then allowed to cool. In the right tube a piece of egg-white was covered with trypsin that had not been heated beforehand. Note that the pre-heated trypsin has failed to digest the egg-white.

The properties of enzymes

Enzymes have five important properties:

1 They are always proteins
This is one reason why we need proteins in our food.

2 They are specific in their action
What this means is that each enzyme controls one particular reaction, or type of reaction. Thus maltase will only act on maltose, and sucrase on sucrose.

3 They can be used over again
This is because they are not altered by the reaction in which they take part. However, an enzyme molecule eventually runs down and has to be replaced.

4 They are destroyed by heating
This is because enzymes, in common with all proteins, are destroyed by heating. This is called **denaturation**. Most enzymes stop working if the temperature rises above about 45°C (picture 4).

5 They are sensitive to pH
The term pH refers to the degree of acidity or alkalinity of a solution. Most intracellular enzymes work best in neutral conditions, i.e. conditions that are neither acidic nor alkaline.

How do enzymes work?

Look at picture 5. This shows in a very simple way how enzymes are believed to work. As you know, molecules are constantly moving about and bumping into each other. Now when a substrate molecule bumps into a molecule of the right enzyme, it fits into a depression on the surface of the enzyme molecule. This depression is called the **active site**. The reaction then takes place and the molecules of product leave the active site, freeing it for another substrate molecule.

The active site of a particular enzyme has a specific shape into which only one kind of substrate will fit. The substrate fits into the active site rather like a key fits into a lock. This is why enzymes are specific in their action.

When an enzyme is denatured by heat, the shape of the active site is changed so that the substrate no longer fits. A change in the pH has a similar effect.

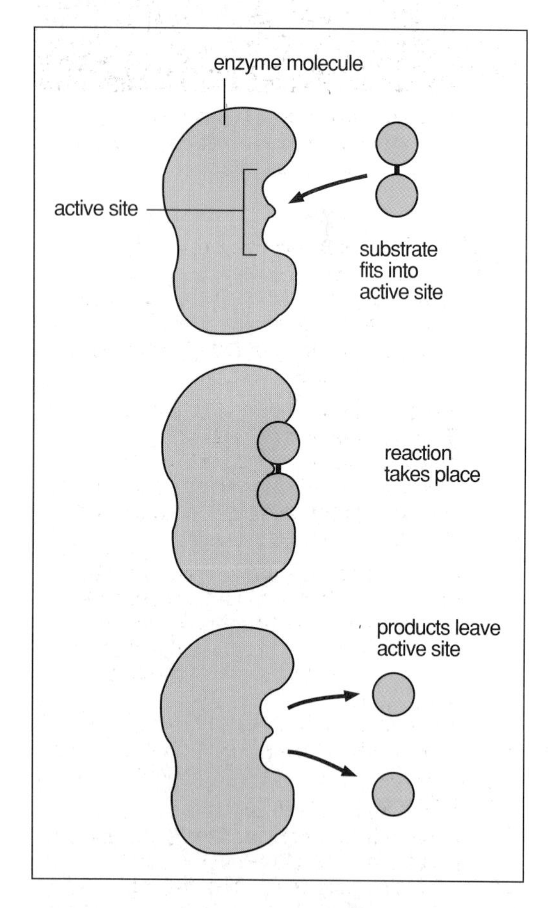

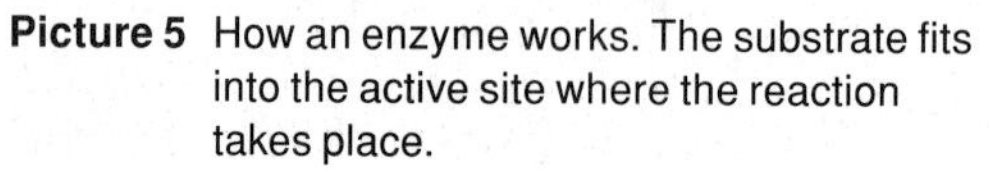

Picture 5 How an enzyme works. The substrate fits into the active site where the reaction takes place.

Helping and hindering enzymes

Anything which helps substrates to come into contact with the right enzymes will make enzyme-controlled reactions go faster. For example, raising the temperature will increase the rate of the reaction: a 10°C increase in temperature can double the rate. Heating increases the random movements of the molecules and increases the chances of substrate and enzyme colliding with enough energy to react. But of course the temperature must not be raised too much or the enzyme will be destroyed.

Certain vitamins and mineral elements also help enzymes by making it easier for the substrate to fit into the active site and react there.

On the other hand various **poisons** such as cyanide and arsenic inhibit enzymes by blocking the active site. Some poisons block the active site permanently, others only temporarily. Either way, the substrate finds it difficult or impossible to enter the active site.

Certain **pesticides** work by blocking the active site of an enzyme. If the enzyme is involved in respiration, the cells cannot transfer energy and so the pest quickly dies.

The uses of enzymes

Enzymes can be extracted from organisms in a purified form and then used in all sorts of scientific and industrial processes. An everyday use in the home is in **biological washing powders**. Various protein-digesting enzymes (proteases) are added to the washing powder, and these are supposed to dissolve protein stains.

The advantage of biological washing powders is that they work at relatively low temperatures. This makes them particularly useful for delicate fabrics, and it saves electricity too. However, some people are allergic to them and they can cause skin trouble.

Biological washing powders are an example of how enzymes can be useful in the home. Enzymes are useful in industry too. Here are some examples:

- **Proteases** are used for tenderising meat, skinning fish, removing hair from hides, and breaking down proteins in baby foods.
- **Amylases** convert starch to sugar in making syrups, fruit juices, chocolates and other food products.
- **Cellulase** breaks down cellulose and is used for softening vegetables, removing the seed coat from cereal grain, and extracting agar jelly from sea-weed.
- **Isomerase** converts glucose into fructose. Fructose is much sweeter than glucose; this makes it useful in slimming foods as only small amounts are needed.
- **Catalase** releases oxygen from hydrogen peroxide and is used in making foam rubber from latex.

Nowadays enzymes for human use are obtained mainly from microbes. The microbes are grown on a large scale in **industrial fermenters** (see page 128) so large amounts of the enzymes can be obtained.

Picture 6 Enzymes are used in manufacturing sweets, in biological washing powders and for tenderising meat.

Questions

1 Give two functions in the human body of each of the following:

carbohydrate, fat, protein, water.

2 Look at picture 4 and read the caption carefully. Describe an experiment which you would do to find out, as closely as possible, the exact temperature at which trypsin is destroyed.

3 A technician carried out an experiment to investigate the effect of temperature on a certain metabolic reaction. Her results are shown in the graph.

a Between which temperatures does the rate of the reaction (i) increase, (ii) decrease?

b At what temperature is the rate of the reaction fastest? (This is called the optimum temperature.)

c Explain the effect of temperature on the rate of the reaction.

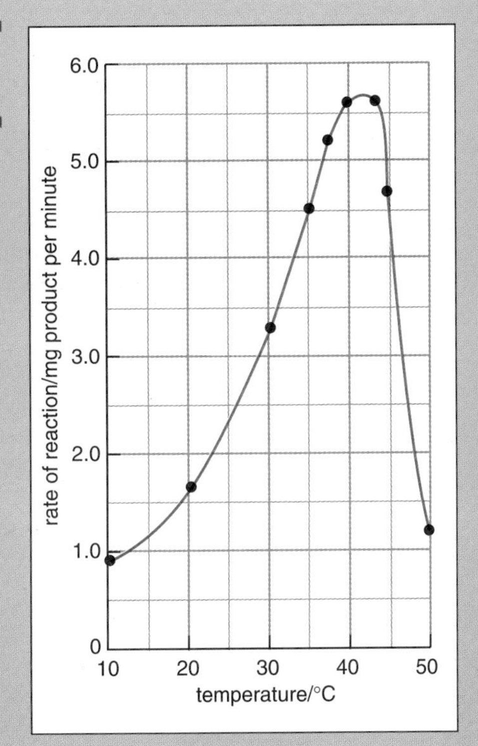

4 Biological washing powders contain protein-digesting enzymes. What advice would you give to the public on how to get the best results with a biological washing powder? Back up your answer with reasons.

5 Paper consists mainly of cellulose. Suggest a way of turning old newspapers into sugar. What problems would you be likely to have, and what precautions would you need to take?

6 An industrial enzyme may be brought in contact with the substrate in one of two ways. In the first method a solution is made of the enzyme and this is then mixed with a solution of the substrate. In the second method the enzyme molecules are permanently attached to an inert surface such as plastic beads which are then immersed in the substrate solution.

a What is meant by an inert surface?

b What are the advantages and disadvantages of each method?

Activities

A What are you made of and how much are you worth?

1 Weigh yourself. Write down your mass in kilograms.

2 Study picture 1. This tells you the percentage of different substances in an average human being.

3 Assuming that you contain these substances in the same proportions, work out the mass of each substance in your body.

4 Find out the cost of each of the substances in the shops. Assume that all the carbohydrate is sugar, all the protein is meat, and the inorganic substances are all table salt. Ignore the 'other organic substances'.

5 Work out the value of your body in pounds and pence.

This is not a valid way of expressing the value of a person. Why isn't it?

B Some important biological chemicals and their properties

1 Dissolve some solid glucose (dextrose) and sucrose in water.

Do they dissolve equally easily?

Explain the differences between them.

2 Try dissolving some starch in water.

What happens?

Why is it more difficult to dissolve starch than sugars?

3 Examine some fat and oil.

What do they feel like and what happens when you mix them with water?

How do their properties fit in with their functions in the body?

4 Examine a soluble protein such as albumen.

Does it mix easily with water?

What happens if you heat it gently?

5 Examine a solid protein such as hair.

How does it differ from a soluble protein such as albumen?

Relate the differences to their functions in the body.

C Watching an enzyme in action

1 Obtain two test tubes. Pour a 2 per cent hydrogen peroxide solution into one of them to a depth of about 2 cm.

CORROSIVE

CARE! Hydrogen peroxide is corrosive and can harm the skin.

Pour water into the other test tube to serve as a control.

2 Drop a small piece of liver into each test tube. Liver contains a lot of the enzyme catalase.

Watch carefully and describe what happens.

Explain your observations.

3 Repeat the experiment using a piece of liver which has been boiled for three minutes.

Explain the result.

4 Carry out experiments to find out if catalase occurs in other things, e.g. kidney, muscle and potato.

How could you make a rough comparison of the rates of the reaction?

5 Repeat the experiment using iron filings.

What happens? The reaction is catalysed by iron. However, to achieve the same reaction rate as the piece of liver you would have to use six tonnes of iron!

D Investigating the effect of pH on the action of catalase

Using the experience you have gained in activity C, plan an experiment to investigate whether catalase works best in acid, alkaline or neutral conditions.

Check your plan with your teacher.

Obtain the materials you need, then carry out the experiment. Report fully on your findings.

How could you find out the exact pH at which catalase works best?

E The action of enzymes on different substances

Your teacher will give you two small beakers labelled A and B. Beaker A contains a solution of the digestive enzyme amylase. Beaker B contains a solution of the digestive enzyme trypsin.

Plan an experiment to investigate which of the following materials each of these enzymes will digest: (1) starch, (2) egg white. Check your plan with your teacher, then carry out the experiment.

What general conclusions do you draw from your results about the properties of enzymes?

F How good are biological washing powders?

Plan an experiment to test the efficiency of a biological and a non-biological washing powder at removing biological stains.

Be sure that your comparison is a fair one. What must you do to ensure that this is so? Think about this before you start.

Check your plan with your teacher, then carry out the experiment.

Report fully on your findings.

Different types of carbohydrates

- **Monosaccharides** consist of single glucose (or glucose-like) molecules on their own. Glucose itself is an example. So is fructose (fruit sugar).
- **Disaccharides** consist of two glucose (or glucose-like) molecules linked together, e.g. sucrose (cane sugar), maltose (malt sugar) and lactose (milk sugar).
- **Polysaccharides** consist of many glucose molecules linked together, e.g. starch, glycogen and cellulose.

These three types of carbohydrate are interchangeable. Thus monosaccharides can be built up into disaccharides and polysaccharides, and polysaccharides can be broken down into disaccharides and monosaccharides.

1 Where and when would you expect glucose to be built up into starch?

2 Where and when would you expect starch to be broken down into glucose?

3 Starch, glycogen and cellulose all consist of chains of glucose molecules, and yet these three substances differ from each other. How would you explain the differences?

How enzymes were discovered

Enzymes were discovered in 1897 by a German scientist called Eduard Buchner. He discovered them by accident.

Buchner was interested in yeast. This little organism had long been known to convert sugar into alcohol. We call this process fermentation (see page 126). However it wasn't fermentation that Buchner was interested in; he had the idea that yeast might contain proteins which could be useful medically. So he gave up a holiday in order to investigate this idea.

He decided to extract the juice from yeast and test it on people to see if it might help to cure certain diseases. He obtained a large amount of yeast and squeezed it. Out of it oozed a brown liquid, rather like treacle. But before he had a chance to try it out on anyone, it went bad.

What could he do to preserve it? His laboratory assistant remembered that sugar was used to preserve fruit, so he suggested that they might add some sugar to the yeast extract. They did this, and to their surprise the brown liquid converted the sugar into alcohol!

Until now people had thought that sugar could only be fermented by *living* yeast cells. Buchner showed that fermentation does not depend on living cells, but is achieved by a chemical substance in the cells. We now know that this substance is a mixture of enzymes.

(Adapted from '*The Science of Life*' by Gordon Rattray Taylor, Thames and Hudson)

1 What do you think caused Buchner's yeast extract to go bad?

Suggest two things that could be done with it today to prevent it going bad.

2 What is the advantage to yeast of being able to convert sugar into alcohol?

Picture 1 Eduard Buchner who discovered enzymes.

3 How could you show that the substance in Buchner's yeast extract was an enzyme and not some other kind of substance?

Using enzymes to manufacture syrups

Enzymes, obtained mainly from bacteria, are used for making sweet syrups from starch. The syrups formed contain two sugars, maltose and glucose.

The starch is obtained from plants from which it is extracted by milling. The milled starch consists of masses of starch grains (see page 23). The grains are broken open by heat treatment so as to release the starch itself.

After cooling, the starch is rather like unsweet jelly. It is now mixed with certain enzymes which liquefy it and turn it into maltose. Treatment with other enzymes turns some of the maltose into glucose. Both these sugars are sweet.

By varying the amounts of the different enzymes, it's possible to produce syrups consisting of mainly maltose with relatively little glucose, or mainly glucose with relatively little maltose.

High maltose syrups are used in the brewing industry, whereas high glucose syrups are used for making jam and confectionery. By further enzyme treatment, glucose syrups can be turned into fructose syrups. Fructose is sweeter than glucose and is used as a sweetener in foods and drinks.

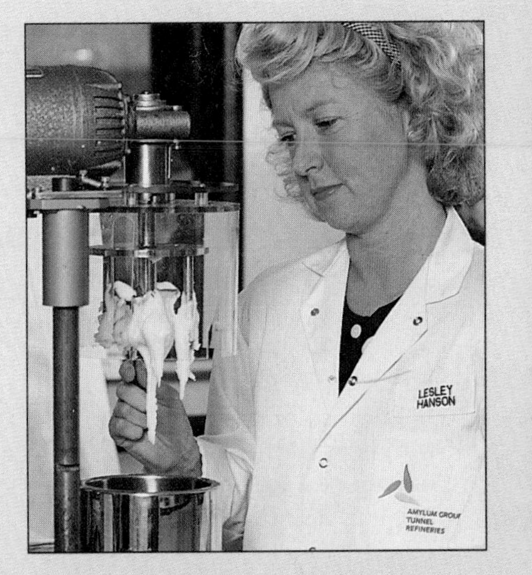

Picture 1 Making fondant from glucose syrup.

Picture 2 Part of the pipework in a refinery where syrups are manufactured.

1 Before enzymes were used, starch was converted into sugars by heating it with an acid. Why is the enzyme method better? Think of as many reasons as you can.

2 Suppose you are starting up a syrup-producing factory. What sort of building would you need, and how should it be equipped? What kind of staff would you employ?

B1 Identifying, naming and classifying

About one and a half million kinds of organisms have been discovered. How do we identify, name and classify them?

Picture 1 Five flowering plants which you might see on a walk.

Picture 2 The sea mouse *Aphrodite*. It is not a mouse at all, but a relative of the earthworm.

How do we identify living things?

Suppose you go for a walk and you see one of the plants shown in picture 1. How can you find out its name? One way might be to compare it with pictures in a book. This is all right if it is a short book, but if it is a long one it can be tedious and it is difficult to know where to start.

It is better to use a **key**. Keys are used by biologists to identify organisms quickly and accurately.

A key for identifying the plants in picture 1 is shown below.

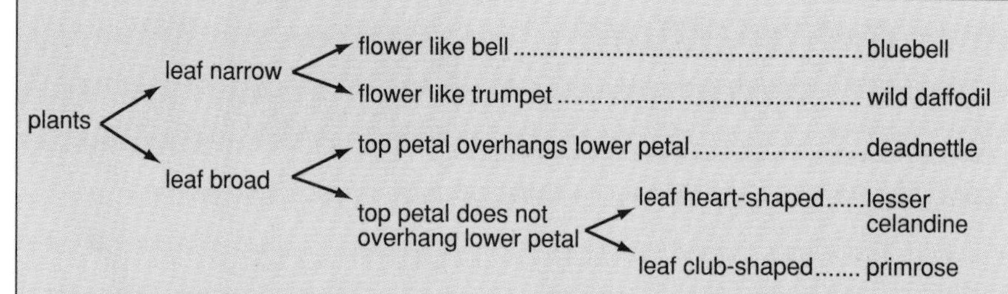

Because of its shape, this is called a **spider key** or **tree key**. The trouble with keys of this kind is that they take up a lot of room. So usually we use a **numbered key**. A numbered key for the plants in picture 1 is shown below. Use it to identify each of the plants.

1 leaves narrow	go to 2
leaves broad	go to 3
2 flower like bell	bluebell
flower like trumpet	wild daffodil
3 top petal overhangs lower petal	deadnettle
top petal does not overhang lower petal	go to 4
4 leaf heart-shaped	lesser celandine
leaf club-shaped	primrose

Picture 3 The marsh marigold, *Caltha palustris*. Its many common names include King cup, golden cup, brave celandine, horse blob, May blob, Mary bud, soldier's button and publicans and sinners!

What's in a name?

The names of the plants given in the key which you have just used are **common names**. They are the names which we use in everyday language like cat, dog, lion and so on.

The trouble with common names is that they can be misleading. One reason is that they are often based on superficial appearance. Look at the animal in picture 2, for example. It is called a 'sea mouse'. But it isn't a mouse at all. It's a relative of the earthworm. It got its name because it looks hairy and reminded someone of a mouse!

Another problem is that an organism may have more than one common name. Take the plant in picture 3, for example. Most people would probably call this plant a marsh marigold. However, it is known by hundreds of other names. Some of them are given in the caption underneath the picture. To make matters worse, in parts of America it is called a cowslip, a name which in Britain is given to a completely different plant.

A better way of naming organisms

Biologists use a standard international system. Each type of organism is given two names. This is called the **binomial system**. For example, the lesser celandine in picture 1 is called *Ranunculus ficaria*.

The first name, *Ranunculus*, is the name of the **genus** to which the plant belongs. The lesser celandine shares this name with other closely related plants such as buttercup and crowfoot.

The second name, *ficaria,* is the name of the **species** to which the plant belongs. This name is possessed only by the lesser celandine; no other plant in the genus has this name.

All organisms are given two names in this way. Thus the family dog is *Canis familiaris*, the cat is *Felis catus*, and you are *Homo sapiens*. These are called **proper names** or **scientific names**. They are written in Latin or have Latin endings. It is usual to start the genus name with a capital letter, and the species name with a small letter, and to print both in italics.

The trouble with scientific names is that they are often long and difficult to remember. For example, there is a certain type of worm which is called *Haploscoloplos bustorus*! So, to make things easier we often call organisms by their simpler common names. This is all right so long as we can be sure that there will be no confusion.

How do we classify organisms?

Classification means puttings things into groups. The smallest group that an organism belongs to is the species. As we have seen, related species are placed in the same genus (plural: genera). Now we can put similar genera into larger groups, and these groups can in turn be lumped together into even larger groups.

If we go on doing this, we finish up with five very large groups called **kingdoms**. Each kingdom contains many different organisms, but they all have certain basic features in common. In the next topic we shall look at the kingdoms and see something of the variety that exists within them.

Questions

1 Write down the common names of five well-known organisms other than the ones mentioned in this topic, then use books to find out their proper names.

2 It is particularly important that organisms should be given their proper names in medicine, agriculture and industry. Why?

3 The following questions are about the key to the plants in picture 1.

a Why were bluebells distinguished from daffodils by their flowers rather than by their leaves?

b If you were to distinguish between the lesser celandine and primrose by their flowers rather than by their leaves, what would you say about them?

4 A scientist visits an uninhabited island and discovers the insects shown in the illustration below. Make up a name for each insect, and devise a key which would enable another visitor to the island to identify them.

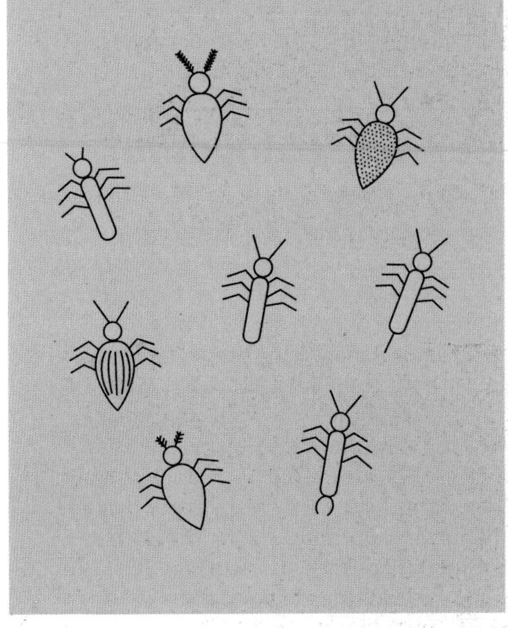

Activities

A Using a key to identify organisms

Your teacher will give you a collection of organisms or pictures of organisms labelled A, B, C, etc. You will also be given a key for identifying them. Use the key to identify each organism. Make a table showing which features enabled you to carry out your identifications.

B Making your own key

1 Your teacher will give you a collection of organisms, or pictures of organisms, together with their names.

2 Make a key, similar to the one on page 28, which would enable a person to find out the name of each organism. To do this, write out a spider key first, and then make a numbered key from the spider key.

3 Ask a friend to identify the organisms using your key.

4 If your friend runs into any difficulties, improve the wording of the key to make it clearer.

Why is a numbered key better than an spider key?

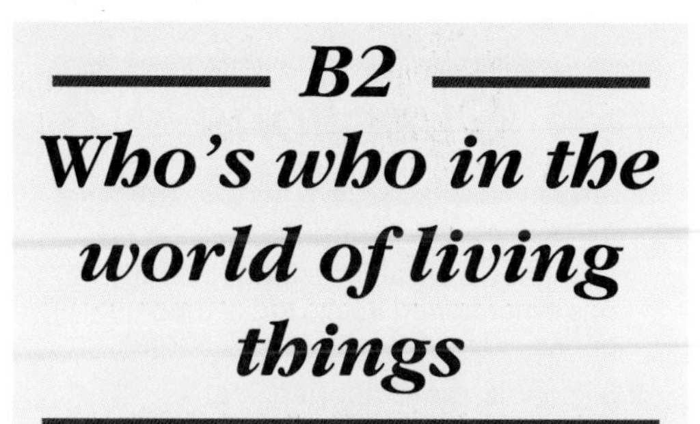

B2 Who's who in the world of living things

Here we shall look at the main groups of living things, and get a glimpse of the variety that is found amongst organisms.

Introduction

There are several different ways of classifying living things. In this book we shall divide them into five kingdoms:

Bacteria Kingdom,
Protoctist Kingdom,
Fungus Kingdom,
Plant Kingdom,
Animal Kingdom.

Most living things belong to the plant and animal kingdoms. It is here that we find the greatest number of species and the widest variety of form. In the lightning tour that follows we can only touch on the tremendous variety that really exists.

The sizes which are given with the pictures are approximate. The smallest organisms consist of only one cell (**unicellular**). The larger ones are made up of many cells (**multicellular**). Those that can only be seen properly by using a microscope are **microorganisms** (**microbes**).

Bacteria Kingdom

Can only be seen with the high power of the light microscope. Consist of a single cell with a wall but no proper nucleus. Varied methods of feeding. Some, e.g. blue-green bacteria, feed by photosynthesis. Occur in air, water, soil or inside other organisms. Many of them cause diseases.

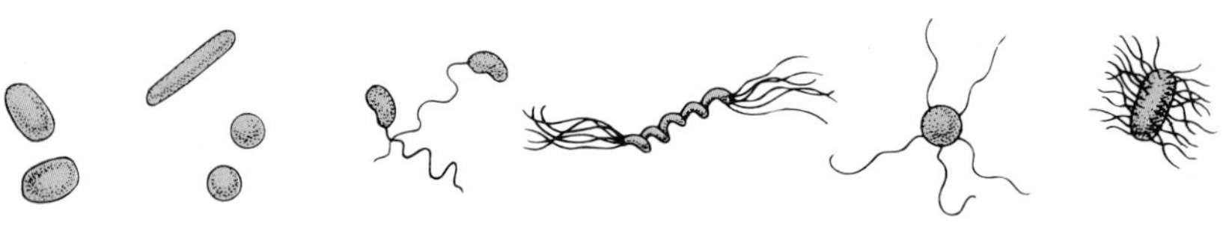
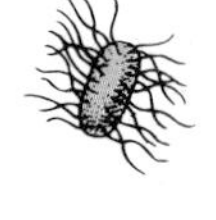

average width 1µm

Protoctist kingdom

A wide range of single-celled and simple many-celled organisms. The single-celled ones have a proper nucleus and can usually be seen with the low power of the microscope. Some feed by taking in organic substances, others by photosynthesis. The photosynthetic ones are called **algae**. Most algae are many-celled but do not have roots, stems or leaves. They are usually green, but sometimes brown or red. Protoctists live mainly in water and soil; some of the single-celled ones are parasites. The many-celled ones include seaweed.

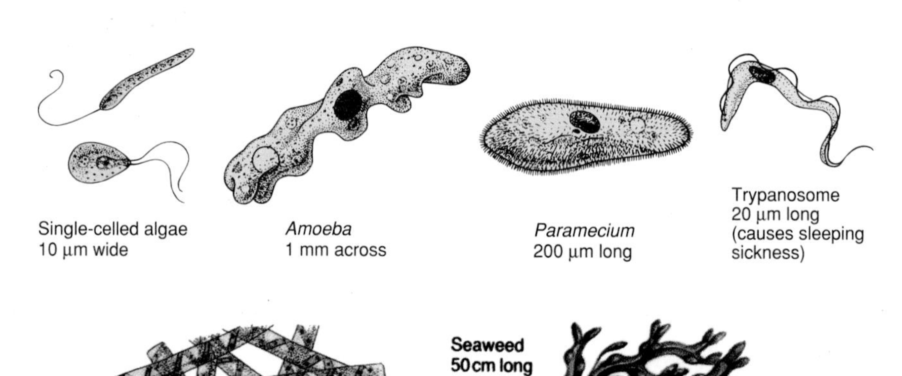

Single-celled algae 10 µm wide

Amoeba 1 mm across

Paramecium 200 µm long

Trypanosome 20 µm long (causes sleeping sickness)

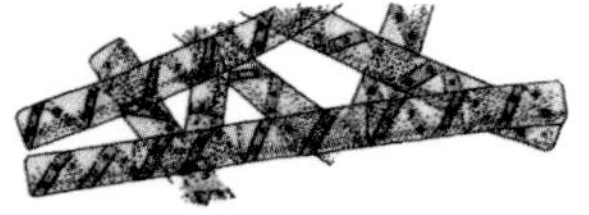

Spirogyra (a thread-like alga)

Seaweed 50cm long

Fungus kingdom

Most are many-celled, consisting of a network (**mycelium**) of fine threads (**hyphae**). The threads may be densely interwoven to form mushrooms and toadstools. Feed by absorbing organic substances from dead or living material. Latter are parasites, especially of plants.

Pin mould

Mushroom 5 cm wide

Yeast each cell 5 µm wide

Potato blight fungus

Lichens
Consist of a fungus and a plant-like protoctist combined together. Grow on rocks and tree trunks. Very resistant to drying.

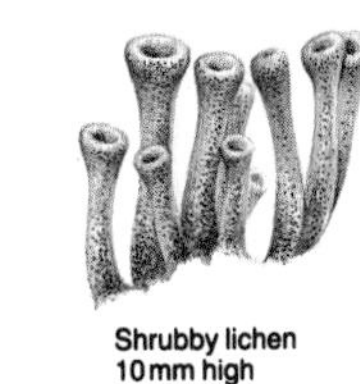

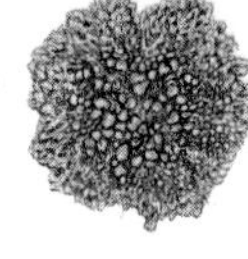

Plant kingdom

Many-celled organisms which contain the green substance chlorophyll and make their own food by photosynthesis.

Mosses and liverworts (Bryophytes)
Have simple leaves or leaf-like form but no proper roots or stems, and no vascular tissue. Found mainly in damp places. Reproductive spores are formed in **capsule**.

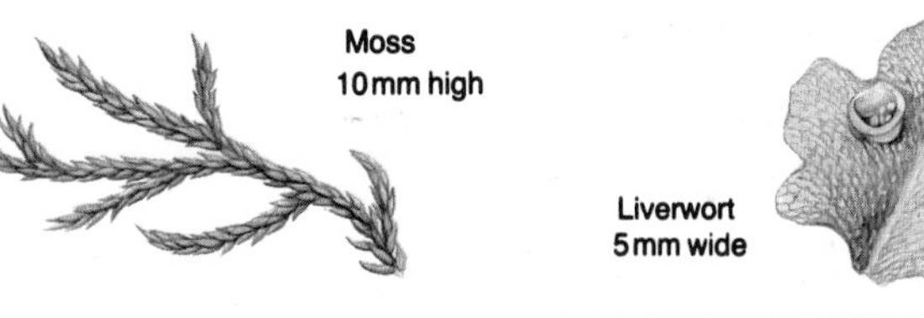

Ferns
Have proper roots and stems, and leaf-like fronds with vascular tissue. Young fronds are coiled into a bud (called a 'fiddlehead'). Found mainly in damp places. Reproductive spores are formed on the undersides of the fronds. The spore bodies are in clusters called **sori** (singular: **sorus**).

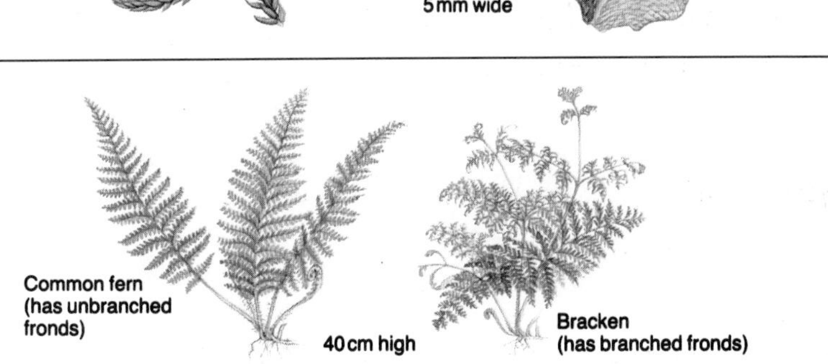

Conifers
Mainly large plants with seed-bearing **cones** for reproduction. Most of them keep their leaves throughout the year (evergreen).

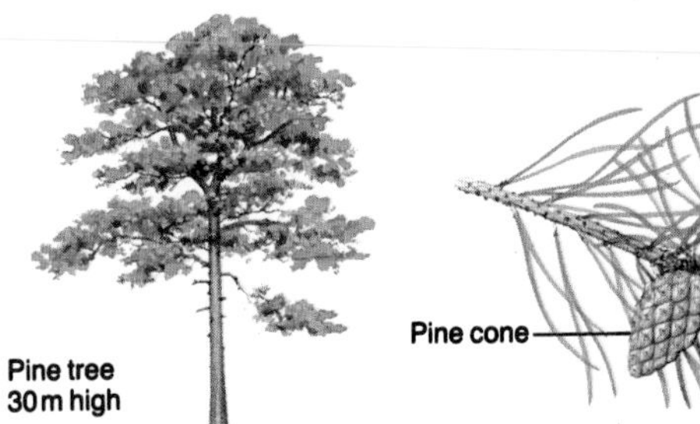

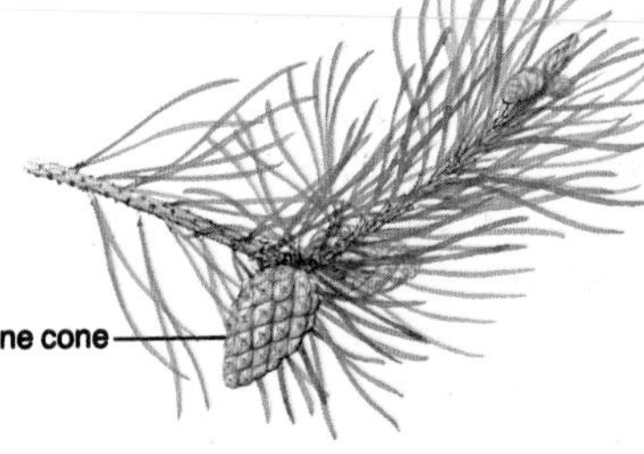

Flowering plants
Wide range of plants with seed-bearing **flowers** for reproduction. Seeds protected inside fruits. Range from small herbs to massive trees. Some are evergreen, others drop their leaves in winter (deciduous).

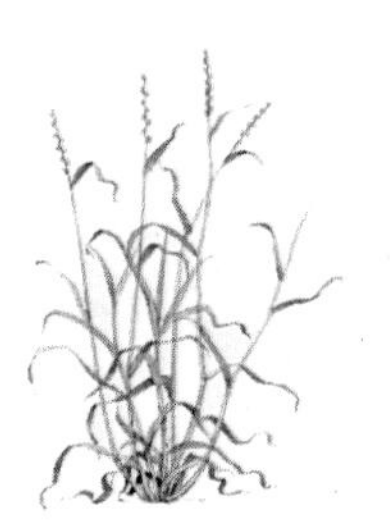

Animal kingdom

Many-celled organisms that feed on other organisms. They all have a nervous system and usually move around.

Animals without backbones (invertebrates)

Cnidarians

Simple sac-like body with tentacles and **stinging cells**. Live singly or in colonies, either attached or floating. May produce a hard external coating (e.g. corals). Most live in the sea, a few in fresh water.

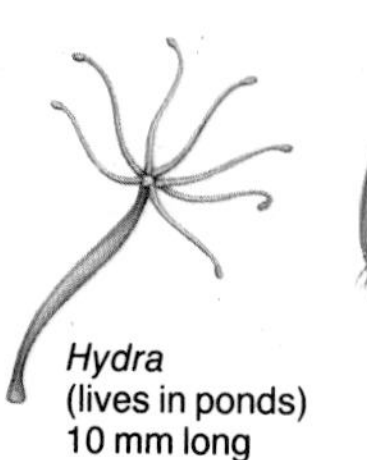

Hydra
(lives in ponds)
10 mm long

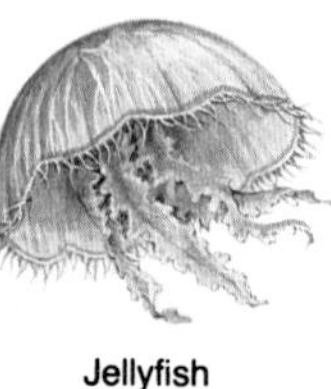

Jellyfish
10 cm wide

Sea anemone
5 cm tall

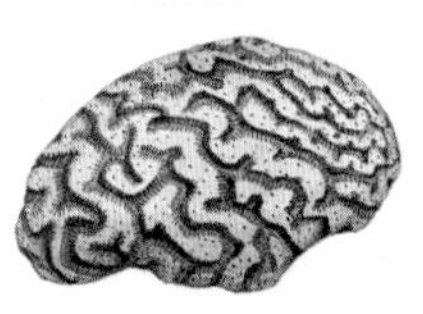

Coral

Flatworms

Body elongated and flat. Some live in fresh water, but most are parasites of animals.

Fresh-water flatworm
10 mm long

Tapeworm
5 m long

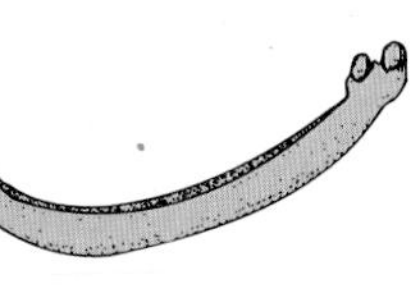

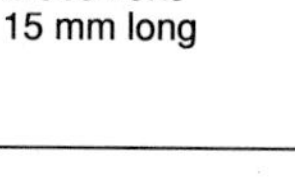

Blood fluke
15 mm long

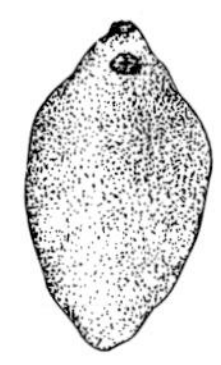

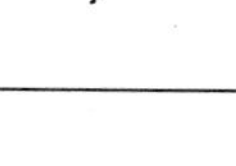

Liver fluke
2 cm long

Roundworms

Body elongated and thread-like, round in cross-section. Some live in soil but most are parasites of plants or animals.

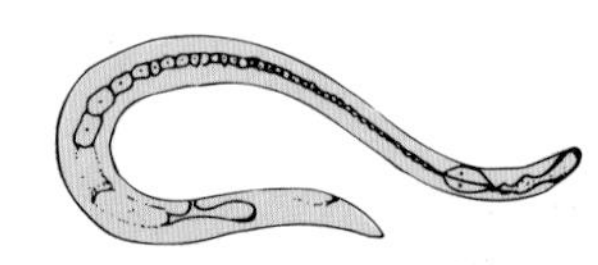

Ascaris 30 cm long
(human roundworm)

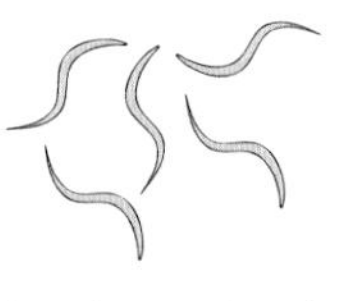

Threadworms 1 cm long
(live in rectum of humans)

Annelids (segmented worms)

Body long and divided by rings into a series of **segments** (annelid means 'ringed'). Lack legs but have bristle-like **chaetae** which may assist in locomotion. Most are aquatic (live in water), but some live in the soil. Some are external parasites.

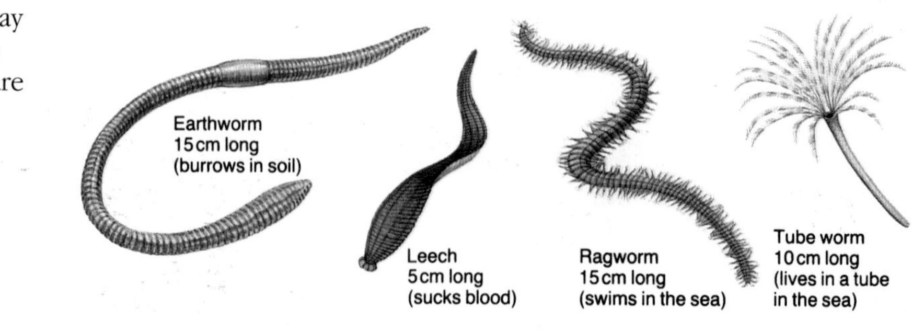

Molluscs

Body soft and unsegmented, usually covered by a **shell**. Most are aquatic, some live on the seashore and on land.

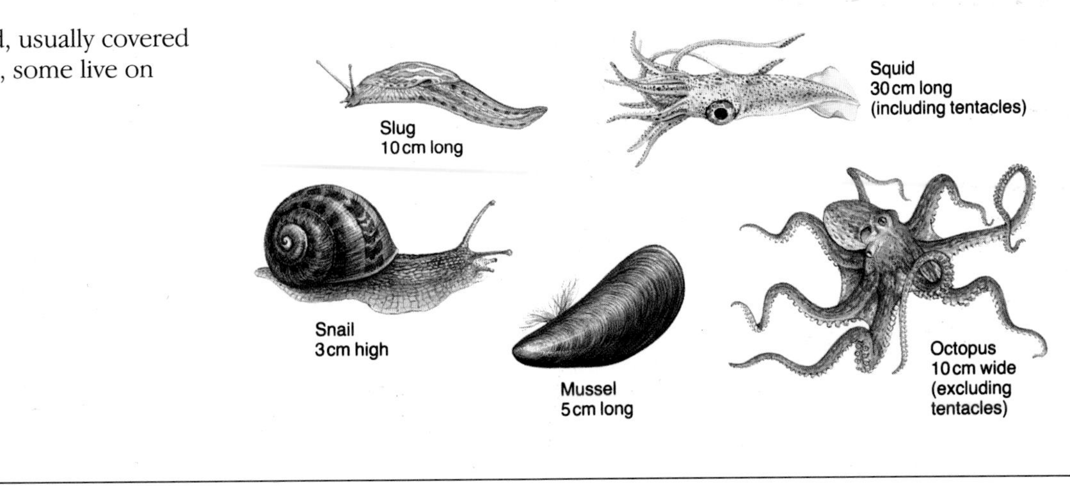

Echinoderms

Body based on a pattern of five parts which typically radiate out like a star. Have a tough skin, often with **spines** (echinoderm means 'spiny skin'). All live in the sea.

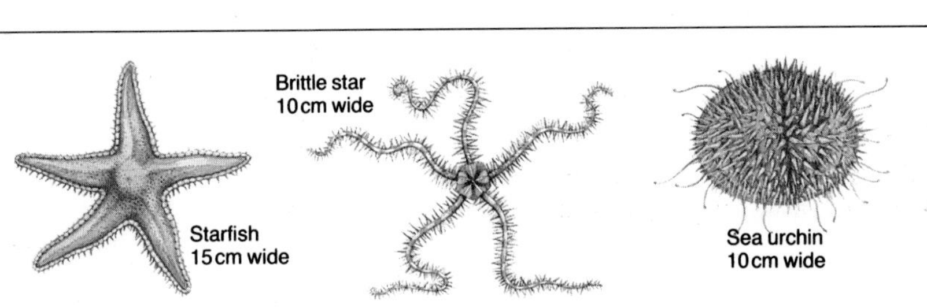

Arthropods

Segmented animals with a hard cuticle (**exoskeleton**) and jointed appendages on at least some of their segments (arthropod means 'jointed foot'). Most have feelers (**antennae**) and **compound eyes**. (Compound eye is made up of many little eyes packed together.) Some, e.g. insects, have air holes (**spiracles**) for breathing.

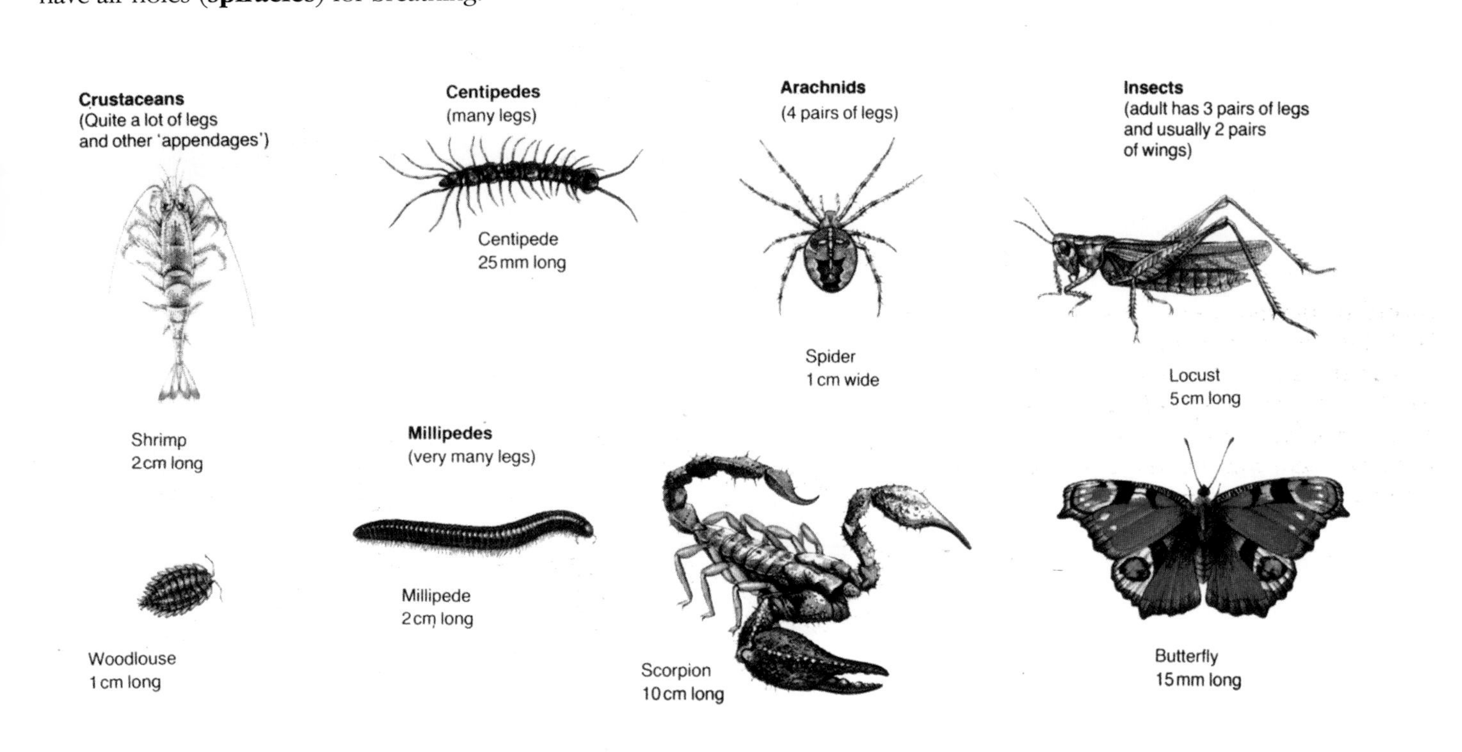

Animals with backbones (vertebrates)

Fish

Skin covered with **scales**. Live in water. Have **gills** for breathing and fins for movement.

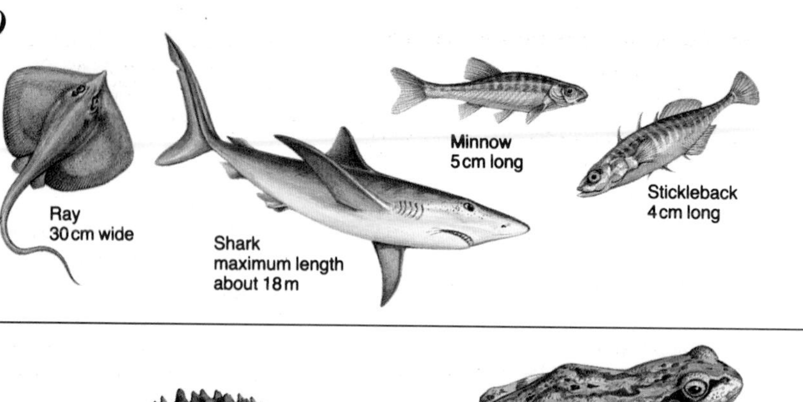

Amphibians

Have soft skin without scales. Live on land but lay eggs in water. Have **tadpole** (larva) which changes into the adult. Tadpole is aquatic with gills, adult is usually terrestrial (land-living) with lungs.

Reptiles

Have hard, tough skin with **scales**. Eggs have a soft shell and are laid on land. Lungs for breathing.

Tortoise
20 cm wide

Common lizard
12 cm long

Crocodile
about 9 m long

Snake
about 10 m
(python etc.)

Birds

Have skin with **feathers**. Eggs have hard shells. Wings for flying, and a beak for feeding. Have lungs and are 'warm-blooded'.

Mammals

Have skin with **hair**. The young develop inside the mother and after birth are fed on milk from her **mammary glands**. Have lungs and are 'warm-blooded'.

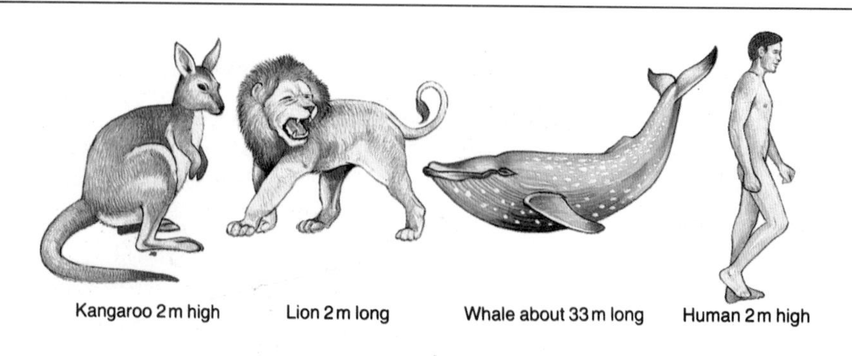

Activities

A Putting some familiar animals and plants into groups

1 Examine various organisms, or pictures of organisms, provided by your teacher. All of them are featured in the classification on pages 30–34.

2 Write down the name of the group to which each organism belongs. Use the classification to help you.

3 Look carefully at each organism.

Which particular feature or features, enabled you to place each organism in its group?

From its structure, what can you say about the sort of place where it lives, and the kind of life it leads?

Give the results of this investigation as a table. Devise the table yourself so as to present all the information in the clearest possible way.

B Putting some unfamiliar animals and plants into groups

1 Examine various organisms which are not illustrated in the classification on pages 30–34.

2 Write down the name of the group to which you think each organism belongs. Do this by relating the characteristics of the organism to the information given on pages 30–34.

3 Look carefully at each organism.

Which specific illustration on pages 30–34 does each organism resemble most closely?

In what ways do they resemble each other?

What special features does each organism have?

Do you have difficulty in placing some of the organisms? If so, why?

Present the results of this activity as a table so as to show all the information clearly.

C Collecting and naming organisms

1 Collect organisms from a habitat near your school or home. Your teacher will show you how to collect the organisms.

2 Examine each organism, using a hand lens or microscope if necessary.

3 Use the classification on pages 30–34 to find out what group each organism belongs to.

4 Return all living organisms to their habitat afterwards.

Questions

1 What group does each of these organisms belong to: moss, jellyfish, turtle, tapeworm, whale, mushroom, pin mould, tube worm, seaweed, newt?

2 What would be the easiest way of telling the difference between:
a an arthropod and a vertebrate,
b an insect and an arachnid,
c an amphibian and a reptile,
d an alga and a fungus,
e a conifer and a flowering plant?

3 From books, try to find out the largest member of each of the following groups: algae, ferns, conifers, flowering plants.

In each case give the proper name and common names of the organism, and state its approximate size.

4 Give the name of an animal which:
a is shaped like an umbrella and has stinging cells,
b lays eggs with a soft shell,
c has a pouch in which the young develop,
d is shaped like a star,
e has two pairs of wings,
f lives on land but lays its eggs in water,
g has long tentacles and belongs to the same group as snails,
h has mammary glands,
i has four pairs of legs,
j has scales and gills.

5 Give the name of an organism which:
a reproduces by means of flowers,
b has frond-like leaves,
c consists of only one cell and is coloured green,
d causes a disease,
e has no chlorophyll.

6 Which of the following features are possessed only by insects, and which ones belong to other arthropods too:
a hard cuticle,
b joints,
c six legs,
d feelers,
e two pairs of wings?

7 The picture below is of a small insect which lives in the soil. Name two important structures, typical of most insects, which it lacks. Why do you think this insect does not need these particular structures?

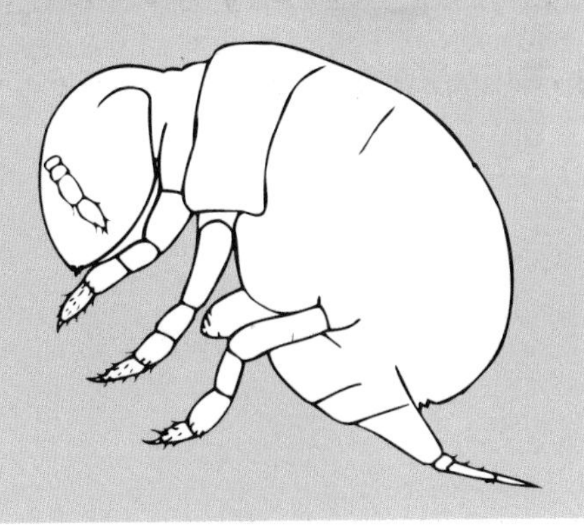

Tops, bottoms, fronts and backs

Most animals move with one end of the body in the lead. This is the front or **anterior** end. The other end is called the **posterior** end. In most animals there is some kind of head at the anterior end. This is where we expect to find the mouth and the main sense organs.

The lower side of the body, the side that's usually closest to the ground, is called the **ventral** side. The upper side is called the **dorsal** side. Humans stand and walk on two legs – we are **bipedal**. In this case the ventral side faces the front.

1 Why is it useful for an animal to have its mouth and main sense organs at the anterior end?

2 Do plants have an anterior and posterior end? If not, why?

3 What are the advantages and disadvantages of being bipedal?

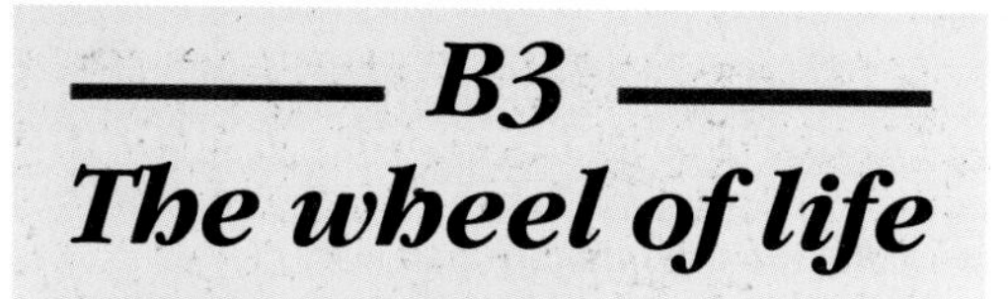

B3 The wheel of life

Chemical elements circulate in nature and can be used over and over again. Here we see how this applies to carbon and nitrogen.

The cycling of carbon

The air around us contains a small amount of carbon dioxide. This is constantly being absorbed by plants and turned into sugar and other complex carbon compounds (**photosynthesis**).

Now after a plant has been eaten by an animal, the carbon compounds get into the animal's cells. Here they are broken down into carbon dioxide and water (**respiration**). The carbon dioxide is breathed out. As a result, carbon dioxide is put back into the air. Plants also put carbon dioxide back into the air, but not all the time (see page 38).

When the animals and plants die, they decay. Bacteria and other decomposers feed on them. They too respire, and so once again carbon dioxide is put back into the air.

So we see that carbon goes round and round in nature. This is known as the **carbon cycle** (picture 1).

Plants do not always decay after they have died. In certain circumstances they form peat which becomes fossilised into coal. When coal and other fossil fuels are burned, carbon dioxide is released (**combustion**). This too is part of the carbon cycle.

The cycling of nitrogen

In the soil there are inorganic nitrogen compounds called **nitrates**. They are dissolved in the soil water. The nitrates are absorbed by the roots of plants, and the plants then turn them into amino acids which are used to build up **proteins**.

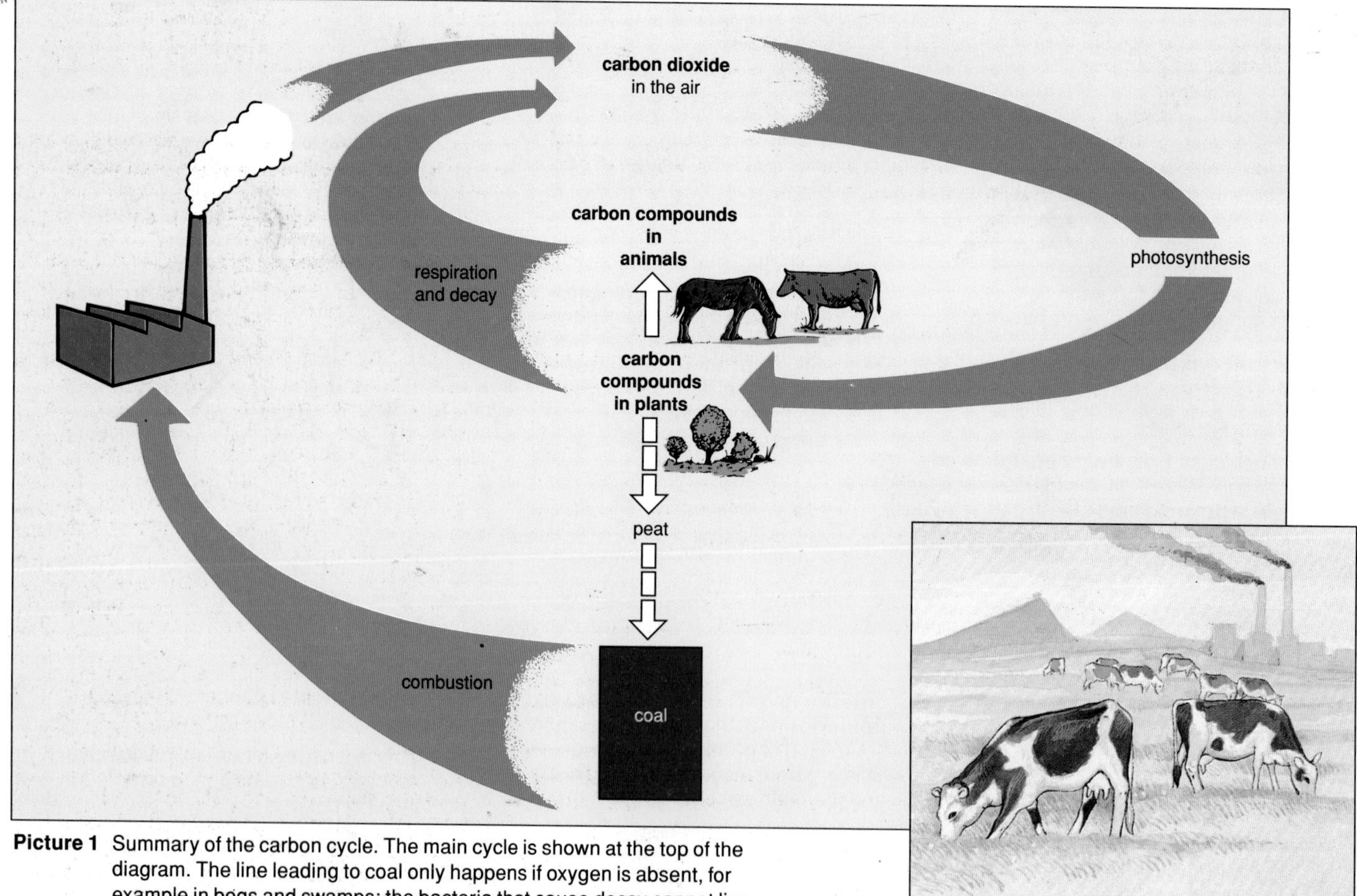

Picture 1 Summary of the carbon cycle. The main cycle is shown at the top of the diagram. The line leading to coal only happens if oxygen is absent, for example in bogs and swamps: the bacteria that cause decay cannot live without oxygen, and so the dead plants pile up, forming soft, black peat. In the course of time the peat gets buried and hardens to form coal.

Picture 2 This scene contains some of the main participants of the carbon cycle.

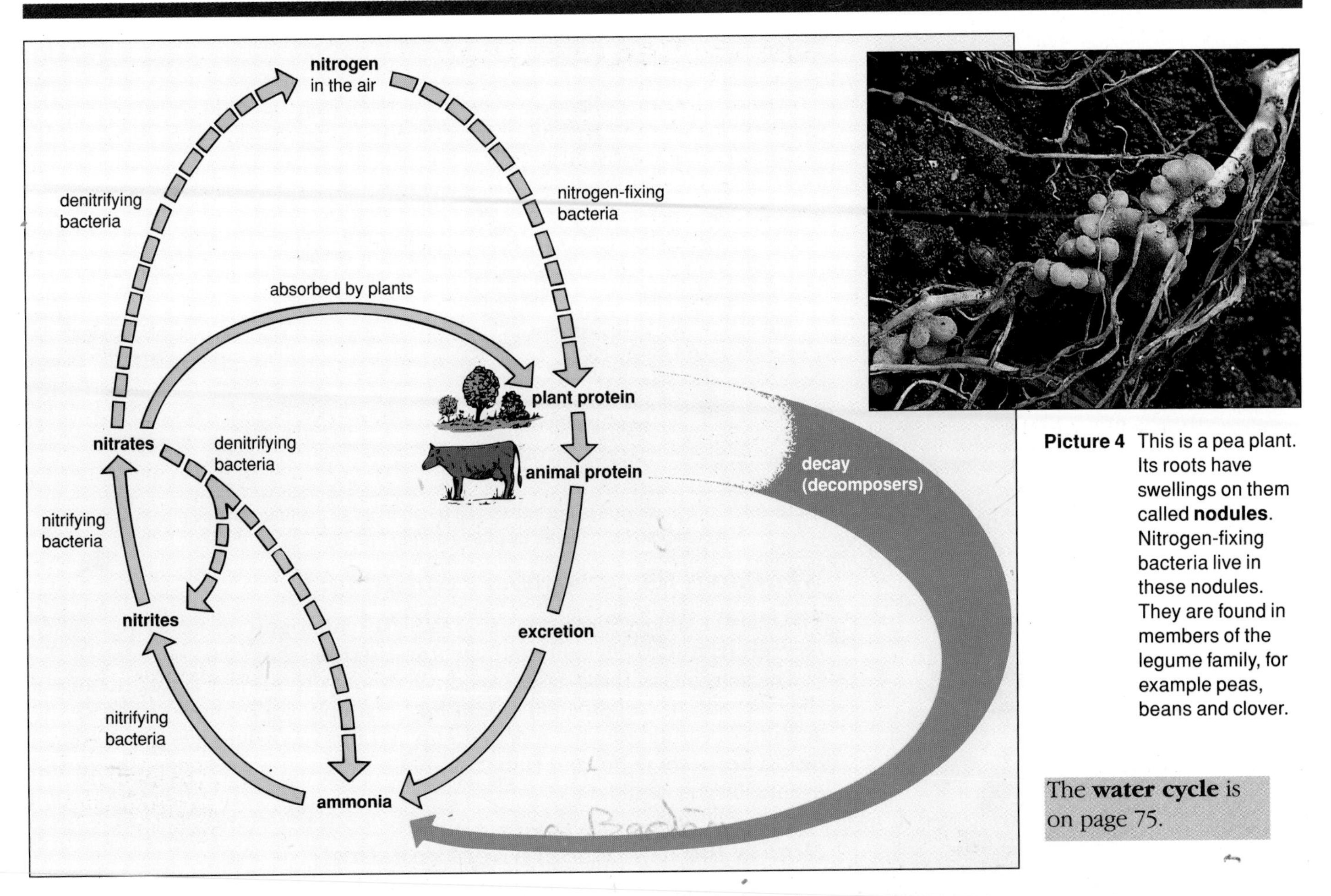

Picture 3 Summary of the nitrogen cycle. The diagram shows how nitrates can be formed by the action of bacteria on animal and plant protein. Some nitrates can also be made by the action of lightning on the nitrogen in the air.

Picture 4 This is a pea plant. Its roots have swellings on them called **nodules**. Nitrogen-fixing bacteria live in these nodules. They are found in members of the legume family, for example peas, beans and clover.

The **water cycle** is on page 75.

Now think what happens when a plant is eaten by an animal. The nitrogen in the plant protein gets into the animal's body and becomes part of its protein.

When the animals and plants die, they decay. Bacteria and other decomposers break down the proteins into ammonia. Ammonia is also formed from the animals' excreta.

Now in the soil there are bacteria which turn ammonia into nitrites, and others which turn nitrites into nitrates. The effect is to put nitrates back into the soil, where they can be used again by plants. These bacteria enrich the soil in nitrates and make it good for plants to grow in. Because of this, they are called **nitrifiers**.

The circulation of nitrogen is known as the **nitrogen cycle** (picture 3). In picture 3 notice that certain bacteria turn nitrates into nitrites, ammonia and even nitrogen. These bacteria lower the nitrate content of the soil and make it less good for plant growth – in fact nitrites are poisonous to most plants. For this reason, we call these bacteria **denitrifiers**.

Plants cannot make use of the element nitrogen. However, certain bacteria can absorb it from the air and use it for making protein. These bacteria are called **nitrogen-fixers**. Some of them are found free in the soil. Others live in the roots of plants belonging to the legume family, for example peas, beans and clover (picture 4). When these plants decay, the nitrogen which the bacteria have fixed goes into the soil where it can be used by plants. Nitrogen-fixing bacteria are therefore very useful because they increase the amount of useful nitrogen compounds in the soil.

Questions

1 Study the carbon cycle in picture 1, then answer these questions:

- a What is photosynthesis and how does it remove carbon dioxide from the air?
- b What is respiration and how does it add carbon dioxide to the air?
- c What is decay and how does it add carbon dioxide to the air?
- d What happens chemically when coal is burned, and how does it affect the carbon cycle?

2 Construct a diagram to show how oxygen circulates in nature. Why is this important to humans?

3 Farmers often plough plants such as clover or lucerne into the soil. Why is this a good thing to do? Explain your answer.

4 The bacteria responsible for decay, nitrification and nitrogen-fixation all require oxygen. However, denitrifying bacteria do not need oxygen. Bearing this in mind, what advice would you give to farmers about how to look after their soil?

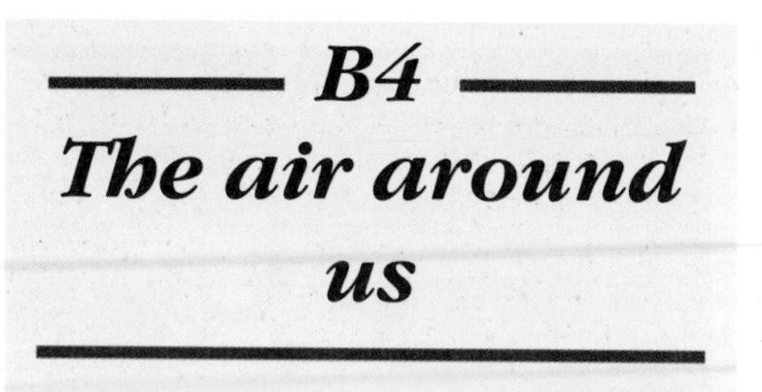

B4
The air around us

How is the atmosphere surrounding the Earth able to support life?

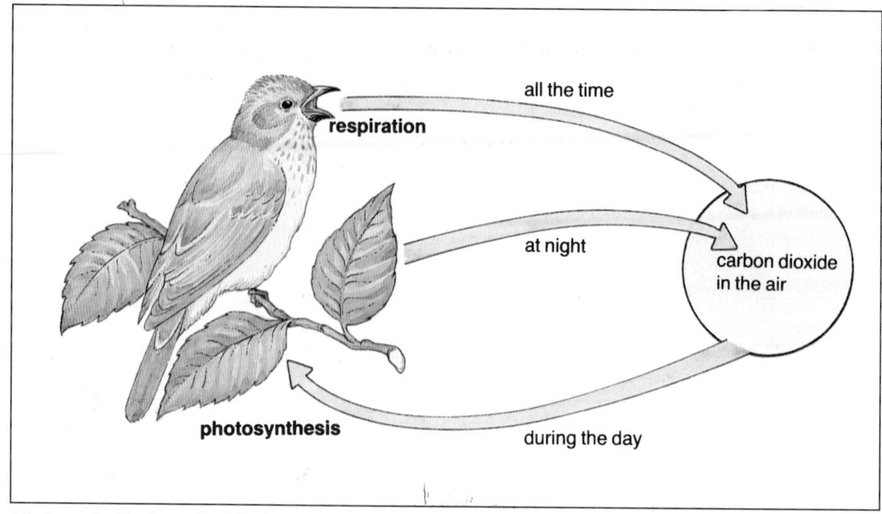

Picture 1 Carbon dioxide is put into the air by animals all the time, and by plants at night. It is removed from the air by plants during the day.

What does the atmosphere consist of?

The Earth's atmosphere consists mainly of **nitrogen** (over 70 per cent), **oxygen** (about 20 per cent) and **carbon dioxide** (about 0.03 per cent).

The amounts of these gases do not vary much. This is just as well because if they did our air could become unfit for breathing. For example, suppose the amount of carbon dioxide was to go up and up. You would feel drowsy, get a headache, become hot and faint. Your brain would stop working properly, and eventually you would die.

How is the atmosphere kept constant?

When living things respire they use up oxygen and give out carbon dioxide. On the other hand, when green plants photosynthesise they use up carbon dioxide and give out oxygen. In other words *respiration removes oxygen from the atmosphere and adds carbon dioxide to it, whereas photosynthesis removes carbon dioxide from the atmosphere and adds oxygen to it.*

In the world as a whole, respiration and photosynthesis are in balance with each other. The result is that the amounts of carbon dioxide and oxygen are kept constant.

A more detailed look at carbon dioxide

Animals give out carbon dioxide all the time. Plants, however, give out carbon dioxide only at night. During the day they take it in (picture 1). Whether a plant takes in, or gives out, carbon dioxide depends on which process is going at the faster rate, photosynthesis or respiration. That depends on how much light there is. And that, in turn, depends on the time of day (picture 2).

The overall effect of animals and plants living together in a normal environment is to keep the amount of carbon dioxide in the atmosphere more or less constant.

Humans interfere

We have seen that respiration adds carbon dioxide to the atmosphere. So does burning things (combustion.) What happens when we put some extra carbon dioxide into the air by, for example, burning coal in a factory or making a bonfire? So long as the amount isn't too great, plants respond by using it up more quickly, so the amount in the atmosphere remains unchanged.

It would be nice if that was the end of the story, but unfortunately it is not. You can find out why in the next topic.

Picture 2 Whether a plant takes in, or gives out, carbon dioxide, depends on which process is going faster, photosynthesis or respiration. This in turn depends on the time of day.

Activities

A How do animals and plants affect the amount of carbon dioxide in the atmosphere?

In this experiment we shall make use of a hydrogencarbonate indicator. This changes colour according to how much carbon dioxide is present:

- Yellow means there is more carbon dioxide than in atmospheric air.
- Purple means there is less carbon dioxide than in atmospheric air.
- Reddish-orange means there is the same amount of carbon dioxide as in atmospheric air.

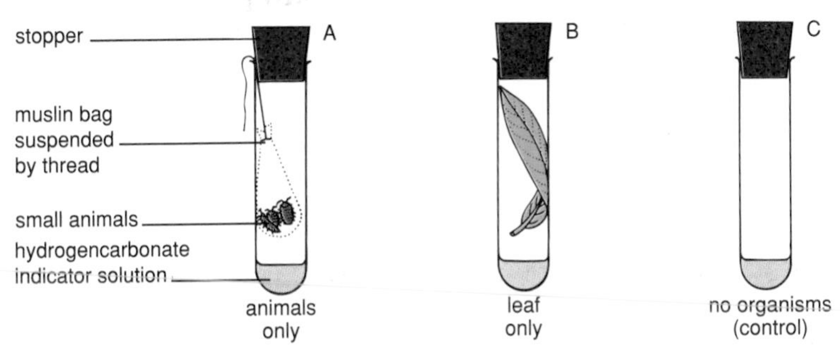

1 Label three test tubes A, B and C. Pour the same amount of indicator solution into each tube.

Notice that the indicator is reddish-orange: this is its colour when it is in contact with ordinary atmospheric air.

2 Set up the three test tubes as shown in the illustration. Seal each tube with a stopper so there is no chance of air getting in or out.

3 Leave the three tubes in a well lit place for about an hour.

4 Now compare the colour of the indicator in the three tubes. It's best to look at the colours against a plain white background.

5 For each test tube, say whether carbon dioxide has been added to, or removed from, the air. Explain the reason in each case.

6 Use the same method to investigate the effect of the following on the amount of carbon dioxide in the atmosphere:

a a leaf in the dark,
b animals in the dark,
c a leaf and animals in the light,
d a leaf and animals in the dark.

What conclusions can you draw about the effect of animals, plants and the two together, on the amount of carbon dioxide in the atmosphere?

B To see if plants will survive in a sealed container

Respiration removes oxygen from the air and adds carbon dioxide to it, whereas photosynthesis removes carbon dioxide from the air and adds oxygen to it.

So respiration provides the carbon dioxide for photosynthesis, and photosynthesis provides the oxygen for respiration. If this is true, plants should be able to survive in a sealed, air-tight container. Test this idea for yourself by making a 'bottle garden'. What would be the effect of introducing animals into your bottle garden?

Questions

1 Why is it a good idea to open classroom windows whenever possible?

2 Scientists measured the carbon dioxide content of the air in a part of the northern hemisphere with a temperate climate. They found that the content in March was 0.02971 per cent, whereas in September it was 0.02905 per cent. Explain the difference. Would you expect the same to be true in the tropics?

3 It has been suggested that a suitable atmosphere might be maintained in a manned space capsule by having some plants inside. Do you think this is feasible? What problems might be encountered in putting it into practice?

4 Plants respire all the time (i.e. day and night), but they only give out carbon dioxide at night. Explain.

5 Imagine there is a catastrophe in which all the plants of the world are suddenly destroyed. Describe in detail the effects which you think this might have.

6 Under what circumstances, if any, would you expect a plant to:

a respire but not photosynthesise,
b photosynthesise but not respire,
c respire and photosynthesise,
d neither respire nor photosynthesise?

7 Can you think of any circumstances in which a non-photosynthetic organism might give out carbon dioxide without taking in oxygen. (The answer to this question is given in the book. If you don't already know the answer, see if you can find it. Try using the index.)

8 At an agricultural research station, a group of scientists measured the amount of carbon dioxide in the air in the middle of a wheat field every three hours for 24 hours. Here are their results:

Time	Percentage of carbon dioxide in the air
24 (midnight)	0.042
3	0.037
6	0.031
9	0.029
12 (noon)	0.028
15	0.030
18	0.032
21	0.035
24 (midnight)	0.042

a Plot these results on graph paper.
b Explain them as fully as you can.
c How would you expect oxygen to change during the same period?

B5 Air pollution

Human activities release polluting gases into the atmosphere. Here we look at their effects and how they may be controlled.

Air pollution is not new. The earliest humans cooked on wood fires which produced polluting gases. People have been polluting the air for thousands of years. However, there are many more people on Earth now. More people mean more homes, more industries, more motor vehicles – and more air pollution.

The gases we release spread rapidly through the atmosphere. Some of them are harmless, but others can damage the environment.

Table 1 summarises the main substances that cause air pollution. Some are **local pollutants**: their effects are felt mainly near the places where they are produced. Others are **long range pollutants**: their effects are felt over long distances, often hundreds or even thousands of miles from where they were produced. In this Topic we shall look at three such long range pollutants.

Picture 1 Industry is one source of air pollution. This smoke is coming from an asphalt plant in Colorado.

The greenhouse effect and global warming

A greenhouse keeps plants warmer than they would be outside. It does this because the glass traps some of the Sun's radiation energy and prevents it getting out.

In a similar way, the atmosphere helps to keep the Earth warm. It traps some of the Sun's radiation energy that would otherwise escape. This is called the **greenhouse effect** (picture 2). It is very useful to us. Without it the average temperature of the Earth's surface would be about 33°C lower than it is, –18°C instead of a comfortable 15°C.

Some gases are better than others at keeping the Earth warm. Oxygen and nitrogen are not very effective, but carbon dioxide is very good – so good, in fact, that there may be a problem. When fossil fuels such as coal and oil are burned, carbon dioxide is released into the atmosphere. People are concerned that we have been putting so much *extra* carbon dioxide into the atmosphere that the Earth could overheat. This is known as **global warming**. Matters are made worse by the mass destruction of natural vegetation which would normally absorb some of the extra carbon dioxide.

There is no doubt that the concentration of carbon dioxide in the air *has* increased (picture 3). The average temperature of the Earth has also shown a small increase. In the past 100 years the average temperature in the USA has risen by about 2°C. It is estimated that if the amount of carbon dioxide in the atmosphere doubles, the average temperature of the Earth will rise by about 3°C. This may not sound a lot, but it will make a big difference to the climate.

Table 1 The main air pollutants.

Air pollutant	Source	Effects	Possible methods of control
Sulphur dioxide (SO_2)	Burning fossil fuels	Causes acid rain	Remove sulphur from fuels before burning. Remove sulphur dioxide from chimney gases of power stations
Nitrogen oxides (NO, NO_2, N_2O)	Vehicle exhausts, burning fuels	Help cause acid rain and photochemical smog	Fit catalytic converters to vehicle exhausts. Modify engines to run on a weaker mixture of fuel and air
Carbon dioxide (CO_2)	Burning fuels	Causes greenhouse effect, affecting Earth's climate	Burn less fossil fuels
Carbon monoxide (CO)	Burning fuels, vehicle exhaust, cigarette smoke	Poisonous to animals, including humans	Ensure vehicle engines are well maintained. Prevent cigarette smoking
Hydrocarbons	Vehicle exhausts, burning fuels	Help cause acid rain and photochemical smog	Fit catalytic converters to vehicle exhausts. Modify engines to run on a weaker mixture of fuel and air
Smoke	Burning fuels	Damages lungs; reduces photosynthesis of plants	Use smokeless fuels. Make sure engines and burners have plenty of air to burn fuel efficiently
Lead compounds	Car exhausts	Damage nervous system of humans	Use unleaded petrol
Chlorofluorocarbons (CFCs)	Aerosol propellants, refrigerators	Destroy ozone in the ozone layer which protects Earth from ultra-violet radiation. Also contribute to the greenhouse effect	Use different substances as aerosol propellants and refrigerants

Some of the effects may be welcome. In places with good rainfall, farmers would be able to grow bumper crops. But there would be more droughts, wetlands would dry up and more land would be turned into deserts. The level of the sea would rise. This is because water expands when heated, and some of the ice at the Poles would melt. This would bring floods to low-lying areas, including parts of Britain.

Carbon dioxide is not the only 'greenhouse gas', but it is the most serious one because there is so much of it. So much that there is no chance of removing it from all vehicle exhausts and other places it comes from. The best way to reduce it is to burn less fossil fuel. But it will take time to put things right. Even if we cut back on using fossil fuels today, global warming would still continue for many years. This is because so much carbon dioxide has already been added to the atmosphere from fuel we have burned in the past.

Destroying the ozone layer

Ozone (O_3) is an unstable form of oxygen. It is a very reactive gas, and when formed near the Earth's surface it can cause pollution problems. But further up in the atmosphere, it does a *useful* job. There is a very thin layer of ozone about 20–40 kilometres above the Earth, in the part of the atmosphere called the stratosphere. This ozone layer acts as a kind of sunscreen. It filters out some of the harmful ultraviolet radiation from the Sun. If this radiation reaches Earth, it can cause sunburn and skin cancer.

In 1984 scientists discovered a large 'hole' in the ozone layer above Antarctica. Since then the hole has got larger. If it spreads to more populated parts of the Earth, there could be serious consequences for people's health.

Damage to the ozone layer is mainly blamed on chemical compounds called **chlorofluorocarbons (CFCs)**. These compounds are used as refrigerants and as propellants for aerosols. CFCs are very stable and unreactive, and they stay in the atmosphere for a long time. Over the years they slowly diffuse upwards until they reach the stratosphere. Here they react with ozone, and destroy it.

Many countries have now agreed to stop using CFCs, and scientists are developing alternative compounds that react more quickly with the air and so never reach the ozone layer. But even if all use of CFCs was stopped today, the ozone layer would still go on being destroyed for at least 20 years because of all the CFCs that are already in the atmosphere.

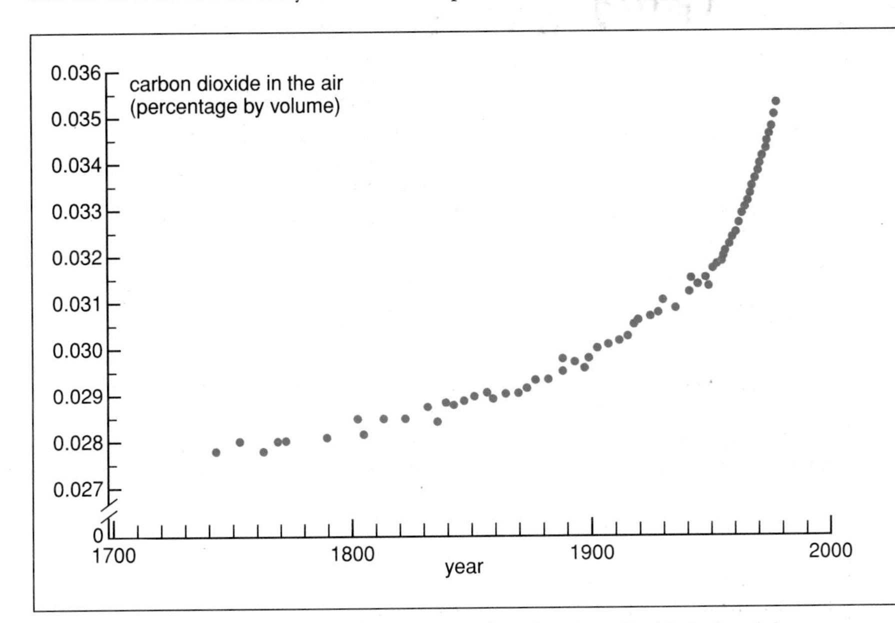

Picture 3 This graph shows how the concentration of carbon dioxide in the air has increased since 1700.

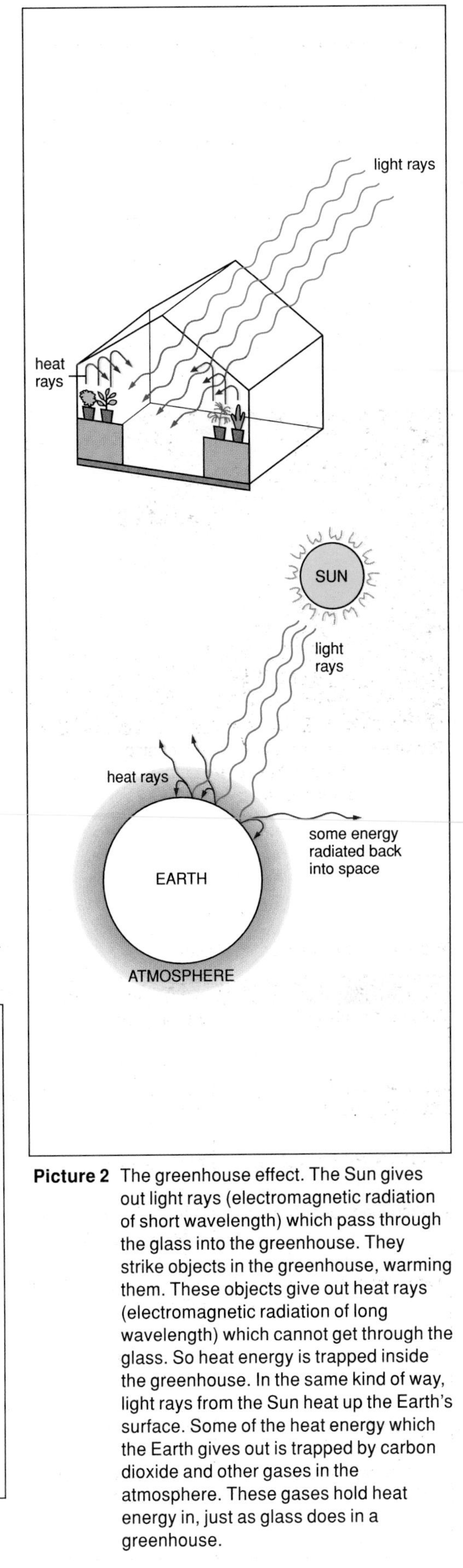

Picture 2 The greenhouse effect. The Sun gives out light rays (electromagnetic radiation of short wavelength) which pass through the glass into the greenhouse. They strike objects in the greenhouse, warming them. These objects give out heat rays (electromagnetic radiation of long wavelength) which cannot get through the glass. So heat energy is trapped inside the greenhouse. In the same kind of way, light rays from the Sun heat up the Earth's surface. Some of the heat energy which the Earth gives out is trapped by carbon dioxide and other gases in the atmosphere. These gases hold heat energy in, just as glass does in a greenhouse.

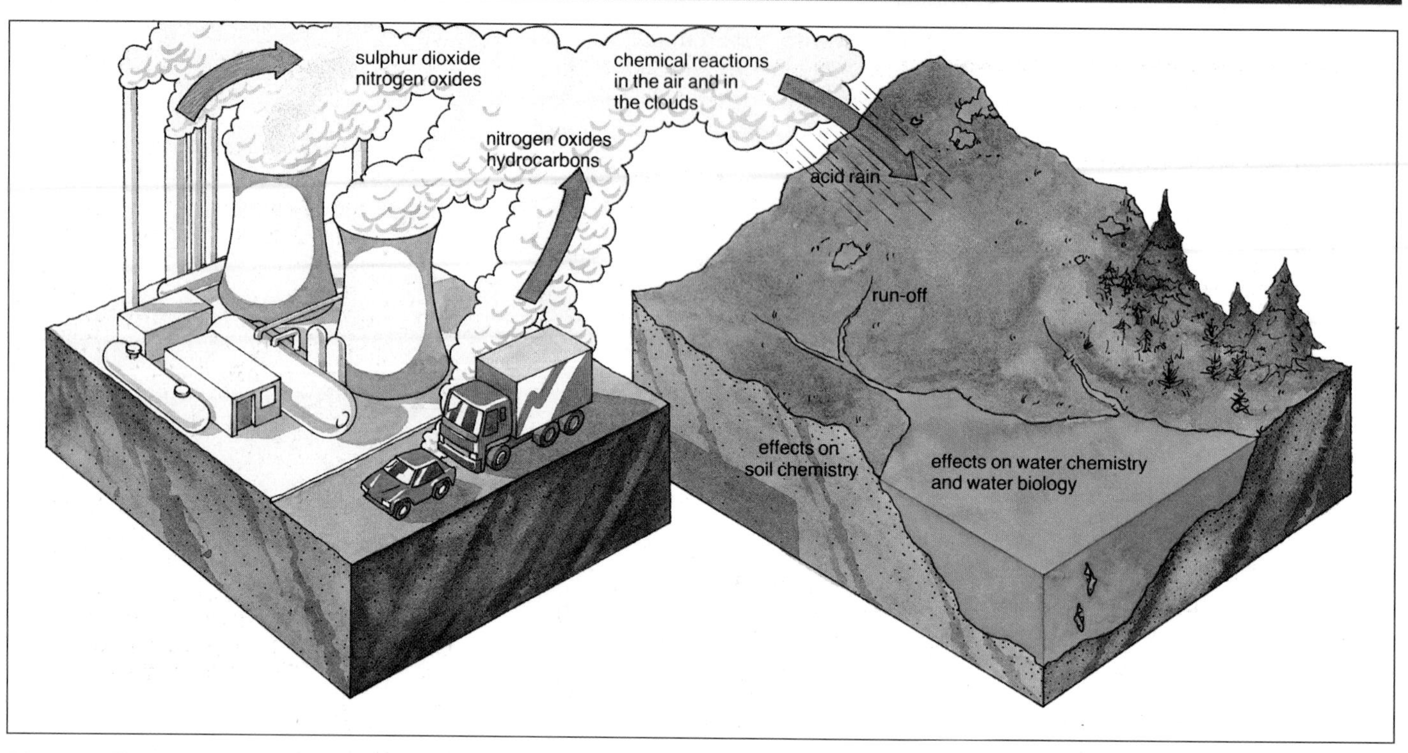

Picture 4 This diagram summarises the formation of acid rain.

The pH scale

The pH of a solution tells us how acidic or alkaline it is. The pH scale runs from 0 to 14, as shown below.

- pH 7 is neutral – neither acidic nor alkaline.
- A pH of less than 7 indicates acidity, and the lower the figure the greater the acidity.
- A pH of more than 7 indicates alkalinity, and the higher the figure the greater the alkalinity

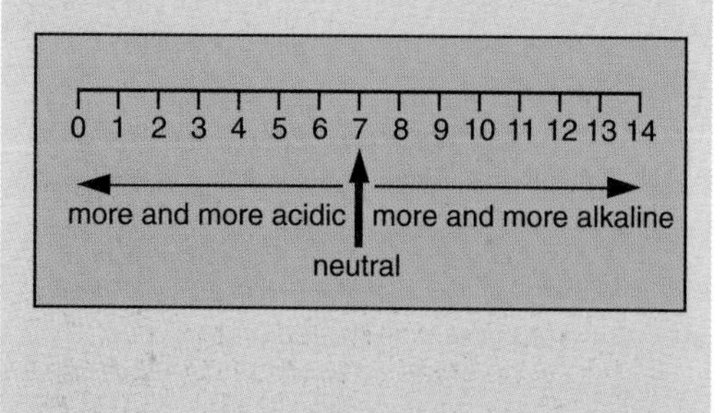

Acid rain

Two of the gases released when fuels are burned, especially in vehicle engines and power stations, are sulphur dioxide and nitrogen dioxide. After they have been released, these gases may undergo chemical reactions in the air and clouds (water vapour) to form acids. For example, sulphur dioxide reacts with oxygen to form sulphur trioxide:

$$\text{sulphur dioxide} + \text{oxygen} \rightarrow \text{sulphur trioxide}$$
$$2SO_2 \quad O_2 \quad 2SO_3$$

The sulphur trioxide then reacts with rain water to form sulphuric acid:

$$\text{sulphur trioxide} + \text{water} \rightarrow \text{sulphuric acid}$$
$$SO_3 \quad H_2O \quad H_2SO_4$$

Reactions like this make rain water acidic, resulting in **acid rain**. Acid rain generally has a pH of between 5 and 2 (pH is explained in the margin). In extreme cases it may be as acidic as vinegar. Picture 4 summarises how acid rain is formed.

When acid rain falls into streams and lakes, it makes them acidic. This may kill fish and other water life. Fish have been dying in lakes in Scandinavia, probably because of acid rain. Some of the acid gases there come mainly from Germany and Britain, carried hundreds of miles on the prevailing wind.

Acid rain also affects the soil. It may have helped kill pine and fir trees in Europe, particularly Germany. It also damages buildings and other structures by corroding metals and wearing away stonework.

It is not easy to control acid rain because scientists do not yet fully understand what causes it and what its effects are. Sulphur dioxide is certainly involved. The chimneys of coal-burning power stations can be fitted with units which remove sulphur dioxide from the chimney gases, but this is very expensive. Nitrogen oxides in car exhaust can be controlled by fitting cars with catalytic converters which remove the gases, or by careful engine design.

Activities

A Estimating the amount of pollution in rainwater

1 Obtain a jar about 15 cm tall, a funnel and filter paper. Set them up as shown in the illustration.

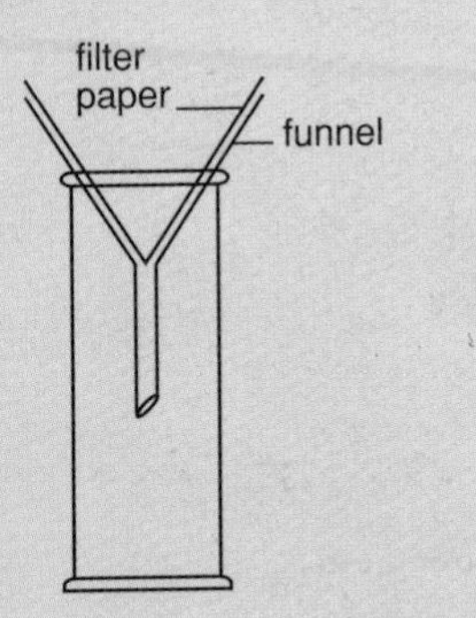

2 When it is raining, place the apparatus in a safe place out of doors and leave it there for at least 30 minutes.

3 Examine the filter paper for particles of dust which have been brought down by the rain.

4 Use universal indicator paper to measure the pH of the rainwater in the jar.

You will probably find that the pH is round about neutral. However, if the air is badly polluted the pH may be acidic. What might the acidity be caused by?

If you have the chance, try doing this experiment in different parts of the country and compare the results.

B To find how much dust is deposited on outside walls

1 Cut off a short length of sticky tape about 8 cm long.

2 Place the sticky side of the tape against a wall out of doors and press it gently.

3 Remove the sticky tape, then hold it against a sheet of white paper. Note how much dust it has picked up.

4 Repeat the experiment on other walls in different places, and compare the amounts of dust picked up.

Suggest reasons why some walls appear to be dirtier than others.

If you have the chance, try doing this experiment in a town and in the country and compare the results.

How could you test the idea that dust is carried by the prevailing wind?

C To find how much dust there is in the atmosphere

Your teacher will give you six microscope slides which have been coated with a thin layer of pure agar jelly.

1 Place the slides in different places out of doors. Make sure they are in safe places where they will not be interfered with. Leave the slides in position for a few days.

2 After a few days compare the amounts of dust deposited on the agar surface of the six slides. Look at the slides under a microscope to see how dense the dust particles are. Can you express the density *quantitatively*?

This investigation could form the basis of a large-scale survey of air pollution in your district. How could you carry out such a survey? Discuss it with your teacher.

D Giving up cars

Hold a debate. The motion is as follows: *People should give up their cars and travel by public transport in order to reduce air pollution.*

Questions

1 In large American cities drivers are encouraged to offer seats in their cars to other people ('pool riding'). The fast lane of highways is reserved for such drivers, and heavy fines are given to drivers of single-occupant cars that use the fast lane. Consider the case for and against such a scheme.

2 The figures below give the estimated total amounts of carbon involved in four processes per year for the whole world.

Process	Amount of carbon (gigatonnes)
Photosynthesis	142
Respiration	136
Burning fossil fuels	5
Deforestation	2

(One gigatonne = 1000 000 000 tonnes)

Comment on these figures.

3 Here are three ways of slowing down the greenhouse effect and global warming:

a Use energy sources that do not produce carbon dioxide.
b Save more energy than we do at present.
c Speed up the rate at which carbon dioxide is removed from the atmosphere.

How might each of these approaches be achieved?

Photochemical smog

Photochemical smog is caused by the action of sunlight on motor vehicle exhaust. This photochemical smog is held close to the ground by a layer of warm air developing above a region of cooler air. The result is an obnoxious brown haze which makes your eyes sting and can cause severe headaches and breathing difficulties.

This type of smog is common in densely populated sunny places like Los Angeles and Tokyo. In Tokyo many people wear masks, and there are fresh air dispensers in offices. The smog also damages plants and is said to have reduced the yield of citrus fruits (oranges and lemons) in the Los Angeles area.

On occasions we get this sort of smog in London and other British cities. When do you think this happens?

Picture 1 Exhaust from motor vehicles is a major cause of photochemical smog.

B6 Feeding relationships

In the natural world, animals feed on plants and on other animals. This is an essential part of the balance of nature.

Picture 1 The larva of the great diving beetle, *Dytiscus marginalis*, is an aggressive predator found in ponds. It sinks its fangs into the prey and sucks up its juices. It may get through as many as 20 tadpoles in an hour. The adult beetle is equally voracious.

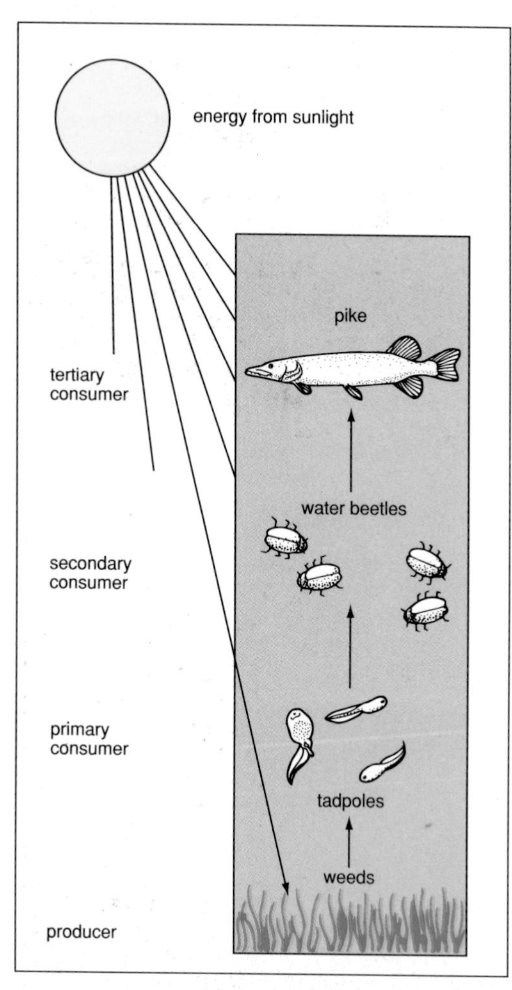

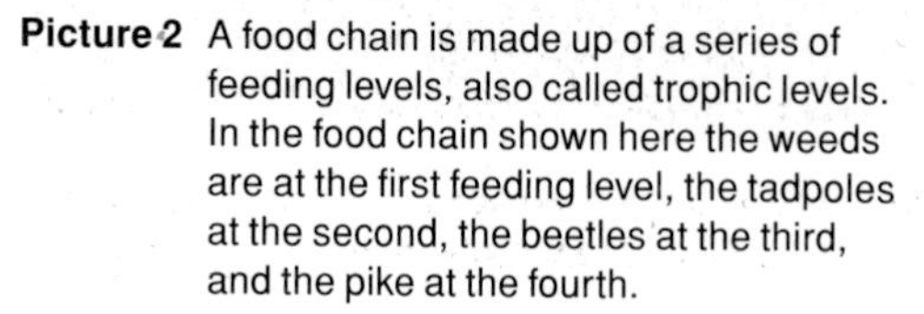

Picture 2 A food chain is made up of a series of feeding levels, also called trophic levels. In the food chain shown here the weeds are at the first feeding level, the tadpoles at the second, the beetles at the third, and the pike at the fourth.

A food chain

Suppose we put some water weeds, tadpoles and a couple of great diving beetles into a jar of pond water, and watch what happens. We find that the tadpoles nibble at the weeds, and the beetles eat the tadpoles. We can sum up the feeding relationship between the three organisms like this:

weeds → tadpoles → beetles

We call this a **food chain**. Tadpoles eat only plants: they are **herbivores**. In contrast, the great diving beetle feeds only on animals: it is a **carnivore** (picture 1).

There are only three organisms in the food chain shown above. However, in a lake there might be some pike. These fish eat water beetles, amongst other things. So in the lake the food chain would be:

weeds → tadpoles → beetles → pike

The pike has been called a 'water wolf'. It is one of the most savage predators found in fresh water. The animal that comes at the end of a food chain like this is called the **top carnivore**.

Producers and consumers

Let's think about this food chain in a bit more detail. The weeds make their own food by photosynthesis; they get the energy for doing this from sunlight. Because they make food, we call them **producers**.

In contrast, the animals in the chain get their food by eating other organisms. For this reason we call them **consumers**.

In this particular chain there are three consumers. The tadpoles are the first consumers, the beetles are the second consumers, and the pike is the third consumer. They are known as the **primary, secondary** and **tertiary consumers** respectively (picture 2).

The stages in a food chain are called **feeding levels** or **trophic levels**.

Pyramid of energy

In picture 2 only a very small fraction of the Sun's energy that falls on the weeds is transferred to the plants' tissues and food stores. And when a tadpole eats a weed only about one tenth of the energy in the plant is transferred to the body of the tadpole. The rest is lost in the tadpole's waste matter (excreta) or in its respiration. The same thing happens when the tadpoles are eaten by the beetles, and again when the beetles are eaten by the pike.

In other words, at each step of the food chain, a lot of energy is lost. Picture 3 shows this as a diagram: it is called a **pyramid of energy**.

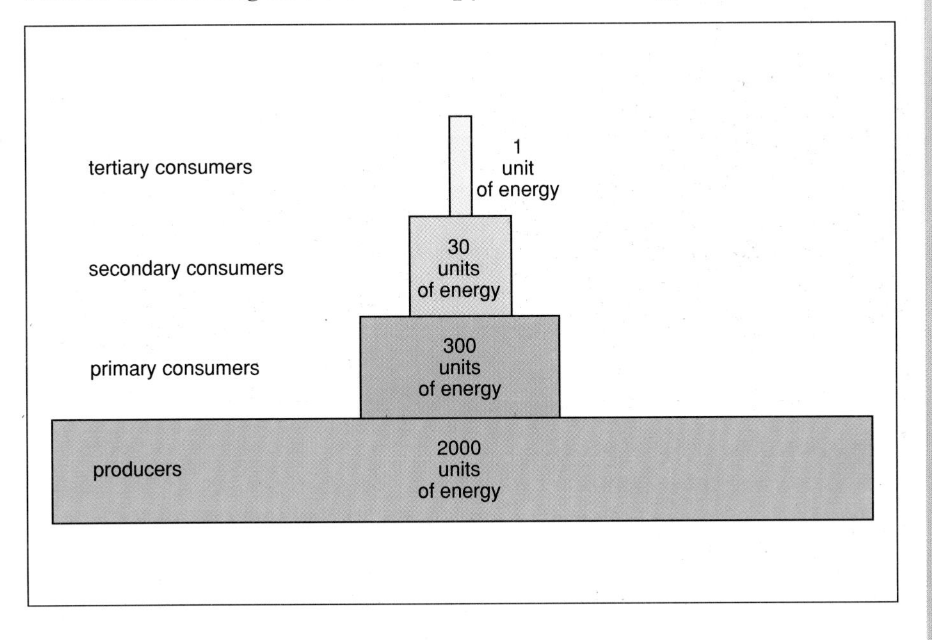

Picture 3 This diagram shows the decrease in energy that occurs at each level in a food chain. The diagram is called a pyramid of energy. The figures are typical of a food chain in fresh water such as a lake.

Pyramids of numbers and biomass

The energy loss just described means that, as you go along the food chain, the number of organisms which can be supported at each level gets less and less. Thus a given number of tadpoles will feed a smaller number of beetles, and these beetles will feed an even smaller number of fish. This drop in numbers at each level in a food chain gives what we call a **pyramid of numbers**.

For the same reason there is also a drop in the total mass of living material at each level of a food chain. This is called the **pyramid of biomass**.

Food webs

In a natural habitat such as a lake or pond, it would be unusual for the organisms to be linked together in a simple chain. Many more species will be present than the ones in picture 2, and each may have several sources of food.

By observing all the organisms in a habitat, you can build up a diagram summarising who feeds on what. This is called a **food web**. A simple food web is shown in picture 4.

In a habitat which contains a large number of different species, the food web may be very complex. A complex food web is shown in picture 5. Food webs can be divided up into a series of feeding (trophic) levels, just as food chains can. Can you recognise the different feeding levels in pictures 4 and 5? As with food chains, there is a drop in energy, numbers and biomass at each level.

Questions

1 In picture 2 which organisms are

a herbivores,
b carnivores,
c predators,
d prey?

2 Fill in the missing organism in each of the following food chains:

a grass → ? → human
b grass → deer → ?
c lettuce → ? → fox
d aphid (greenfly) → ladybird → ?

3 The following is a food chain that ends with the human:

plant → bee → human

Explain precisely how plants provide food for bees, and how bees provide food for humans.

How does this food chain differ from the ones in question 2?

Do you think it should be regarded as a food chain? Explain your answer.

4 This question is about the food web in picture 4.

a How many food chains can you see in this picture? Write them out separately.
b What would happen to the numbers of the other organisms in the web if all the fish were destroyed?
c A food chain is more easily destroyed than a food web. Why? Use picture 4 to illustrate your answer.

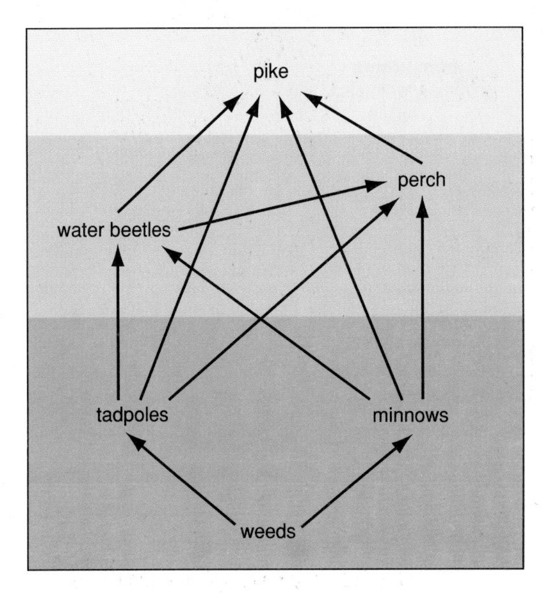

Picture 4 A simple food web in a pond. The colours are different trophic levels.

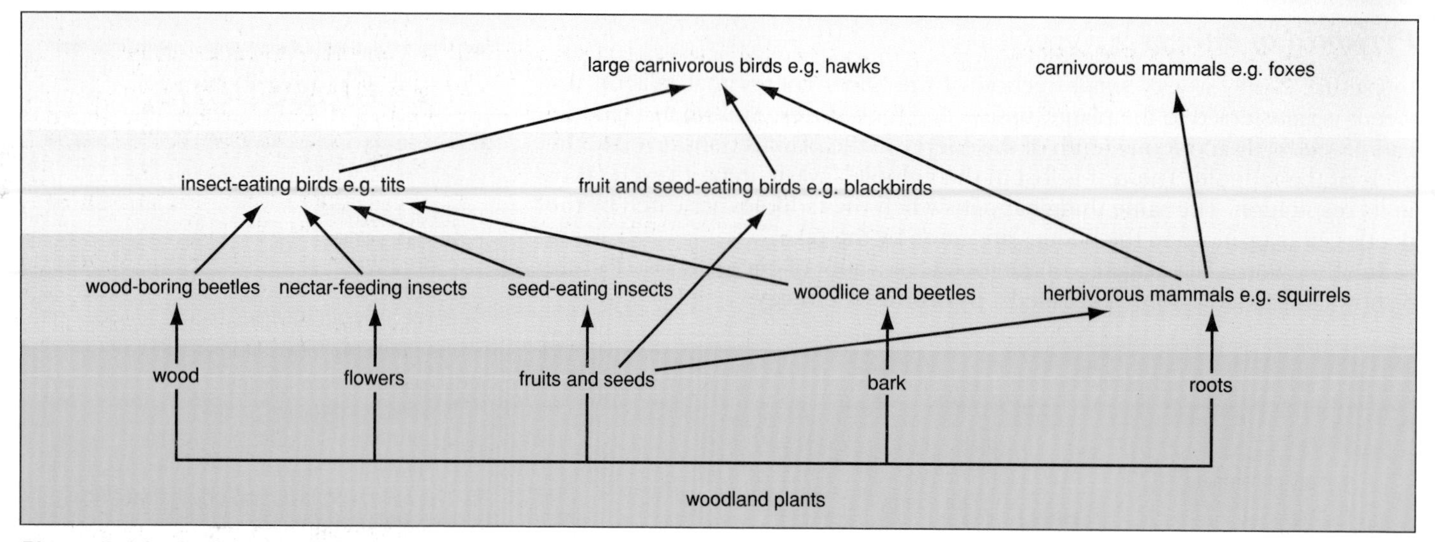

Picture 5 A food web in a wood. The colours are different trophic levels.

Picture 6 Humans are at the end of many food chains.

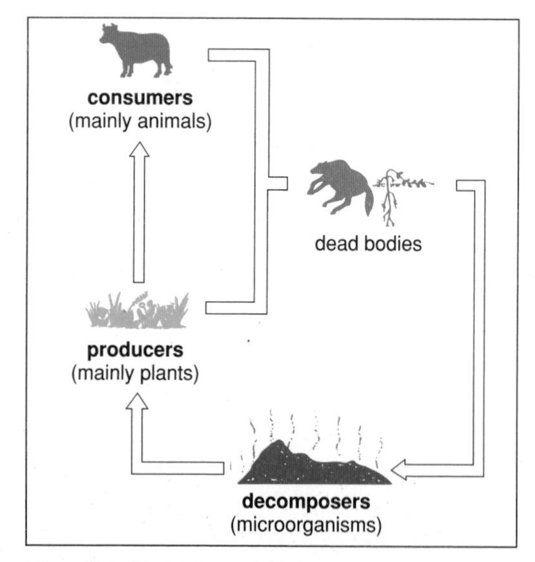

Picture 7 Decomposers enable the chemical elements in the bodies of the producers and consumers to be used again. The arrows show the flow of materials from one organism to another.

Food chains in the service of humans

Look at picture 2 again. The pike that ate the beetles that ate the tadpoles that ate the weeds might be caught by a fisherman for his supper. The food chain would then be:

weeds → tadpoles → beetles → pike → human

So food chains can provide us with food.

Some of the most important food chains occur in the sea. In the surface water where light can penetrate, there are millions of microscopic organisms called **plankton**. Some of these organisms are like plants and feed by photosynthesis. Others are like animals and feed on the plant-like ones. The animal plankton, in turn, is eaten by fish such as the herring, giving us the following food chain:

plant plankton → animal plankton → herring → human

Our fishing industry depends on this and other similar food chains. On land, farming involves several important food chains such as:

grass → sheep → human
grass → cattle → human

Of course humans eat plants as well as animals: we are **omnivores**. When we eat such things as bread or cornflakes, we are the primary (and only) consumers in this very simple food chain:

wheat → human

The decrease in energy which occurs at each level of a food chain is a very important consideration in growing crops and raising livestock. This is dealt with in the next topic.

Decomposers

When animals and plants die their bodies **decay**. This is because they are fed upon by bacteria and other microorganisms which break them down.

The organisms which bring about decay are called **decomposers** or **saprobionts**. As a result of their activities, simple substances are released from the dead bodies. These substances are absorbed by plants, and can go through the food chain all over again (picture 7).

Decomposers thus enable chemicals to be re-cycled and used again. They play an important part in the cycling of elements such as carbon and nitrogen (page 36).

Activities

A Building up a food web

1 Set up an aquarium in your laboratory.

Use a large transparent container. Wash it thoroughly, then put in some clean sand to a depth of about 2 cm. Root some water weeds in the sand. Slowly pour in some pond water until the container is approximately three-quarters full. Put in some floating plants like duckweed and *Spirogyra*. Now add as large a variety of animals from a local pond or stream as you can. If possible include water beetles and their larvae, dragonfly and caddis fly nymphs, mosquito larvae and pupae, shrimps, water snails and a few small fishes such as carp, minnows and sticklebacks. Don't put in so many carnivores that they eat up all the other animals!

2 Using a simple identification key find out the names of the animals and plants in the aquarium.

3 Observe the animals, and see if you can find out what each one feeds on. If necessary use books to help you.

4 Write down the names of:

a the producers,
b the herbivorous consumers,
c the carnivorous consumers.

Which of the carnivorous consumers are not eaten by any other organism?

What do you think would happen if you removed the consumers from the aquarium?

5 Construct a food web similar to the one in picture 4, showing the feeding relationships of the organisms in your aquarium.

B A food web in a natural habitat

1 Choose a habitat. It might be a freshwater pond, a hedgerow, a patch of grass, or a rock pool on the seashore.

2 Find as many animals and plants in the habitat as you can.

Whenever you find an animal, try to see what it feeds on.

Write down the name of each animal and its food in your notebook.

3 Construct as many food chains as you can for your habitat.

4 Now try to construct a food web for the habitat.

If you are not sure what a particular animal eats, try to find the answer in books.

Questions

1 Study the food web in picture 5, then answer these questions:

a Give one example of a predator in the food web, and write down the name of an animal which it preys on.
b The food web does not include the leaves of the woodland plants. Write down a food chain which might lead from the leaves.
c A chemical substance, poisonous to animals but not to plants, leaks onto the ground in the wood. Explain how this might affect the food web.
d What might happen to the wood eventually if all the foxes were destroyed? Explain your answer.

2 As one proceeds along a good chain, each organism tends to be larger than the one before.

a Give an example of a food chain which illustrates this.
b Why do you think this is true?
c Give an example of a food chain which is an exception to this.

3 The loss of energy at each level of a food chain or web can be shown as a pyramid of numbers, a pyramid of biomass or a pyramid of energy.

Which do you think is the best way of showing the energy loss, and why?

4 The diagram below shows the total number of organisms per square metre at each level of a food web in a river.

1	tertiary consumer
24 000	secondary consumer
400 000	primary consumer
1000 000 000	producers

a What process enables the producers to provide food for the primary consumers?
b Why are there fewer secondary consumers than primary consumers?
c What can you say about the kinds of organism that comprise the producers and primary consumers?
d Suggest one change in the environment that might greatly increase the number of producers in this pyramid.

5 The following figures show the total mass of body material, excluding water, formed in one square metre of grassland during one year:

plants	470.0 g
herbivores	0.6 g
carnivores	0.1 g

a Why does the total mass of body material decrease at each step of the chain?

b These figures were obtained by measuring the 'dry mass': that is the mass after all traces of water have been removed by drying. Why is this better than measuring the 'wet mass'?

Funny pyramids

If we construct pyramids of numbers for various food chains, we find that some of them have unusual shapes so that they are not really pyramids. Here are three examples of 'funny pyramids':

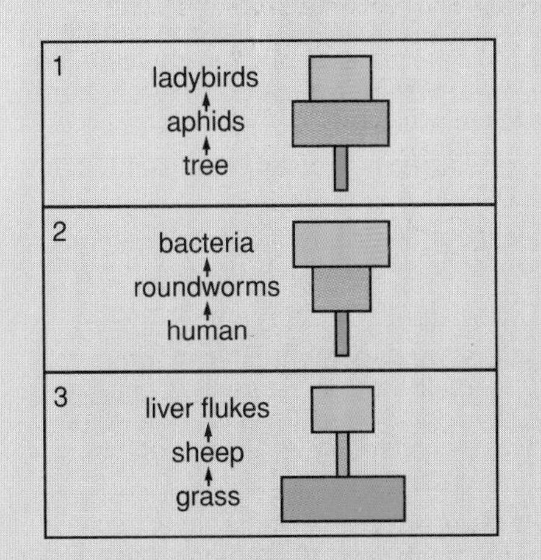

1 Explain in your own words why each of these food chains gives a funny pyramid.

2 For each food chain, draw what you would expect to get if you constructed (a) a pyramid of biomass and (b) a pyramid of energy.

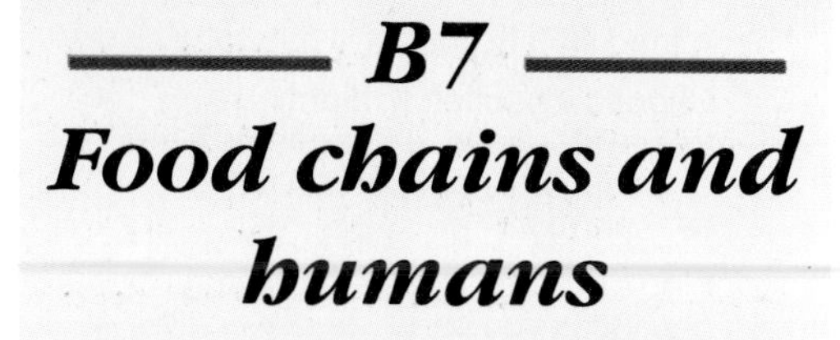

B7 Food chains and humans

Food chains play a vital part in our lives, as we shall see here.

Picture 1 Wheat, one of our most widely grown crops. From its seeds (grain) we get flour for making bread, cakes and breakfast cereals.

Growing plants for human use

The aim is to transfer the maximum amount of energy from the Sun into the part of the plant you want to harvest. This is what **plant farming** (**plant husbandry**) tries to achieve.

Agriculture is growing crop plants such as wheat, barley and oats.

Horticulture is growing mainly garden plants out of doors or in glasshouses.

Both involve:

- **Soil management**. The soil must contain the right nutrients, if necessary by adding a **fertiliser**, and it must be adequately watered by rain and/or irrigation.
- Making sure the plants get enough **light** and **warmth**, by natural or artificial means.
- **Controlling pests** so that they do not harm the plants or reduce their yield.
- **Breeding plants** so as to ensure a supply of new ones for future use. Plant breeding aims to produce improved varieties.

All these aspects of growing plants are dealt with in detail in other parts of the book.

Energy from plants

Think of a crop plant such as wheat (picture 1). It captures light energy from the Sun and transfers the energy to starch and other organic substances by photosynthesis. However, only a small proportion of the light energy that strikes the plant is transferred in this way. Most of it is reflected, or passes straight through the leaves, or is used up in various chemical reactions inside the plant. The rest of the energy gets transferred to sugar, but most of this is used in respiration and only a small proportion gets into the starch and tissues of the plant. This is summed up in picture 2.

In picture 2, only five per cent of the total light energy that strikes the leaf finishes up in the starch and tissues. We call this the **energy conversion efficiency.** An energy conversion efficiency of five per cent is the best that can be achieved by a crop plant growing in really good soil in an ideal climate. In practice the average figure for a crop plant like wheat is about one per cent. Even sugar cane, grown in the best possible conditions in the tropics, rarely exceeds two and a half per cent.

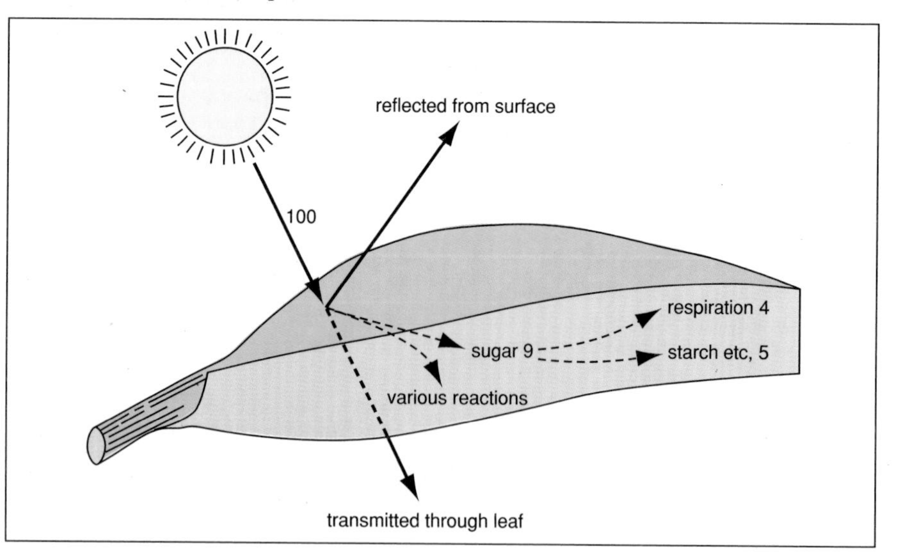

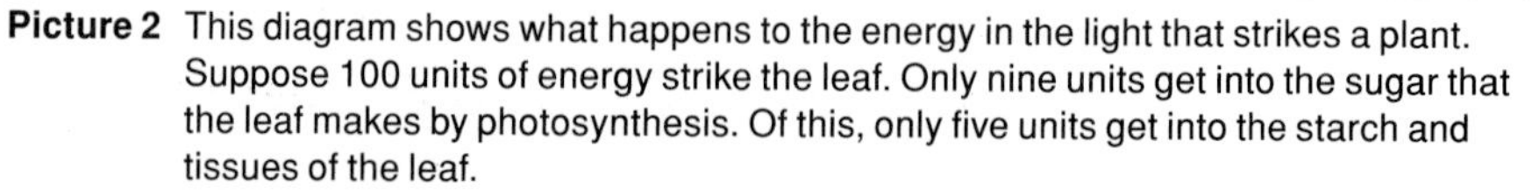

Picture 2 This diagram shows what happens to the energy in the light that strikes a plant. Suppose 100 units of energy strike the leaf. Only nine units get into the sugar that the leaf makes by photosynthesis. Of this, only five units get into the starch and tissues of the leaf.

Picture 3 Cattle transfer energy from plants such as grass to their tissues.

Energy from meat

Think of a cow eating grass in a lush meadow (picture 3). The grass contains energy in its starch and tissues. These are eaten by the cow. However, only a small fraction of the energy in the grass is transferred to the cow's tissues.

Picture 4 shows what happens to the rest of the energy. Most of it passes straight through the gut without being absorbed and is lost in cowpats, or in methane gas given off by microbes in the cow's stomach. Someone has worked out that there is enough energy in the gas given out by a cow to light a small house! Further energy is lost in urine, or used up in respiration.

Altogether only four per cent of the energy taken in by the cow gets into its tissues. Still more energy is lost when the cow is eaten by humans, because we only eat the meat. Non-edible structures like the bones, horns and skin get left behind.

Plant or animal food, which is better?

Humans are **omnivores** – that is, we eat animals *and* plants. From the energy point of view, it is more economical to eat plants than animals. This is because so much energy is lost as it passes from plants to animals in the food chains.

For this reason, plant crops like wheat, maize and rice will feed more people than livestock such as cattle and sheep. This is particularly important in densely populated countries: an area of land with crops can support more people than the same area of land with cattle.

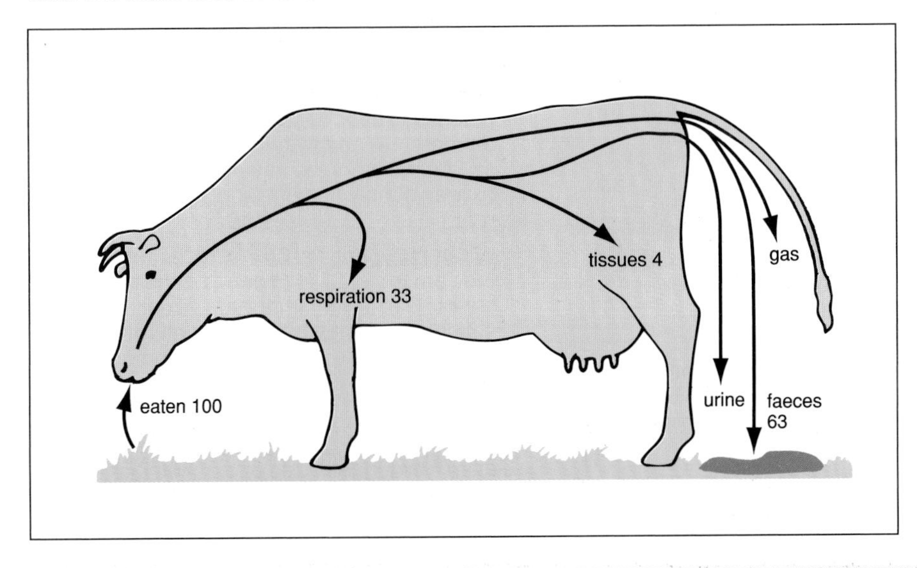

Picture 4 Here we see what happens to the energy in the grass eaten by a cow. Suppose 100 units of energy are ingested. Only four units get into the cow's tissues. The rest is lost in respiration, urine, faeces and gas.

Raising animals for human use

The aim is to transfer the maximum amount of energy from plants into the tissues of the animals. This is what **animal farming** (**animal husbandry**) tries to achieve.

Cattle are farmed for beef and milk, sheep for meat and wool, pigs for meat, and poultry for meat and eggs.

Animal farming involves:

- Providing the animals with the **right environment**, either out of doors or, when necessary, under cover.
- Giving the animals the **right kind of diet** of natural food, e.g. grass, and/or artificial food containing the right combination of nutrients.
- **Keeping the animals healthy** by good hygiene and having them immunised against infectious diseases.
- **Breeding animals** so as to ensure a supply of new stock. Animal breeding, like plant breeding, aims to produce improved varieties.

These topics are dealt with in detail in other parts of the book.

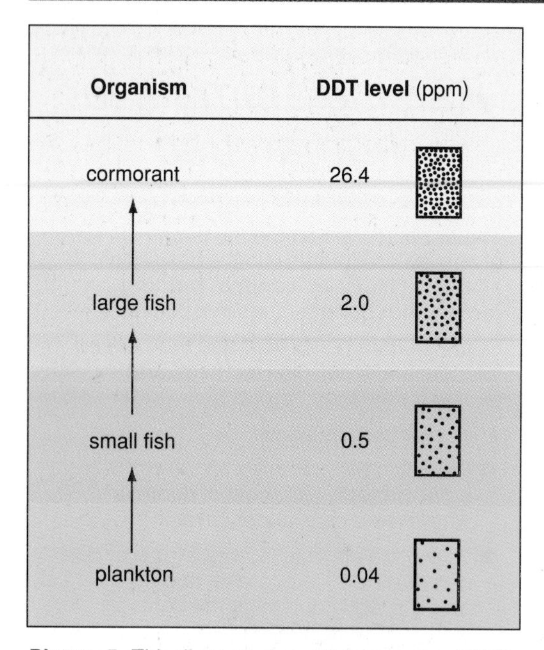

Picture 5 This diagram shows the amounts of DDT in four organisms in a food chain. The amounts of DDT are expressed in parts per million (ppm). Notice how the DDT gets more and more concentrated as it goes from one level of the food chain to the next. This is because it is held in the organisms' bodies instead of being excreted.

Getting other things from food chains

We don't just get energy from food chains. In addition we get all sorts of chemical substances needed for life.

Some of these substances get more and more concentrated as they pass along the food chains. Take vitamin D, for example. This is made by plant plankton in the sea. Then it gets into the animal plankton. Eventually it gets into the fish that eat the plankton. The fish store and concentrate it in their livers. This is why cod liver oil is such a good source of vitamin D.

Food chains and poisons

Unfortunately it is not always useful substances that get into food chains. Sometimes poisons do so too. Such is the case with the insecticide DDT.

DDT keeps its poisonous properties for a long time after it has been released into the environment (see page 85). If it gets into a river or lake, it is taken up by the plankton. It then passes along the food chains, becoming more and more concentrated as it does so (picture 5). DDT damages animal tissues (see page 51). It may also be a danger to humans. For this reason it has been banned in many countries, including Britain.

Unfortunately DDT isn't the only poisonous substance to get into food chains. Some years ago over sixty people died in Japan from eating fish whose bodies contained mercury. The mercury had been discharged into the sea from a factory and had passed right through the food chains.

Germs, too, may pass along food chains. For example, if *Salmonella* bacteria are present in the food given to hens, the hens and their eggs may become infected. The bacteria may then get into humans, causing Salmonella poisoning. Matters are made worse if the excreta and/or remains of dead hens are fed to the hens. This closes the food chain and results in repeated infection of the hens.

Activities

A What food chains do you enter into?

Make a list of all the foods you can think of which you eat during a typical day. For each one write down the food chain which it is part of, putting yourself as the final consumer.

B Comparing the cost of meat and vegetables

Visit a supermarket and find out the cost of a selection of

1 meats, e.g. beef, lamb, pork and chicken,

2 vegetables, e.g. cabbage, cauliflower, carrots, potatoes.

Express each cost as price per gram. Calculate the average price of meats and vegetables.

Can you relate the difference in the prices of meats and vegetables to the food chains invlved? Bear in mind that other factors besides food chains help to set the prices of specific foods. Can you suggest what these factors are?

Which are the cheapest and the most expensive meats? Suggest reasons for the difference. Which are the cheapest and most expensive vegetables? Again, suggest reasons for the difference.

C The economics of human food chains

Think about these two food chains:

wheat → cow → human

wheat → human

1 You have been asked by Her Majesty's Government to compare the cost to society of these two food chains. Make a list of all the information which you would need in order to carry out this task. In each case suggest where the information might be obtained. Say how you would present your comparison.

2 Cost does not just mean money. It also means cost to personal health and cost to the environment. Compare the two food chains from these points of view.

Fish farming

The aim in fish farming is to cultivate fish so that they breed rapidly enough to produce populations that can be harvested profitably.

Fish farming is becoming more and more important. In 1993 there were nearly two thousand fish farms in Britain. About one third of the total value of all fish produced or landed in Britain comes from fish farms.

Two main types of fish are farmed in Britain, salmon and trout. The eggs are hatched in special hatcheries where the temperature and water flow are controlled so that the embryos have an adequate oxygen supply and develop at the optimum rate.

When the young fish are large enough they are transferred to ponds, tanks or cages. Diet is very important. Trout and salmon are carnivorous, and natural food (small fish, etc) may be supplemented by scientifically formulated pellets.

Disease can spread quickly through the fish population, particularly if the fish are overcrowded. All sorts of precautions are taken to prevent this.

Questions

1 Imagine you are a wheat farmer. You are determined to grow wheat so efficiently that you achieve the maximum yield of energy per hectare. how could you do this?

2 The cow in picture 4 has an energy conversion efficiency of only 4 per cent. However, the perch, a fresh water fish, has an energy conversion efficiency of 22.5 per cent.

a What is meant by the term energy conversion efficiency?
b Suggest reasons why the perch is a better energy converter than the cow.

3 Suppose the leaf in picture 2 is eaten by the cow in picture 4. What proportion of the Sun's energy that strikes the leaf will get into the cow's tissues. Show your reasoning.

4 It is better for people in densely populated countries to concentrate their efforts on growing crops rather than raising livestock. Why? What might make it difficult?

5 Some types of chicken-feed include the ground-up remains of dead hens.

a What are the advantages of using dead hens for making chicken feed?
b Some health officials think that it is a bad practice. Why?
c If the practice was discontinued, suggest an alternative use for the dead hens.

6 You are faced with having to choose between a beefburger and a vegeburger.

a Sketch the probable pyramid of energy for each, with you at the top.
b Which is more likely to be the more expensive, and why?
c What are the advantages of being a vegetarian from the point of view of the environment?

7 The following figures show the concentration of mercury in sea water and in various organisms in a particular area. The measurements are in parts per million (ppm).

sea water	0.00003 ppm
algae	0.03 ppm
fish	0.3 ppm
water birds	2.0 ppm

a How many times more concentrated is the mercury in the fish than in sea water?
b Suggest an explanation for these figures.
c Eating fish with a mercury level higher than 0.5 ppm is potentially dangerous to humans. A level of 0.2 ppm in the human body causes symptoms of mercury poisoning which, in extreme cases, may lead to kidney damage, paralysis and death. Use this information to explain why the figures given above are important to us.

The case of the Peregrine falcon

The Peregrine falcon is a swift, deadly predator – one of the most beautiful of all birds of prey. It nests on cliff ledges and rocky crags and is frequently found in coastal areas where it feeds mainly on sea birds.

After the end of the Second World War, a dramatic decline in the number of Peregrine falcons was noticed in both America and Europe. No reason for this decline could be seen, until in 1958 a British ornithologist, Derek Ratcliffe, observed that many of their nests contained broken eggs and that the parents often ate their eggs.

Ratcliffe noted that the decline in the number of Peregrines had begun at about the time that DDT and other pesticides were being used a lot. At first he thought that egg-eating by the parents was a behavioural disorder caused by the pesticides attacking the nervous system. But later he showed that the eggs were not deliberately broken by the parents, but were crushed during incubation because they had abnormally thin shells. They were eaten as a natural response to the fact that they were broken.

Since then a great deal of research has been done into the way that DDT affects birds. It has been found that it interferes with the hormones that control the deposit of calcium in eggshells. This results in the eggs being thin-shelled and fragile. DDT also causes a reduction in the amount of the sex hormone oestrogen, with the result that the bird lays fewer eggs than usual.

(Adapted from Tim Halliday, '*Vanishing Birds*', Sidgwick and Jackson)

1 DDT is particularly harmful to birds of prey, that is birds which eat other animals. Why do you think this is?

2 Although DDT in food chains harms birds, there is no evidence that it has ever harmed humans. Suggest a reason why humans have not been harmed by it?

3 DDT has been banned in Britain but is still used in certain developing countries. Why do you think developing countries still want to use it?

4 How do you think new pesticides should be tested to see if they are safe?

Picture 1 A peregrine falcon.

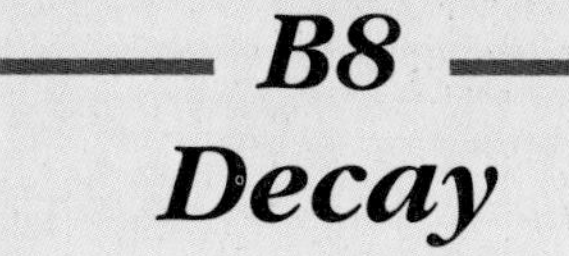

B8 Decay

If no decay had occurred since the year 1600, dead bodies would cover the Earth to a depth of a kilometre!

Picture 1 The carcass of this buffalo will eventually decay.

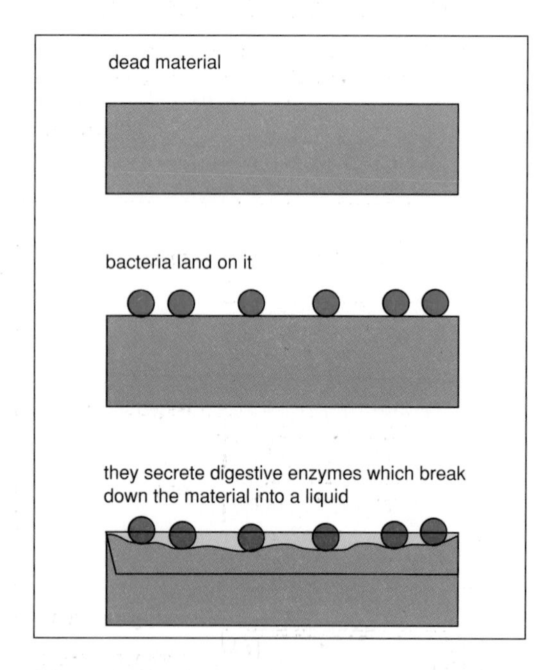

Picture 2 How bacteria bring about decay. Fungi do the same kind of thing.

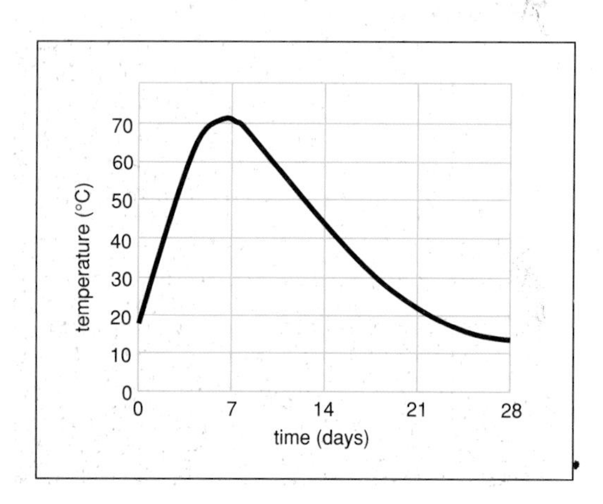

Picture 3 This graph shows the temperature changes which occurred in the middle of a heap of vegetation.

What is decay?

After an organism dies, it becomes dead organic matter (detritus). This gradually disintegrates and eventually becomes a liquid consisting of nothing but inorganic substances (picture 1). This is what we mean by **decay**.

Three main things cause decay to happen:

- Immediately after death, the organism's own enzymes start breaking down the body. The organism literally digests itself.
- Various natural processes break the body up. For example, **scavengers** such as birds peck at it, maggots and roundworms wriggle through it and rain softens it.
- Certain microbes, mainly bacteria and fungi, feed on the dead remains and break it down. These **decomposers** are the main agents of decay.

How do decomposers bring about decay?

After an animal or plant has died, it isn't long before some spores of bacteria and fungi land on it. The spores give rise to new individuals. These grow and multiply, spreading quickly through the dead material. A teaspoonful of rotting vegetation may contain over a thousand million bacteria!

To feed on the dead material, the microbes must first break it down into a soluble form, just as we have to digest our food before we can absorb it. Microbes do this by releasing digestive enzymes into the surrounding material (picture 2). The enzymes dissolve the material, and the microbes absorb the soluble products.

Soft remains like skin and muscle decay more quickly than hard structures like bone and wood. A skeleton may remain intact for years after the rest of the body has decayed. Indeed, in certain circumstances bones may become **fossilised** and remain indefinitely (see page 286).

As decay gets under way, the rotting material may get warm (picture 3). This is because of the heat given out by the millions of respiring microbes. It may also have a foul smell. This is because some of the microbes have special methods of respiration and produce smelly gases such as hydrogen sulphide.

What's needed for decay to occur?

Experiments tell us that these conditions are needed for complete decay to occur:

- **Moisture must be present**
  This is needed for the spores to germinate, and for the microbes to grow and multiply. If a dead body is kept dry, it loses moisture, the skin shrinks and decay does not occur. This process is called **mummification**. The ancient Egyptians used it to preserve the bodies of their kings. Closer to home, it is how hay is made; hay is simply dry grass.
- **It must be warm enough**
  Microbes thrive, and multiply fastest, in a warm environment. Under these circumstances decay will occur quickly. If it is cold, decay is slowed down, and if it is well below freezing it won't happen at all. Extinct mammoths, which died in Siberia thousands of years ago, have been dug out of the ice with their flesh intact and local people cooked the flesh as steaks.
- **Oxygen must be present**
  The microbes which bring about decay need oxygen for their respiration. If oxygen is lacking, they respire without it – that is, **anaerobically** (see page 126). The end products of this process are acids and they stop further decay taking place. So when there is no oxygen present, decay is incomplete. This is how peat is formed. A body left in peat may resist decay (see page 286). It is also how silage is made (see page 92).
- **Chemicals which kill the decomposers must not be present**
  A biologist who wants to preserve a specimen will put it in a chemical preservative such as alcohol. In the past this sort of thing has happened naturally when animals have fallen into a tar pit, that's a lake full of an oily liquid. In California there are tar pits containing the undecayed skeletons of extinct sabre-toothed tigers which died about a million years ago.

Picture 4 These are the remains of a baby mammoth which died thousands of years ago. Because of the extreme cold in Siberia, where it was found, it has been very well preserved.

Making decay occur

To bring about decay, all we need to do is to put some dead material in a place which has all the right conditions for decomposers to flourish. This is what a gardener does when making a **compost heap** (picture 5).

Why is decay important?

If it wasn't for decay, the dead remains of organisms would simply pile up. What a thought!

Decay is important for another reason too. It enables chemical elements such as carbon and nitrogen to circulate in nature. This means that they can be used again and again (see page 36). It's why decaying matter in the form of compost or manure is so good for plants: the locked-up nutrients are released and can be used by plants.

We make use of decay in a number of ways. For example, in a **sewage works** and in **making cheese** (see page 92).

Decay is also important in **getting rid of rubbish**. Things like left-over food, potato peelings and tea leaves can be broken down by decomposers. We call this sort of rubbish **biodegradable**. Thanks to decomposers, the chemicals in these materials can be recycled by nature and used again.

Other kinds of rubbish will not decay, because they are made of substances which microbes cannot live on. They are **non-biodegradable.** They include plastic, polythene and many other artificial materials. If you throw an apple core into a hedge, it will eventually decay and disappear. But if you throw a polythene bag, it will remain there indefinitely – unless some worthy citizen removes it.

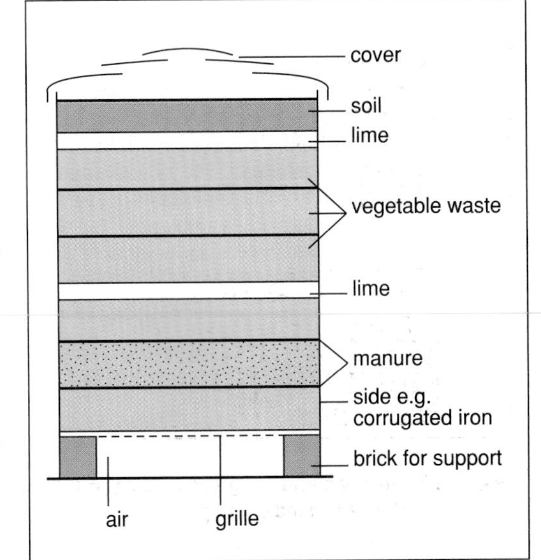

Picture 5 The diagram shows a compost heap in section. In the photograph a gardener tends his compost heap.

Picture 6 Plums being put in a jar for bottling. The lid is fixed tightly in place by the clip to prevent air getting in. The lady will then heat the jar to kill any microbes present. As the jar cools down afterwards, the air inside shrinks and creates a vacuum which pulls the lid on tightly.

Preserving food

Preserving food means preventing it from decaying (going bad). How can this be done? The principle is to kill the microbes, or make conditions unsuitable for them. Some microbes, such as the bacterium *Salmonella*, can cause food poisoning. It is particularly important that these microbes should be killed.

There are five main ways of preserving food:

- **Heating** to a high enough temperature kills microbes. So heating can **sterilise** food (see page 100). The heat-treated food must be sealed in a container so that microbes cannot get at it. This is done by **canning**, **vacuum packing** or **bottling** (picture 6). Milk is heat-treated by two different methods which are described on the opposite page.
- **Freezing** does not kill microbes, but it slows down their action and stops them multiplying. As soon as the food thaws, the microbes start up again and the food begins to go bad. So frozen food should be eaten as soon as possible after it has thawed.
- **Drying** removes most of the moisture from the food. This inactivates any microbes and stops them multiplying. It also stops any spores germinating if they land on the food.
- **Chemical treatment** adds to the food a substance which kills microbes but is harmless to humans. Examples are **pickling** in a preservative such as vinegar, **smoking** over a wood fire (substances in the smoke kill the microbes), and **salting**. In salting, the food is soaked in a solution of salt, or salt is rubbed into it. The salt causes water to flow out of the microbes by osmosis (see page 15). This kills them.
- **Radiation**. Microbes can be killed by ionising radiation, for example gamma rays. This provides a modern way of preserving food. One snag, though, is that radiation may kill bacteria such as *Salmonella* but leave their harmful toxins behind. So radiation is used with discretion.

Activities

A What conditions are needed for decay to occur?

In the text it says that the following three conditions are necessary for decay to occur: moisture, warmth and oxygen.

Plan an experiment to find out if these conditions really are needed for an object such as a piece of bread to decay. Be sure to take any necessary safety precautions. Check your plan with your teacher, then carry out the experiment.

BIOHAZARD

Decaying food is a breeding ground for microbes some of which might be harmful.

B Finding the best temperature for storing fruit

Storing fruit so that it does not decay is important for fruit-sellers. Plan an investigation to find the best temperature for storing a particular type of fruit, e.g. grapes, apples or bananas. Check your plan with your teacher. If possible carry out the investigation. You will need to control your variables carefully. This investigation is probably best done as a project.

Questions

1 Decay is brought about mainly by microbes such as bacteria and fungi. However it is helped by other agents, physical as well as biological. Name five such agents.

2 A human body was found in a remote cave in the Sahara Desert. Forensic experts estimated that it had been there for well over a hundred years. The skin, though dry and shrivelled, was still intact. Suggest a reason why the body had not decayed.

3 Study picture 3 on page 52. Suggest why the temperature inside the heap of dead vegetation rose and then fell.

4 A family of litterbugs left the remains of their picnic in a secluded part of a wood. A year later, their left-over food had disappeared but their plastic plates were still there.

a Suggest two things that may have happened to the food.

b Why didn't the plastic plates disappear?

c The father left his newspaper screwed up in a bush. Do you think it was still there after a year? Explain your answer.

5 Food which has been frozen and then thawed should not be re-frozen. Suggest a reason for this.

Making a compost heap

Compost is decaying vegetable matter. It makes a good fertiliser for the garden. Any rottable material can be used to make compost: old cabbages, potato peelings, tea leaves – you name it.

In making a compost heap, the following rules should be followed:

1 Choose a place for your compost heap which is sheltered from the wind and Sun.

2 Enclose the heap in a container with sides and a rainproof lid.

3 Support the heap on a grille so that air can get into it from underneath.

4 As you pile up the dead material, add a thin layer of soil or manure every now and again.

5 Add a little lime occasionally.

6 If you put in old cabbage stalks, bash them up with a mallet beforehand.

7 If you put in grass cuttings, mix them with bulkier material first.

8 Keep the heap moist, but don't saturate it.

9 Don't throw on sticks, plastic bags or materials tainted with oil, paint or creosote.

10 Turn the heap occasionally with a fork, so that the inside is moved to the outside and the outside to the inside.

Suggest a scientific reason for each of these rules.

How is milk heat-treated?

There are two ways of heat-treating milk.

UHT milk is superheated at about 150°C for a few seconds, then sealed (UHT stands for ultra high temperature). This sterilises the milk: virtually all the microbes are killed, and the milk will keep for months. However, its flavour is altered slightly.

Pasteurised milk is usually heated at a lower temperature for a longer time (e.g. 70°C for 15 seconds). This kills the *harmful* microbes, but leaves some of the others behind. These make the milk go off after a few days. However, the flavour is not affected by the treatment.

The use of heating to destroy microbes in milk was first described by Louis Pasteur in 1885. Later it was introduced into Britain and other countries to protect people from several serious diseases, including tuberculosis (TB). The bacteria that cause TB were often present in cow's milk.

How Louis Pasteur discovered the cause of food spoilage

At one time people thought that when food went bad, microbes were created out of the food material itself. In 1850 Louis Pasteur did an experiment which disproved this idea. He showed that the microbes which spoil food come from the surrounding air.

Pasteur's experiment is shown in the pictures below.

First he poured some clear broth into a flask. He heated the neck of the flask so as to soften the glass, and pulled it out into a long S-shape like a swan's neck. Then he boiled the broth to kill any microbes present. Pasteur found that the broth remained clear and fresh, and microbes did not develop inside it.

After some days he broke the neck off the flask. Only then did it go bad – it went cloudy and microbes developed in it.

How would you interpret the result of this experiment? Pasteur interpreted it like this. Any microbes that entered the swan-necked flask from the air got caught on the sides of the long neck and so never reached the broth. But once the neck was broken, microbes were able to reach the broth and turn it bad.

We now know that Pasteur's interpretation was right. Today, in the Pasteur Institute in Paris, you can see one of Pasteur's original swan-necked flasks containing broth which he put in it over a hundred years ago. It is as clear now as it was then.

Pasteur's discovery had a great influence on the way we treat our food today. Pasteurisation of milk is just one example.

1 Imagine you are Pasteur talking to a colleague. Your colleague firmly believes the old idea that microbes are created out of the food itself. How would you convince him otherwise?

2 Did Pasteur *prove* that the microbes which cause food spoilage come from the air? Explain your answer.

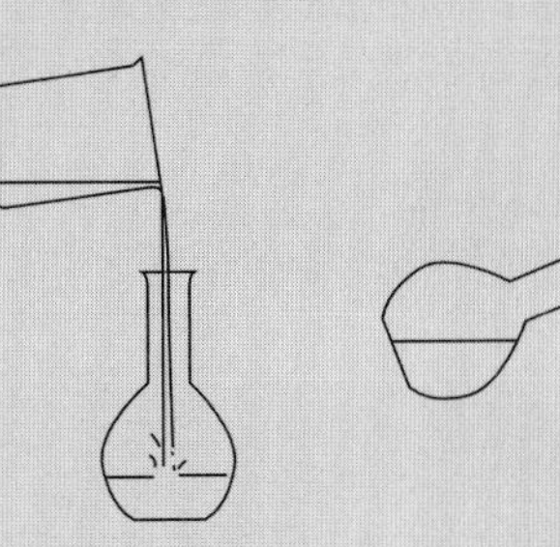

1 Nutrient broth poured into flask

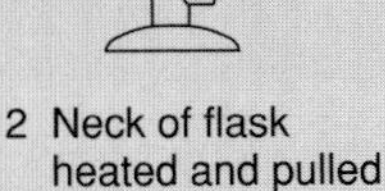

2 Neck of flask heated and pulled out into S shape

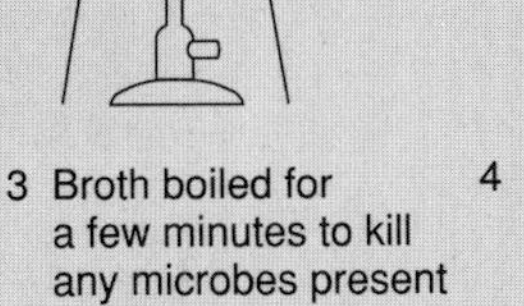

3 Broth boiled for a few minutes to kill any microbes present and drive out air

4 Broth allowed to cool. **It stayed clear and fresh for months.**

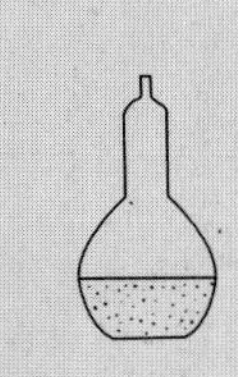

5 Neck of flask broken off without touching it with hands. **Broth went bad within a few days. It became cloudy and was soon teeming with microbes.**

B9
Habitats and communities

Usually lots of different species are found living together in particular places.

Picture 1 A freshwater pond in England.

Picture 2 An English wood in autumn.

Where do living things occur?

Living things are found almost everywhere: on land and in the air, in water and underground. They occur in the soil, under logs and stones, in grass and in trees. Various pests may share our homes with us, and some live as parasites inside us.

The place where an organism occurs is called its **habitat**. Examples of habitats are shown in pictures 1 to 4. Within a habitat organisms may live in a particular place such as under a stone or log. These are called **microhabitats**.

Environment and adaptation

The conditions in an organism's habitat make up the **environment**. The environment includes such things as the temperature and humidity, and the presence of certain other organisms. Every habitat has its own particular environment, and it is this that decides what sort of organisms can live there.

Organisms are suited or **adapted** to their environments. For example, many animals are cleverly camouflaged, or can move very quickly to escape from their enemies. The caterpillar in picture 5 is a particularly striking example of adaptation. However, all living things are adapted to the environment, though it may not always be obvious. They must be adapted to survive. There is more about this on page 291.

Picture 3 A grassland in East Africa.

Picture 4 Desert plants in the Californian desert.

Picture 5 This is not a twig, but the caterpillar of the peppered moth. By being camouflaged like this it avoids being eaten by birds. When alarmed the caterpillar sticks out rigidly from the plant and does not move. This is essential if the camouflage is to work. So it is adapted in its behaviour as well as in its structure.

Habitat restrictions

Look again at picture 5. Obviously this caterpillar's habitat has to be where the right sort of twigs are found, for example woodlands and hedgerows. If it was sticking out from a rock on the side of a mountain, its camouflage would be less effective.

Many organisms are restricted to particular habitats because they are not adapted to live anywhere else. For instance, fish are adapted to breathe in water, but not on land. On land their gill surfaces stick together, greatly reducing the surface area over which gaseous exchange can occur. Similarly, earthworms are adapted to burrowing through soil, but not to moving on a smooth surface.

It is interesting to compare the range of environments in which different groups of organisms can live. For example, mosses and liverworts are restricted to damp places because they do not have the necessary adaptations for living in dry places. On the other hand, flowering plants are found almost everywhere, including the desert. We shall see how they manage to do this in the next Topic.

Communities and ecosystems

In a habitat such as a pond or wood, the various organisms can be divided up into producers, consumers and decomposers (see page 46). Together these organisms make up a **community**. Every habitat has its own typical community. The main species characteristic of the habitat are called **indicator species**. Which are the indicator species in pictures 1 to 4?

Within the community each species feeds on, or is fed on by, other species. In other words each species occupies a particular position in a food web. We call this its **ecological niche**.

The organisms in a community are influenced by environmental conditions such as temperature, humidity and rainfall. The organisms interact with these conditions, and with each other, to make up an **ecosystem**. An ecosystem is a stable, balanced community in which matter is recycled (see page 36).

Changes through the year

Habitats don't stay the same all the time, at least not usually. They change with the seasons. For example, think of the changes that occur in a wood in the autumn. The deciduous trees and shrubs lose their leaves and their buds become **dormant** – they go to sleep, as it were. Birds such as the swallow migrate to warmer places. Other animals such as the dormouse **hibernate**. Many herbaceous plants die back and rest in the soil as **storage organs** adapted to survive the winter. Seeds which have been produced during the summer remain dormant until the next spring.

When spring arrives the migrant birds return, animals come out of hibernation, buds open, herbaceous plants send up new shoots and seeds germinate. This is the **breeding season** when many animals produce offspring. They are stimulated to do so by an increase in the amount of sex hormones in the body. This in turn is brought on by the longer days as spring approaches. Increases in the day-length are also responsible for making many plants flower. The warmer temperatures help too.

Daily changes

As well as changing with the seasons, habitats change every day. For example, natural habitats are light during the day but dark at night, and usually the temperature changes at the same time. How do you think organisms are affected by this? The seashore and estuaries change every time the tide goes in and out, and this influences the types and distribution of organisms that live in these

Plants in a community

Plants are the backbone of any community. They provide a home for many animals and, as producers, they also provide food. There are many kinds of plants. Here are the main kinds:

- **Herbs** These do not contain much wood and are generally small(ish), ranging in height from a few centimetres to about a metre.
- **Shrubs** These contain a lot of wood and may reach several metres in height. They branch close to the ground and so usually have a bushy appearance.
- **Trees** These are larger still and contain a great deal of wood in a thickened main stem or trunk.
- **Annuals** are herbs which grow from seeds, produce flowers and seeds, then die all in one year.
- **Biennials** are herbs which send up a leafy shoot in the first year but don't produce flowers and seeds till the second year, after which they die.
- **Perennials** are plants which continue to live year after year, producing flowers and seeds every year.

  Woody perennials (trees and shrubs) keep their stems and branches through the winter.

  Herbaceous perennials die down during the winter and send up new shoots the following spring.
- Trees and shrubs which drop their leaves before winter comes are described as **deciduous**. Those that keep their leaves during the winter are **evergreen**.

Look at plants around your school or home and decide which of the above categories they belong to.

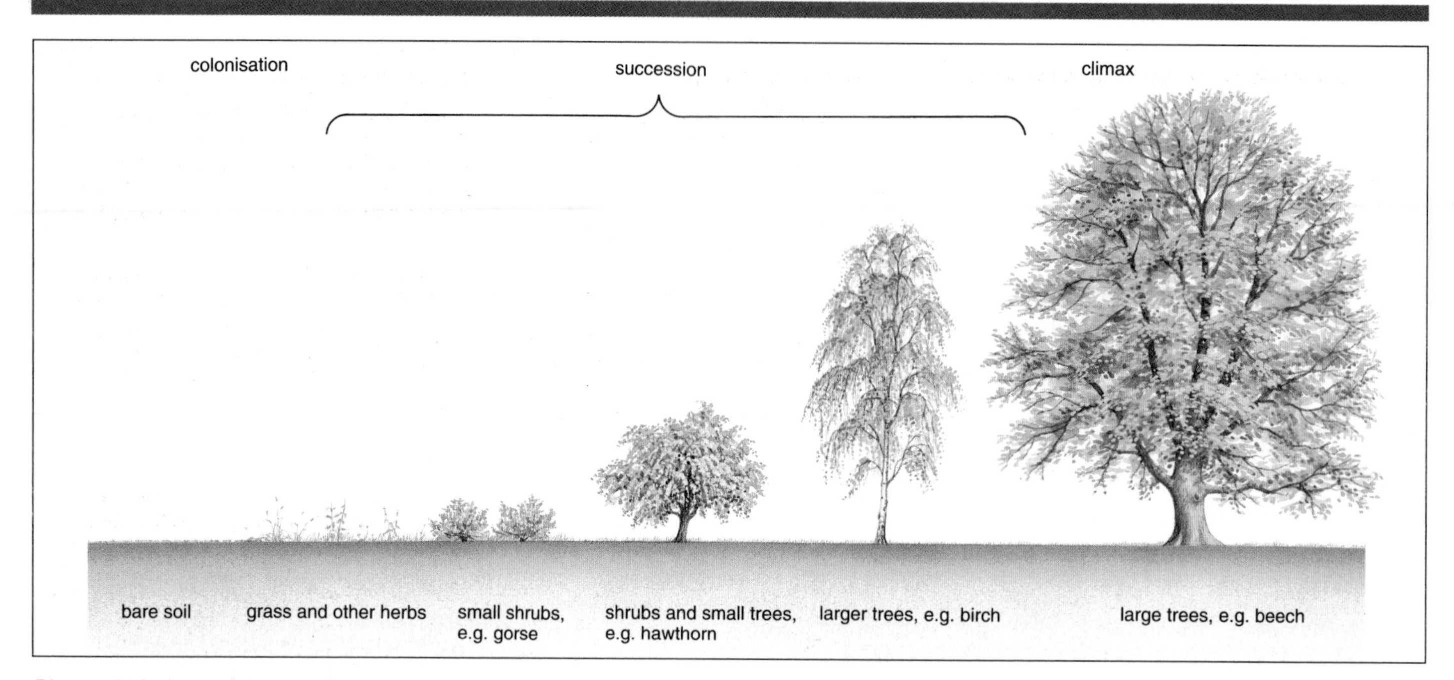

Picture 6 A plant succession. Only the main (dominant) species are shown at each stage.

places. If you study a rocky shore, you can see this for yourself.

Colonisation and succession

Imagine an area of bare soil which is left alone. Quite quickly it becomes **colonised**. Grasses and other herbs appear, forming a covering of low vegetation. Then larger plants such as gorse and broom appear, turning the area into a scrub. Next, shrubs and small trees such as hazel, elder and hawthorn grow up, followed by larger trees such as birch. Finally, large trees such as oak and beech may become established. A series of changes like this is called a **succession**. The succession just described is illustrated in picture 6. We can sum it up like this:

grassland → scrub → woodland

In a succession, the plants present at one stage alter the environment in such a way that new species can move in. In other words, each lot of plants prepares the habitat for the next lot. For example, oak and beech are able to move in at the end of the succession because the shrubs and small trees shade the ground and prevent other plants from developing; oak and beech seedlings are adapted to survive being shaded, so they can grow up without competition from other species.

While changes occur in the vegetation, changes also occur in the animals present. For example, once the shrubs and trees grow up, birds and squirrels can move in and make their homes in them. Each newcomer provides food for further species. So, as time goes on, a complex ecosystem builds up.

Once the large trees are established no further changes occur; the ecosystem has reached a **climax**. Birth, death, growth and decay go on at the same rate, so a kind of equilibrium is reached. The only thing that can change it now is a disaster such as a forest fire, or humans chopping down the trees.

Sometimes an animal may prevent a succession from developing. For example, grazing animals such as rabbits and sheep will keep an area in a grassy state and prevent shrubs and trees from growing up. This has happened on the chalk downs of Britain. On a larger scale the prairies were produced by herds of grazing buffalo.

Questions

1 Explain the difference between an organism's habitat and its environment.

2 Write down one special adaptation which would be needed to enable:
   a a small mammal to feed on the fruits from the top of a tree,
   b a lizard to avoid being eaten by birds in a sandy desert,
   c a parasitic worm to live in the human small intestine,
   d the leaves of a plant to avoid being eaten by cows,
   e a fish to be equally at home in fresh water and the sea.

3 Describe the main features of the habitats in pictures 1 to 4.

4 Some organisms move from one habitat to another in the course of their lives. Give two examples of such organisms and explain why they change their habitats.

5 A new motorway is built through a cutting in open countryside. The local authority has no money for landscaping the sides and planting shrubs, so they leave the soil bare. What do you think the sides will look like after (a) one year, (b) five years and (c) 20 years?

6 How could scientists prove that it's the longer days which makes an animal mate in the spring?

Activities

A Some examples of adaptation

1 Examine various organisms, or pictures of organisms, provided by your teacher.

Examine each organism carefully. From its appearance try to say where it lives and what sort of life it leads. Check with your teacher to see if your suggestions are correct.

For each organism write down one way in which it appears to be well adapted to its environment. Explain how the adaptation may help the organism to survive.

2 Search for organisms around your home or school. Look carefully because they may be camouflaged. Good places to look are in leaf litter, under stones and logs and at the foot of trees and shrubs.

Examine each organism and note how it is adapted to living in its particular habitat.

B Colonisation of water

1 In the spring half fill an aquarium tank with cold tap water. Place the tank in a safe place out of doors where it won't be interfered with.

2 In the following autumn examine the contents of the tank.

Examine drops of the water under a microscope. Are any organisms present? What sort of organisms are they? How do you think they got into the tank? What do you think would happen if you left the tank for several years?

C Colonisation of land

Plan a way of investigating how a patch of bare land becomes colonised. First decide exactly what questions you want to ask, and how they can best be answered. The investigation need not be carried out on a large scale; you are more likely to obtain useful information from a small patch of land than from a large patch.

When your teacher has approved your plan, carry out the investigation.

D Describing a habitat

Choose a habitat. It could be the same one that you chose for activity B on page 47. If possible, visit your habitat in winter as well as summer, and analyse it in as much detail as you can. Write a report on it.

Plan what you intend to do before you start. Your teacher will help you with this. Much will depend on how much time you have. A full report would take years. However, a lot can be achieved in quite a short time.

E Life on the seashore

Study a rocky seashore when the tide is out. Working with one or more friends, make a line or belt transect between the high and low tide marks so as to show which sort of organisms occur where (see page 64). What do you think decides where particular organisms occur?

Write down the changes which occur between the low and high tide marks in the course of 24 hours. What sort of adaptations would you expect to see in the organisms that live here? Do the organisms which you have observed have these adaptations?

Krakatau, an example of colonisation

Krakatau is a volcanic island in the Indian Ocean. On 26 August 1883 its volcano erupted with such force that a cubic mile of rock was thrown up, and dust rose to a height of 17 miles. More than 40 000 people were killed by tidal waves, and weather changes were experienced all over the world.

The island itself, 40 kilometres from the nearest land, was rendered lifeless.

After nine months a spider was seen on the island, and three years later there were blue-green bacteria, 11 species of fern and 15 species of flowering plants. Within ten years coconut palms were growing on the island, and 15 years later there was dense forest with at least 263 animal species. Fifty years after the eruption, there were 47 species of vertebrates: 36 birds, 5 lizards, 3 bats, a rat, a crocodile and a python!

(Adapted from T.J. King, *Ecology*, Nelson)

Picture 1 The island of Krakatau with its now quiet volcano peak. The island lies in the Sunda Straits between Sumatra and Java in the Indian Ocean.

1 How do you think the various organisms listed above got to the island?

2 To what extent did each new species make it possible for others to follow?

3 What might cause Krakatau to return to its original lifeless condition?

B10
The environment and survival

Why do organisms thrive in certain places but not in others? It's because of their environment.

Picture 1 An oasis in the Sahara desert with a pool and palm trees. The water brings life to an otherwise barren landscape.

The physical features of the soil are dealt with on pages 74–76.

Palm trees in the desert

The Sahara Desert is a barren landscape except here and there where clusters of palm trees can be seen. One such cluster is shown in picture 1.

Why do palm trees grow in these particular places? The answer is that here there is water, whereas the rest of the desert is extremely dry. In these oases, as they are called, water is part of the **environment**.

The environment as a whole can be divided into two parts, the **physical environment** and the **biological (biotic) environment**. Let us look at each in turn.

The physical environment

This includes all the physical features of the environment such as temperature, light and so on. We shall illustrate them by seeing how they affect life on land.

In general the physical features of the land environment are closely related to the climate.

Water

The palm trees in picture 1 can grow only where there is water. Like many other organisms, they are dependent on a regular supply of water.

Some plants have special ways of cutting down water loss so they do not need as much. For example, their stems and leaves may be covered with a thick waterproof cuticle. Their leaves may be small, reducing the surface area from which water evaporates, or they may be folded or have fewer air pores (stomata) than usual. The stomata may even be sunk down into pits in the epidermis. These kinds of plants are called **xerophytes** (picture 2).

Many plants that live in dry places have swollen stems or leaves in which water is stored. Such plants are called **succulents** (picture 3).

Animals too have ways of reducing water loss. For example, reptiles such as lizards and snakes have a scaly, waterproof skin. In contrast, amphibians such as frogs and newts have a thin, moist skin from which water readily evaporates – such animals are normally found only in wet places. The same applies to mosses and liverworts; they have been called the amphibians of the plant kingdom.

Tropical organisms have ways of surviving the dry season equivalent to the ways that temperate organisms survive the winter. For example some species go into a dormant state, rather like hibernation, until it rains.

Picture 2 This Joshua tree has narrow, spiky leaves with a thick cuticle for reducing water loss. It grows in the hot, dry desert of North America.

Picture 3 These Saguaro cacti in the North American desert store water in their thick stems and branches.

Picture 4 This rattlesnake has many adaptations for living in the desert. How many adaptations can you see in this picture?

Humidity

This is a measure of the amount of moisture in the air. A high humidity means that there is a lot of moisture in the air; a low humidity that there is relatively little moisture in the air – in other words the air is dry.

The drier the air, the faster water evaporates from the surfaces of animals and plants. Organisms that are well waterproofed can live in the dry atmosphere of the desert. Organisms that are not well waterproofed are restricted to humid places.

Although a dry atmosphere can cause water-loss, evaporation can be useful because it cools the organism.

Picture 5 Paulet Island in the Antarctic. The air temperature is nearly always below freezing, even in the summer.

Air movement

Wind and breezes speed up the rate at which water evaporates from the surfaces of animals and plants, adding to water loss but increasing the cooling effect. Have you noticed how much cooler it is on a hot day when there is a breeze?

The speed and direction of the wind are important in pollination and in the dispersal of fruits, seeds and spores. Some harmful pests such as locusts and the potato blight fungus are carried long distances by the wind.

Temperature

The temperature of the environment may vary greatly between day and night, from day to day and from one season to another. In the desert, for example, the temperature may reach 50°C during the day, but fall to below 10°C at night.

Cold-blooded animals are particularly affected by fluctuations in temperature. Some cold-blooded animals, for example lizards and snakes, bask in the sun to keep warm, and cool off in the shade (picture 4). This is an example of the way an organism's **behaviour** can help it to survive.

Warm-blooded animals are better able to cope with very low temperatures. The penguins in picture 5, for example, thrive in the freezing conditions of the Antarctic.

Picture 6 The plant on the left was grown in the light and looks normal. The one on the right was grown in the dark and is etiolated. The etiolated plant is tall and spindly and its leaves have not expanded. It is yellowish because plants need light for making chlorophyll.

Light

Plants need light for photosynthesis, so they occur only where there is sufficient light. In the darkness they become **etiolated** (picture 6). Different species differ in the amount of light they need. For example, chrysanthemums like bright sunlight, whereas many species of orchids prefer shady places.

Plants tend to grow towards light when it comes from a particular direction. On the other hand, some animals avoid light. The light acts as a stimulus and they move away from it.

Picture 7 Caterpillars of the privet hawk moth live almost entirely on privet leaves.

Picture 8 Epiphytic mosses and ferns growing on tree branches.

The biotic environment

Look at the caterpillar in picture 7. This animal feeds almost entirely on privet leaves. The privet is part of the caterpillar's biotic environment.

An organism's biotic environment consists of all the other organisms which affect its life, directly or indirectly. These include any organisms that it feeds on – and any organisms that feed on it.

Now let's take another example. Bees get nectar and pollen from flowers, so flowering plants (of the right sort) are an essential part of the bees' biotic environment. It works the other way round too: the plants depend on the bees for pollination, so bees are part of the plants' biotic environment.

Many organisms live inside, or sometimes on the surface of, another species. The latter is called the **host**. If the host is harmed in some way, we call the organism a **parasite**. The blood fluke and the head louse are examples of parasites. They feed on the host's blood.

The blood fluke lives inside its host. It is an **endoparasite**. The head louse, on the other hand, lives on the surface of the host. It is an **ectoparasite**. Every parasite is adapted to live in a particular place in or on its particular host, and to spread from one host to another. The host is an essential part of the parasite's biotic environment.

Organisms which live in or on other organisms don't always harm their hosts. For instance, mosses and lichens grow on the trunks and branches of trees. However, they don't feed on the tree's tissues and they don't cause any damage. Plants which grow on the surface of other plants like this are called **epiphytes** (picture 8).

In some cases organisms which live together may help each other. Such is the case with the nitrogen-fixing bacteria which live in the root nodules of leguminous plants (see page 37). The bacteria get shelter from the plant, and also carbohydrate which the plant has made by photosynthesis. In return, the plant gets some of the nitrogen compounds which the bacteria have made from atmospheric nitrogen. This kind of association, where both species benefit, is called **mutualism**. Each partner is part of the biotic environment of the other.

Activities

A Comparing the physical environments of two habitats

Choose two contrasting land habitats, e.g. a dense wood and an open meadow. Compare the physical environments of the two habitats. Measure the physical features as follows:

- **Water**. Place a **rain gauge** (picture 1) in each habitat, and measure the height of the water after a period of time.

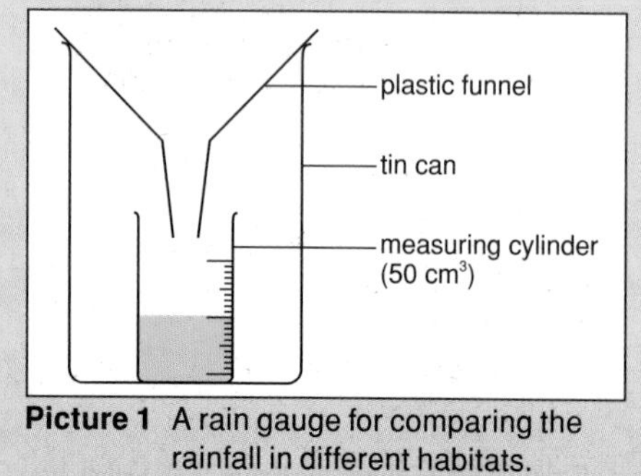

Picture 1 A rain gauge for comparing the rainfall in different habitats.

- **Humidity**. Use **cobalt chloride paper**. This is blue when dry and pink when moist. Time how long it takes for a piece of the paper to change from blue to pink in each habitat.
- **Air movement**. Compare the wind speed in the two habitats with an **anemometer** (picture 2). Count the number of times the arms swing round in a certain time.

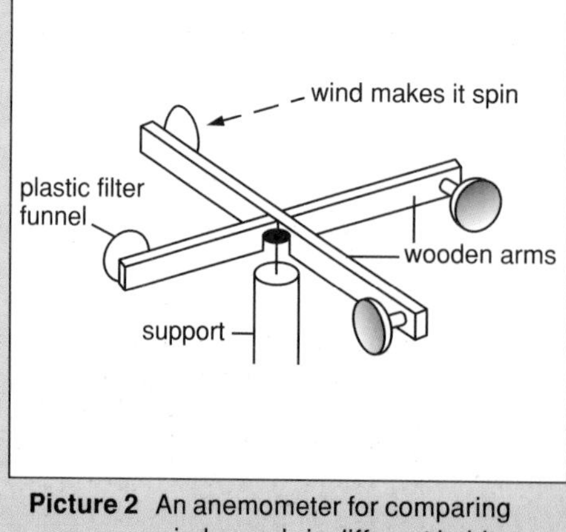

Picture 2 An anemometer for comparing wind speeds in different habitats.

- **Temperature**. Use a **thermometer** to measure the temperature at different times of the day, and/or leave a **maximum–minimum thermometer** in each habitat for 24 hours to obtain the temperature range.
- **Light**. Use a **light meter** to measure the light intensity in each habitat.
- **Soil features**. See Activity B on page 77 for how to estimate soil water, humus, air and acidity/alkalinity (pH).

Can you think of any other physical features that might be measured?

Decide how often you should make your measurements, and at what times of the day. Be sure you have enough measurements to make your comparison of the two habitats valid.

How should you express the measurements? Should you draw graphs or take averages – or both?

1 Summarise the main differences between the physical environments of the two habitats.

2 How may the differences affect the types of organisms found in each habitat?

B Do woodlice prefer dry or humid conditions?

For this investigation you will need a **choice chamber**. This is a transparent box which is divided into several interconnected compartments. The box has a removeable top and a perforated floor. Substances can be placed beneath the perforated floor to create different conditions in the various compartments.

1 Set up a choice chamber as follows. Under the perforated floor of one compartment place anhydrous calcium chloride. Under the perforated floor of another compartment place some water.

IRRITANT

Calcium chloride is an irritant.

Anhydrous calcium chloride absorbs moisture, so the air in this compartment will become dry. In contrast, the air in the other compartment will become humid. Wait at least ten minutes to give time for these conditions to develop.

2 Place ten woodlice in the centre of the choice chamber. Observe them at intervals during the next half hour. Every five minutes count how many there are in each compartment.

Which part of the choice chamber do the woodlice tend to move into? Which conditions do they prefer? Where do woodlice live? How does this fit in with your results?

C Do woodlice prefer to be in the light or dark?

Use a choice chamber to investigate whether woodlice prefer light or dark places. You will need to cover one of the compartments of the choice chamber with aluminium foil.

What conclusions do you draw from your results?

D Do blowfly larvae respond to light?

1 Lay a sheet of white paper, approximately 24 cm long, on a table.

2 Place a lamp at one end, and direct it vertically downwards towards the paper.

3 Switch the lamp off and darken the room.

4 Place six blowfly larvae on the paper, beneath the lamp.

5 Switch the lamp on and observe the blowfly larvae.

How do the larvae respond to your switching on the lamp? Why is this response useful to them in their natural environment? (Hint: find out where blowfly larvae live and what they do there.)

What could you do to make sure that the response is caused by the light and not heat from the lamp?

E What, where and why?

Have you noticed a green powdery substance on the bark of tree trunks? It is an organism called *Pleurococcus*.

Carry out your own investigations to answer these questions:

1 What kind of organism does the 'green powder' consist of? (Hint: use a microscope.)

2 On which sides of tree trunks does it occur? (Hint: find the percentage cover on different parts of a tree trunk, using a very small quadrat.)

3 Which environmental factors might influence its distribution on a tree trunk? (Hint: in what way do the conditions differ on different sides of the trunk?)

F The environment in water

This topic has been all about the environment on land. In this activity you can apply what you have learned to a different kind of environment: water.

Choose any watery (aquatic) habitat. It may be a pond, stream, river, or a rock pool on the seashore.

First, make a list of all the physical features of the environment which you can think of. Then make a list of the problems that you would expect the organisms to face. Finally, suggest adaptations which might enable the organisms to overcome each problem.

What you have done is to make a series of predictions about the habitat, based on your general understanding of the environment. You must now visit the habitat and see if your predictions are correct. Check with your teacher for any safety precautions you will need to take. In the course of doing this,. you may discover various things that you would like to investigate in more detail.

Questions

1 Name as many organisms as you can think of which form the biotic environment of the following:

a a lion,
b a tadpole of the common frog,
c a mosquito,
d a cabbage white butterfly (adult and larva),
e you.

2 The tapeworm, *Taenia solium*, lives in the small intestine of the human. What can you say about its physical environment? What problems does it face?

3 In the hot dry desert, plants have two types of root system. Some species have very deep roots which grow straight down; others have superficial roots which spread out close to the soil surface. How do these two systems help the plants to survive? (Hint: in the desert the water table is very low, and short sharp showers occur at certain times of the year.).

4 A brick wall runs east–west. There are mosses on the north side, but not on the south side. Suggest two possible reasons for this. Describe experiments which you could do to test your suggestions.

5 If you place an earthworm on the surface of some soil in daylight, it soon burrows into the soil. Here are two possible explanations of why the worm does this:

a it is repelled by light,
b it is attracted to the soil.

Describe experiments you could do to test each of these suggestions.

If (b) is correct, what might it be about the soil that attracts the worm? If neither (a) nor (b) is correct, suggest another explanation.

6 The human is more adaptable than any other species. This has allowed us to survive in a wide range of environments. How do humans manage to survive in the following situations?

a the hot desert,
b the Poles,
c the top of Mount Everest,
d under the sea,
e in space.

7 Give examples of how the behaviour of animals and plants can help them to survive in their environments.

B11 Finding out where organisms live

The places where a species occurs make up its distribution. How can we investigate distribution?

Picture 1 These students are using a quadrat to estimate the number of plants of a particular species in a meadow.

Investigating distribution

To study a species' distribution we need to find out roughly how many individuals occur in different places.

The simplest way of doing this is to look at the habitat and describe in words what occurs where. Each species is given one of these descriptions:

- **Dominant**. The species that has the greatest effect on the environment, e.g. a tree.
- **Abundant**. Species that are hardly ever out of sight, e.g. grass.
- **Frequent**. Species that are constantly found but are fairly spread out, e.g. thistles in a field.
- **Occasional**. Species that are seldom found.
- **Rare**. Species that are hardly ever found.

This is not an accurate way of assessing distribution, but it may enable you to spot something which you can then investigate more carefully if you want to.

Sampling

Suppose you suspect that there are more thistles in one field than in another field of the same size. What can you do to make sure? One way would be to count all the thistles in each field. However, this would probably drive you mad. So you count the number of thistles in, say, ten squares, chosen at random. Then you work out the average. We call this process **sampling**.

Sampling is often carried out in biology, not just in studying habitats but in other situations as well. For example, the number of blood cells in a person's circulation can be estimated in the same kind of way.

Quadrats

One of the commonest ways of sampling a habitat is to use a **quadrat**. This is a square frame made of wood or metal. You lay the quadrat on the ground and count the individual plants inside it (picture 1). Activity A tells you how to use a quadrat to find the **density** of weeds in a field.

Grids

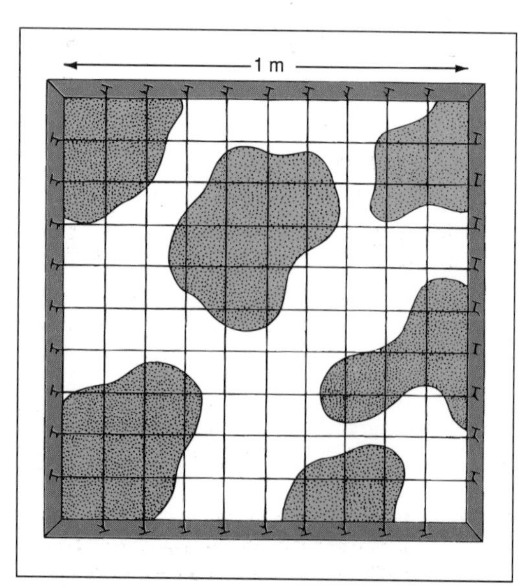

Picture 2 A grid made by dividing a one-metre quadrat into 100 smaller squares. The green areas represent patches of grass. You estimate the percentage of the quadrat which is occupied by grass.

Suppose you want to find out how much grass is growing on a piece of waste ground. In this case you can't count the plants individually because it's impossible to tell where one ends and the next one begins. Instead you estimate the area of the ground covered by grass.

To do this you use a **grid**. This is a quadrat which has been divided into 100 small squares (picture 2). You lay the grid on the ground and count the number of small squares in which the grass occurs. Activity B tells you how to use a grid in this way in order to find the **percentage cover** of grass on a piece of ground.

Transects

You may have noticed that in some places the types of organisms gradually change as you go across a habitat. In such cases it is useful to record where each type of organism occurs. You can do this by making a **line transect**. A length of tape, marked at regular intervals, is stretched across the habitat. You then record the positions of all the plants that are touching the tape (picture 3).

The trouble with a line transect is that it only gives you the organisms that are right on the line. A better method is to lay out two parallel tapes a metre apart and record the plants between them. We call this a **belt transect**. If you lay a grid between the two tapes, it will help you to put the different plants in their right positions.

You may want to see how the numbers of a particular species change as you

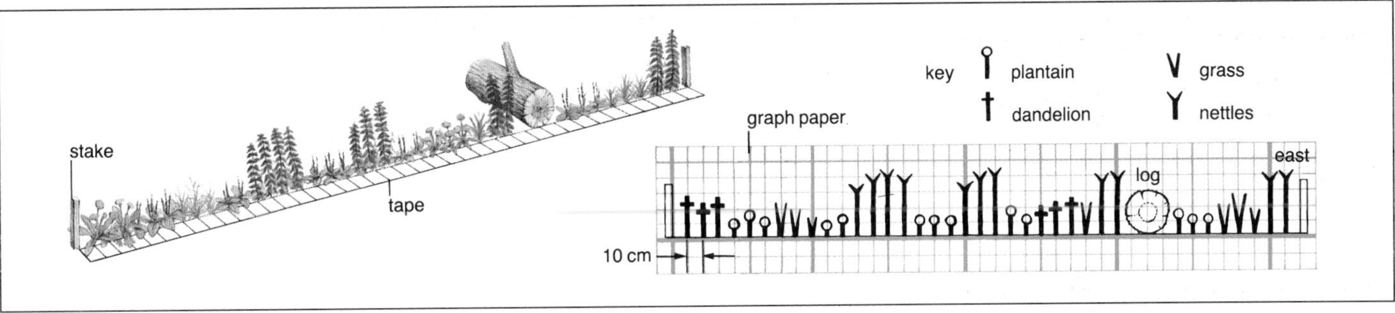

go across a habitat. To do this you lay a quadrat or grid at regular intervals in a straight line across the habitat. You then count the number of plants, or estimate the area occupied by them, in each square. What would be the best way of presenting your results?

Picture 3 How to make a line transect. On the left are the plants which might be touching a tape that you have attached to the ground. On the right is the way you should record the positions of the various types of plant.

Collecting and trapping

The methods described so far are suitable for stationary organisms, e.g. plants and sessile animals such as mussels on the sea shore. But what about organisms that move around – insects for example?

Here again, sampling comes to the rescue. Without harming them, you simply collect, or trap, the organisms over a given period of time, and count them.

Of course this does not tell you the total number of individuals in the habitat, but it is a good way of comparing their relative numbers in different places.

When making such a comparison, you must be sure that the method used is a random one (picture 4). The same method should be used in the different places, and carried out at the same time of day for the same length of time. Why do you think this is necessary?

If you want to estimate the *total* number of individuals in a habitat, you can use a special technique called the **capture–recapture method**. This is explained on the next page.

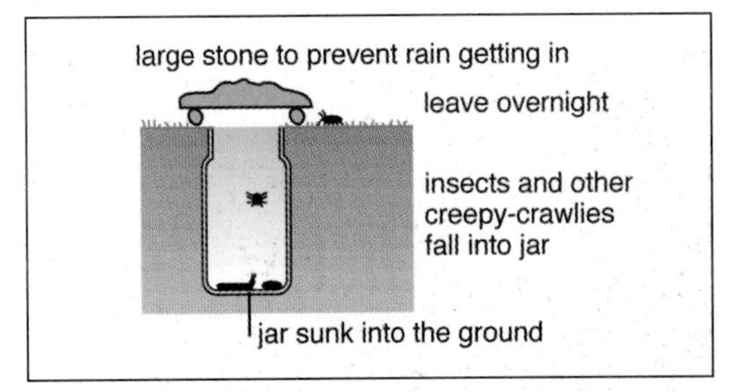

Picture 4 A pitfall trap, an example of a random trapping device. It is pure chance as to whether or not an animal falls into the jar.

Relating an organism's distribution to the environment

There is a large beech tree in the middle of my lawn. I have noticed that the number of daisy plants increases as one goes further from the trunk. Why is this?

The most obvious reason that springs to mind is that there is more light further from the trunk. To test this idea you estimate the number of daisy plants per square metre at different distances from the trunk, using a quadrat. You also find the average light intensity at each position, using a light meter of the kind that photographers use.

You then plot graphs of the number of daisies and the average light intensities against the distance from the trunk, as shown in picture 5. You can see at once that the two curves are very similar. This suggests that they are related in some way. In mathematician's language, there is a **positive correlation** between them.

Biologists often look for relationships, particularly when investigating the environment. But remember: an apparent relationship between two things does not prove that they are connected. For example, the change in the number of daisies on my lawn might be caused by some factor other than the light intensity, or by several factors acting together. The only way of proving that light intensity is involved is to rule out all the other possibilities.

It's often difficult to find out why organisms are distributed in a particular way simply by making measurements in their natural habitats. Sometimes you have to do **controlled experiments** indoors as well. You put the organism in different conditions and find out which one suits it best.

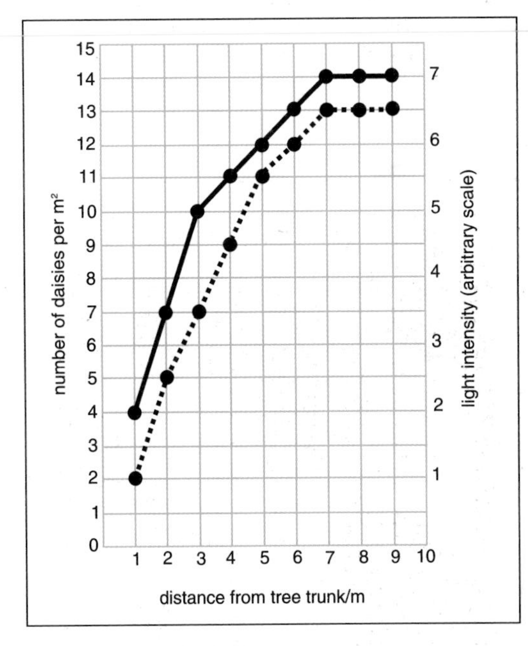

Picture 5 This graph shows the number of daisy plants, and the average light intensities, at different distances from a tree trunk. The number of daisy plants is represented by the solid curve, the light meter readings by the dotted curve. Notice that the two curves follow each other exactly. This suggests that they *may* be related in some way.

Activities

A Estimating the number of weeds in a field

1 Obtain a quadrat, one metre square.

2 Select a field, and decide what particular weed you wish to investigate.

3 Lay the quadrat on the ground, and count the number of weeds inside it.

If a weed is touching the frame, include it in your count if more than half of it is inside the quadrat.

4 Repeat the above procedure with the quadrat in at least five different places, chosen at random.

5 Work out the average number of weeds per square metre in the field. This is the **density** of weeds.

Do you think this is a good method of finding out how many weeds there are in the field? If not, why not?

What are the main reasons for any inaccuracies in the results?

What could be done to improve the method?

B Estimating the area of ground occupied by grass

1 Obtain a grid. This should be a one-metre quadrat that has been divided into 100 squares.

2 Find a suitable area of ground that has patches of grass.

3 Lay the grid on the ground, and estimate the number of squares which contain grass. If a square is only partly filled with grass, take this into account in making your estimate. For example, four squares that are each a quarter full count as one square.

The final figure you arrive at is the **percentage cover.**

4 Repeat the above procedure with the grid in at least five different places, chosen at random.

5 Work out the average percentage cover of grass in the area.

Do you think this is a good method of finding out how much grass there is in an area. If not, why not?

C Investigating distribution

Look around your local countryside, a park or the grounds of your school and find a wild plant whose distribution looks interesting. If you're stumped your teacher will help you.

Choose one of the methods described in this topic and use it to investigate the plant's distribution.

Do your results show anything interesting about the plant's distribution? If they do, what further investigations might you carry out?

Questions

1 When using a quadrat to investigate the number of plants in a field, you should place the quadrat on the ground randomly. Why is this important, and how can it best be achieved?

2 Look at picture 2 on page 64. Work out the percentage cover of the grass in this illustration.

3 Your friend maintains that there are more minnows in his pond than in yours. What could you do to find out if he is right?

4 A student counted the number of daisy plants in thirteen one-metre quadrats on a lawn. Here is a summary of her results:

Number of daisies	4	5	6	7	8	9
Number of quadrats in which the above number of daisies occurred	1	2	4	3	2	1

a Plot these results on graph paper.

b Calculate the average number of daisy plants per square metre.

c What sort of graph have you drawn? Why is it the best way to present the results?

Capture–recapture method

Suppose you want to estimate the total population of animals in a particular area. You can do this using the **capture–recapture method**.

First you capture a sample of individuals, and count them. You then mark them in some way that does not harm them. With beetles, for example, a small dab of non-toxic paint on the back will do.

You now release all these marked individuals back into the habitat, and give them time to mix with the rest of the population. Then you capture another sample of individuals. You count the individuals in this second sample; you also count the marked individuals in the sample.

To find out the total population, you use this neat little formula:

$$\text{population size} = \frac{n_1 \times n_2}{n_m}$$

where:

n_1 is the number of individuals captured in the first sample (the ones that were marked and released),

n_2 is the number of individuals in the second sample,

n_m is the number of marked individuals in the second sample.

Using this method, you can compare the total populations in different areas, or you can compare the populations in the same area at different times – day and night, for example, or different times of the year.

Suppose you want to estimate the size of the population of ground beetles in a particular habitat using the capture–recapture method. Explain *in detail* how you would do this. What precautions would you take to ensure that your estimation is as accurate as possible?

Two examples of adaptation

Here we look at the main features of two animals that live in very different environments. We shall see how the features of these animals help to adapt them to their environments.

The camel

Look at the Arabian camel in picture 1. Although there were wild camels at one time, the camel is now a domestic animal. It lives in the desert areas of north Africa and the Middle East where temperatures are generally high and rainfall low. Although the desert can be extremely hot, the temperature may vary greatly between night and day and between winter and summer. Camels are herbivorous, feeding on desert plants.

Here are the main features of camels which enable them to live in the hot, dry desert:

- The nostrils are narrow slits and are lined with hair. They can be almost completely closed. They filter dust from the air during sand storms.
- The long thick eyelashes protect the eyes from sand and grit.

Picture 1 The Arabian camel, also known as a dromedary, has a single hump and long legs. It is a fast runner and is used for riding. An Arabian camel has been known to cover 150 miles in 11 hours.

- The feet, instead of having hooves, are splayed out and have leathery soles for walking on soft shifting sand.
- Camels can go for a long time without eating or drinking. Water is stored in the stomach. The hump contains fat, which serves as a food store from which water can be made by metabolism. When a camel does drink, it drinks a lot (as much as 40 litres in ten minutes!).
- The tissues can tolerate large temperature swings. The body temperature may vary from 34°C at night to 41°C during the day.
- The tissues can tolerate dehydration. As more and more water is lost, the body fluids become more and more concentrated.
- The body is cooled by sweating (see page 186). However, the camel sweats only when the body temperature is high and there is a reasonable amount of water in the body.
- Being large, camels have a relatively small surface–volume ratio. This reduces heat energy loss.
- The skin has a covering of fur which insulates the body against the loss or gain of heat energy.

The polar bear

There is only one species of polar bear. It lives in the Arctic, in areas where the temperature is below freezing for most of the year. Polar bears spend much of their time walking on floating ice, and swimming in the water (picture 2). They are carnivorous, and feed mainly on fish, seals and walruses.

Here are the main features of the polar bear which help it to live in the Arctic:

- The fur is white, like the surroundings.
- The forelimbs have enormous claws and are very powerful.
- The feet have hairy pads which assist movement over ice.
- Eyesight is good, and so is the sense of smell.
- The animal is an excellent swimmer, propelling itself through the water with its powerful forelimbs.
- The thick fur insulates the body against heat energy loss when the animal is on land.
- A thick layer of fat (blubber) beneath the skin insulates the body against heat energy loss when the animal is in the water.
- Although carnivorous, polar bears will eat plant food if necessary.

Picture 2 Polar bears are good at walking on slippery surfaces such as ice. They are also good at swimming.

1. Why is it helpful for the polar bear to eat plant food as well as animal food?
2. Polar bears are white. What might be the advantage of this? Can you think of any possible disadvantages?
3. The polar bear's fur is a good insulator on land, but not in the water. Why the difference?
4. The polar bear is so well adapted to living in the cold that it does not need to hibernate. Why is hibernation useful to a cold-dwelling animal?
5. Why is it useful to a camel to be able to tolerate wide fluctuations in its body temperature?
6. Desert mammals face a conflict: they need to save water but they also need to lose it (by sweating, for example) to cool themselves. How does the camel resolve this conflict?
7. Do you think camels would survive in the Arctic and polar bears in the desert? Give reasons for your answer.

B12 Populations

People, people everywhere! This topic is about populations: how they grow and what happens when they get too big.

Picture 1 Oxford Street, London, on a busy day.

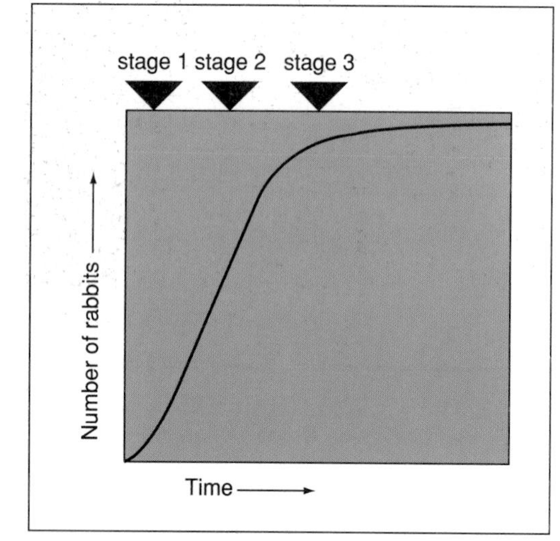

Picture 2 This graph shows how a population of rabbits may grow. In stage 1 the population grows slowly at first and then gradually gets faster. In stage 2 the population grows exponentially at the maximum rate. In stage 3 the population growth gradually slows down, and eventually the population stops growing altogether.

How do populations grow?

Suppose you put 100 rabbits onto an unpopulated island. The rabbits reproduce and the population increases. If you counted the rabbits at intervals and plotted their numbers against time, you would find that the population rises as shown in picture 2.

One of the most noticeable things about populations is that they increase very quickly. This is because the numbers go up by *multiplication*, like this:

$$100 \xrightarrow{\times 2} 200 \xrightarrow{\times 2} 400 \xrightarrow{\times 2} 800 \xrightarrow{\times 2} 1600 \xrightarrow{\times 2} 3200 \text{ etc.}$$

In other words, the total number *doubles* at regular intervals. This type of increase is described as **exponential**. It is how populations tend to grow, whether they are rabbits, flies or humans – unless something happens to stop the natural increase.

Why do populations grow?

In any community new individuals are born and older ones die. The rate at which individuals are born is called the **birth rate**, and the rate at which individuals die is called the **death rate**. Birth rate and death rate are explained in more detail in the box at the foot of this page.

Populations increase because new individuals are born at a faster rate than older ones die. In other words, the birth rate is greater than the death rate.

Populations are also affected by individuals entering or leaving the community, that is **immigration** and **emigration**.

What stops populations growing?

Look again at the graph in picture 2. Notice that eventually the curve flattens out. In other words, the population growth slows down and the numbers level off.

Why does this happen? In the case of our rabbits there are a number of possibilities. Here are some of them:

1 The food (grass and so on) begins to run out, so some of the rabbits starve.
2 There are so many rabbits that there is no room for any more burrows.
3 The rabbits are so overcrowded that diseases spread rapidly and many die.
4 Being overcrowded, the rabbits suffer from stress and this interferes with their reproduction.
5 Predators such as foxes and birds of prey eat more of the rabbits because, being commoner, they are easier to catch.

How to work out population growth from the birth rate and death rate

The birth rate and death rate are expressed as yearly percentages.

If the birth rate is 10 per cent, it means that for every 100 individuals at the beginning of the year, there are 10 more (i.e. 110) at the end.

If the death rate is 3 per cent, it means that for every 100 individuals at the beginning of the year, there are 3 less (i.e. 97) at the end.

The whole population therefore shows a net increase of 10 – 3 = 7 per cent.

This means that for every 100 individuals at the beginning of the year, the actual number at the end is 107.

These are the kinds of checks which stop populations growing for ever. The first two involve **competition**: the rabbits compete with each other for scarce resources such as food and living space. There may also be competition between the rabbits and other species. Which other species might compete with rabbits, and for what?

If there are no predators, humans may step in and take control. At one time Australia had no rabbits, but in 1859 some domestic rabbits escaped from their pen when it was swept away by a flood. These rabbits ran wild and bred at such a rate that parts of Australia were soon overrun with them. They did a lot of damage to crops and gardens. Unfortunately there were no predators to keep them under control. Eventually the virus disease myxomatosis was deliberately introduced to destroy them – an example of **biological control** (see page 85). The disease swept through the rabbit population, and their numbers fell.

Picture 3 Foxes feed on rabbits and help to keep their numbers down.

The human population

Picture 4 shows the population of Britain over the last 7000 years. Notice how it has increased. In fact the graph looks rather like the one in picture 2. In particular, notice that the population has grown more during the last 200 years than in the whole of the previous 5000 years.

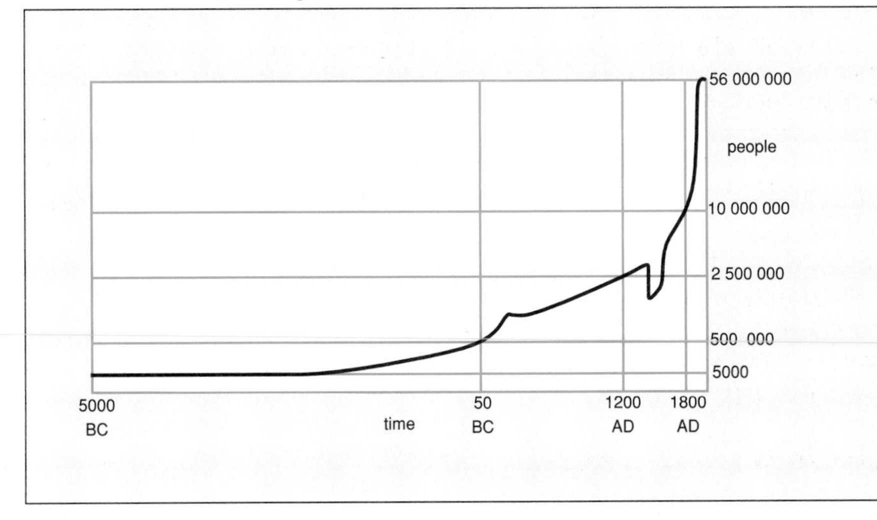

Picture 4 The graph shows how the population of Britain has grown since 5000 BC.

What has caused this recent increase? It's mainly because the death rate has fallen due to better health. Fewer infants die and old people live longer.

Obviously this increase can't go on for ever. One way of stopping it is by **birth control** (see page 234). Birth control certainly seems to be having an effect in Britain because the population is now showing signs of levelling off. But this is not true of many developing countries where the populations are continuing to rise rapidly.

In the world as a whole the population is increasing by about 80 million people each year (picture 5): that's 9000 an hour, or 150 a minute. Someone has worked out that if the human population were to go on rising uncontrollably, the whole of the Earth's surface would be covered with people standing shoulder to shoulder within 600 years!

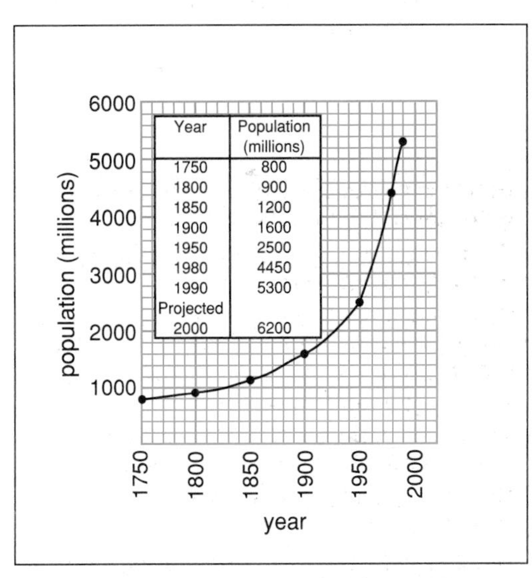

Year	Population (millions)
1750	800
1800	900
1850	1200
1900	1600
1950	2500
1980	4450
1990	5300
Projected 2000	6200

Picture 5 World population growth from 1750 to 1990.

Populations under control

In a natural community of organisms, such as a pond or wood, the population sizes of the various species stay more or less constant. Numbers are kept under control by all sorts of checks, and there is a natural balance between predators and prey. Uncontrolled growth of the sort that we see in the human population is the exception rather than the rule.

Questions

1 Study picture 2, then suggest two reasons why the rise in population is slow to begin with and then speeds up.

2 If all its offspring survived, a single greenfly could produce 600 000 000 000 offspring in one season, with a mass of over 600 000 kg – roughly equivalent to 10 000 men. What prevents this happening?

3 Study the graph in picture 4 and then answer these questions.

a How can we say what the population of Britain was in 5000 BC?

b Suggest a possible reason why the population was rising around 50 BC.

c What do you think caused the sudden fall in the population between AD 1200 and 1800?

d Why has the population risen so quickly since AD 1800?

e Suggest two reasons why the population appears to be levelling off now.

4 The table shows the birth and death rates for four countries.

Country	Birth rate %	Death rate %
UK	1.71	1.19
USA	1.76	0.96
China	2.9	1.3
India	4.2	1.7

Calculate the percentage yearly increase in the population of each country. Suggest reasons why the population is increasing at different rates in the four countries.

5 'Most of the ecological and social problems facing the human race are caused by over-population.' Think about this statement and discuss it with your friends. Start by making a list of the problems in order of urgency.

6 Population growth is kept under control by competition for scarce resources. Illustrate this with reference to (a) rabbits and (b) humans.

What other forces besides competition keep population growth under control?

The environment under strain

Suppose you are a parent of a large family. What would be your main concern? Surely it would be to make sure that your family is well fed, clothed and housed. This means spending your money wisely, and making certain that you don't spend more than you earn.

The same thing applies to a whole country. It's part of **economics**. One of the main tasks of a country's government is to look after the economy. The government tries to do this in such a way that everyone enjoys a high standard of living.

With a large population, goods – including food – have to be produced on a massive scale. This requires **industry** and **farming**, and that in turn requires energy and raw materials. The larger the population, the greater the strain on our natural resources. Between 1924 and 1985 the number of new houses built in England and Wales since 1919 rose from less than half a million to over 14 million. In one year the British Sunday newspapers alone swallow up paper from about five million trees.

No wonder there is not much natural countryside left. No wonder so much of our land is covered with houses, factories and farms.

Since the Second World War the number of licensed vehicles in the United Kingdom has increased from less than 4 million to over 24 million (see picture 1). No wonder there is so much **pollution**.

To satisfy the needs of a rapidly growing population that wants more and more, we have changed our world.

What exactly has changed? Not just the surface of the Earth where people live, but the whole of that part of the Earth and its atmosphere where there is life. This is called the **biosphere**.

Here are two particularly important ways in which the biosphere has changed:

- We have increased the amount of carbon dioxide in the air. This is raising the temperature of the world as a result of the **greenhouse effect.**
- We have released chlorofluorocarbons (CFCs) which destroy the **ozone layer** in the upper atmosphere. The ozone layer protects us from harmful ultraviolet rays which can cause skin cancer.

These are both aspects of pollution. You will find more about them on pages 40–41. All we shall say here is that they are made much worse by the size of the human population. When you squirt an aerosol can, the amount of CFCs that you put into the air is tiny. But millions of people using CFCs all over the world can have a tremendous effect on the atmosphere.

It's only relatively recently that these momentous changes have started taking place on our planet. That's because only in recent times has the world had to support such a huge and demanding population.

Humans have been on Earth for about a million years; we have started ruining it in less than two hundred. It could take thousands of years to put right the harm that we have already done.

1 Do you think it would be possible to solve our environmental problems without reducing the population?

2 Refrigerators are now manufactured on a large scale in many developing countries. These refrigerators release CFCs but the countries concerned say that they cannot afford to change their factories to produce non-CFC refrigerators. They say to the developed countries: 'For years you've been producing refrigerators that give out CFCs, and now you tell us not to!' How would you answer this?

3 To what extent does greed add to our environmental problems?

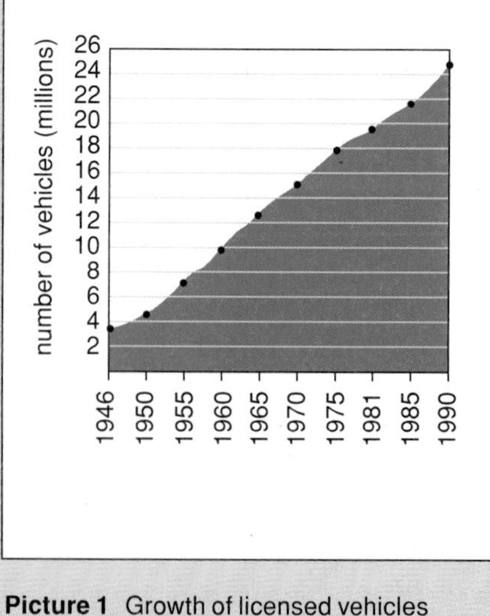

Picture 1 Growth of licensed vehicles 1946–1990. From Guiness UK Data Book 1992.

Population structure

Look at the picture below. These diagrams are called **population pyramids**. They show the population for different age groups in Great Britain at various intervals of time from 1891 to 1980.

In 1891 the most numerous people were the youngest ones. However, by 1947 the proportion of people in the 35 to 39 age group had increased, due to better health, but the proportion of teenagers had decreased. Notice the increase in the number of children in the nought to four age group: this was caused by an increase in the birth rate just after the second world war. It is known as the **post-war bulge**.

You can see the post-war bulge again in the 1956 and 1980 pyramids. These later pyramids also show a marked increase in the proportion of elderly people compared with the 1891 pyramid.

Population pyramids are useful because they enable us to forecast the population structure in the future. This is important in planning things like schools, housing needs and medical services.

Examine the 1980 population pyramid below.

a In what age group is the post-war bulge evident?

b How would you account for the post-war bulge?

c Why do you think the proportion of teenagers was so high in 1980?

d Why do you think the proportion of under tens was relatively low?

What do you think the population pyramid for Great Britain will look like in the year 2000? Try drawing it, and give reasons for your prediction.

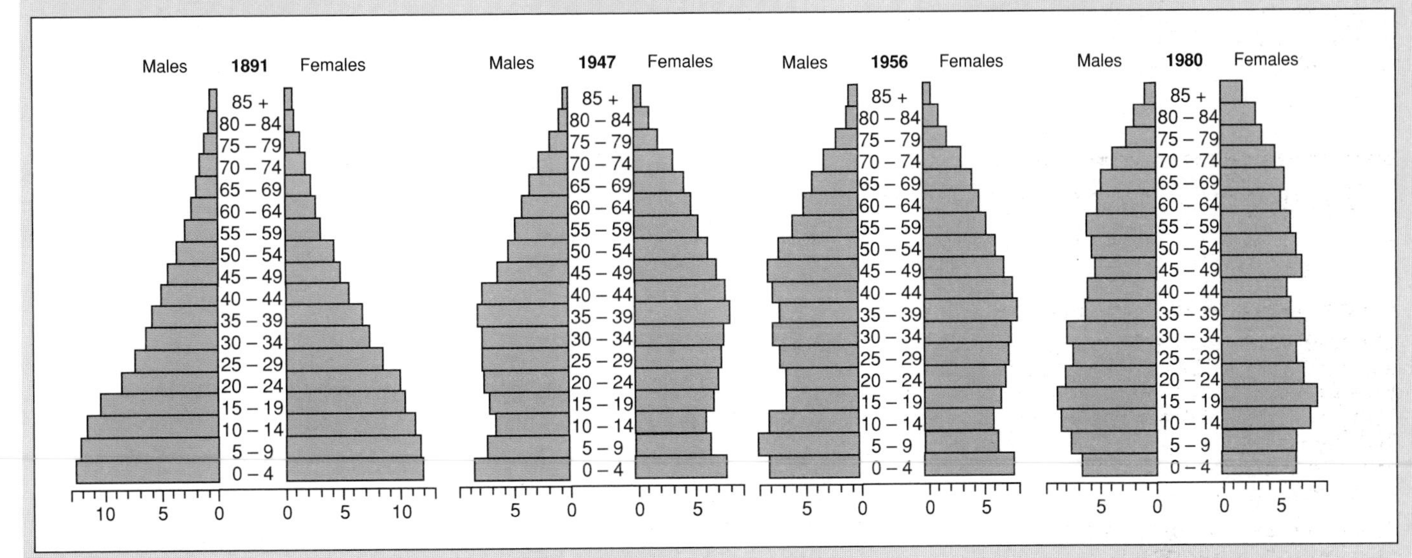

Picture 1 Population pyramids for Great Britain. The length of each horizontal bar represents the percentage of people, male or female, in a particular age group in the population.

£27,000 war on island's rabbit army

By George Turnbull

THE tiny Orkney island of Rousay (pop. 200) is under attack by thousands of marauding rabbits.

It is estimated that a rabbit army of at least 100,000 – and increasing rapidly – is eating about 40 tons of grazing grass a day on the 16 square-mile island.

Sixteen farmers have formed a Rabbit Clearance Society and mounted a £27,000 three-year campaign to fight the invaders.

A full-time rabbit control operator has been hired to fight the nibbling pests with gas, snares and nets and take the battle to their burrows in the heather-clad hills in the middle of the island.

Night raids

Mr Christopher Soames, 36, who has been raising beef in Rousay for 12 years and is the secretary for the society, said the burrows were difficult to get at or even spot. "And it is at night that the rabbits come down and raid the farms."

Because of the amount of grass the rabbits eat, farmers are not able to carry as much stock as they might.

"I farm 170 acres of beef, but I estimate the rabbits are costing me £1,500 to £2,000 a year," said Mr Soames.

The local authority and the highlands and Islands Development Board are contributing to the rabbit clearance scheme.

1 Why do you think rabbits suddenly started being a problem on the island?

Suggest *two* possible reasons.

2 The rabbit control operator might have considered introducing foxes onto the island. Why might this have been a bad idea?

3 Consider the pros and cons of infecting the rabbits with myxomatosis.

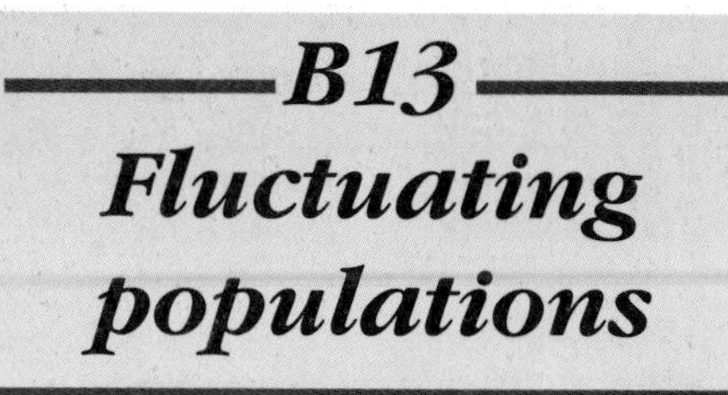

B13 Fluctuating populations

Populations don't stay the same but change with time.

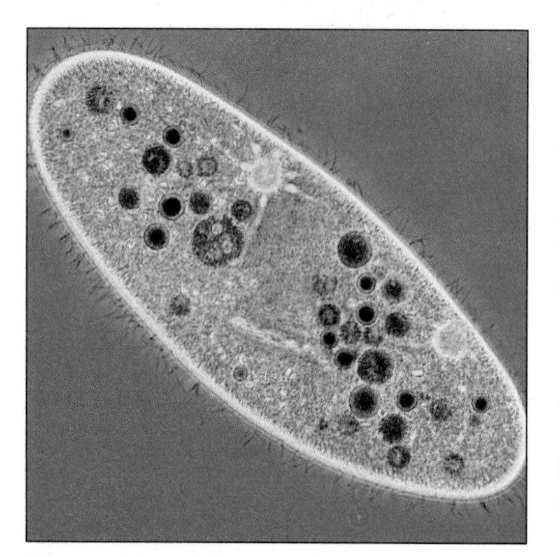

Picture 1 *Paramecium* digesting yeast cells (the red blobs). *Paramecium* is a single-celled organism whose body is covered with hair-like cilia. Beating of the cilia drives the organism through the water. They also draw the yeast cells into a 'mouth' at the side of the body.

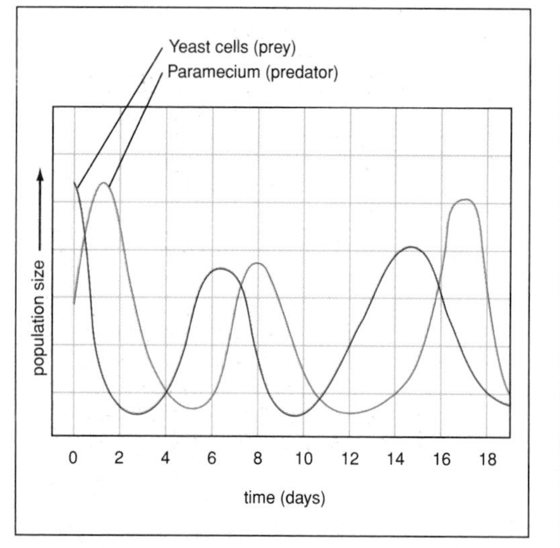

Picture 2 The effect on a population of yeast cells of adding *Paramecium*. The numbers of yeast cells and paramecium cells go up and down (fluctuate) as shown.

Population growth of yeast

Yeast is a single-celled fungus which is ideal for studying population growth. If you put a few yeast cells in a flask containing a suitable medium (e.g. sugar solution) at the right temperature, the yeast cells reproduce rapidly by budding (see page 248). As a result, the yeast population increases exponentially, doubling every hour or so.

The effect of adding a predator

Paramecium is a single-celled organism that is fond of eating yeast cells (picture 1). Now suppose we add a few paramecia to a flask of yeast cells. Picture 2 shows the result of doing this. The yeast and paramecium populations go up and down (fluctuate) together. The fluctuations in the paramecium population follow the fluctuations in the yeast population, lagging slightly behind.

How can we explain this? We can suggest that the following cycle of events occurs. As soon as the paramecia start feeding on the yeast cells, the yeast population falls. Meanwhile the paramecium population rises. After a while, the yeast population falls so low that there aren't enough yeast cells for the paramecia to feed on, so the paramecium population falls. With fewer paramecia feeding on them, the yeast population rises again. The cycle is then repeated.

Similar fluctuations are found in other predator–prey populations. It is nature's way of controlling the populations of predators and prey. It is also the basis of how pest numbers are kept down by **biological control** (see page 85).

The hare–lynx story

Some years ago a scientist called Charles Elton carried out an interesting study on the populations of various mammals in Canada. He found out from the Hudson Bay Company how many skins had been obtained from different animals each year from 1860 to 1935.

Two of the animals that he studied were the snowshoe hare and its predator, the lynx. The population fluctuations for these animals are shown in picture 3. You will see that the numbers of hares and lynx fluctuate together. The most obvious explanation is that the lynx feed on the hares, causing fluctuations similar to those that are known to occur in other predator–prey populations.

But there is a snag with this explanation. On islands off the east coast of Canada there are hares but no predators. It has been found that on these islands the hare populations still fluctuate!

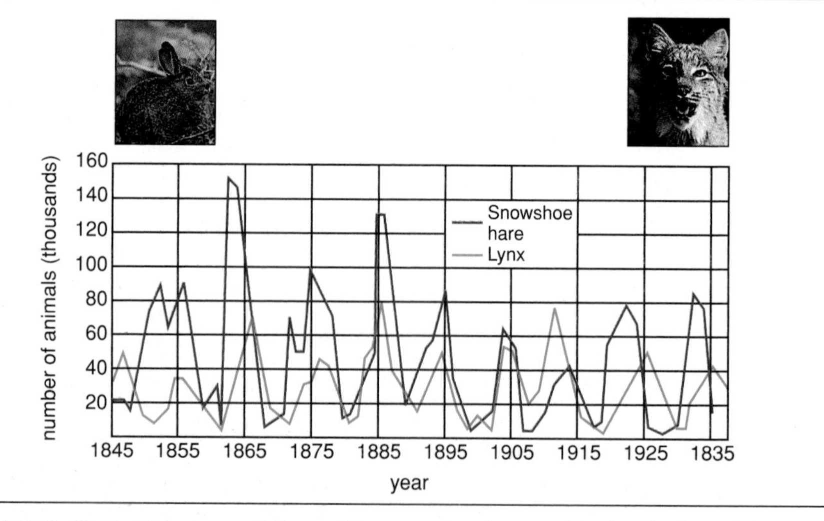

Picture 3 Fluctuations in populations of the snowshoe hare and the lynx in Canada between 1845 and 1935. The photographs show the hare (left) and the lynx (right).

Fluctuations and adrenal stress

Many animals show regular fluctuations in their numbers, but the reason is not always clear. Take the lemming, for example (picture 4). This small Arctic rodent shows a regular four year cycle in its numbers. When the population reaches a peak the animals start suffering from severe stress. This makes them run all over the place, jumping off cliffs and falling into rivers. The population then falls.

Other animals suffer from stress when overcrowded. For example, rats kept in overcrowded cages show all sorts of antisocial behaviour such as fighting with each other and killing their young. Scientists have shown that these rats have abnormally high concentrations of the hormone adrenaline in their bloodstream. They are suffering from **adrenal stress**.

It is possible that adrenal stress may occur in humans in overcrowded conditions. It might help to explain the aggressive behaviour shown by people who live close together in high-density housing. However, the causes of human aggression are very complex and many factors are involved.

Picture 4 The lemming, an Arctic rodent, has an unusual way of controlling its numbers.

Counting yeast cells

Suppose you want to study the way a population of yeast cells increases with time. How do you go about it?

First you put a small number of yeast cells (100, say) in a suitable medium (e.g. sugar solution) and leave the flask at a constant, fairly warm temperature. The yeast cells multiply quickly. After a period of time (e.g. half an hour) you swirl the flask to ensure that the yeast cells are spread evenly through the medium. You then remove a small sample of the contents of the flask with a pipette, place a drop of it on a special microscope slide and cover it with a coverslip. This slide enables you to count the number of yeast cells in a known volume of the medium.

You repeat this procedure every half hour for several hours. You then plot the results on graph paper: the number of yeast cells on the vertical axis, and time on the horizontal axis. The graph in picture 1 was constructed from data obtained in this way.

1 The method described here is an example of *sampling*. What does the term sampling mean, and why is it important in biology?

2 How could you use this method to find the effect of varying the temperature on the rate at which yeast cells multiply?

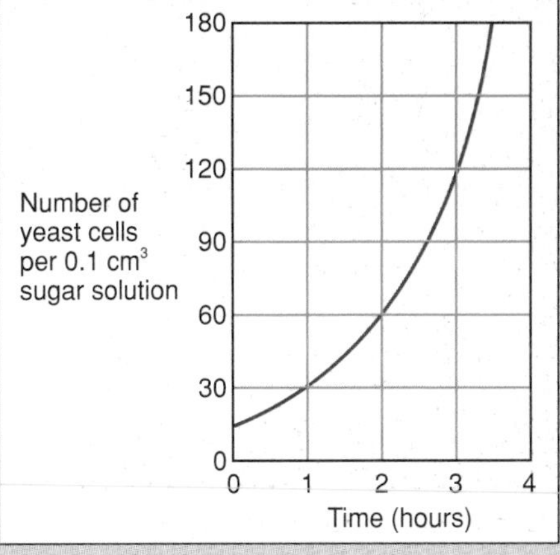

Picture 1 Population growth of yeast cells in a flask of sugar solution. Note that the increase in numbers is exponential, the number of cells doubling every hour.

Questions

1 In an experiment a scientist set up a culture of *Paramecium*. Two days later he added a small sample of another single-celled organism called *Didinium* which feeds on *Paramecium*. Changes in the populations of the two species are shown below.

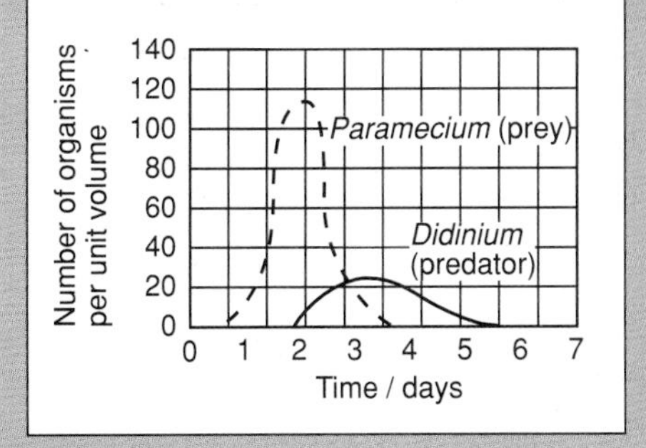

a Explain the population changes of the two species.
b What do you think would have happened to the *Paramecium* population if *Didinium* had not been added to the culture?

3 How would *you* explain the fluctuations in the populations of snowshoe hares on those Canadian islands where the hares have no predators?

4 The predatory mite (*Typhlodromus*) feeds on the spotted mite (*Eotetranychus*) which feeds on oranges. A scientist placed a small batch of spotted mites in a flask, together with some oranges. Three days later he added a small batch of predatory mites to the flask. He then estimated the numbers of both species of mites at weekly intervals for eight months. His results for the first two months are shown here.

Week	Numbers	
	Spotted mite	Predatory mite
1	210	100
2	920	340
3	1400	1250
4	750	1900
5	300	950
6	170	750
7	250	360
8	580	130

a Plot the results on graph paper.
b Suggest an explanation for them.
c Predict how the numbers of mites changed during the next six months. Give reasons for your predictions.

B14 Soil

Soil affects plant growth and is important for the welfare of humans.

Picture 1 Notice the different layers of soil in this cutting.

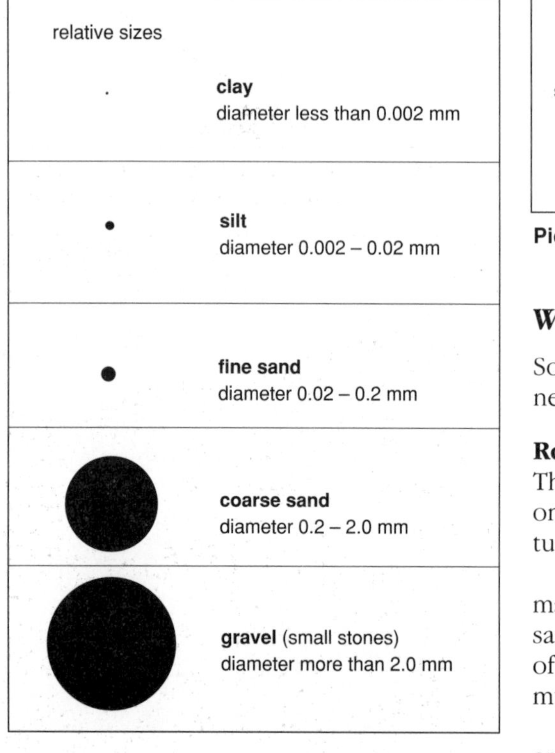

Picture 3 The different kinds of particles which make up soil. You need a microscope to see clay particles. In contrast, gravel is so large that it can be separated from the rest of the soil by sieving.

How is soil formed?

Land which is now covered with soil started off as bare rock. In time the surface of the rock got broken up by rain, wind, frost and snow into small particles. This process is called **weathering**. The particles gradually piled up on top of the rock to form soil.

If you look at a cliff or a new motorway cutting, you will see that the soil is made up of layers (picture 1).

Picture 2 explains the layers. At the top is a dark layer where plants and other organisms live. We call this the **topsoil**. It's formed by surface weathering and the activities of the many organisms that live in it. It may be covered with dead leaves (**leaf litter**), and it is usually rich in nutrients.

Beneath the topsoil is a lighter-coloured layer of stones, gravel, clay and so on. This is called the **subsoil**. It contains the deeper roots of large plants, like trees, but otherwise not much lives there.

Further down still is solid **rock**. This may not let rain through, so water gathers above it. The surface of this water is called the **water table**.

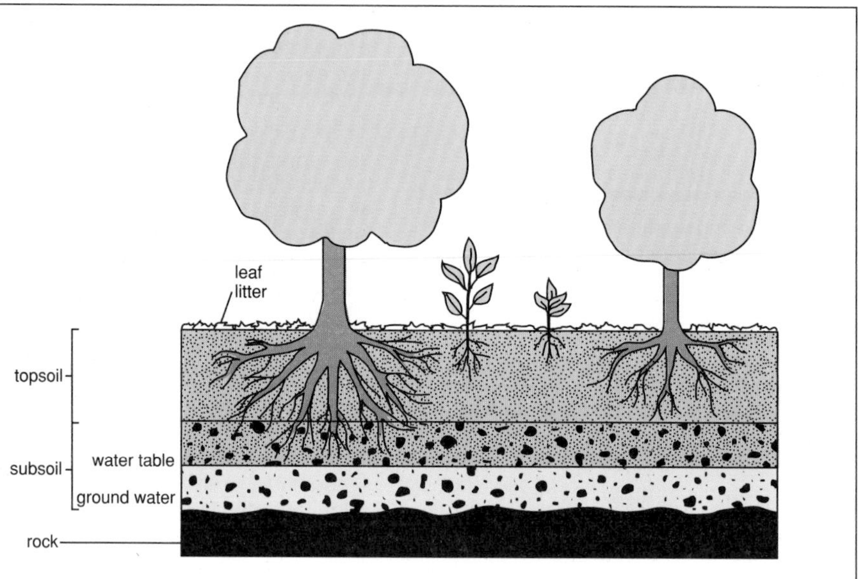

Picture 2 Sectional view of the Earth's crust to show the different layers of soil and other minerals.

What does soil consist of?

Soil contains six main things. For organisms living in the soil, these components make up their **physical environment**. Let's look at them in turn.

Rock particles

These make up the framework of the soil, its 'skeleton' as it were. Depending on their size, the particles are classified into **clay**, **silt**, **sand** and **gravel** (picture 3).

Clay and sand are especially important. Clay holds onto water firmly; this makes clay sticky and helps to bind the rest of the soil together. In contrast, sand is looser and more easily penetrated by air and water. Good soil consists of a mixture of the two. This is called **loam**. Loam contains roughly twice as much sand as clay.

If you look at a sample of good garden soil, you will notice that the particles are stuck together in small clumps. These are called **soil crumbs**. They give the soil its structure, helping air to get into it and water to drain through it.

The roots of plants grow down between the soil crumbs, gripping them as they do so. This gives plants a firm anchorage.

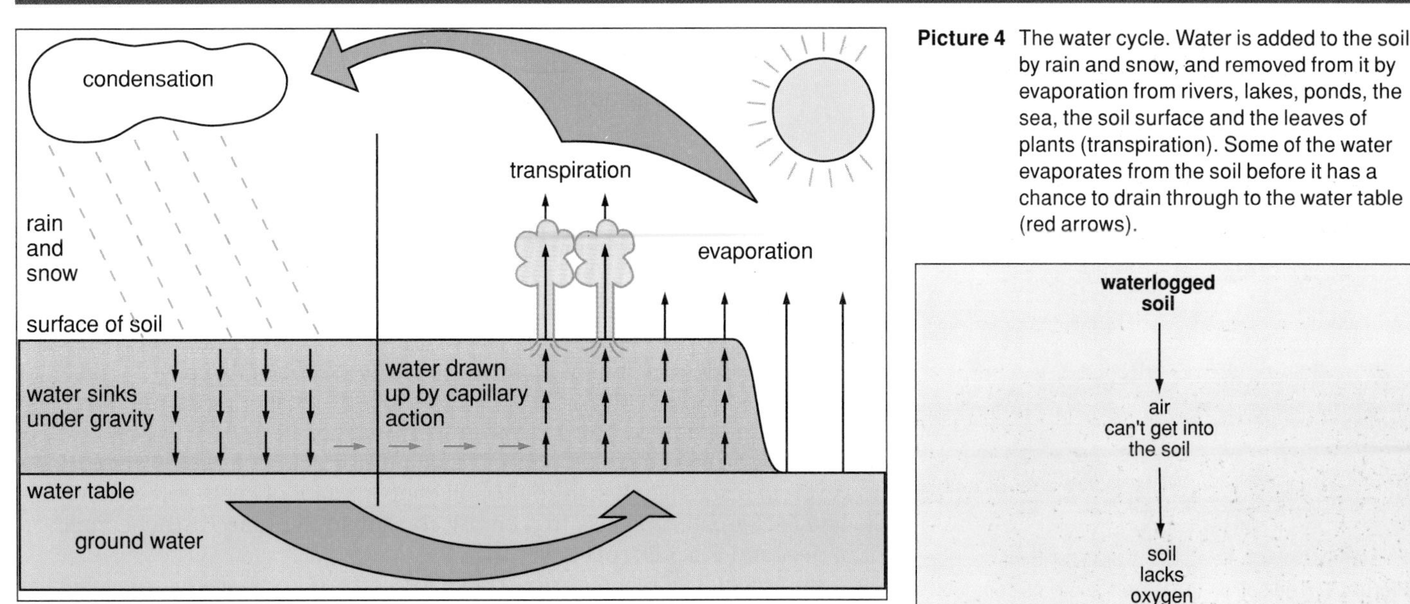

Picture 4 The water cycle. Water is added to the soil by rain and snow, and removed from it by evaporation from rivers, lakes, ponds, the sea, the soil surface and the leaves of plants (transpiration). Some of the water evaporates from the soil before it has a chance to drain through to the water table (red arrows).

Soil water

Soil particles are usually surrounded by a thin film of water. Plant roots take up all the water they need from these films.

What ensures that these films of water are present? Thanks to the rain and sun, water constantly moves through the soil, keeping the soil particles wet. This movement of water through the soil is part of the **water cycle** (picture 4).

For water to move through the soil, the soil particles must be the right size. if they are too large, water will flow straight through the soil and useful nutrients get washed out of it. This is called **leaching**. If the particles are too small and tightly packed, water cannot get through – it just stays on top or flows off the surface.

Soil air

In good soil there are spaces between the soil particles, filled with air. The oxygen in this air is needed for respiration by plant roots and by the microbes that bring about decay.

If there is heavy rain and the drainage is poor, the soil may become **waterlogged**. The air spaces get filled with water and the soil becomes short of oxygen. This results in poor plant growth. The reason is explained in picture 5.

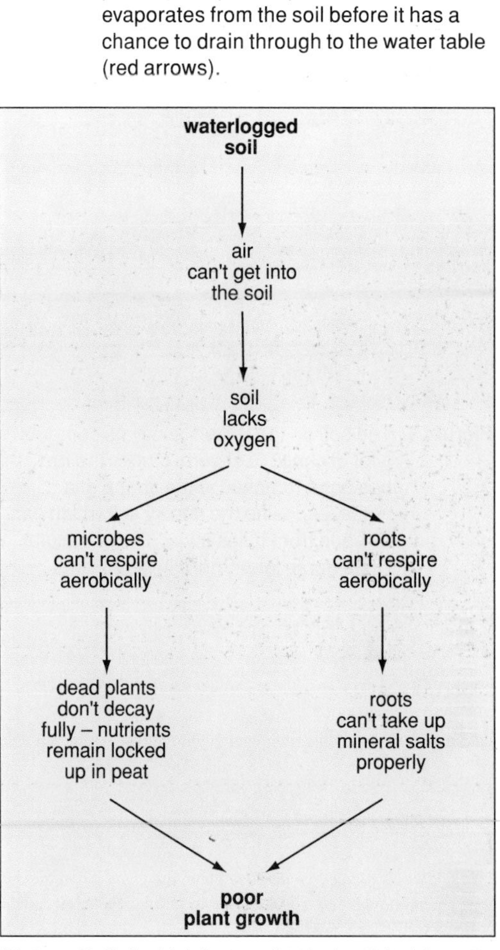

Picture 5 Soil which is waterlogged results in poor plant growth. This flow chart tells you why.

Mineral salts

Dissolved in the soil water are various **mineral salts**. These provide plants with important elements such as nitrogen, phosphorus and potassium which are essential for growth (see page 79).

Some mineral salts come from the rock which formed the soil. An example is given in picture 6. Other minerals, such as nitrates, come from dead organisms when they decay.

Picture 6 A red sandstone cliff in Devon. The red colour of Devon soil is caused by the presence of a lot of iron compounds.

Humus

When animals and plants die in the soil, they normally decay into a dark, sticky substance called **humus**. Humus is found mainly in the topsoil, and much of it comes from the leaf litter on the surface. For the gardener, one of the best sources of humus is **compost** (see page 55).

Humus makes the soil rich in nutrients which are needed for plant growth. It also forms a sticky coating round the soil particles, helping them to clump together into soil crumbs. Humus holds onto water and prevents valuable nutrients being washed out of the soil when it rains.

For dead material to decay completely, oxygen is needed. This is because the microbes that bring about decay need oxygen for their respiration. If there isn't enough oxygen, dead plants build up into a thick carpet of partly decayed

Picture 7 The soil on the Wiltshire downs contains a lot of chalk. In several places the turf has been removed in the shape of a horse, exposing the chalky soil which can be seen from miles away. This particular white horse is at Cherhill, near Calne.

Picture 8 Heather and other moorland plants like living in acid soil.

material called **peat**.

If peat is added to well-aerated soil, it will decay into humus. Peat and humus tend to make the soil acidic.

Lime

Lime comes from a type of rock (limestone) which contains chalk. Chalk is calcium carbonate. Lime is important for three main reasons:

- Calcium is one of the elements which all plants need for proper growth and development.
- Lime causes the smaller soil particles to clump together into soil crumbs.
- Calcium carbonate neutralises acids and this prevents the soil being too acidic: in gardeners' language, it prevents the soil being 'sour'.

We can find out how acidic or alkaline the soil is by measuring its **pH** (see page 42). Soils from different places differ in their pH.

In many places the soil is neither alkaline nor acidic: it is neutral (pH 7). However, in places like Wiltshire the soil contains a lot of chalk and is therefore alkaline (pH greater than 7). In contrast, the soil in the Lake District contains a lot of peat and is acidic (pH lower than 7).

Certain species of plants thrive in alkaline soil but not in acidic soil, whereas others thrive in acidic soil but not in alkaline soil. However, most plants grow best in soil which is round about neutral.

The best soil contains approximately 50 per cent sand, 30 per cent clay, 10 per cent humus and 10 per cent lime.

Life in the soil

Many organisms live in the soil. Some are harmful, others helpful. Harmful species include the animals in picture 9. They eat the roots of plants and do a lot of damage. Also harmful are denitrifying bacteria because they lower the nitrate content of the soil (see page 37).

One of the most helpful species is the earthworm which – through its burrowing activities – turns over, drains and aerates the soil. Decomposers, such as bacteria and fungi, are also helpful because they cause decay and thus make the soil fertile. And of course nitrogen-fixing and nitrifying bacteria add nitrates to the soil.

Grass is a helpful plant because its roots hold the soil together. Overgrazing by sheep and other herbivores can destroy the grass and cause **erosion**. The animals crop the grass so close to the ground that they kill it. This is a particular problem in many developing countries.

Picture 9 Four animals which live in the soil. Their approximate lengths are given in brackets.

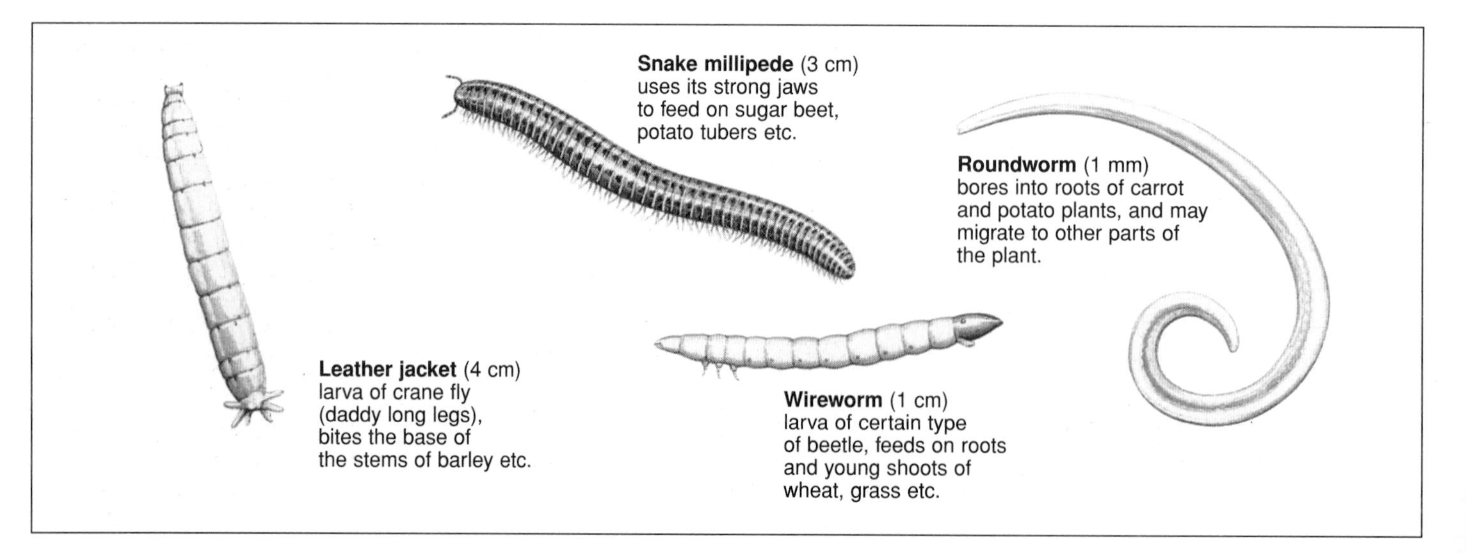

Activities

A Separating the components of soil

1 Quarter fill a large test tube with soil.

2 Add water until the tube is three-quarters full. Notice that air bubbles are given off: what does this tell you?

3 Put your hand over the open end of the tube, and shake well.

4 Put the tube in a rack, and let the soil settle. The heaviest components of the soil will sink to the bottom, and the lighter ones will float at various levels.

Does the appearance of your test tube agree with the illustration?

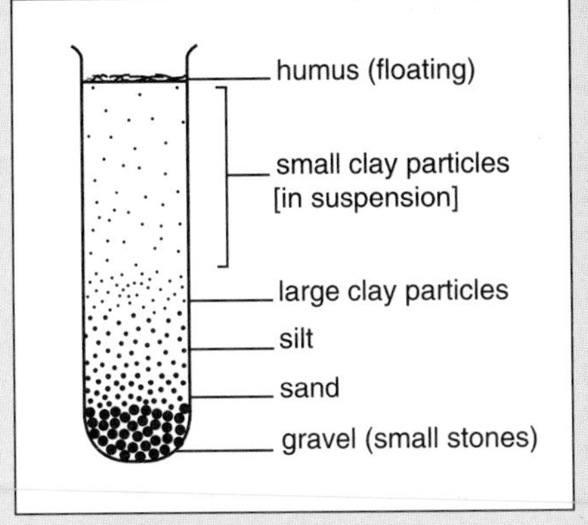

B Investigating soil

Collect some good garden soil. Investigate its properties, using the following methods.

1 *Air content*

Take a sample of the soil. Carefully place it in a measuring cylinder, then cover it with water. Stir it so as to dislodge the air. Measure the fall in the level of the water.

2 *Water content*

Take another sample of the soil and weigh it. Dry it to evaporate all the water. Then re-weigh the sample.

3 *Humus content*

Weigh the dried soil sample from step 2. Heat it strongly to burn off the humus. Then re-weigh the sample.

WEAR EYE PROTECTION

4 *Acidity/alkalinity (pH)*

Shake another soil sample with distilled water. Let the water settle. Find the pH of the water by dipping a strip of pH paper into it and noting the colour.

These notes only tell you the principles behind each method. You must plan the details for yourself. In particular ask yourself these questions: Will my method give accurate results? What is the best way of expressing my results? Is my method safe? Can I improve it?

Discuss your plan with your teacher before you start work.

C Comparing different types of soil

You often hear people say that the soil in their garden is dreadful. Your teacher will give you one of the following types of soil. All of them give problems to gardeners.

Sandy, clayey, chalky, peaty soil.

Using whatever means you have available, investigate the properties of your particular soil. Don't feel you have to use complicated methods. Do obvious things like looking at it and feeling it. If you carry out any experiments, make sure you check them with your teacher first.

Write a report on the soil, describing its bad qualities and suggesting ways of improving it.

D Collecting small organisms from soil

1 Obtain a sample of good garden soil.

2 Set up the apparatus shown in the illustration, spreading the soil out on the perforated tray. (This apparatus is called a **Tullgren funnel**.)

3 Switch the lamp on, and leave the apparatus for between one and three days. The light and heat from the lamp should drive the organisms downwards out of the soil into the beaker.

4 Observe the contents of the beaker. Are there any organisms in it? What kind of organisms are they? Use a simple key to identify them.

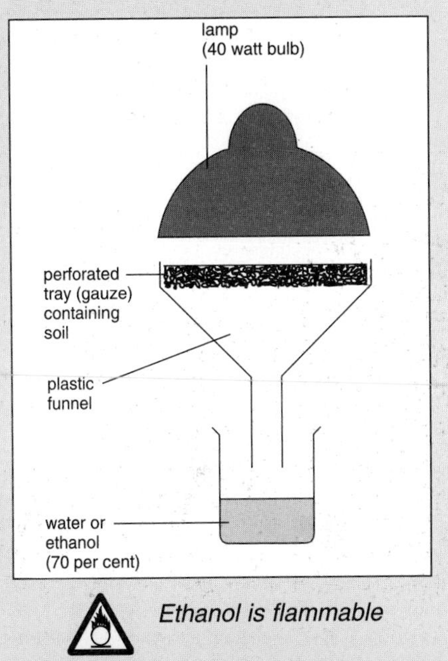

Ethanol is flammable

FLAMMABLE

Questions

1 It's better to water a garden in the evening than in the middle of the day. Why?

2 You have a potted plant and you pour some water onto the soil. However, none of the water comes through the hole at the bottom of the pot. What has happened to it?

Some people water their potted plants by standing the pot in a saucer of water. Will this work? Explain your answer.

3 What sort of soil

a shifts easily beneath your feet,
b sticks to your shoes,
c looks very black,
d has a whitish appearance,
e is very heavy to dig?

4 On a certain mountain top the soil is only a few centimetres thick, whereas in a forest at the foot of the mountain the soil is about 20 metres thick. Suggest two reasons for the difference.

5 Sometimes a gardener will spread a layer of peat or manure or bits of bark over the surface of the soil. This is called *mulching*. Suggest four ways in which this might be useful.

6 A student carried out an experiment to determine the amount of water in a sample of soil. He placed the soil in a small dish, and weighed it before and after drying. These are his results:

mass of dish on its own	10 g
mass of dish plus damp soil	15 g
mass of dish plus dry soil	20 g

a How might the student have dried the soil?
b How could he be certain that the soil was completely dry?
c What percentage of the soil consisted of water?

7 Suggest two reasons why a gardener might add lime to the soil.

B15
Making the land fertile

With so many mouths to feed, it's essential that the land should be fertile.

Picture 1 Lucerne belongs to the legume family and has nitrogen-fixing bacteria in its roots. It raises the nitrogen content of the soil, especially when ploughed into the soil as green manure.

Picture 2 The wheat seedlings on the left were given all the nutrients they needed. Those on the right were given everything they needed except nitrogen. Notice the poorer growth shown by the seedlings that were deprived of nitrogen.

Soil nutrients get lost

Think of a natural plant community – a forest if you like, or grassland. When one of the plants dies, it decays. As a result, the various chemicals in its body are set free and returned to the soil. They can then be absorbed and used again by other plants.

Now think what happens in a field with a crop in it, such as wheat. The crop is harvested and the plants removed. The chemicals are not returned to the soil, so the soil becomes poor. The soil is made even poorer if heavy rain washes useful nutrients out of it.

How can we overcome this problem? There are five main ways.

- **Grow crops in different places**
  You grow your crops on one piece of land for several years and then move somewhere else. This is what nomadic tribes do in certain parts of the world. It's called **shifting cultivation**. It is only possible where there is plenty of land and not many people.
- **Leave fields fallow**
  Every few years you leave a field empty of crops, that is **fallow**, for a year or so. This gives the soil a chance to regain the nutrients that it has lost. Where do you think the nutrients come from? What other advantages might be gained by leaving a field fallow for a year?
- **Rotate crops**
  Some plants take more of certain chemicals out of the soil than others. If the same crop is grown in a field year after year, a particular element – nitrogen say – may get very low. So you grow a different type of crop in the field every few years. This is called **rotation of crops** and it has been practised since Roman times. Crop rotation on its own won't prevent a particular nutrient getting used up, but it can delay it.
- **Make use of nitrogen fixers**
  Every now and again you include a crop of clover, or some similar plant, in the rotation (picture 1). These plants belong to the legume family. Their roots contain **nitrogen-fixing bacteria** which absorb nitrogen from the air and turn it into nitrogen compounds which the plants can use. Some of these nitrogen compounds get into the soil, enriching it for other plants.
- **Add fertilisers to the soil**
  A fertiliser is a substance which contains chemical elements needed by plants. *When you add fertilisers to the soil, you are putting back the chemicals which the plants have taken out*. This is the surest way of preventing the soil becoming short of nutrients.

Different kinds of fertilisers

There are two kinds of fertiliser, organic and inorganic. **Organic fertilisers** are generally obtained from animals and plants, and are therefore natural. **Inorganic fertilisers** are produced by manufacturing processes, and are therefore artificial.

Organic fertilisers

One of the best known organic fertilisers is **farmyard manure**. This consists of the faeces and urine of farm animals, mixed with straw. It is spread on the ground where it decays. As it decays, nitrates and other inorganic nutrients are released from it into the soil. These can then be absorbed and used by plants.

Another organic fertiliser is **compost**. This consists of the rotting remains of vegetable matter (see page 55). As with manure, the decay process releases inorganic nutrients which can be used by plants.

Because of their colour, compost and farmyard manure are referred to as **brown manure**. Sometimes a farmer will grow a crop of green plants and then plough them into the soil. This is called **green manure**. Once ploughed in, the plants decay and the inorganic nutrients are set free. The advantage of green manure is that the plants absorb and hold onto nutrients which might otherwise be washed out of the soil by the rain.

Plants belonging to the legume family make particularly good green manure. Why do you think this is?

Inorganic fertilisers

These consist of simple substances which can be absorbed by plants straight away.

General inorganic fertilisers contain **nitrogen**, **phosphorus** and **potassium** (**N**, **P** and **K**). These are the three main elements which plants need from the soil in order to grow well. Picture 2 shows the effect of depriving wheat seedlings of just one of these elements (nitrogen), so you can see how important they are.

Inorganic fertilisers are manufactured either from natural materials such as bones and horns, or by special chemical processes. They are produced in chemical forms that can be absorbed by plants, such as nitrates or ammonium salts.

Inorganic fertilisers may be sprayed onto the soil in liquid form. Alternatively, they may be scattered as pellets or powders which are then dissolved and washed into the soil by the rain (picture 4). They must not be too concentrated, otherwise the plants may be harmed and water may pass out of their roots as a result of osmosis. Also, putting too much fertiliser on the soil can cause pollution (see page 81).

Picture 4 An agricultural helicopter spraying fertiliser over a vineyard in California's Napa Valley.

Picture 3 Inorganic fertilisers, prepared and packaged by the manufacturers, provide a convenient and effective way of adding nutrients to the soil.

What elements do plants need?

In addition to nitrogen, phosphorus and potassium, plants need sulphur, magnesium, calcium and iron. All these elements are needed in quite large amounts and are therefore called **major elements**. Normally they are obtained from the soil as mineral salts, such as nitrates and phosphates, which are absorbed by the roots.

Here are the main reasons why plants need the major elements:

- Nitrogen is in proteins and DNA.
- Phosphorus is in DNA, cell surface membranes and in chemicals involved in respiration and photosynthesis.
- Potassium helps enzymes in respiration and photosynthesis.
- Sulphur is in proteins.
- Magnesium is in chlorophyll.
- Iron is needed for chlorophyll formation.
- Calcium is needed for forming cell walls.

Apart from poor growth, one of the commonest signs that a plant is short of a major element is yellowing of the leaves. This is called **chlorosis**.

Plants also need certain other elements in much smaller amounts. These are called **trace elements** or **micronutrients**. They include copper, molybdenum, sodium and manganese. If one of these trace elements is absent from the soil, plants are likely to show poor growth.

In certain parts of Australia crops grew very badly until it was discovered that there was no molybdenum in the soil. The soil was then sprayed with a very dilute solution of molybdenum, and the plants grew splendidly! Very little molybdenum was needed: one teaspoonful was enough for an area the size of a tennis court. Too much of a trace element can have a damaging effect on plants.

Picture 5 The rice plants on the left were given a general fertiliser. Those on the right were not. Notice the much greater growth shown by the plants that were given the fertiliser.

Do fertilisers work?

Look at picture 5. This shows the result of a trial with rice, carried out in the Far East. The results speak for themselves.

In England there is an experimental research station at Rothamsted in Hertfordshire. Here there are plots of soil where wheat has been grown year after year for over 100 years (picture 6). In one plot no fertiliser has ever been added to the soil. Its first crop was harvested in 1843. Since then the annual yield of grain has fallen to less than half what it was originally.

In other plots different kinds of fertiliser have been added to the soil every year. In some of these plots the yield is more than twice that of the unfertilised plot (picture 7). So fertilisers certainly help.

Which is better, organic or inorganic fertilisers?

The Rothamsted results suggest that it doesn't matter much. High yields have been obtained with both.

An important feature of organic fertilisers is that they have to decompose before plants can make use of them. Small amounts of nutrients are therefore released slowly and steadily. This is a disadvantage if you want quick results. However, in certain circumstances it can be a good thing. For example, if you put manure on the soil in the autumn, it won't start acting until the following spring. Organic fertilisers also improve the structure of the soil, aerating it and helping the soil particles to stick together. They also help the soil to hold water (see page 75).

Inorganic fertilisers can be used by plants straight away, so they act more quickly. Large amounts of nutrients are therefore made available to the plants just when they want them, that is during the period of maximum growth. However, if inorganic fertilisers are used year after year, and no organic fertilisers are used at all, the soil loses its structure and becomes fine and dusty. It then gets blown about by the wind and this can lead to **erosion**. Leaving fields fallow every few years gives the soil a chance to regain its structure, particularly if an organic fertiliser such as manure is ploughed into it.

Picture 6 The Broadbalk field at Rothamsted Experimental Research Station, showing the plots of wheat. Wheat has been growing on these plots since 1843. Some experiments take a long time!

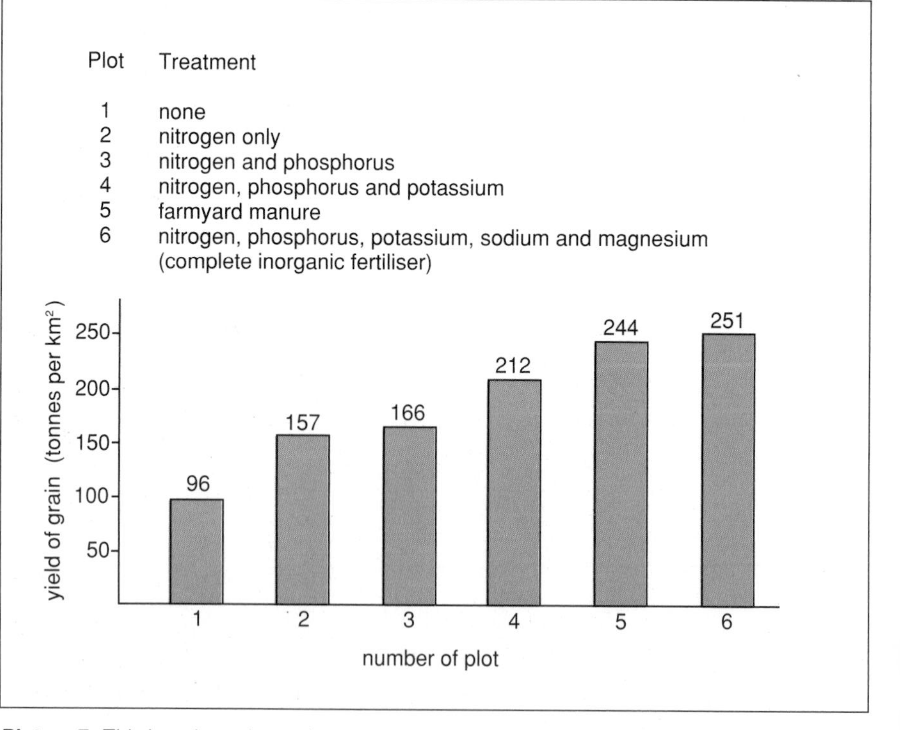

Picture 7 This bar chart shows the average daily yield of grain given by six plots of wheat in the Broadbalk field at Rothamsted between 1852 and 1967.

Fertilisers and the environment

If a lot of fertiliser is added to the soil, it may drain into lakes and rivers from the surrounding land. With so much nitrate and other nutrients available to them, the algae multiply rapidly and the water turns green. The enrichment of the water with nutrients resulting in this great growth of photosynthetic organisms is called **eutrophication**. When the organisms die, they decay. The microbes that bring about decay (decomposers) multiply so fast that they use up all the oxygen. As a result, the fish and other organisms suffocate and die.

And there is another problem. Some of the nitrate may get into the drinking water. A high nitrate level in drinking water can be a health hazard. It causes 'blue baby syndrome' and may be linked with stomach cancer. This is an important aspect of **water pollution** (see page 83).

Picture 8 A lake showing severe eutrophication. Notice the duckweed covering the surface of the water. The extra nutrients which caused this may have come from fertilisers, sewage or farm slurry.

What is 'blue baby syndrome'?

If nitrate gets into a baby's gut, it is turned into nitrite. The nitrite is absorbed into the baby's bloodstream. Once it gets into the bloodstream, the nitrite makes it difficult for the baby's blood to carry oxygen. It does this by interfering with the haemoglobin, the red pigment in the blood (see page 156). As a result, the baby's tissues suffer from shortage of oxygen. In severe cases the skin goes blue – hence the name 'blue baby syndrome'.

Doctors think that babies might be at risk if they are fed on bottled milk made by mixing powdered milk with drinking water. The danger period is the first two months of life. By the time the baby is two months old, it has made an enzyme in its blood which counteracts the effect of nitrite.

Activities

A Finding out about fertilisers

Visit a local gardening shop or garden centre. Find the fertiliser section. Read the labels on the various brands of fertiliser to see what substances they contain. Make a list of the substances. Write down the chemical elements each substance provides. From books find out why plants need these particular elements.

B Investigating the beneficial effects of fertilisers on plants

Set up two groups of seedlings. Barley or broad bean seedlings will do. Grow one group in water or moist sand which contains no fertiliser, and the other group in sand or water to which a fertiliser of your own choice has been added. Make sure that both groups of plants are given exactly the same conditions except for the presence or absence of the fertiliser.

Compare the two groups of seedlings over several weeks. Record any differences in their appearance, making measurements when appropriate. What conclusions do you draw?

C Investigating individual elements

Plan an experiment to investigate the effect of depriving seedlings of

a nitrogen
b phosphorus and
c potassium.

Discuss your plan with your teacher, then carry out the experiment.

Questions

1 Gardeners often plant a particular type of vegetable in a different part of the garden each year. Why is this a good idea?

2 Some farmers go in for 'organic farming'. This involves using only organic fertilisers. Inorganic fertilisers aren't used at all. Discuss the advantages and disadvantages of organic farming. Try to distinguish between advantages which have a scientific basis and those that are just hunches.

3 A new magazine called *Best Buy* has asked you to investigate which brand of garden fertiliser is best, 'Kwikgrow' or 'Bioshoot'. Outline the method that you would use. How would you make sure that your investigation was fair to both products?

4 Look at picture 7, then answer these questions:

a Express the yield of grain given by each of the plots 2 to 6 as a percentage increase over that given by plot 1.

b From the data you might conclude that artificial fertiliser is better than farmyard manure. Do you think this conclusion is justified? Give reasons for your answer.

5 It is said that inorganic fertilisers, if used intensively, can spoil the structure of the soil, making it fine and dusty. Describe an experiment to find out if this statement is true.

The pros and cons of fertilisers

Fertilisers illustrate the conflict between having to feed a large population and conserving the environment. Here we consider the main arguments for and against fertilisers. We shall confine ourselves to artificial (inorganic) fertilisers, particularly nitrates, though many of the points apply to organic fertilisers as well.

Pros

- Fertilisers give crop plants such as wheat the nutrients they need for healthy growth. Wheat will produce more grain if it is grown in a well fertilised field than if it is grown in an unfertilised field (see page 80).
- Fertilisers enable farmers to grow their crop plants very close together. This increases productivity and means that the land can be used more intensively.
- Intensive farming means that less land is needed for crop-growing, so more countryside can be left in its natural state.

Cons

- Fertilisers get into rivers and lakes, causing eutrophication. Severe eutrophication kills the fish and other animals (see page 81).
- Fertilisers get into our drinking water. Nitrate in drinking water can cause 'blue baby syndrome' (see page 81) and possibly cancer of the stomach. High nitrate levels also occur in certain foods, particularly vegetables.
- Manufacturing fertilisers requires energy from fossil fuels. As well as using up precious fuels, this adds to pollution of the atmosphere.

Can you think of any other advantages or disadvantages of fertilisers?

Some facts and figures

Here are ten facts which you should bear in mind when weighing up the pros and cons of fertilisers:

1 In Britain the amount of nitrate fertiliser put onto the land by farmers increased ten times between 1945 and 1989.

2 Only half the nitrate put onto the land by farmers is taken up by plants. The rest runs off the surface, or seeps through it into the ground water and then into rivers, ponds and lakes.

3 The nitrate moves down through the soil very slowly – at about one metre per year on average. Typically the ground water level is about 20 metres below the surface. This means that our present nitrate probem is the result of fertilisers that were applied about 20 years ago.

4 The upper safe legal limit for nitrate in drinking water, set by the European Commission, is 50 mg per litre. (The World Health Organisation has recommended that it should be 45 mg per litre.) Britain has said that it will meet the European Commission regulations by 1995 at the latest.

5 In 1989 it was estimated that in Britain 1.6 million people had received drinking water with nitrate levels of more than 50 mg per litre at some time during the previous two years.

6 Only 14 cases of 'blue baby syndrome' were reported in Britain between 1950 and 1987. Most of these were in places where freak nitrate levels reached 100 mg per litre or more. Of course there may have been many mild cases that were unreported. In extreme cases 'blue baby syndrome' can be fatal.

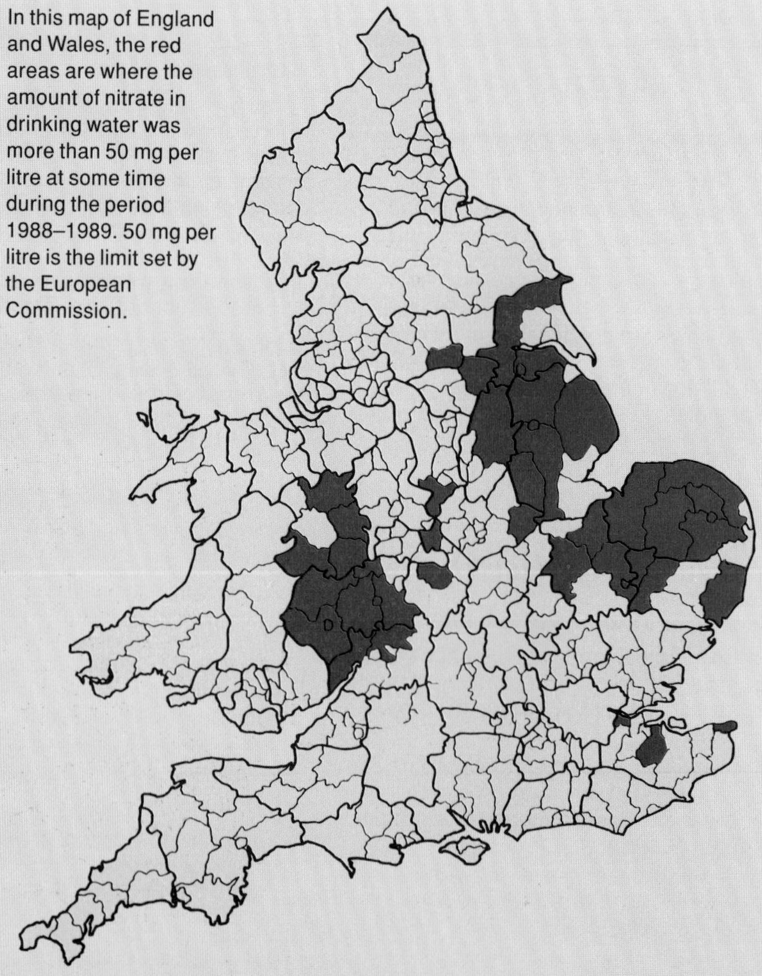

7 Nitrate is known to cause stomach cancer in laboratory animals, but there is no evidence that the same happens in humans.

8 In Britain, the number of cases of stomach cancer is lower than average in East Anglia, although the nitrate levels in drinking water are higher there than in other parts of the country. The number of cases of stomach cancer in the country as a whole has fallen in the last 20 years.

9 It may take many years for the over-use of nitrate fertilisers to have a serious effect on people's health. Like other aspects of pollution, the time scale may be very long.

10 Some of the disadvantages of artificial fertilisers may be overcome by using organic fertilisers such as manure. But remember that organic fertilisers have to be broken down into inorganic substances before they can be used by plants, so the advantages may not be as great as they seem.

Picture 1 In this map of England and Wales, the red areas are where the amount of nitrate in drinking water was more than 50 mg per litre at some time during the period 1988–1989. 50 mg per litre is the limit set by the European Commission.

Things you can do yourself

When you are considering two sides of an argument, it's a good idea to obtain some of the evidence from your own experience. This can be done by reading articles in newspapers and magazines, and by talking to people – for example, farmers, market gardeners and conservationists.

It is also helpful to carry out your own practical investigations. If you have done activity B on page 81, you will already realise how beneficial fertilisers can be.

Here are three investigations which you might carry out on fertiliser pollution:

1 Look for evidence of eutrophication in rivers, lakes and ponds in your area.

BIOHAZARD

Poisonous gases may be released from eutrophic water when disturbed.

First decide how you are going to tell if eutrophication is taking place. Next decide which particular rivers, lakes and ponds you are going to investigate. Use an ordnance survey map to help you – this will show you where farms are too. Check your plans with your teacher. Then carry out a systematic survey, preferably in Spring or Summer, recording your observations in whatever way you think best. If you do find evidence of eutrophication, suggest possible causes. How could you test your suggestions?

2 Measure the amount of nitrate in your drinking water, and also in local lakes, ponds, rivers and streams. Your teacher will give you some special reagent strips for doing this.

Record your results. Does the amount of nitrate exceed 50 mg per litre anywhere? You may be able to relate particularly high levels of nitrate with eutrophication, but beware of jumping to conclusions.

3 Suppose a farmer comes along and says that he does not believe that fertilisers cause eutrophication. Plan a laboratory experiment to show him that they do.

The purpose of this experiment is to demonstate the truth of an established theory. It should be set up very carefully, and the results and conclusions should be presented as convincingly as possible but without bias.

Discuss your plan with your teacher. When you have had it approved, carry it out.

To think about

1 Do you think fertilisers, as used at the present time, pose an environmental and/or health problem? If you do, what recommendations would you make for reducing the problem?

2 How would you test the suggestion that nitrates cause stomach cancer without experimenting on animals?

3 Suppose inorganic fertilisers were banned and everyone used organic fertilisers. Do you think this would solve the problem of nitrate pollution?

4 'With millions of mouths to feed, some degree of nitrate pollution is inevitable'. Do you think this is a fair statement?

5 The map opposite shows high nitrate levels in East Anglia. Suggest why high nitrate levels should occur here.

Water pollution

Eutrophication, caused by fertiliser getting into lakes and rivers, is an example of **water pollution**. Eutrophication is likely to occur when *any* source of nitrates and other nutrients gets into a river, lake or pond. So it is also caused by **sewage** and **farm slurry**. Slurry is the liquid and solid waste from farm animals.

Eutrophication is made worse if hot water from power stations is discharged into rivers. This is called **thermal pollution**. The extra warmth makes the algae and decomposers multiply even faster.

Eutrophication is just one of many types of water pollution. Here are some others.

From time to time **oil** is spilled into the sea from a tanker or off-shore oil rig. The oil forms a thin layer, or **slick**, which floats on the surface of the water. The slick may be carried by currents to the coast and deposited on the shore.

Oil ruins beaches for local residents and holiday-makers. It also kills fish and sea birds (see picture).

Oil slicks can be sprayed with detergents which break up the oil into drops. However, the detergents may be even more deadly than the oil. Nowadays less destructive methods are used.

Picture 1 This cormorant was the victim of a spillage of oil from a damaged oil tanker.

Finally **chemical waste products** from factories are sometimes discharged into seas and rivers. They may be so concentrated that the fish are killed straight away. But sometimes they are taken up into food chains and concentrated, like DDT (see page 50).

Some years ago, over 60 people in Japan died from eating fish whose bodies contained mercury that had been discharged into the sea from a local factory. The mercury had got into the food chains and become concentrated in the fish.

1 Can you think of any other examples of water pollution besides the ones mentioned here? You may know of some examples from where you live.

2 Find out how oil slicks are got rid of, both when they are floating on the sea and when they get onto beaches.

3 It is said that there is much more pollution in the Mediterranean Sea than in the English Channel.

a How could the truth of this statement be investigated?

b Suggest possible reasons for it.

B16 Controlling pests

Certain organisms are harmful to humans. How can we keep them under control?

Picture 1 A locust eating a leaf.

Picture 2 Rust on the surface of a leaf.

What sort of pests do we need to control?

There are two kinds of pest: those that harm humans and domestic animals, and those that harm plants, particularly crops.

The first group includes flies which contaminate food and spread diseases and mosquitoes which suck blood and spread malaria. The second group includes locusts and wireworms which feed on crop plants, fungi such as brown rust which infests wheat and barley, and weeds such as thistles which compete with crops (pictures 1 and 2).

Two different approaches

We get rid of pests in two main ways:

- We can put into the environment a chemical substance which kills the pest. This is called **chemical control**.
- We can put into the environment another organism which kills the pest. This is called **biological control**.

Chemical control

A substance which kills pests is called a **pesticide**. Those that kill insects are called **insecticides**, those that kill fungi are called **fungicides**, and those that kill weeds are called **herbicides**. Pesticides are usually applied as pellets, powders or sprays (picture 4).

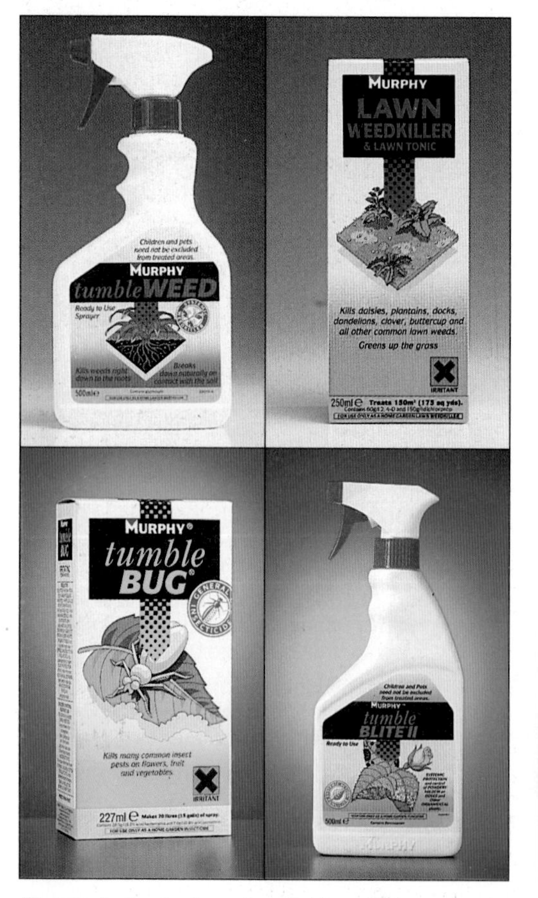

Picture 3 A selection of pesticides which are used for getting rid of specific types of pest.

Picture 4 A pear orchard being sprayed with a fungicide.

Picture 5 shows how herbicides work. Some kill the leaves as soon as they touch them. Others are carried from the leaves or roots to some other part of the plant, which they then kill.

One of the most widely used herbicides is the **hormone weed killer** which many gardeners use to clear their lawns of weeds. If you spread it over the lawn in the right concentration, it will kill all the broad-leaved weeds such as dandelions and plantain, but leave the grass unharmed. It is therefore selective in its action, and is called a **selective weed killer** (see page 258).

Twenty years ago, DDT (which stands for dichlorodiphenyltrichloroethane) was the most widely used insecticide. This was used against mosquitoes and many other insect pests, especially in the tropics. Worldwide it saved about 15 million lives through the destruction of harmful insects.

The trouble with pesticides like DDT is that they may kill not only harmful insects, but useful ones too. Such pesticides are non-selective. Their victims can include the caterpillars of butterflies and moths, and insects such as bees which pollinate flowers. And sometimes they kill other insects that kill other pests. More modern products are much more specific to the target pest.

And there is another problem. If a particular pesticide is used a lot, the pest may eventually become resistant to it (see page 291). The pesticide no longer kills the pest, and a new one has to be developed to take its place.

Some pesticides lose their killing action quite quickly after they have been released. They are **non-persistent**. An example is the insecticide **pyrethrum**. Pyrethrum is obtained from the roots of a type of chrysanthemum and has the merit of being a natural substance. It is used against whitefly and other insects which attack plants.

Other pesticides remain active for a long time after they have been released. These **persistent** pesticides may sometimes get into food chains and harm other organisms further along the chain. Such is the case with DDT (page 50).

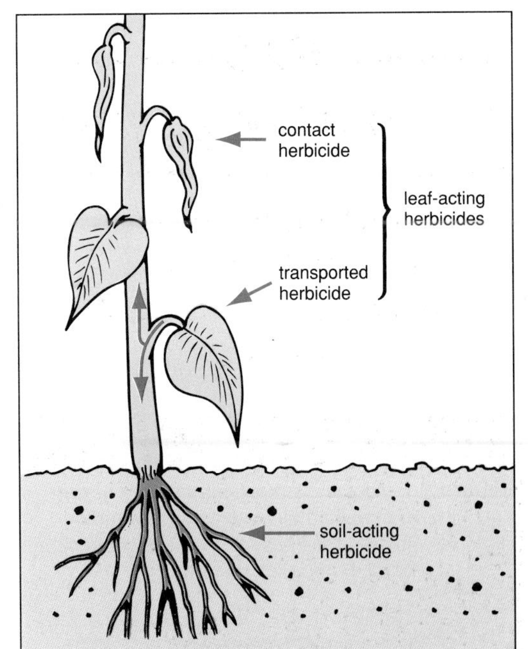

Picture 5 Some herbicides act on the leaves of the plant, others on the roots. Leaf-acting herbicides either kill the leaf on contact, or are transported to the site of action in some other part of the plant.

Biological control

Suppose your house was overcome with mice. How could you get rid of them? You could put down some mouse poison. That would be chemical control. Alternatively you could keep a cat. That would be biological control. The cat would be serving as a **biological control agent**.

Here is a more serious example. Greenflies damage plants. Now greenflies are eaten by ladybirds and their larvae. So if ladybirds are released into an area they will help to keep down the greenflies.

The advantage of biological control is that it avoids polluting the environment with poisonous substances. Instead nature gets rids of the pest for us. Biological control agents are available as insect eggs or pupae. They can be put on crops, garden plants or greenhouse plants (picture 6).

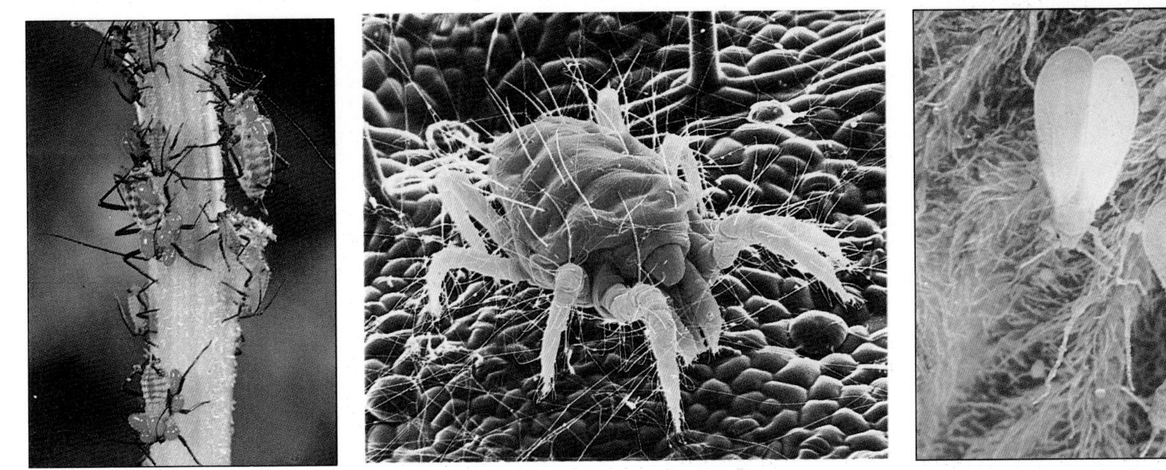

Picture 6 The weevil (top right) and whitefly (bottom right) are serious pests that attack plants. So is the red spider mite (centre). However, other species of mite can be used to control pests. The picture on the left shows mites attacking aphids.

Picture 7 Biological control gone wrong. The cat that was obtained to catch unwanted mice in the house has turned its attention to the birds in the garden.

But even biological control has snags. For example, suppose you bring in a predator to get rid of a certain pest. The predator may eat up all the pests and then start eating a useful species. The control agent then becomes a pest itself. What you have done is to alter the food web with results that you had not thought about. An example of a biological control agent becoming a pest is given on the opposite page.

Making the decision

Suppose you are a farmer or market gardener and you are trying to decide whether to use chemical or biological control. In reaching a decision you have to take four things into account:

- How efficient each method is at controlling the pest.
- The cost.
- Damage that might be caused to the environment.
- Possible health hazards.

In practice, a combination of chemical and biological methods is often the answer. This would be part of a programme of management aimed at making the farm or market garden as productive as possible with minimum damage to the environment. There is more about this in the next topic.

Activities

A Finding out about pesticides

Visit a local gardening shop or garden centre. Find the pesticide section. Read the labels on the various brands of pesticide to see what they consist of, and whether they are selective or non-selective, persistent or non-persistent, and what sort of organisms they kill. From books, find out why these particular organisms are harmful. Present your results as a table.

B Greenflies and ladybirds, an example of biological control

Greenflies, or aphids, are well known garden pests. They suck plant juices, attacking the tender young shoots and leaves. A single rose plant may be covered with hundreds of them and they can cause a lot of damage.

It is said that greenflies can be kept under control by ladybirds. Plan a way of finding out if this is true. Check your plan with your teacher before you start work. In particular try to answer these questions:

1 Observe a greenfly and a ladybird under a lens or binocular microscope. What sort of animals are they?
2 Are greenflies eaten by ladybirds, and if so are they eaten by adult ladybirds, larval ladybirds, or both?
3 How do greenflies feed on plants, and how do ladybirds feed on greenflies?
4 Do ladybirds eat greenflies quickly enough to keep their numbers under control?

C A clever way of getting rid of insect pests

A modern and rather clever method of controlling certain insect pests is to release into the population very large numbers of males which have been sterilised by radiation. The result is that the size of the insect population gradually falls.

Find out as much as you can about this method of pest control. Where is it carried out, and against what sort of insects? How does it work? It is effective? What are its advantages over other kinds of insect control? Does it have any snags? Could it be used against other animal pests.

Questions

1 Give an example of an insect pest which spreads a disease, and name the disease. How is the insect controlled?

2 You are a manufacturer and you intend to produce the perfect insecticide. Make a list of all the features which it should have.

3 Explain the difference between chemical control and biological control. List the advantages and disadvantages of each.

4 Pyrethrum occurs in the roots of a certain type of chrysanthemum. Suggest why it might be useful to the chrysanthemum.

5 Suppose the woodlands of Britain became overrun with an unpleasant species of insect which was damaging the trees. It is discovered that a certain species of lizard from New Zealand likes eating these insects – in fact it gobbles them up at a fantastic rate. A proposal is therefore put forward to introduce this lizard into British woodlands. Outline the research that would need to be carried out before the lizard was introduced.

6 In some species of moth, the female secretes tiny amounts of scent which attract males and stimulate them to mate with her. A male may detect a female 10 km away.

In some cases males have tried to mate with pieces of blotting paper on which a few drops of the scent have been absorbed.

Use this information to suggest a way of controlling the apple codling moth. This moth lays its eggs on apples and the larvae (caterpillars) bore into the fruit and feed on it.

Biological control gone wrong: the cane toad

Watch your step, killer toad ahead

Families are living in terror of a plague of killer toads.

Parents have slapped a curfew on children and pets as millions of the plate-sized monsters march across Australia spitting poison.

The cane toads, measuring a foot across and weighing up to 6lbs, have swamped Brisbane and are now heading for Sydney.

"They'll kill a dog in 30 seconds. They devour everything in their path including mice, rodents even snakes. They can even attack humans," said one biologist.

Emergency services are on full alert as the evil green giants threaten to overrun residential areas.

In some suburbs people are literally tripping over the brutes.

Tape-recordings of the male toad's mating call are being played to attract females so that trained trappers can shove the toads into plastic bags and then freeze them to death.

Australian Robyn Foyster, now living in England, says a cane toad killed her pet dachshund Sheba in Brisbane.

"One day Sheba was out in the garden playing and she came across one of these toads. They're huge, like giant rats, and they spit poison over a few feet," she said. "Unfortunately, we didn't realise what was happening until it was too late. Within no time at all she was dead."

The toads were introduced from South America in 1935 to eradicate a beetle plague destroying sugar cane crops in Cairns, Queensland.

As well as moving south to Brisbane and Sydney, the wart-covered toads have started migrating north towards Darwin.

Now conservationists fear they are threatening Crocodile Dundee country and the beautiful Kakadu National Park.

One forestry official said last night: "There seems to be nothing we can do to stop them."

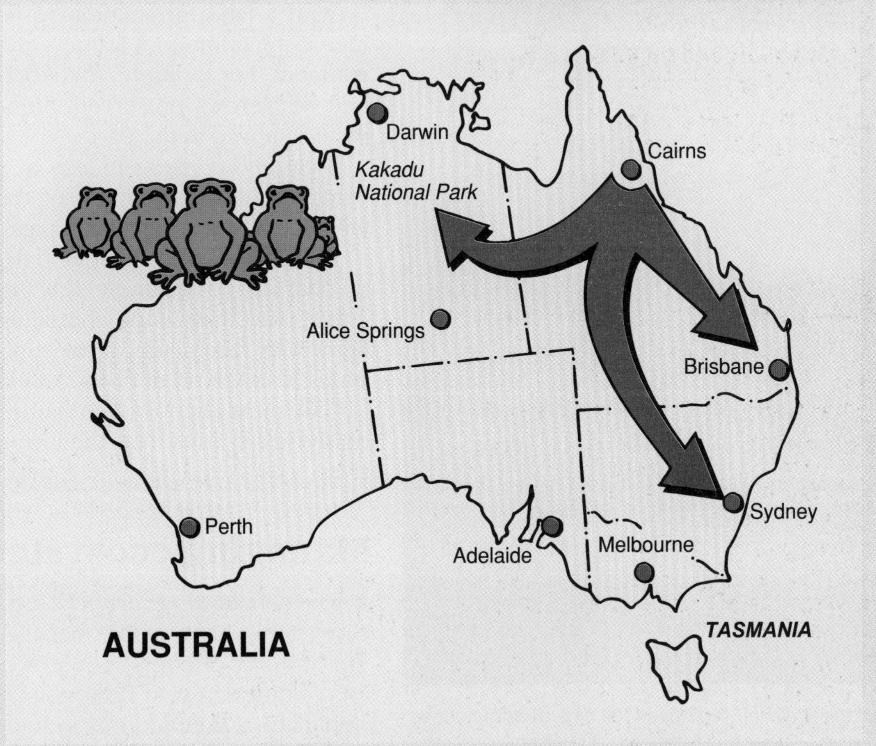

B17 Management and conservation

We can control the environment to serve our needs, but we must be careful how we go about it.

Picture 1 Harvesting wheat on a large scale in North America. In Canada there is a single fenced wheat field with an area of 140 km². That's larger than the city of Sheffield!

Picture 2 Since 1947 over 100 000 miles of hedges have been ripped up in Britain – enough to go round the Earth four times.

Modern farming

Nowadays farmers usually fill their fields with just one type of crop, wheat for instance, which they grow year after year. This is called **monoculture**. Fertilisers enrich the soil with nutrients, and herbicides ensure that no weeds are present to compete with the crop. With plenty of food and no competition, the crop plants can be grown very close together. All this has the effect of increasing the yield. Modern machinery for ploughing, sowing the seeds and harvesting means that fields can be enormous (picture 1).

The disadvantages of monoculture

Monoculture enables the land to be used intensively. This means more food for more people. However, when carried out on a large scale it involves cutting down trees and removing hedges. This destroys the natural environment, kills wildlife and makes the countryside look dull. What's more, with repeated ploughing and the use of artificial fertilisers, the soil crumbs break up and the soil becomes fine and dusty. It then gets blown about by the wind, causing **erosion**. The structure of the soil can be regained by leaving it fallow for a year, or by growing grass on it, or by adding manure to it.

There is another drawback too. Suppose a pest happens to get into the crop. Any natural predators that might control it will have been killed by pesticides. For example, if greenflies get into the crop, there will be no ladybirds to control them. With the plants so close together, the pest will spread quickly through the crop. To some extent the risk of this happening can be reduced by rotating crops; different types of crop are attacked by different pests, and rotation may prevent a pest completing its life cycle. The same thing may be achieved by varying the times when particular crops are sown.

Producing without destroying

A wheat field is an **artificial ecosystem**. It is created and managed by farmers with the sole purpose of mass-producing one species of plant – wheat. Any other species which interfere with this are destroyed.

With a large population to feed, this may seem inevitable. However, an increasing number of farmers are trying to reduce damage to the natural environment. For example, the edges of the wheat field can be left uncultivated, and hedgerows preserved. Some farmers have even replanted hedges that were removed in the past.

Trees and copses can also be preserved, and marshy areas left undrained. Sympathetic management of this kind can do a lot to save our dwindling wildlife, as well as providing pleasure for hikers and tourists (picture 3).

Even within the wheat field itself care can be taken to reduce environmental damage. For example, it isn't necessary to destroy all the weeds, only the larger ones that would compete with the wheat or shade the seedlings from light. With careful use of selective pesticides and biological control agents, it is possible to get rid of pests but leave other species unharmed.

Care must also be taken in the use of fertilisers. There is no point in putting on more fertiliser than the crop requires. Too much can lead to pollution of rivers, ponds and lakes (see page 81).

Managing ecosystems

A wheat field is an artificial ecosystem, but even natural and semi-natural ecosystems have to be managed. Consider the chalk downs of Southern England, for example. The downs are covered with short grass and support lots of different wild flowers, many of them rare. The downs are kept in this semi-natural state by grazing animals such as rabbits and sheep. In fact for generations farmers have used the downs for summer grazing. Not only do the grazers keep the grass short, but they also eat the seedlings of larger plants which are thus prevented from growing up.

Now think what would happen if the grazers suddenly disappeared. The grass would become long, and the seedlings of the larger plants would survive. The wild flowers would be stifled, and in time the downs would become covered with shrubs and small trees. This actually happened in the 1950s when most of Britain's rabbits were killed off by the virus disease myxomatosis. Animal life was affected too. For example, birds of prey like the buzzard, which fed on rabbits, fell in number until they found another source of food.

Of course rabbits can be a nuisance if there are too many of them and they start eating farm crops. However, they are normally kept under control by natural predators such as foxes and birds of prey. If there are more rabbits on the downs, more buzzards are likely to appear. *Good management involves making sure that there is the right balance between predators and prey*.

If rabbits are scarce, much of their job can be done by sheep. But if hawthorns and brambles start growing on the downs, they should be removed either by hand or by means of a herbicide.

Of course over-grazing can be damaging too. It can destroy the wild flowers and may cause erosion (see page 76). Over-grazing can be prevented by keeping the rabbit population under control and by making sure that farmers don't put too many sheep on the downs.

Picture 3 The copse in the centre of this wheat field is a haven for wildlife.

When management is lacking

There are places in the world where ecosystems are not managed properly. An example is the tropical rain forest. Here vast numbers of trees have been felled for timber, or burned to clear the forest for growing crops, raising livestock or industrial development (picture 4). In the Amazon alone, an area the size of France has been destroyed in a single year.

Not only does this kill the wildlife on a massive scale, but it also changes the pattern of rainfall and ruins the soil.

Picture 4 Part of a tropical rain forest being cleared.

It's a matter of need

It is easy to condemn governments and developers in tropical countries for ruining the rain forests, but they are only doing what we in Britain have done – though more slowly – for centuries. At one time Britain was a land of forests. Gradually the trees were cleared to provide fuel, timber for building, and space for growing crops.

Today Britain has very little natural woodland, let alone forests. This is not just because wicked people have destroyed them. There is a real conflict between our desire to have unspoiled places, and the needs of a modern industrial society. The problem is made worse by the fact that our country is small and densely populated (picture 5).

This makes it even more important that the land should be managed properly. In practice this means reaching a compromise between conservation on the one hand, and food-production, industry and housing on the other. To reach a compromise, there has to be cooperation between government, local authorities and individual people.

The place where you live

Think about the place where you live. Has the environment been spoiled or is it cared for? Is there any natural countryside and wild life, and if so is it under threat? What sort of development has taken place in your area, and has it all been necessary? Have attempts been made to improve the environment, and if so by whom? How could you improve it?

Answering these questions helps to bring home the conflict between the opposing forces of urbanisation and conservation. To understand this more fully we need to know exactly what conservation involves.

Picture 5 A small, densely populated country like Britain needs land for urban development.

Picture 6 Wildebeest and zebra in Serengeti National Park in East Africa. The numbers of these herbivores are kept at the right level by natural processes, including predation by lions. The ecosystem as a whole is carefully managed by game wardens. But even in game parks poachers threaten animals like the gorilla and elephant.

Picture 7 Slimbridge Wildfowl Trust Centre in Gloucestershire is a sanctuary for many species of migratory birds.

What is conservation?

To conserve something means to protect it and keep it in a healthy state. Applied to our environment, it means protecting ecosystems so that the organisms in them can flourish. Perhaps the most successful, and certainly the largest, examples of conservation are seen in some of the game parks of Africa (picture 6).

Sometimes special steps have to be taken to help endangered species to survive. Such was the case with the Hawaiian goose. This unfortunate bird was shot by hunters and its chicks were preyed on by animals introduced by humans. By 1950 it was almost extinct. Then a few were brought to the Wildfowl Trust Centre at Slimbridge in Gloucestershire. Here, in one of the best-managed semi-natural ecosystems in the world, they bred successfully (picture 7).

Soon there were more Hawaiian geese at Slimbridge than there had been in Hawaii! In the early 1960s some of them were taken back to Hawaii and released. Since then their numbers have gradually increased. This is a success story in conservation, but sadly there are not many like it. There is more about threatened species on pages 288 amd 289.

Conserving food resources

One reason for conserving ecosystems is that they may contain animals we (and future generations) need for food. Such is the case with sea fish such as herring and cod.

One of the main ways of catching fish is by **trawling** (picture 8). The modern trawler is a very efficient vessel. The nets are large and special echo-sounding equipment is used for locating large shoals of fish. In **factory trawlers** there are machines for gutting, skinning, filleting and freezing the fish on board.

This looks like a success story in technology. However, we have to make sure that we do not take so many fish out of the sea that their numbers begin to fall. This means that *we must not remove them faster than they can be replaced by their natural processes of reproduction*.

In the European Union there are agreed quotas regulating the number of tonnes of the various types of sea fish that may be caught by each member country each year. The quotas are set annually by negotiation, and are based on scientific advice from bodies concerned with fisheries management.

We must also make sure that no damage is done to the food chains which fish depend on (see page 46). These food chains start off with **plankton**, and so it is particularly important that no harm comes to the organisms that make up the plankton.

Picture 8 Many of the fish we eat are caught by trawling. The net here is one of several different kinds that are commonly used. To prevent over-fishing, various regulations limit the sizes of nets and their mesh.

Activities

A Finding out about a market garden

A market garden is a place where plants such as lettuces and tomatoes are grown for selling. This is called horticulture, and it is a good example of an artificial ecosystem.

Visit a market garden. Find out what sort of crops are grown, and who the main customers are.

Choose one crop, and try to find answers to these questions:

1 How is the crop protected from unsuitable weather conditions?

2 What sort of pests have to be controlled, and how is this done?

3 What kind of fertilisers are used, and how are they applied?

4 In what form are the plants sold? What steps are taken to ensure that they are free of disease and do not present a health hazard?

Write an account of the way the crop is grown, concentrating on the extent to which it forms part of a managed ecosystem.

B Finding out about a farm

1 Choose a particular type of farm, e.g. one that grows cereal crops such as wheat or barley, or one that rears livestock such as cattle, sheep or pigs.

2 From books, find out as much as possible about this type of farm. Try to get interested in one particular aspect – such as how the farmer achieves maximum productivity without damaging the local environment.

3 If possible, talk to a farmer or, better still, visit a farm. Write down a list of questions beforehand which you would like to ask the farmer.

4 Write a report on your farm, concentrating on the aspects that interest you most.

C The pros and cons of monoculture

Hold a debate. The motion is as follows: *Monocultural crop-growing on a large scale damages the environment unacceptably and should be stopped.*

D Conservation in your area

1 Try to find one example of a conservation project which has been carried out in your area. Find out as much about it as you can, in particular how it is looked after and managed. Write an account of it, and say how it has benefited (a) wildlife and (b) people.

2 Suggest a conservation scheme which has not been carried out in your area but which, in your opinion, should be. Outline what might be done, and give reasons for your suggestions.

Questions

1 You have been appointed Minister of Conservation. Write down ten laws you would like to introduce, in order of priority. Give reasons for each one, and say what problems you might have in getting it through Parliament.

2 An area of natural woodland in Britain is destroyed and replaced by a very productive wheat field. As a result, the ecosystem of the area is changed.

a What is meant by the word ecosystem?

b In what ways does the ecosystem of the wheat field differ from that of the woodland?

3 Why don't the wild flowers on the downs get eaten by the grazers? Suggest two possible hypotheses and, describe how each could be tested.

4 Why do you think clearing the tropical rain forest

a alters the pattern of the rainfall,

b ruins the soil.

What other effects might it have?

5 Even the best run farms are at the mercy of the climate. Illustrate this with reference to a wheat farm. Suggest two events which can ruin a harvest.

6 The figures show the agreed fishing quotas in tonnes for cod and herring in 1994 for the United Kingdom and the whole of the European Union.

	UK	EU
Cod	56 700	205 720
Herring	106 360	833 200

a What sort of information is needed to set the quotas?

b Why do you think the quota for herring is much larger than the quota for cod?

c The quotas for herring and cod could be quite different in a few years time. Suggest a reason for this.

d How do you think the UK keeps to its quota? (Hint: the Ministry of Agriculture, Fisheries and Food (MAFF) publishes statistics every month on UK fish catches.)

e Can you suggest ways of managing fisheries other than by setting quotas?

Killer worm threatens farms

Read the newspaper extract on the right, then answer these questions.

1 What lesson does this story teach us about conservation?

2 Does it matter if earthworms get eaten? What use are they?

AN UNWELCOME visitor from New Zealand has colonised Northern Ireland and is killing off the native earthworms that are essential to soil fertility.

The visitor is a flatworm called *Artioposthia triangulata*, which can grow up to six inches long and weigh two grams. It can eat a whole earthworm in 30 minutes.

The killer worm has been found in every Ulster county and, according to the Department of Agriculture, the only check to its increasing numbers appears to be the availability of food.

'The potential impact on local populations of earthworms cannot be overestimated,' said a department spokesman.

By Godfrey Brown

Agriculture Correspondent

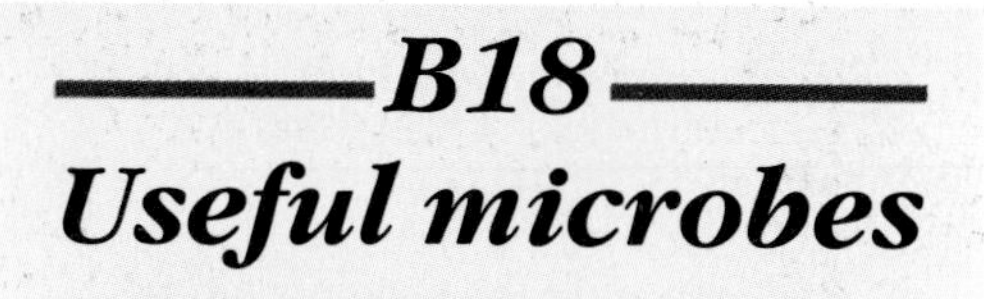

B18 Useful microbes

Humans make use of microbes in many ways. Here we take a look at some of them.

Picture 1 Filter beds in a sewage works. Liquid sewage is sprinkled from long arm-like pipes which rotate slowly over a bed of stones/clinkers. The stones/clinkers are coated with a slimy film of aerobic bacteria and other microbes.

Treatment of sewage

Sewage treatment is carried out in a **sewage works**. The process makes use of bacteria that normally bring about decay. They break down the waste matter into harmless products. Some of the bacteria are aerobic: they need oxygen for respiration. Others are anaerobic: they do not need oxygen (see page 126).

On arrival at the sewage works, the sewage is pumped into a large tank. Here the solids sink to the bottom, forming a sludge. The sludge is broken down by anaerobic bacteria, after which it may be dumped out at sea or disposed of on land. It can also be dried and used as a fertiliser.

Meanwhile the liquid part of the sewage is acted on by aerobic bacteria. They break down any organic material in the liquid, giving off carbon dioxide gas. This process is carried out by mixing the liquid with bacteria in special aerated ponds, or by filtering it through a bed of stones or clinkers coated with bacteria (picture 1).

The liquid that results from this treatment may have lots of bacteria in it. They settle out, forming more sludge.

The liquid is now fairly pure water and may be discharged into the sea or a river. If it is discharged into a river, the water can be collected, purified and used again for household supplies. So sewage works enable water to be recycled. If you drink a glass of water in London, it has probably already been through at least three other people!

A simple diagram of a sewage works is shown in picture 2.

Making silage and compost

Silage is undecayed vegetable matter, mainly grass. Farmers use it for feeding livestock in winter. Silage is made by packing the material tightly in a container so that air cannot get in. Without oxygen, bacteria in the container respire anaerobically and produce lactic acid (see page 127). The acid prevents decay and preserves the grass.

Compost is *decayed* vegetable matter. In this case air is allowed to get in while the compost is forming. With oxygen available, aerobic bacteria break down the material into a semi-decayed state. It is then used as an organic fertiliser (see page 78).

Making yoghurt and cheese

Certain bacteria convert milk sugar (lactose) into lactic acid. The lactic acid curdles the milk, making it go lumpy. **Yoghurt** is made by solidifying milk in this way. It is then made more creamy and tasty by further bacterial action. Fruit juices may then be added. These stabilise the yoghurt and give it particular flavours.

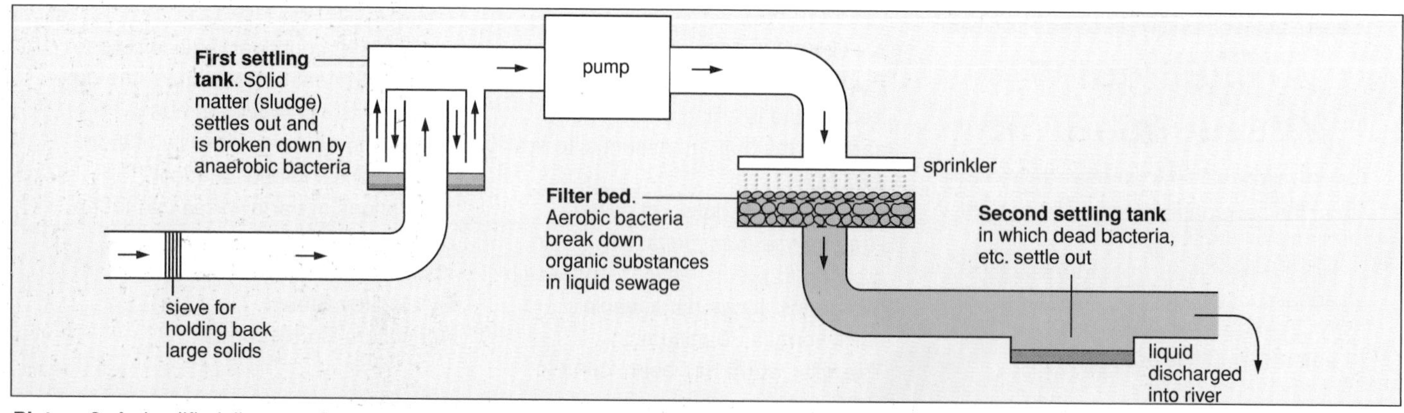

Picture 2 A simplified diagram of a sewage works.

To make **cheese**, the solid part of the curdled milk (the **curds**) is separated from the fluid part (the **whey**). One way of doing this is to put the curdled milk in a muslin bag and squeeze out the fluid whey. The paste-like substance left behind is cheese.

At this stage the cheese is white and tasteless. It must now be ripened. This is carried out by microbes, particularly fungi – the sort that bring about decay. They break down the cheese, softening it and giving it its characteristic smell, flavour and appearance (pictures 3 and 4).

Picture 3 Blue cheese is ripened by a mould (fungus) which grows on it. This machine makes holes in the unripe cheese. The holes allow oxygen to get in and help the mould to spread through the cheese.

Brewing, wine-making and baking

These processes make use of yeast, a single-celled fungus. They depend on the fact that yeast can respire anaerobically, producing carbon dioxide and alcohol (**alcoholic fermentation**). You will find details on page 126.

Manufacturing vinegar

Certain bacteria turn ethanol into acetic acid. They are used for producing vinegar which contains acetic acid. The bacteria are aerobic and therefore need oxygen.

Vinegar can be made by trickling wine, beer or cider through tall cylinders packed with wood shavings coated with a film of aerobic bacteria (picture 5). The cylinders have holes at the bottom so that air can circulate freely inside them.

Picture 4 Different types of cheese are ripened by different microbes. Blue cheeses are ripened by moulds which are visible as a network of blue threads. The cheese with large holes in it was ripened by bacteria which gave out carbon dioxide gas which could not escape.

Producing edible protein

Bacteria and fungi, grown in bulk in special chambers, are used as a source of protein food for farm animals and humans. Protein produced this way is called **single cell protein (SCP).**

In the case of fungi, the fungal threads are collected and compressed to form a material called **mycoprotein**. Fungal 'chicken' and 'ham' can be made from flavoured mycoprotein, and mycoprotein pies can be bought in food shops.

An advantage of mycoprotein is that it has a high protein and fibre content but no cholesterol. Also the fungi can be grown on unwanted material such as flour waste or wood and paper shavings.

Making antibiotics

An **antibiotic** is a substance produced by a microbe which kills, or inactivates, other types of microbe. Microbes produce antibiotics to prevent other microbes competing with them.

The most famous antibiotic is **penicillin**. Its story is told on page 95.

Many other antibiotics besides penicillin are now manufactured from microbes. Some come from fungi, others from bacteria. They are used for treating people with bacterial and fungal diseases (see page 102).

Picture 5 The cylinders used in vinegar production.

Manufacturing enzymes and other useful compounds

Many commercially important substances, including enzymes, are obtained from microbes, particularly bacteria and fungi.

Some of these compounds are natural products which the microbes make for themselves. For example, microbes have the same sort of enzymes as we do. These can be extracted and used for various purposes (see page 25).

It is now possible to make bacteria produce compounds that they do *not* normally produce. This is achieved by **genetic engineering**. In this process genes of another species, e.g. humans, are transferred to bacteria. The bacteria then make useful substances for human use, as instructed by the human genes. An example is the hormone insulin which prevents diabetes. Genetic

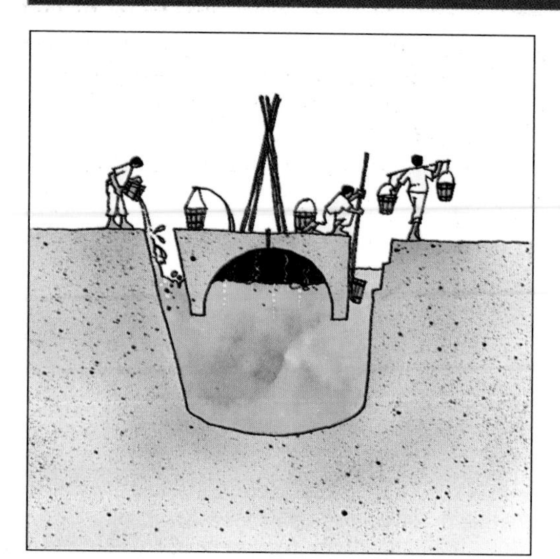

Picture 5 An underground biogas digester. What do you think the various people in the picture are doing?

engineering is explained more fully on page 282.

Producing fuel

The anaerobic bacteria in a sewage works give off a gas which is mainly methane. This **biogas** burns well and is used to drive the machinery in modern sewage works.

In developing countries animal and plant waste matter is fermented by anaerobic bacteria in special **biogas digesters** (picture 6). The gas is collected and used for cooking, lighting and other energy-requiring purposes.

In some countries, such as Brazil, alcohol is used as a fuel for cars, either on its own or mixed with petrol as **gasohol**. The alcohol is obtained by fermenting sugar with yeast. The sugar comes mainly from sugar cane plantations, and the fermentation is carried out in factories close by.

The advantage with alcohol fuel is that it is a renewable energy source and does not cause pollution. However, it involves using a lot of land which could be used for growing food crops.

Activities

A Culturing bacteria

CARE Work with bacteria should be carried out only under strict supervision by your teacher.

1 Obtain a Petri dish containing sterile nutrient agar. Keep the lid on whenever possible.

2 Your teacher will give you a plugged tube containing a particular type of bacteria.

3 Sterilise a wire loop by passing it quickly through a small bunsen flame.

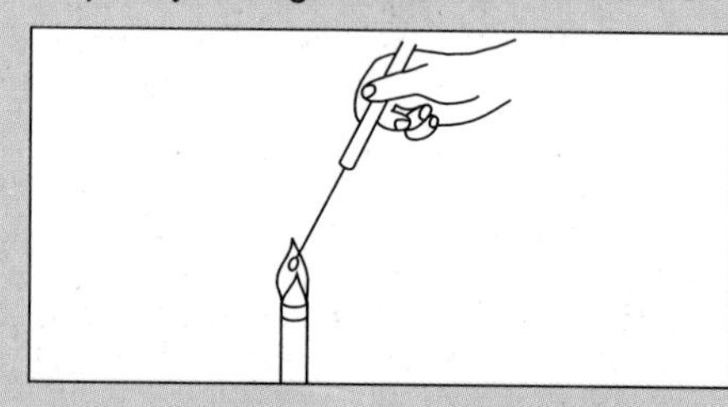

4 Keep hold of the wire loop, and unplug the tube with the same hand. With your other hand, sterilise the neck of the tube by passing it through the flame.

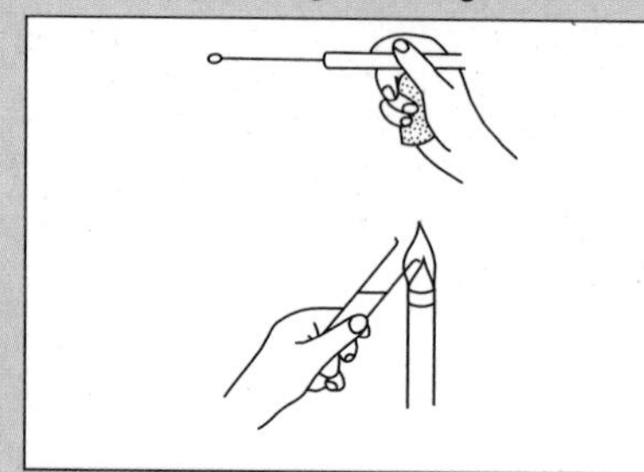

BIOHAZARD

Always wash your hands after working with bacteria.

5 Collect a sample of the bacteria on the wire loop.

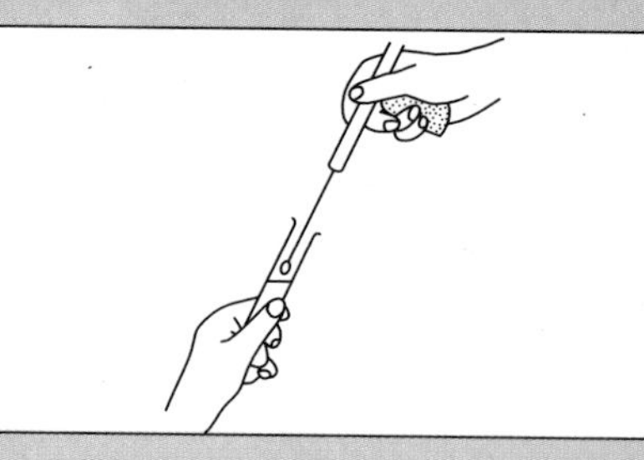

6 Pass the neck of the tube through the flame again, then replace the plug.

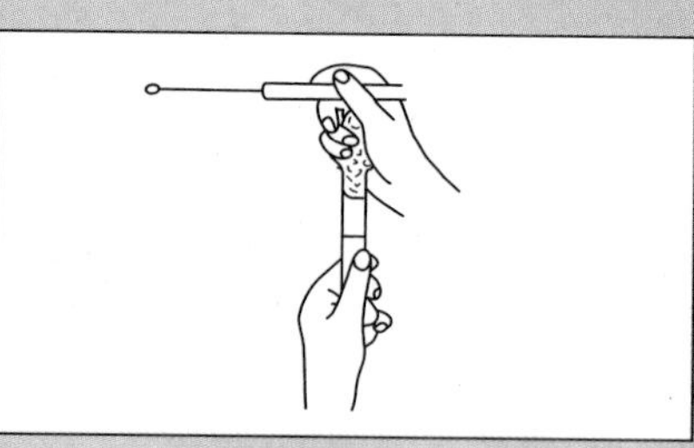

7 Remove the lid of the Petri dish. With the wire loop make a zig-zag streak on the surface of the agar.

8 Pass the loop through the flame again. Then replace the lid of the Petri dish and fix it firmly with sellotape.

9 Place the Petri dish upside down in an incubator at 30°C.

10 After a day or two examine the dish without removing the lid.

The bacteria should have multiplied to form **colonies**. Each colony consists of thousands of bacteria clumped together.

Different types of bacteria can be recognised by the size, shape and colour of their colonies. How would you describe the colonies in your Petri dish?

Sterilising the wire loop and the neck of the tube helps to prevent contamination with other microbes. What other precautions might you take against contamination?

Why was it necessary to put the Petri dish in an incubator?

The tube which your teacher gave you at the start of this activity contained one particular type of bacteria. This is called a pure culture. Suppose you were given a Petri dish containing colonies of several different types of bacteria. How could you produce a pure culture of one of these types?

Why is it useful to produce pure cultures of bacteria?

B Making yoghurt and cheese

First read on pages 92 and 93 about how yoghurt and cheese are made. Then try to make some yoghurt and cheese yourself, using the materials and instructions provided by your teacher.

What part is played by microbes in making these products palatable?

Questions

1 List three useful chemicals which are made by bacteria, and one which is made by a fungus. Explain why each is useful.

2 A sewage works saves water by recycling it. Explain.

3 Certain bacteria convert ethanol into acetic (ethanoic) acid. Under what circumstances are these bacteria (a) a nuisance, and (b) useful?

4 It is better to use microbes to produce useful substances rather than synthesise them from raw materials in chemical factories? Why?

5 A politician makes a speech in which he urges scientists to find a way of getting rid of all bacteria. Write a letter to a newspaper airing your views on this idea and explaining the probable consequences.

6 What are the pros and cons of using alcohol from microbes as a fuel for cars? Consider the cost of manufacture, effects on the environment, and any other problems you can think of.

Using rennet to make cheese

The first step in cheese-making, the solidifying of the milk, can be carried out by bacteria, as described on page 92. However, milk can also be solidified by treating it with **rennet**, the same substance that is used for turning milk into junket. For years rennet has been used for solidifying milk in cheese-making.

Rennet is the commercial name for the enzyme **rennin** which is found in the stomach of calves (see page 138). Until recently rennin for the cheese industry was obtained from calves' stomachs. But, thanks to genetic engineering, we now have bacteria which can make calves' rennin. Bacterial rennin is used increasingly in making cheese.

1 What are the advantages of using rennin from bacteria rather than calves' rennin in making cheese?

2 Cheese which has been made using bacterial rennin is sometimes called 'vegetarian cheese'. Is this a good name for it? If so, why?

The story of penicillin

In the late 1920s a Scottish bacteriologist called Alexander Fleming was working at St. Mary's Hospital, London. He was growing bacteria on dishes of agar, rather like you may do in Activity A.

Normally Fleming covered his bacterial colonies with a lid to prevent them getting contaminated. But one day he accidentally left a dish uncovered. When he examined the dish later, he found that a mould fungus was growing on the agar. But the really interesting thing was that close to the mould the bacterial colonies were absent or had degenerated (picture 2).

Picture 1 Alexander Fleming examining one of his culture dishes in the laboratory.

The mould turned out to be *Penicillium notatum* which grows on the surface of fruit. Spores of the mould had strayed onto the agar. The mould had then prevented the bacteria from growing.

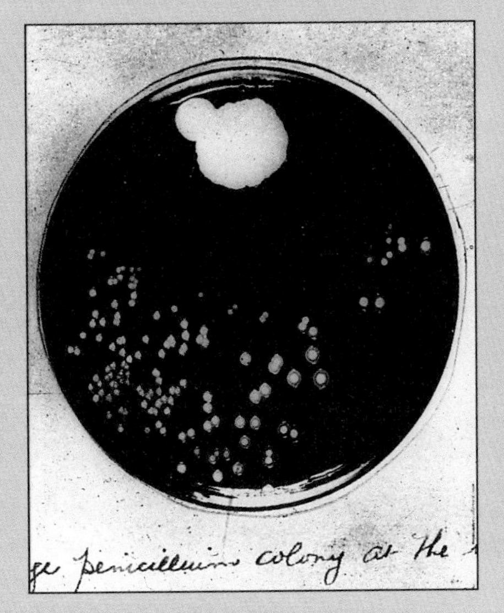

Picture 2 Fleming's culture dish. The bacterial colonies are the small white blobs towards the bottom; the fungus is the much larger white blob at the top. Notice that there are no bacterial colonies close to the fungus. The notes were made by Fleming himself.

Fleming at once saw the significance of his discovery. The mould must have produced a substance which acted against the bacteria. If this substance could be extracted from the mould, it might be possible to use it to cure people of bacterial diseases.

It took scientists 12 years to obtain the substance in a usable form. This was achieved by two biochemists, Howard Florey and Ernst Chain. In the early 1940s penicillin, as it came to be called, was tried out on patients in hospital. The results were dramatic. People who were dying of bacterial infections started to recover almost immediately. Since then, penicillin has saved millions of lives.

Penicillin works by preventing bacteria making their cell walls. The result is that newly-formed cells burst open and colonies are prevented from spreading. Penicillin has this effect only on bacteria; it does not affect human cells. This is one of its great advantages.

Today vast amounts of penicillin and other antibiotics are produced in industrial fermenters rather like the one illustrated on page 128.

1 It is often said that Fleming discovered penicillin by luck. But it wasn't *only* luck. What else was involved?

2 Why do you think it took so long for penicillin to become available in a usable form?

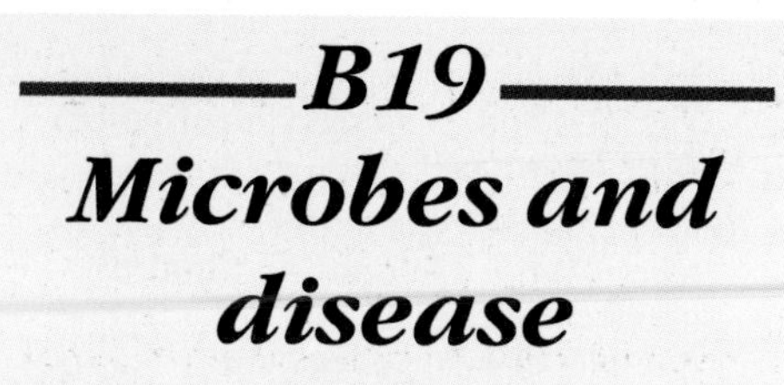

B19 Microbes and disease

Some microbes cause diseases. We look at some of these diseases and how they are spread.

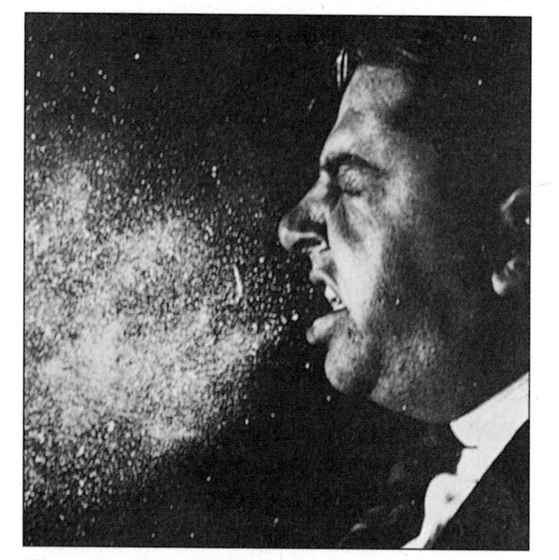

Picture 1 Flash-photo of a sneeze. Thousands of droplets of moisture, containing germs, shoot out of the man's mouth and nose. The droplets may travel at up to 70 miles an hour! A person with tuberculosis may cough up as many as four thousand million bacteria in 24 hours.

What causes what?

These are the main agents of disease, and some of the diseases they cause:

- **Viruses** cause the common cold, influenza ('flu'), measles, chicken pox, hepatitis B, AIDS and yellow fever.
- **Bacteria** cause Salmonella poisoning (a type of food poisoning), impetigo, tuberculosis, diphtheria, whooping cough, typhoid, tetanus,bacterial dysentery, cholera and syphilis.
- **Protoctists** cause amoebic dysentery and malaria.
- **Fungi** cause athlete's foot and thrush.

  In addition, certain diseases are caused by parasitic flatworms and roundworms. For example, bilharzia, a serious disease of the tropics, is caused by a type of flatworm.

How do we know that microbes cause disease?

In the middle of the nineteenth century Louis Pasteur discovered that microbes in the air make food go bad (see page 55). Pasteur suggested that microbes might also cause diseases.

Some years later a German doctor called Robert Koch showed that diseases such as cholera and tuberculosis are caused by certain bacteria. Similar investigations were carried out by other scientists, and by the end of the nineteenth century the particular microbes responsible for many diseases were known.

Which microbes cause diseases?

Most human diseases are caused by bacteria and viruses. However, diseases are also caused by certain types of fungi, protoctists and parasitic worms. Organisms which cause diseases are called **pathogens**.

In this topic we shall be concerned mainly with bacteria and viruses – or **germs** as they are generally called.

When you get a disease, certain signs usually enable the doctor to tell what it is. The signs are called **symptoms** and they allow the doctor to make a **diagnosis**.

What do germs do to us?

Germs get into the body through the nose and mouth, and through cuts and wounds. Once inside, they may multiply very quickly. This early stage is called the **incubation period**, and several days or even weeks may go by before the person starts feeling ill.

Germs harm us in two ways. Some of them attack and destroy our cells. Others release poisonous substances (**toxins**) which get into the bloodstream. For example, the bacteria which cause cholera never leave their victim's gut, nor do they invade the cells lining it. The harm they do is caused entirely by poisons which they give off.

Germs may be passed from person to person

In 1918 there was an outbreak of flu in Spain. Within a few months it had spread all over the world. Well over 20 million people died of it, twice as many as were killed in the whole of the First World War. When a large number of people go down with a disease, we call it an **epidemic**. If it's worldwide, it's called a **pandemic**.

Diseases spread because the germs that cause them get passed from one person to another. A healthy individual catches the disease from someone else, or maybe from an animal. Such diseases are described as **infectious**.

Sometimes a person may have germs in his or her body without showing any signs of the disease. We call the person a **carrier**. Carriers may be infectious even though they are not ill themselves.

How are germs spread?

Here are the main ways that germs are spread.

■ By droplets in the air

When you cough or sneeze, thousands of tiny drops of moisture shoot out of your mouth and nose (picture 1). If you have a disease, the droplets may be teeming with germs. If other people breathe them in, they are likely to catch the disease. The common cold and flu spread rapidly this way, particularly in crowded places.

■ By dust

Germs may stick to dust particles and float through the air. Eventually they settle on surfaces which may be a long way from where they came from.

Diphtheria and tuberculosis can be spread this way. People catch the disease by breathing in the dust, or getting it in their mouths from contaminated food.

■ By touch

Impetigo is a skin disease which breaks out in schools from time to time. You can catch it by touching an infected person or by sharing his or her hairbrush or towel.

Another skin disease is athlete's foot. This can be picked up from the floor of changing rooms and showers. Diseases which are spread by touch are described as **contagious**.

■ By faeces

The faeces of a person with a disease may be swarming with germs. If the germs get into food or drinking water, the disease will quickly spread to other people. Epidemics of typhoid, cholera and dysentery have been caused this way.

Food can become contaminated with bacteria if it is handled by a person with dirty hands. This is why you should always wash your hands after going to the toilet, particularly if you are about to prepare food for other people.

Drinking water may become contaminated if sewage is not got rid of properly. We tend to think of this as happening only in backward countries. However, it can happen in the cleanest of countries if there is a disaster such as an earthquake or flood.

■ By animals

Germs can be brought onto food by animals such as rats and mice, cockroaches and flies. Take flies, for example. These little animals are equally happy feeding on dung or food. Their legs and proboscis may be covered with germs (see page 107).

Many diseases are spread by animals which suck blood. An example is the mosquito which transmits malaria and yellow fever (see page 107).

Bubonic plague, the Black Death of the Middle Ages, is caused by bacteria which are carried from rats to humans by fleas. Occasional outbreaks of this disease still occur in dirty places where rats are common.

An animal which transmits disease-causing microbes from one animal (or plant) to another is called a **vector**.

A number of diseases are spread by pets such as dogs and cats. The most serious is rabies which is caused by a virus. Fortunately it is very rare, at least in Britain. Humans can catch it by being bitten or even licked by an infected dog. The family dog may look innocent, but its tongue may be covered with germs. It is unwise to let it lick your face.

Some nasty diseases can be picked up from the faeces of dogs. One affects the eyes and can cause blindness. Dogs should not be allowed to foul footpaths and parks where children may be playing.

■ By blood

Suppose Jean scratches herself with a needle, and then the needle scratches Ann. Certain germs can be passed from Jean's blood to Ann's blood in this way.

Two diseases which are known to be spread by blood are viral hepatitis and AIDS (see page 238). Drug addicts who share the same needle are at particular risk of getting these diseases.

In the past, AIDS has been spread by people being given contaminated blood in a transfusion. Nowadays blood that is to be used in a transfusion is tested beforehand to make sure it is safe.

Finding the cause of a disease

Tuberculosis (TB) is a serious disease of humans, thankfully now rare in Britain and other developed countries. Suppose you want to find out what causes this particular disease. How could you do it?

You take a small amount of phlegm that has been coughed up by an individual with TB, and examine it under the microscope. You find that it is teeming with a certain type of bacteria. You then examine the phlegm of lots of TB patients, and you discover that they *all* have this type of bacteria. On the other hand, healthy people, or people with other diseases, do *not* have them. This leads you to suspect that these particular bacteria cause TB.

But you are not absolutely sure, so you do a further experiment. You take a sample of phlegm from someone with TB, and grow the bacteria on their own (see page 94). You then inject some of these bacteria into a mouse. If the mouse gets TB, you can be pretty sure that these bacteria are the cause of the disease.

One of the first people to do experiments of this kind was the German doctor Robert Koch (1848–1910). He discovered the bacteria responsible for many diseases including tuberculosis, diphtheria, plague, typhoid and pneumonia.

Picture 1 Robert Koch (1848–1910).

Activity

Finding out about a disease

Choose one infectious disease of humans and find out as much about it as you can. Use books, but also ask a doctor if you have the chance. Doctors can often provide interesting bits of information from their personal experience which you cannot get from books.

Write a detailed account of the disease, the organism that causes it and its effects on humans.

Also find out if, and how, it may be cured.

Questions

1 Give five examples of places where diseases are likely to spread by people coughing and sneezing.

2 What part is played by each of the following in spreading disease:

a flies,
b rats,
c mosquitoes,
d needles,
e aeroplanes?

2 Write down the missing words, (a) to (d) in the following passage. It is what a doctor says to a patient.

'David,. you have chicken pox. The ____(a)____ are quite clear, especially the spots on your chest. I'm going to give your mother a ____(b)____ so that she can get you some lotion from the pharmacy. It will stop you itching so much. The disease is caused by a ____(c)____. I'm afraid it's very ____(d)____ so you won't be able to mix with your friends for the time being, otherwise they may catch it.'

3 The next topic is about how we can protect ourselves against infectious diseases. Before reading it, make a list of all the ways you can think of by which infectious diseases may be prevented.

What are bacteria?

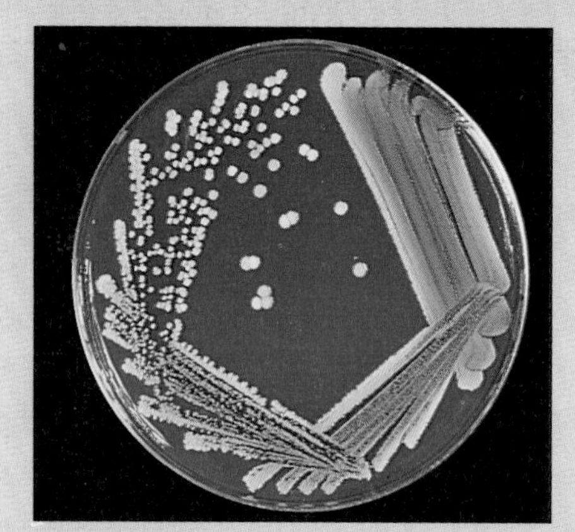

Picture 1 Bacterial colonies growing on an agar plate.

Bacteria are amongst the smallest organisms. You find them almost everywhere: in air, water, soil and inside other organisms. So they are very much part of our environment. Many of them are useful, but some cause serious diseases.

Scientists need to grow bacteria in the laboratory. This is necessary if we are to investigate them, and find ways of fighting the diseases they cause. They can be grown on the surface of a jelly-like material called **agar** to which various food substances are added. In warm conditions, the bacteria multiply to form **colonies** (picture 1).

Each colony consists of thousands of bacteria. If you look at part of a colony under a good light microscope, you can see the individual bacteria as little rods or dots (picture 2). You can see them more clearly if you look at them under the much more powerful electron microscope (picture 3). Different kinds of bacteria have different shapes. Some have whip-like 'hairs' which lash from side to side, propelling the organism along.

If bacteria are grown in a clear liquid medium, they make the liquid go cloudy.

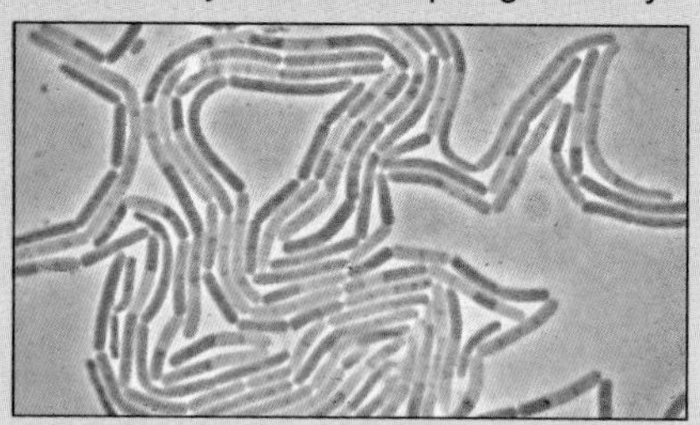

Picture 2 Bacteria under a light microscope.

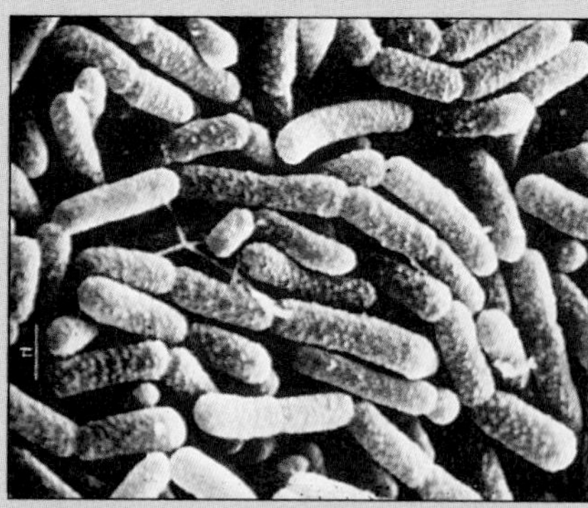

Picture 3 The same type of bacteria under an electron microscope.

Scientists have also studied the *internal structure* of bacteria. They are single cells, but the cell is simpler than those of other organisms and there is no proper nucleus.

Many bacteria are good at surviving bad conditions such as drought, heat, cold and even poisons. They do this by forming a protective coat around themselves. They are then known as **spores**. Inside the spore the bacterial cell becomes dormant – it goes to sleep, as it were. When conditions return to normal, the spore splits open and the bacterial cell comes out (picture 4). The spores of some bacteria can survive for more than fifty years.

dormant bacterial spore
germinating spore
spore coat

Picture 4 On the left is a bacterial spore with a thick protective coat. On the right the coat has split open and the bacterial cell is coming out.

In good conditions bacteria reproduce very quickly (page 248). Their spores and rapid reproduction make disease-causing bacteria difficult to get rid of.

1 In medical research it is important to be able to grow particular kinds of bacteria on their own. Why do you think this is necessary?

2 It is dangerous to eat food which has been left lying around for some time. Which part of the above text helps to explain this?

What are viruses?

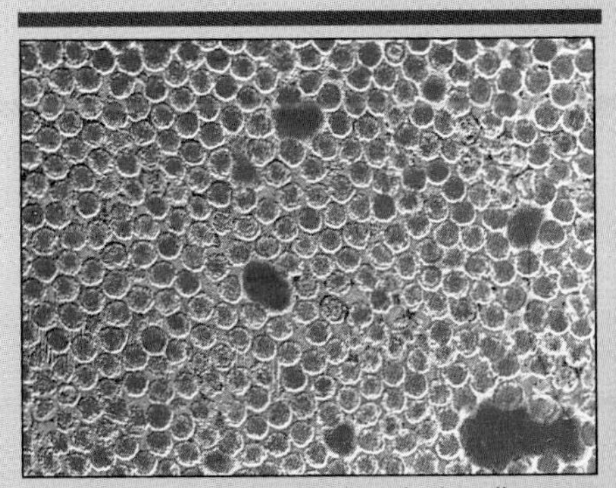

These viruses cause the paralysing disease poliomyelitis.

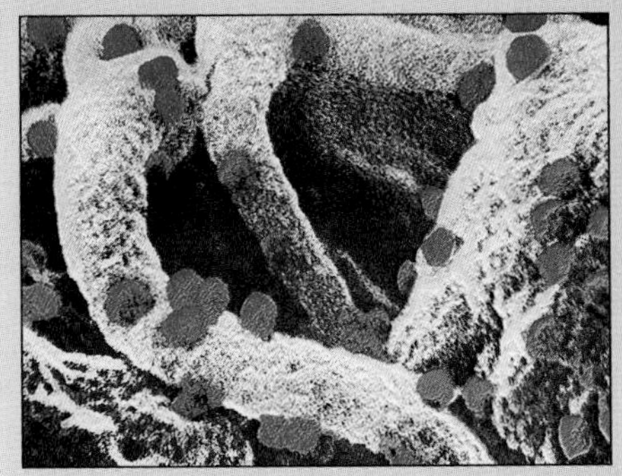

The red objects are the viruses that cause AIDS.

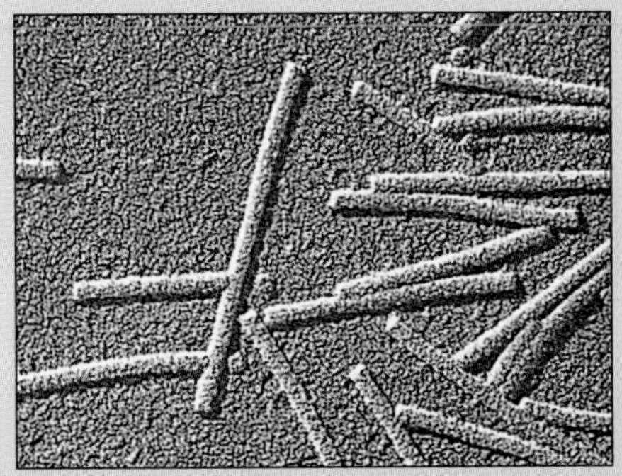

These viruses destroy the leaves of tobacco.

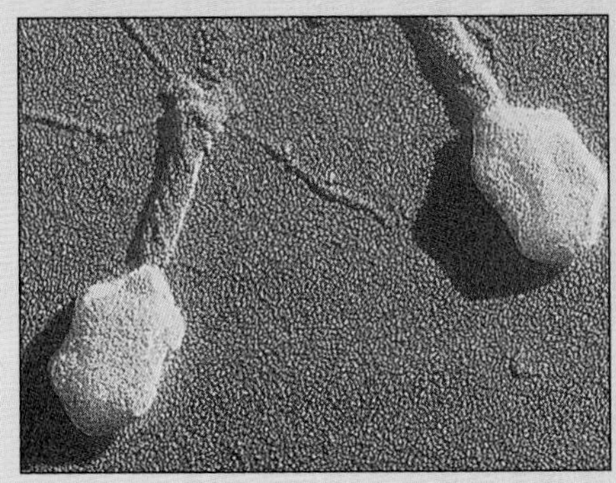

These viruses, known as phage, attack bacteria.

Picture 1 Some examples of viruses.

Viruses are much smaller than bacteria. A typical virus has a width of about a ten thousandth of a millimetre. If you lined them up in a row across this page, there would be over two million of them.

Despite their small size, viruses can cause immense harm. All sorts of diseases in humans, domestic animals and crop plants are caused by viruses. The virus disease which has been most in the news in recent years is AIDS, but viruses also cause less serious diseases such as flu and the common cold.

Because viruses are so small, you can only see them with an electron microscope. Four examples are shown in picture 1. Different kinds of viruses have different shapes and this is one way of recognising them.

A virus has a simple structure. It consists of a coiled string of genes surrounded by a wall made of protein. A famous biologist, Sir Peter Medawar, has described them as a piece of bad news wrapped up in protein.

Viruses can only reproduce inside the cells of living organisms. If they are outside the body they usually survive for only a short time.

Inside the victim's cells, the viruses reproduce and make new viruses. Picture 2 shows how they do this. The materials for making the new viruses come from the victim's cell itself. So the virus is a thief, robbing the cell of its contents and killing it in the process. Thousands of new viruses may be released from one cell, and they then attack more cells. No wonder we feel ill when we've got flu.

Different viruses attack different cells. For example, the common cold virus attacks cells in the nose and throat. The much more serious virus that causes AIDS attacks a certain type of blood cell which helps us to fight disease.

Not all viruses reproduce as soon as they get into our cells. Some just stay there and wait, maybe for years. Then suddenly they become active and start multiplying. Until then you don't even know they are there, unless a test is carried out to show that they are.

1 Viruses can be grown in hen's eggs (picture 3) or in tissue cultures (see page 20). What do we need them for, and why can't they be grown on agar jelly, like bacteria can?

2 Most scientists regard viruses as non-living. Do you think viruses should be regarded as living or non-living things?

3 If viruses survive for only a short time outside the body, how do you think people catch colds from each other?

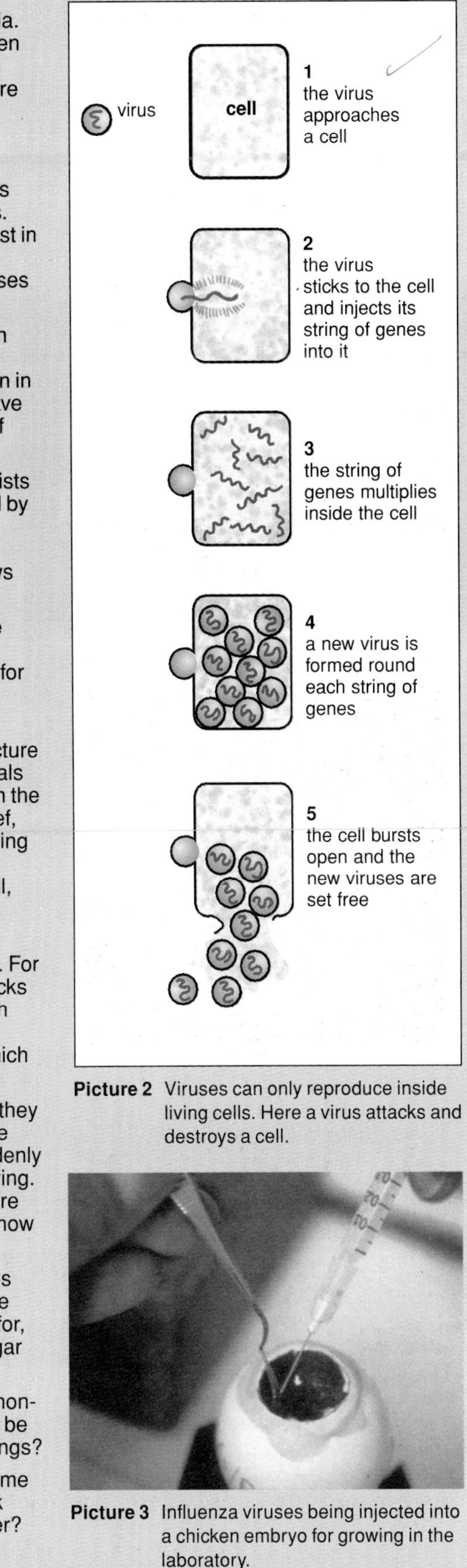

Picture 2 Viruses can only reproduce inside living cells. Here a virus attacks and destroys a cell.

Picture 3 Influenza viruses being injected into a chicken embryo for growing in the laboratory.

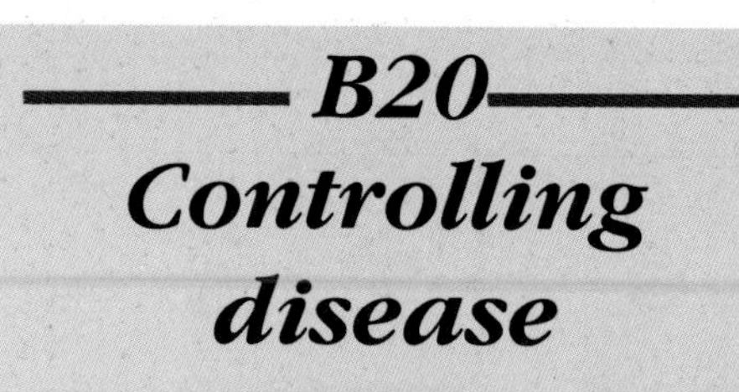

B20 Controlling disease

In this topic we look at the various ways we try to protect ourselves from disease.

Keeping germs out of the body

The surest way of keeping germs out of the body is to get rid of them from our environment.

When you rid an object of germs, it becomes **sterilised**. One way of sterilising is by **heating**. However, the temperature must be high enough. Heating at 120°C for 15 minutes is enough to kill most germs. To achieve such high temperatures, a **pressure cooker** or **autoclave** must be used. Autoclaves are used to sterilise pre-packed foods and hospital instruments. Your school may have an autoclave for sterilising equipment for certain experiments.

Places where we particularly want to get rid of germs, such as kitchens and toilets, should be cleaned with a **disinfectant** such as Dettol. Disinfectants are chemical substances which kill germs.

One of the most germ-free places is the hospital operating theatre (picture 1). Before entering the theatre, the air passes through a special filter which takes out any dust. The surgeons and nurses wear sterilised gowns, head covers and face masks, and all the instruments are sterilised beforehand. There is therefore little risk of the patient becoming infected. A totally germ-free environment is described as **aseptic**.

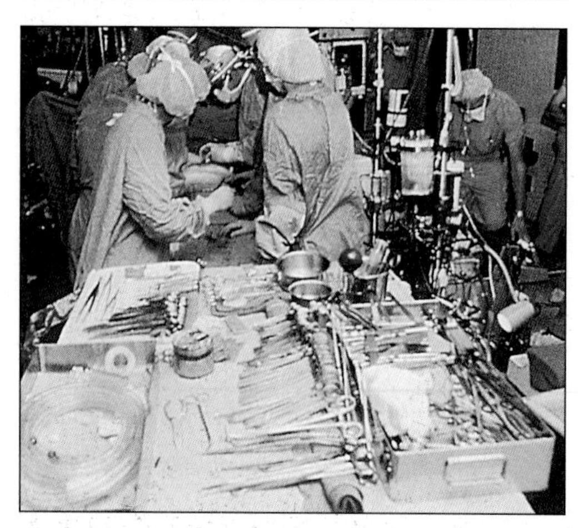

Picture 1 A modern hospital operating theatre – out of bounds to germs!

Animals which carry germs are killed

Great efforts have been made to get rid of disease-spreading animals such as rats, fleas, lice and mosquitoes. The battle against insects has been helped enormously by **insecticides** (see page 84).

One of the most useful insecticides has been DDT. It was first used during the 1939–45 war to get rid of lice. Now it has been banned in many countries because it may be dangerous in the environment (see page 85). However, it has been extremely useful in the fight against malaria and yellow fever, both of which are spread by mosquitoes.

Natural defences against disease

We are so used to the benefits of modern medicine that it is easy to forget that the human body has its own *natural* defences against disease. Our main defences are summarised in the picture. Some of them prevent germs getting into the body; others destroy germs once they have got in.

In defending the body against disease the blood plays a very important part. The role of the blood in defence is explained on pages 157–158.

Thanks to science and medicine, our natural defences can be helped by all sorts of artificial procedures. They are the subject of this topic.

Which of the defences in the picture
(a) prevent germs entering the body, and
(b) destroy germs after they have got in?

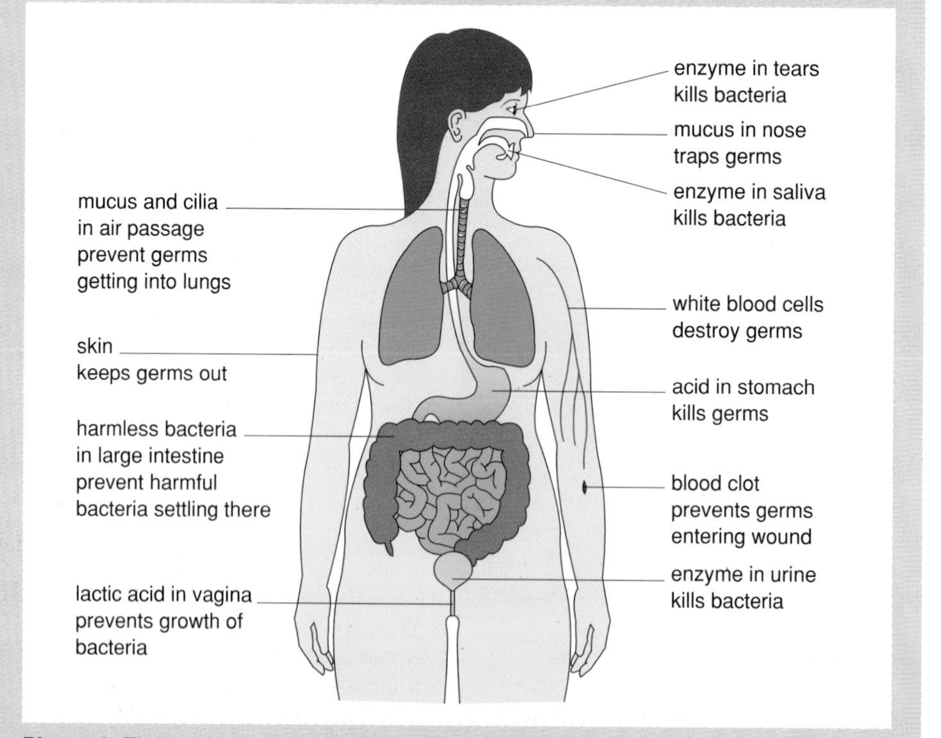

Picture 2 The human body's main defences against disease.

Infectious individuals are isolated

People who have a serious infectious disease, or are carriers of it, must be kept away from other people. So they are isolated until they are no longer infectious. This is called being put in **quarantine**.

Occasionally a person entering a country is placed in quarantine because it's thought that he or she might be carrying a serious disease. This rarely happens nowadays as so many infectious diseases have been brought under control. However, it is always done in Britain with animals. Cats and dogs have to be put in quarantine for six months. This is to make sure that they don't bring in rabies. Rabies is a very serious disease of humans as well as animals. It is caused by a virus which attacks the brain, causing convulsions and death. Anyone who breaks the quarantine law and smuggles a pet into the country could start a rabies epidemic.

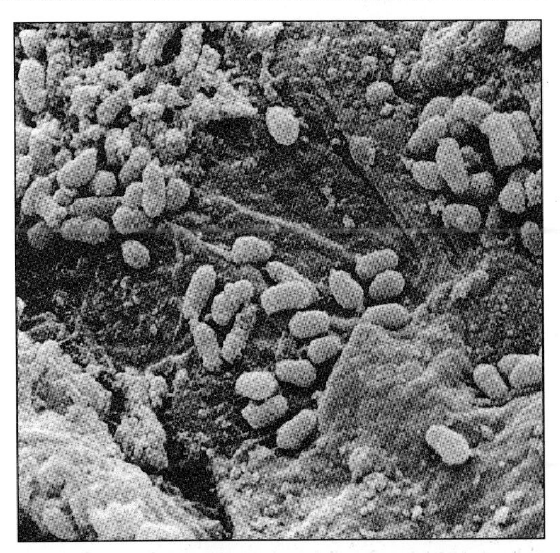

Picture 3 This greatly magnified picture of the surface of human skin shows groups of bacteria living in the crevices.

The skin should be kept clean

The surface of the skin is very uneven, and thousands of microbes make their homes in its nooks and crannies (picture 3). Some of these organisms are useful to us: they kill germs and help to protect us against disease. However, others are harmful and may cause unpleasant skin diseases. For good health it is important to **wash the skin** regularly with soapy water.

If you cut yourself, you open a door to germs and the cut may go septic. You can stop this happening by applying a substance such as 'surgical spirit' which kills germs. These substances are called **antiseptics**.

Antiseptics were discovered in the 1860s by the English surgeon Joseph Lister. In Lister's day more than half the patients in hospital who had had operations died. Many of them got a bacterial infection of their wounds called gangrene. Lister discovered that if he sprayed the patient's wound with carbolic acid during the operation, it did not go septic (picture 4). Thanks to Lister, the number of people who died after operations was greatly reduced.

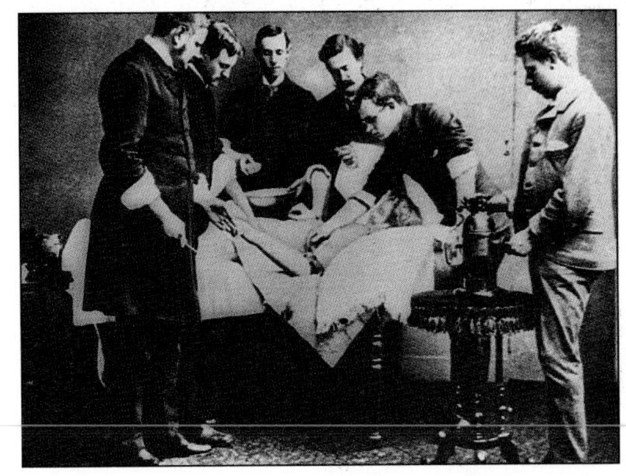

Picture 4 How operations used to be carried out in the nineteenth century. In those days the surgeon and his staff wore their everyday clothes in the operating theatre. Notice the carbolic spray. This helped to create antiseptic conditions in the area of the wound.

If you cut yourself, the cut should be cleaned then covered with sticking plaster or a bandage. This is called a **dressing**. The dressing stems the flow of blood, helps the blood to clot and prevents germs getting in. It also brings the cut surfaces of the skin close together, which speeds up the healing process.

Immunisation

When you get a disease, measles say, substances called **antigens** on the surface of the germs cause certain white blood cells in your body to produce chemicals called **antibodies**. The antibodies then attack the germs, and the germs are destroyed (see page 158).

Now suppose some dead germs are put into your blood before you ever get the disease. What effect will this have? Even though the germs are dead, their antigens still cause you to make antibodies. Antibody-producing cells then remain behind and are ready and waiting to leap into action if you get the disease later. So receiving the dead germs has protected you against the disease. This is what doctors do when they **immunise** you against a disease.

The first person to immunise someone against a disease was the English doctor Edward Jenner. He immunised a young boy against the dreaded disease smallpox. That was in 1796. Since then immunisation has been extended to many other diseases, both viral and bacterial.

When a doctor immunises you, he or she puts a small quantity of dead or inactivated germs, called the **vaccine**, into your bloodstream (picture 5). Normally this is done by injection, using a hypodermic needle, or by scratching the skin. In some cases the vaccine can be taken by mouth.

Nowadays chemicals can be extracted from germs and used as vaccines; or genetically engineered bacteria can be made to produce antigens for use as vaccines.

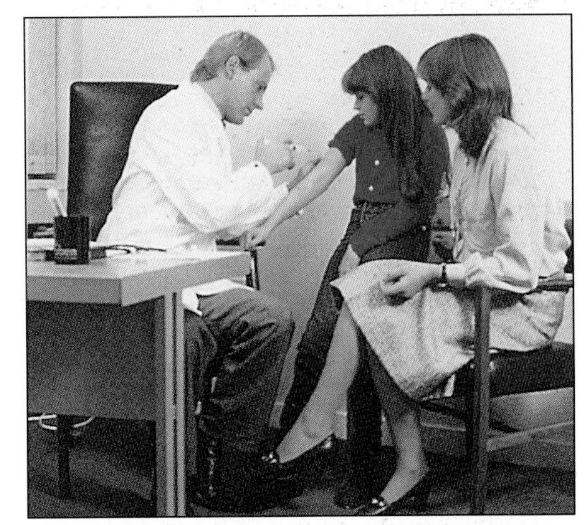

Picture 5 The doctor is immunising this girl against German measles. He is injecting the vaccine into the girl's arm with a hypodermic syringe. This process is called inoculation.

Vaccines

How are vaccines made?

The aim is to produce a substance that contains the antigens of a germ. When introduced into the body, the substance must give the person immunity against the disease without *causing* the disease. There are five main ways of producing such vaccines:

- Kill the germs by heat or chemical treatment, then use the killed germs as the vaccine, e.g. typhoid, cholera (bacterial diseases).
- Inactivate (attenuate) the germs by culturing them in special conditions so that they become harmless (non-virulent), then use them as the vaccine, e.g. tuberculosis (bacterial disease), rubella (German measles, virus disease).
- Extract the toxin from the germ and modify it chemically so that it is no longer harmful, then use that as the vaccine, e.g. diphtheria,tetanus (bacterial diseases).
- Extract antigens from the germ and use them on their own as the vaccine, e.g. influenza (virus disease).
- Use genetically engineered bacteria to mass-produce antigens for use as vaccines. e.g. hepatitis B (virus disease).

Is vaccination safe?

Think of it like this. When you are vaccinated against a disease, the doctor deliberately puts germs (or their antigens) into your body. Isn't there a risk that the vaccine might give you the very disease that it's supposed to protect you against? The answer is no. This is because of the way the vaccine has been produced.

Nevertheless, people sometimes show a slight reaction to a vaccine, and on very rare occasions the reaction may be serious. For example, it has been suggested that the measles and whooping cough vaccines *may* have been responsible for a few cases of brain damage – though this has never been proved.

In deciding whether or not to have a child vaccinated against a particular disease, parents have to weigh up the very slight chance of a bad reaction against the much greater chance of the child suffering from the disease if he or she is not vaccinated. For example, the number of cases of brain damage following measles immunisation is 1 in 87 000; the number of cases of brain damage following an attack of measles is 10 to 20 times greater

Picture 6 Medical drugs are manufactured on a huge scale.

Babies are immunised against various serious diseases such as diphtheria and polio. Teenagers are usually immunised against tuberculosis unless a simple skin test shows that they are already naturally immune to it. Young girls may be immunised against German measles (rubella). This is a mild disease but if a woman gets it in the early stages of pregnancy it may damage her baby.

These immunisations should protect us for the rest of our lives. However, protection against some diseases lasts for only a limited time.

In such cases you need to be given further doses of vaccine from time to time to keep up your protection. These are called **boosters**. Boosters are needed for protection against diseases such as typhoid and cholera which people sometimes get immunised against if they are going abroad.

For immunisation to be effective, you must be given the vaccine before you get the disease. It will not cure you of the disease once you have got it.

Receiving ready-made antibodies

Tetanus is a serious bacterial disease which kills about 100 people every year in Britain. The muscles, particularly those working the jaws, go into painful spasms – the disease is sometimes called 'lockjaw'. Tetanus germs can be picked up if you cut yourself with a dirty instrument such as a penknife.

Suppose you cut your finger with a dirty knife. The doctor wants to make sure you do not get tetanus. It is too late to give you an injection of vaccine; by the time your body had made the necessary antibodies you might well be dead! How can the doctor protect you quickly?

The answer is to give you some antibodies which have already been made by someone else, or by an animal. This is called **anti-tetanus serum**. The doctor injects some of this into your arm, and sends you home. The serum prevents you getting tetanus.

Giving a person ready-made antibodies like this is useful in an emergency. However, the protection does not last long. This is because the antibodies are gradually broken down and got rid of from the body. For long-lasting protection we need to make our own antibodies.

Antibiotics and drugs

Some microbes produce substances which defend them against other microbes. For example, the fungus *Penicillium* produces a substance which acts against bacteria. The substance was discovered by Alexander Fleming in 1928, and is called **penicillin**. The story of penicillin is told on page 95.

Penicillin is just one of a very large number of **drugs** which today are used by doctors to treat people with diseases. They are manufactured on a huge scale by pharmaceutical companies (picture 6). Some, like penicillin, are obtained from microbes. They are called **antibiotics**. others are made by special chemical processes. They exert their effect either by killing germs or inactivating them. Over the years these substances have saved countless millions of lives.

Antibiotics are very successful against bacteria. Unfortunately they do not work against viruses. So they will not cure people of the common cold and flu. Some of the more serious virus diseases are treated with special anti-viral drugs. However these are often not very effective.

One problem with antibiotics is that new kinds of bacteria keep arising which are resistant to them (see page 291). For this reason doctors do not prescribe antibiotics unless they are really necessary.

Is the battle won?

Look at picture 7. This shows the number of people in Britain who died of diphtheria each year between 1911 and 1961. You will see that there has been a tremendous fall in the number of deaths from this disease. The same is true of smallpox and many other infectious diseases. In fact, the World Health Organisation has declared that smallpox is now extinct.

This happy state of affairs has been brought about partly by immunisation and antibiotics, but also by improvements in cleanliness and diet. If you are well fed, clean and healthy, you are less likely to succumb to disease.

Unfortunately the situation is not so good in many developing countries where infectious diseases still claim many lives. This is due to poor food, overcrowding and lack of hygiene, and to a shortage of doctors, nurses and medical supplies.

In developed countries the main problems are now diseases which may be linked with diet or lifestyle – diseases like cancer of the lung and heart disease. One of the most serious infectious diseases which remains to be conquered is AIDS (see page 238). AIDS is a new disease, only discovered around 1980, and it is on the increase. So the battle against disease is not yet over.

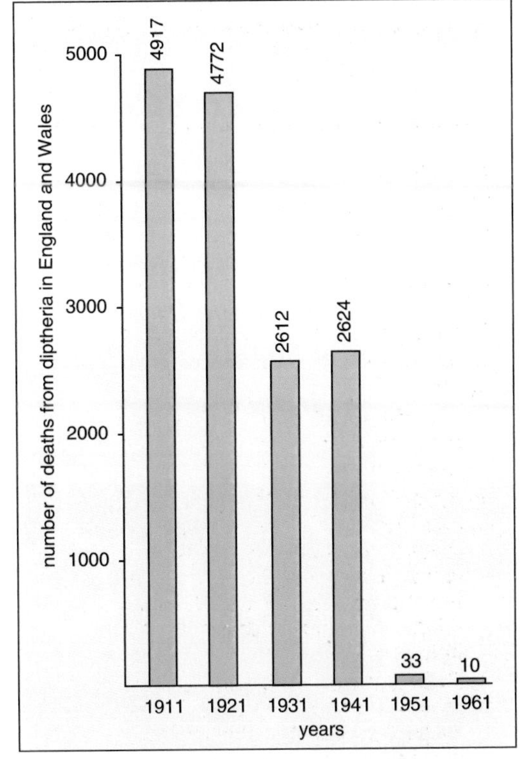

Picture 7 This bar chart shows the annual deaths from diphtheria in England and Wales at ten year intervals between 1911 and 1961. During this period the number of deaths each year fell from 5000 to only ten. Diphtheria is a bacterial disease and used to be a major cause of death among children.

Questions

1 Explain the reason for each of the following:

a A pet which is brought into Britain from overseas is put into quarantine for six months.
b If you graze your knee it is sensible to wash it immediately and put a dressing on it.
c A surgeon wears a mask over the mouth and nose.
d Many of the food items in a supermarket are wrapped in plastic film.
e Chlorine is often added to drinking water.

2 Why is it particularly important that the following places should be as free of germs as possible:

a operating theatres,
b public lavatories,
c hotel kitchens,
d swimming pools,
e doctors' surgeries?

3 Until the late 1940s there were special 'isolation hospitals' for patients with infectious diseases such as tuberculosis. Few such hospitals exist in Britain now because they are not needed any more.

a Why were isolation hospitals necessary in the old days?
b Why are they no longer needed?
c Write down three particular difficulties which you think there might have been in running an isolation hospital.

4 In general bacterial diseases are easier to cure than viral diseases. Why?

5 Consider the effect of each of the following developments on the health of the nation:

a mains drainage,
b mains water supply,
c pasteurisation,
d immunisation,
e antibiotics.

What other developments have improved our health?

Activity

Preventing the growth of bacteria

CARE: Work with bacteria should be carried out only under strict supervision by your teacher.

Always wash your hands after working with bacteria.

BIOHAZARD

Before starting this activity, be sure you have done activity A on page 94. This shows you how to culture bacteria on agar jelly.

Now plan an experiment to find out how good these agents are at preventing the growth of bacteria: a **disinfectant** (e.g. Dettol), an **antiseptic** (e.g. iodine) and an **antibiotic** (e.g. penicillin).

Discuss your plan with your teacher, then carry out the experiment.

You might also compare the efficiency of different disinfectants or antiseptics.

B21 A useful insect: the honey bee

Insects help us in a number of ways. Here we look at the honey bee as a useful insect.

Picture 1 Worker bees on one of the combs in a hive.

Picture 2 In a beehive the workers construct their combs inside wooden frames. The top ones, where the honey is stored, can be taken out easily.

The honey bee lives in an organised society or colony (picture 1). Wild bees live in a nest in, for example, a hollow tree trunk. Bee-keepers rear them in specially constructed **hives** (picture 1) so as to get **honey** from them. A large hive may contain more than 50 000 bees.

Who's who in the beehive

Bees have a remarkable **social organisation**. There are three different kinds of individual in the colony:

- The **queen** is head of the colony. She is a fertile female whose job is to lay eggs and produce all the other bees in the hive.
- **Drones** are fertile males. There are several hundred of them in the colony. Their only job is to mate with a queen. The queen only mates with one drone; she uses his sperm to fertilise most of her eggs.
- **Workers** are sterile females and cannot reproduce. Their job is to look after the hive, rear the young and collect food from flowers.

Each type of bee has its own job to do. There is thus a strict **division of labour** within the hive. An individual bee does what it is programmed to do. The idea of a bee showing original or imaginative behaviour is unthinkable.

Structure of the hive

Inside the hive the workers make **combs** out of **wax**. The wax is produced by glands on the lower sides of the abdomen. The worker takes wax from the abdomen with its hind legs, and moulds it into shape with its **mandibles** (jaws).

A comb consists of numerous small chambers called **cells**. They are hexagonal in shape and fit together neatly, as you see in picture 1. Honey is stored in cells towards the top of the comb, pollen in cells lower down. In the cells towards the bottom the queen lays eggs – several thousand a day in the summer, one in each cell. The young (larvae) develop inside the cells. Workers are reared in the smallest cells, drones in slightly larger cells. Queens are reared in special cone-shaped cells near the edge of the comb.

What is honey?

Honey is made by the workers from **nectar**, the sugary fluid found in flowers (see page 240). When a worker visits a flower, it sucks up the nectar with its tongue-like **proboscis** and stores it in its **stomach**. Here the nectar undergoes slight chemical changes and is turned into honey.

When the bee returns to the hive, it regurgitates the honey into one of the comb cells. The worker then closes the cell with a wax lid.

Worker bees use the stored honey, together with pollen, for feeding the larvae. They also produce a special substance from the mouth called **royal jelly** which is fed to the larvae in varying amounts.

We humans can take honey from bees because normally they make much more than they need (picture 2). In a good summer a hive can produce enough honey to fill 100 pots (picture 3).

Picture 3 Different kinds of honey are made from the nectar of different types of flower. Blended honey is made by mixing different kinds together.

How worker bees communicate

On returning to the hive from a food-collecting trip, a worker bee can tell other workers where to find the food. This was discovered by an Austrian scientist, Karl von Frisch.

When the bee gets back to the hive, it does one of two types of dance:

- If the food source is less than about 80 metres from the hive, the bee dances in a circle, reversing the direction of movement after each circuit. This is called the **round dance**. The closer the food, the faster the dance.
- If the food source is more than about 80 metres from the hive, the bee dances in a figure-of-eight, waggling its abdomen as it does the straight run in the middle. This is called the **waggle dance**. As with the round dance, the closer the food, the faster the dance.

The waggle dance tells other bees not only how far away the food is, but also its direction. This is explained in picture 4.

While a worker is dancing, other workers gather round in an excited manner and after a few seconds they fly off to the food source.

The way bees communicate by dancing is one of the best known theories in biology. But it is only a theory. There is no doubt that bees do dance, and the speed and orientation of the dances can be correlated with the positions of food sources. But it does not follow that bees *communicate* this way. Communication may be by some other kind of signal that accompanies the dances. For example, an American scientist has discovered that bees make a wide range of sounds while they are dancing. It is possible that bees communicate with each other by these sounds.

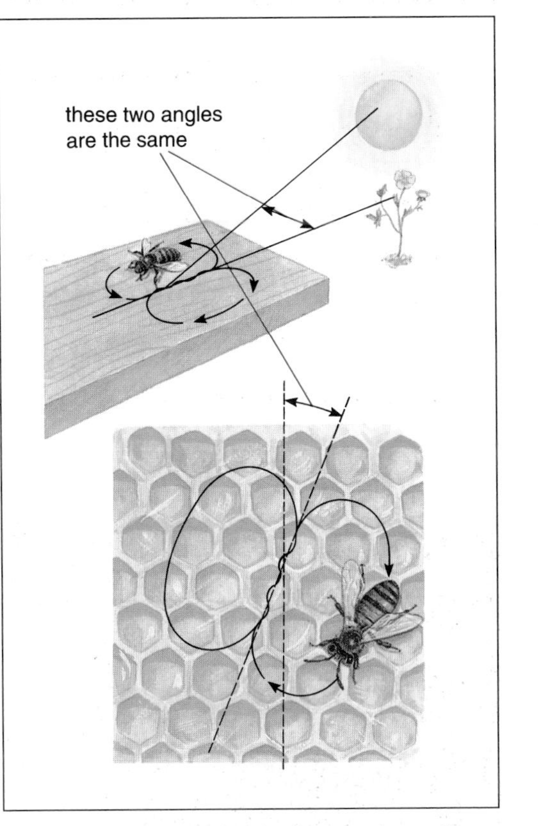

Picture 4 The figure-of-eight dance performed by worker bees, and how it tells other bees where to find food. The angle which the straight run part of the dance makes to the vertical is the same as the angle that the food source makes to the Sun.

Activity

How is a worker bee adapted to do its jobs?

These are the jobs that a worker bee does in the hive during its life of about two months: cleans out the cells and feeds the larvae, builds new combs, defends the hive against intruders (e.g. mice), ventilates the hive, collects nectar and pollen from flowers, places honey and pollen inside cells.

Examine a preserved worker bee, using a hand lens and/or binocular microscope. Also watch bees visiting flowers. In this way find out how bees are adapted to carry out their various jobs.

Be careful you are not stung.

Compare the worker bee with a queen bee and a drone. Are the worker's adaptations absent in these other types of bee?

Questions

1 Suggest an explanation for each of the following:

a In a poor summer bees make less honey than usual.
b Worker bees often sit on comb cells which contain developing larvae.
c A worker bee will often visit only one kind of flower, ignoring others.
d After a worker bee stings someone, it dies.

2 What causes an egg, laid by the queen, to develop into a particular type of individual (i.e. a queen, drone or worker)? The text of this topic contains *three* clues as to *possible* causes. Find the clues and suggest the possible causes.

3 What part do plants, bees and humans play in making a pot of honey?

4 Bees and humans both live in societies. Compare the organisation of a beehive with that of human society.

5 From more specialised books try to find out:

a how the queen bee maintains her supremacy of the colony,
b how and when new colonies are formed,
c what bees do in the winter,
d how and when a drone mates with a queen,
e what happens to the drone after it has mated with a queen.

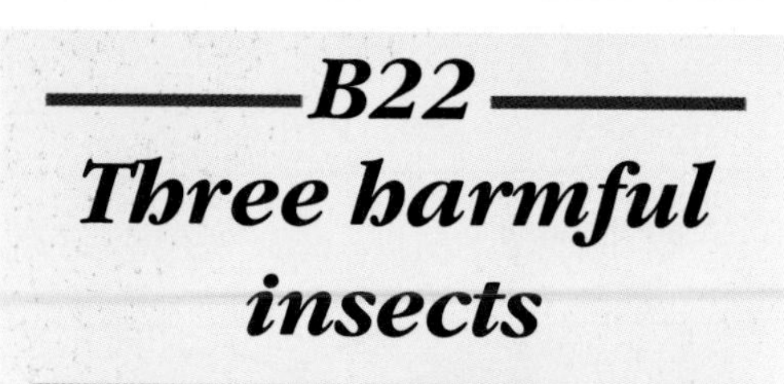

B22 Three harmful insects

Insects can cause a lot of trouble. Here we look at three culprits.

Picture 1 A fly feeding on a lump of sugar. Notice the fleshy lobes of the proboscis in contact with the sugar.

The housefly

Everyone knows this annoying animal and the way it tries to land on our food (picture 1). The housefly belongs to a group of insects called dipterans. Its structure and life cycle are shown in picture 2.

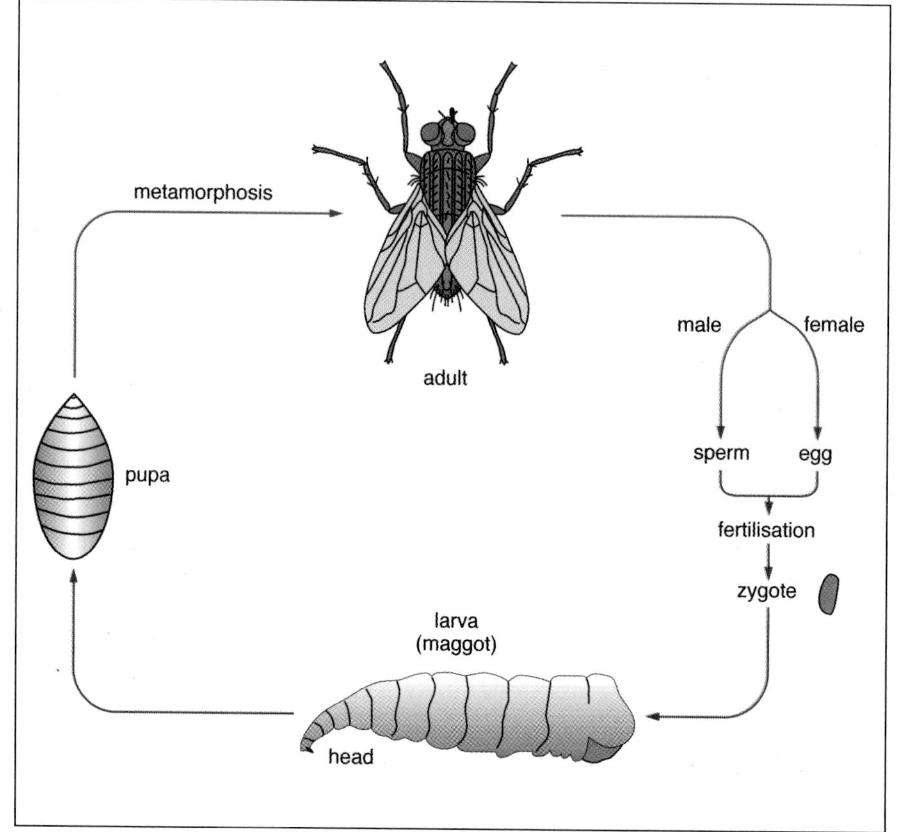

Picture 2 The structure and life cycle of the housefly.

The housefly, in common with other dipterans, has only **one pair of wings** (most insects have two pairs). The second pair is replaced by a pair of **balance organs**. The wings and balance organs make the housefly an agile flyer.

You know how difficult it is to swat a fly. This is because it is very sensitive to stimuli and reacts quickly. Its **feelers** (**antennae**) have hair-like receptors sensitive to air currents. Its **compound eyes** cover a large area of its head, giving the animal a wide range of vision.

You will have noticed how flies can walk on walls and ceilings. The feet have two little hooks with a pad between them. When the pad is pressed it secretes a sticky substance which enables the feet to stick to the smoothest surface.

Flies are equally at home feeding on cake, rotting matter or faeces. They feed by means of a **proboscis** (picture 3). The end of the proboscis is expanded into two fleshy lobes. The undersides of the lobes are traversed by narrow grooves which connect to a tube that runs up the proboscis to the gut.

When the fly feeds on solid food, it places the lobes of its proboscis on the surface. **Saliva** then flows down the proboscis onto the food. The saliva dissolves the food, and the fly sucks up the liquid. The narrowness of the grooves prevents solid matter getting into the proboscis and blocking it.

During the summer the females lay their eggs on decaying vegetable matter, rotting meat or faeces. The eggs hatch into white **maggots** (larvae) which burrow into, and feed on, the decaying material. The maggots then leave their feeding material and pupate. After a few days or weeks, the adult flies emerge from the pupae. The pupae can survive the winter and give rise to adults the following spring.

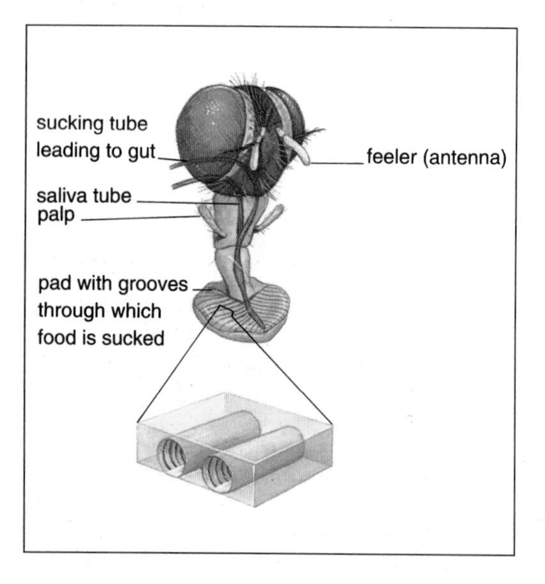

Picture 3 The head of a housefly showing details of its proboscis.

Flies contaminate food with bacteria carried on their bodies, particularly the proboscis and feet, and from their saliva and faeces. A single housefly may have as many as 500 million bacteria on its body. Fortunately most of these are harmless, but flies *can* spread serious diseases such as typhoid and dysentery.

Controlling houseflies

Flies can be kept under control by **wrapping or covering food**, killing them with **special lamps** to which they are attracted, spraying them with a fly spray (**insecticide**) and getting rid of, or covering, any dead material on which they might lay eggs.

The mosquito

The mosquito is another dipteran fly. It is much smaller than the housefly, and its proboscis is needle-like for piercing skin and sucking blood (picture 4). After the proboscis has pierced the skin, a drop of saliva is injected into the wound. This prevents the blood clotting as it flows up the proboscis.

Mosquitoes need water for breeding: ponds, lakes, water tanks – any place where the water is still. Tropical swamps are ideal.

The life cycle of a mosquito is shown in picture 5. The female lays her eggs on the surface of the water. Each egg hatches into a small wriggling larva which hangs onto the surface film by a **breathing tube** at the back end. It breathes air through this tube. Later the larva turns into a pupa which hangs onto the surface film by a pair of breathing tubes on the head. After a few days the pupa splits open and the adult mosquito emerges.

In the tropics mosquitoes transmit two very serious diseases: **malaria** which is caused by a single-celled parasite, and **yellow fever** which is caused by a virus.

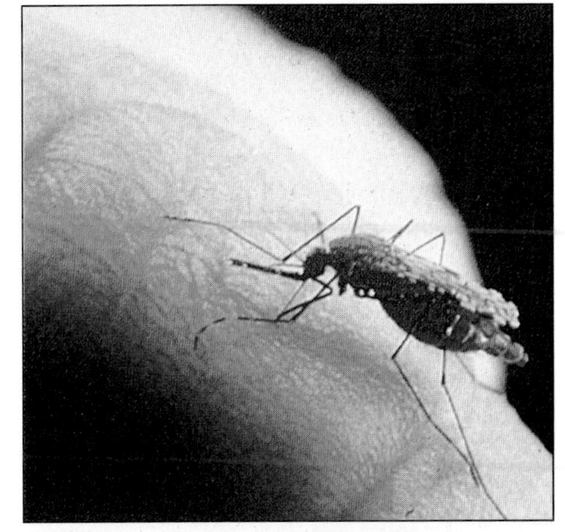

Picture 4 A mosquito sucking blood from a person's arm. This type of mosquito carries malaria. It is called *Anopheles*. Only the female mosquito sucks blood, so only the female carries malaria.

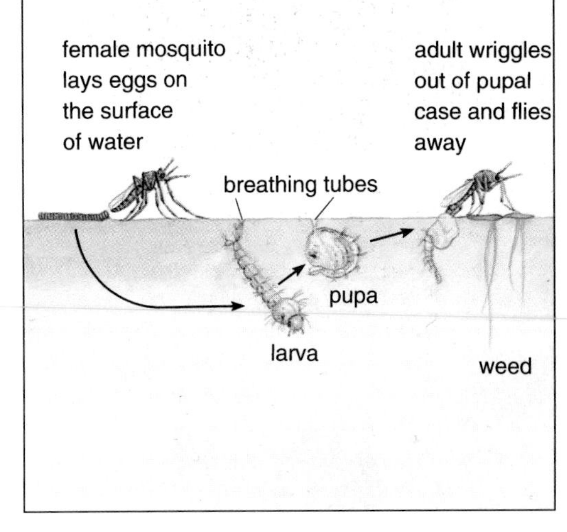

Picture 5 Egg-laying and subsequent development of the mosquito *Culex*.

Controlling mosquitoes

Mosquitoes can be kept out of buildings by covering the windows with a **fine-mesh screen. Mosquito nets** may be used to protect people while they are asleep, particularly campers.

Adult mosquitoes can be killed by spraying them with insecticides. The larvae and pupae can be killed by spraying **oil** on the water. The oil lowers the surface tension of the water. As a result, the animals can no longer hang onto the surface film and water enters the breathing tubes, so they drown. Usually the oil is mixed with an insecticide to make sure they die.

Another approach is to stock up lakes and ponds with fish that eat the larvae and/or pupae. This is an example of **biological control**.

Finally, you can **drain swamps** so as to destroy the mosquito's main breeding areas.

Insect life cycles

Insects have two different types of life cycle. Both are illustrated in this Topic.

- In some insects the egg develops into the adult via a series of **nymphs**. These look like miniature adults, but lack wings. Each nymph sheds its cuticle and grows a bit to give rise to the next nymph. The shedding of the cuticle is called **moulting**. The locust has this kind of life cycle. We can summarise it like this:

  egg → nymph → adult

- In other insects the egg develops into a **larva**. This looks completely different from the adult, and usually feeds on a different type of food. In time the larva changes into a dormant **pupa**. Eventually the adult emerges from the pupa. The change from larva to adult is called **metamorphosis**. The housefly and mosquito have this kind of life cycle. We can summarise it like this:

  egg → larva → pupa → adult

1 Read the account of the honey bee on page 104. Which kind of life cycle does it have?

2 Why is *metamorphosis* a good word for the change from larva to adult? (Hint: use a dictionary.)

3 Why might it be helpful to an insect species to have a larval stage in its life cycle?

Picture 6 Locusts have a huge appetite and can devastate a plant in a very short time.

The locust

Locusts thrive in warm parts of the world where they feed on grass and other vegetation. They possess a pair of powerful **mandibles** (jaws) which cut and crush the food (picture 6).

The locust's life cycle is shown in picture 7. The female lays her eggs in the sand. The eggs hatch into miniature locusts called **hoppers**. The hoppers have no wings, so they cannot fly. As their numbers build up, they crowd together in an excited state and start 'marching' in bands, eating as they go. In the course of the next six weeks they moult several times and grow. Eventually the wings expand and they become adults.

They now start to fly, moving across the country in a vast **swarm**. A single swarm may contain ten thousand million locusts. With the aid of the wind, they fly about 80 km a day, stripping the land of its vegetation. They may travel several thousand kilometres before settling down to breed.

Controlling locusts

In the old days farmers tried to drive locusts away by lighting fires or beating drums. The hoppers were driven into trenches, then buried or burned, or poisoned bait was put in their path. Where possible, the eggs were dug up and burned.

Nowadays crops are sprayed with powerful insecticides from vehicles or aeroplanes. Spray from an aeroplane on one flight can kill as many as 180 million locusts.

Constant watch is kept on locusts, using satellites and other modern methods. This enables scientists to forecast when and where swarming is likely to occur. Crops can then be sprayed in good time. This requires cooperation between different countries, and much of the work is coordinated by the United Nations.

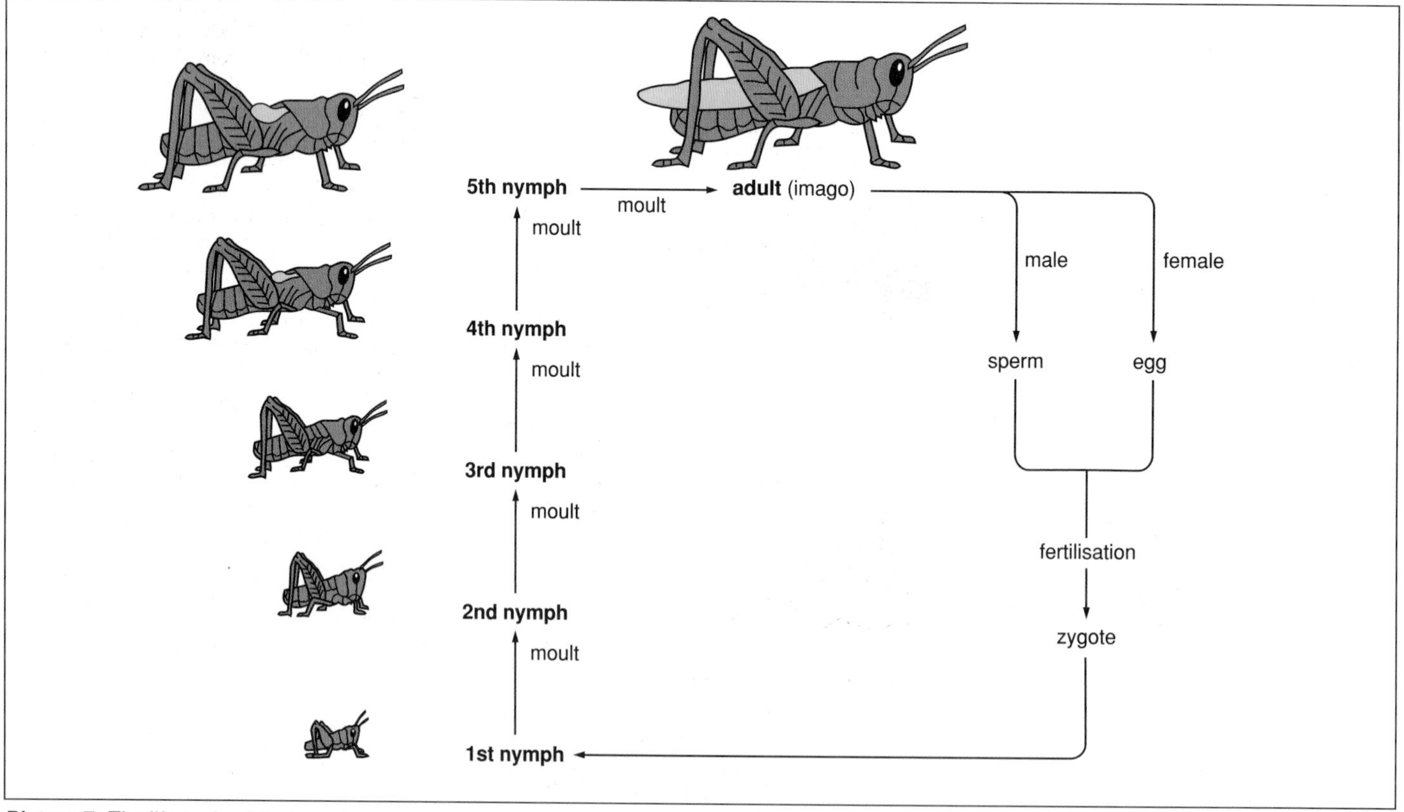

Picture 7 The life cycle of the locust.

Activities

A Observing houseflies

1 What exactly do flies do when they land on a piece of food, or on some other surface? Try to find out by observing houseflies at close quarters. Concentrate particularly on how, and where, they feed, defaecate and clean themselves.

2 It is claimed that flies have taste receptors on the last joint of their legs (the tarsus). Do any of your observations support this idea?

3 Observe a preserved housefly under a binocular microscope. Look for any special features that enable the fly to behave in the way it does.

B Controlling mosquitoes

1 Obtain a dish containing mosquito larvae and/or pupae.

2 Notice that the larvae and pupae hang onto the surface film. The slightest vibration causes them to let go and dive down into the water.

3 Wait till the larvae and pupae have settled at the surface. Then very gently run some oil or paraffin onto the surface of the water.

What happens to the larvae or pupae? Explain your observations.

Do you think this is a good way of getting rid of mosquitoes?

Can you think of any disadvantages?

4 Obtain a preserved adult mosquito and put it on a piece of white paper.

5 Measure its length, and width with wings outstretched.

6 Examine the preserved mosquito under a hand lens or binocular microscope. Make a list of the features which help to adapt it to its way of life.

What advice would you give to the manufacturers of mosquito netting to make sure their product is effective?

Questions

1 The fly in picture 1 is feeding on a lump of sugar.

a How does the fly dissolve the sugar so that it can ingest it?

b If a person put this lump of sugar in his or her tea, what might be the outcome? Explain your answer fully.

2 Do you think it is correct to regard houseflies as 'filthy'?

3 Houseflies are instantly disturbed by a sudden movement of the hand, but they are less disturbed if they are behind a sheet of glass. Suggest an explanation.

4 Read how mosquitoes are controlled on page 107. Assess the advantages and disadvantages of each method, and suggest how each one could be made to work efficiently. Can you think of any other possible methods?

Malaria

Every year about 200 million people get malaria, and about 2 million die of it.

Malaria is caused by a tiny single-celled parasite called *Plasmodium*. There are four different species of *Plasmodium* and they all cause different types of malaria.

Suppose you are bitten by a malaria-carrying mosquito. When the mosquito puts its proboscis into your skin, it injects a drop of saliva into your bloodstream. The saliva contains lots of tiny, worm-like parasites.

The parasites first attack your liver. They invade the liver cells and reproduce asexually inside them (see page 248). Once the parasite has got into a liver cell, it splits up into as many as 1000 offspring.

After about two weeks, the parasites start invading the red blood cells. Each little parasite penetrates a red blood cell and then reproduces asexually inside it. After a few days (the time varies according to the type of malaria) the red blood cell bursts, releasing a batch of new parasites. They then repeat the process.

This procedure takes place in lots of blood cells at the same time. The mass release of parasites from the blood cells is accompanied by a sudden attack of high fever with violent bouts of shivering and sweating. In one type of malaria this happens at regular 48 hour intervals.

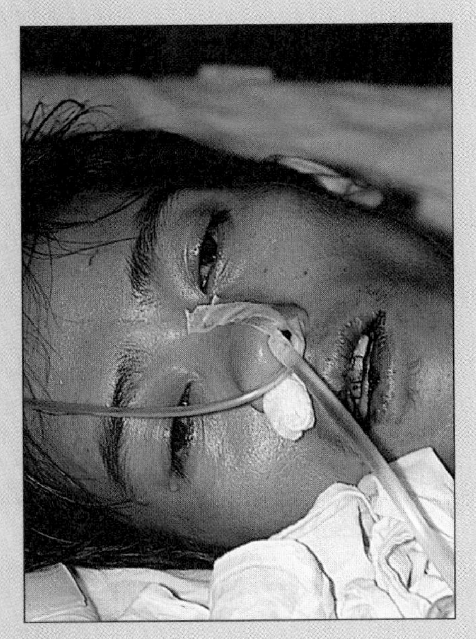

Picture 1 A patient being treated for malaria in a hospital in Vietnam. The disease damages the liver and causes mass destruction of red blood cells.

When a mosquito bites an infected person, parasites are drawn up with the person's blood into the mosquito's stomach. Here the parasites reproduce sexually, and this is followed by further asexual reproduction in the wall of the stomach. The parasites then make their way to the mosquito's salivary glands where they wait until the mosquito bites another victim.

1 Typically, 24 parasites are released every time an infected red blood cell bursts. Assuming that *one* red blood cell in your body is attacked by a parasite at midnight on Monday, and that fever occurs every 48 hours, how many red blood cells will have been destroyed by midnight the following Monday?

2 Malaria can be treated with drugs that kill the parasites in the body. However, this treatment is not always successful. Suggest reasons for this.

3 It has proved very difficult to develop a vaccine against malaria. Why do you think this is?

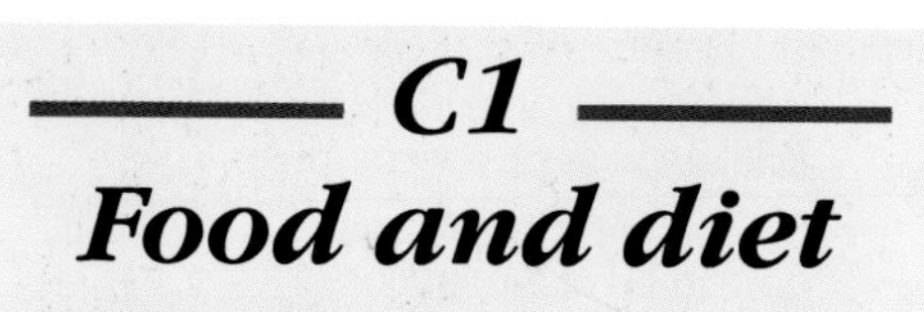

C1 Food and diet

The study of food is called nutrition. The next two topics are about nutrition.

Picture 1 In the course of a lifetime a person may eat 100 tonnes of food.

Why do we need food?

We need food for four main reasons:

- It serves as a fuel, giving us energy and warmth.
- It provides materials for growth.
- It enables us to repair and replace tissues.
- It keeps us healthy and helps us to fight disease.

Our diet

The food we eat each day makes up our **diet**. Whatever we choose to eat, our diet must include the following substances: **carbohydrates, fats, proteins, water, minerals** and **vitamins**.

A diet which contains all the necessary substances, but not too much of any of them, is called a **balanced diet** (picture 2). The food substances themselves are called **nutrients**.

The chemistry of these substances is dealt with on page 22. Here we shall concentrate on the part they play in our diet. We shall deal with the substances that we need in bulk in this topic, and those that we need in only small amounts in the next one.

Carbohydrates

■ Sugar

Sugar gives us energy, so we call it an energy food. Different foods contain different kinds of sugar. For example, the sugar in fruit is **fructose** or **glucose**, and in milk it is **lactose**. Ordinary table sugar is **sucrose**.

Most of the world's sugar comes from sugar cane. This is a giant grass, rather like bamboo, which may reach a height of six metres. It is grown in hot countries. Sugar, in the form of sucrose, is stored in its thick stem. The juice is extracted from the stem, and then purified (refined). After that, water is evaporated from it, so sugar crystals are formed. This is **white sugar**.

Brown sugar is less refined than white sugar. It contains various impurities which give it its brown colour and make it slightly sticky. These impurities do us no harm – in fact, when used in cooking, they add flavour to things like cakes and biscuits.

Another source of sugar is sugar beet which is grown in temperate countries including Britain. Sugar beet stores sugar in thick swollen roots.

■ Starch

Starch is found particularly in bread, potatoes and cereals. Like sugar, starch gives us energy.

Starch occurs in plant cells as **starch grains**. Each grain contains many tightly packed starch molecules, and each starch molecule consists of a chain of sugar molecules. So starch is a concentrated store of energy.

Each starch grain is surrounded by a membrane, rather like a little envelope. When you cook a starchy food such as a potato, the starch grains swell up and burst. This releases the starch molecules, which can then be easily digested.

■ Cellulose

Cellulose is the substance which plant cell walls are made of. It is tough and rubbery. The cellulose walls of neighbouring cells are stuck together by a sort of glue. This is why raw plants are often difficult to chew. Cooking dissolves the glue, with the result that the cells come apart. This makes the plant softer and easier to eat. The same sort of thing happens when fruits ripen.

Humans cannot digest cellulose – we don't have the necessary enzyme in our gut for breaking it down. This means that we cannot get energy from it. However, it still has a useful function: it is our main source of **dietary fibre** (**roughage**). This keeps food moving along the gut, and helps to prevent con-

Picture 2 A balanced diet is a *varied* diet. By eating lots of different kinds of food, you are likely to get all the substances you need.

stipation and cancer of the colon (bowel) (see page 141).

Unrefined foods such as wholemeal bread, bran cereals and fresh fruit and vegetables contain plenty of fibre. This is one reason why such foods are good for us.

Fats

Fats occur in both animal and plant foods. Butter, dripping and lard are animal fats obtained from cattle and pigs. These fats are solid at room temperature, though if you heat them they become liquid.

Plant fats, on the other hand, are normally liquids at room temperature – we call them **oils**. Two well known examples are corn oil and olive oil. Both are used in cooking.

Margarine, unlike butter, consists mainly of plant oils. These are obtained from peanuts, soya beans and so on, and are then turned into solid fat by chemical treatment.

Fats give us energy, so – like carbohydrates – they are energy foods. In humans and other mammals, fat is stored under the skin; this helps to keep the body warm, as well as serving as an energy store.

There are many kinds of fat. Each contains particular **fatty acids** (see page 22). Now some fatty acids are **saturated**, whereas others are **unsaturated**. (An unsaturated fatty acid is one which contains less than the maximum amount of hydrogen possible.) Fatty acids which are very unsaturated are called **polyunsaturates**.

In general animal fats contain a high proportion of saturated fatty acids, whereas plant oils contain a high proportion of polyunsaturated fatty acids. For good health we should eat mainly polyunsaturates. The reason is that polyunsaturates reduce the amount of another substance in the body: **cholesterol**.

Picture 3 Table margarines usually contain a high proportion of polyunsaturated fat.

Picture 4 Doctors recommend us to eat about 70 grams of protein each day for a healthy life. This amount of protein is present in the piece of meat and glass of milk in this picture.

What is cholesterol?

Cholesterol is a fat-like substance which the body needs for a number of purposes. So basically cholesterol is a useful substance. We get it from various foods, particularly egg yolk. The trouble is that if we take in too much, it can cause **heart disease** (see page 154).

Now scientists have found that saturated fats raise the concentration of cholesterol in the body, whereas polyunsaturates lower it. So for good health it makes sense to eat foods that are high in polyunsaturates and low in saturates. Most plant foods are of this kind. Table margarines and spreads with a high polyunsaturate content are generally better for us than butter (picture 3).

The harmful effects of saturated fats and cholesterol start early in life. So children should have a diet with not too much animal fat.

Proteins

A certain amount of protein is present in most foods, but it is particularly plentiful in milk, eggs and meat. In milk and eggs the protein is in liquid form. In meat it consists of solid thread-like fibres – the animal's muscle.

Proteins form the main structures of the body like muscles and skin. So we need proteins for growth and body-building, and for repairing tissues. We also need proteins to make enzymes. In addition, we can get energy from proteins.

How much protein do we need each day? Picture 4 shows what doctors recommend. In practice most people could manage with a lot less than this. In rich countries people tend to eat much more protein than they need. On the other hand, in poor countries many people get very little. A growing child who does not get enough protein may develop **protein deficiency disease** (**kwashiorkor**) (picture 5).

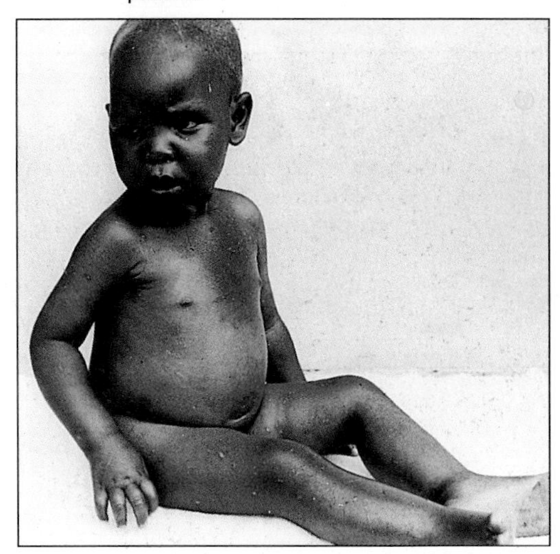

Picture 5 This child is suffering from lack of protein in the diet. The child is weak and listless. He looks fat because fluid has collected in the tissues. This is one of the commonest types of malnutrition in poor countries. It is often seen in babies who are being breast-fed by an undernourished mother, particularly if she is carrying another child.

Getting the right kinds of proteins

Proteins are composed of amino acids (see page 22). We can make some of these for ourselves, so we do not need them in our food. Others we cannot

Table 1 In this table, each food is given marks out of ten, depending on how good it is at giving us all the amino acids we need. A high mark means that the protein in the food contains all the essential amino acids in the right proportions for humans. A low mark means that it is short of certain essential amino acids.

Type of protein	Marks out of 10
Mother's milk	10
Eggs	10
Fish	8
Meat	8
Cow's milk	7½
Potatoes	7
Liver (beef)	6½
Rice	5½
Soya beans	5½
Maize	5½
Wheat (white flour)	5
Peas	4½
Beans	4½

Picture 6 Drought like this may lead to famine.

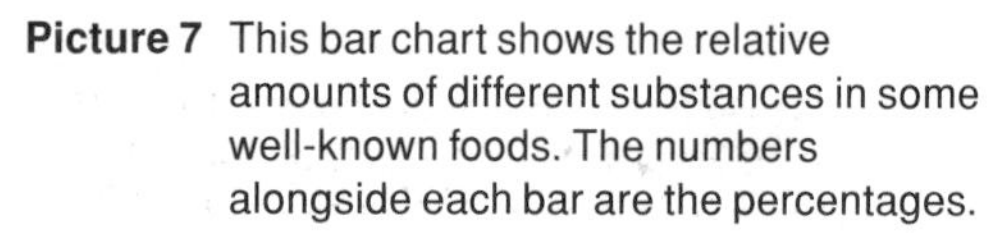

Picture 7 This bar chart shows the relative amounts of different substances in some well-known foods. The numbers alongside each bar are the percentages.

make, so they must be included in our diet. These **essential amino acids** are vital for good health, and absence of just one of them can have serious consequences.

Now look at table 1. This shows how good different proteins are at giving us the amino acids that we need. As you can see, animal proteins come out on top. Plant proteins come lower down. This does not mean that plant proteins are no use. They are, provided that a mixture of different ones is eaten. Different plant proteins contain different essential amino acids, so eating the right mixture of plants can give you all the amino acids you need. By doing this you are also avoiding animal fats which may be harmful. Wheat protein and bean protein provide a good balance of amino acids, so beans on toast can be a nutritious meal.

Notice that one of the best plant proteins is soya bean protein. What's more, the total amount of protein in soya beans is greater than in most plants. Soya beans are therefore used for manufacturing artificial meat. It is called **textured vegetable protein**.

Artificial meat is also made from bacteria and fungi. The protein is textured and flavoured to make it taste and feel like chicken or ham (see page 93).

Water

Water is essential for life (see page 22). It must therefore be included in our diet. A person can go without food for several weeks, but would die in a few days from lack of water.

We take in water mainly by drinking. However, there is plenty of water in most solid foods. A lettuce or cabbage is 90 per cent water, and even bread contains about 40 per cent. Some animals get all their water from solid food and never drink, but humans normally need to drink about a litre of liquid every day.

Shortage of water in our environment (drought) is one of the main causes of famine. It kills livestock, and causes crops to fail.

Finding out what's in our food

Experiments can be done to find out how much of each substance is found in different foods (picture 7). This information is particularly useful to people who plan meals for schools and hospitals, because it tells us what each kind of food is useful for. For example, maize contains mainly carbohydrate and is therefore a good energy food. On the other hand, meat contains mainly protein, which makes it good for growth and body-building.

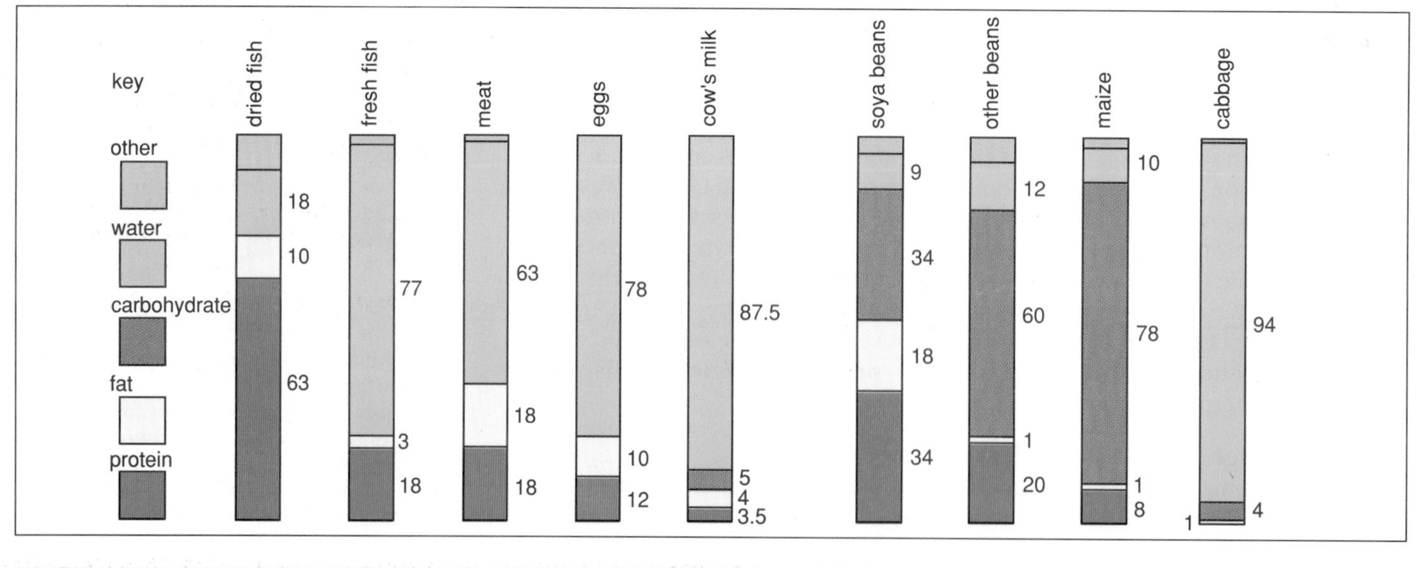

Activity

Finding out what substances are present in various foods

You can do this by carrying out the following tests. Try them on orange juice, banana, bread, milk, egg white, butter or margarine, a breakfast cereal, baby food and other foods of your own choice.

Sugar

1 If the food is not already in liquid form, mash it up with a pestle and mortar, and add a little water to make a suspension.

2 Pour about 2 cm^3 of the food into a test tube.

3 Add about 2 cm^3 of Benedict's solution to the test tube, and shake.

4 Wear eye protection. Boil some water in a beaker over a bunsen burner.

EYE PROTECTION MUST BE WORN

5 Put the test tube in the beaker of boiling water, and leave it there for a minute or two.

If a precipitate develops, sugar is present. (The precipitate is usually green or brown).

Starch

1 Obtain a small quantity of the food: it can be in liquid or solid form.

2 With a pipette add 2 or 3 drops of dilute iodine solution to the food.

If a blue-back colour develops, starch is present.

Fat

Simple test

1 Rub the food onto a piece of thin paper.

2 Hold the paper in front of a light, so light shines through it.

If the food has left a mark on the paper which lets the light through, fat is present.

More complicated test

1 Pour about 1 cm^3 of absolute ethanol into a test tube. Keep it well away from flames.

Absolute ethanol is highly flammable.

FLAMMABLE

2 Add a small amount of the food to the ethanol. (If the food is a liquid, just add one or two drops; if it is solid, cut it up into very small pieces first.)

3 Shake the test tube.

4 Add about 2 cm^3 of water to the test tube.

If the mixture turns cloudy white, fat is present.

Protein

Biuret test

1 If the food is not already in liquid form, mash it up with a pestle and mortar, and add a little water to make a suspension.

2 Pour about 2 cm^3 of the food into a test tube.

3 Wear eye protection. Add a little dilute sodium or potassium hydroxide solution till the mixture clears.

Sodium and potassium hydroxide are corrosive.

CORROSIVE

4 Add a few drops of dilute copper sulphate solution and shake.

If the solution goes purple, protein is present.

Present your results as a table, summarising which substances occur in each food.

If possible try to decide what each food is more suitable for, supplying energy or body-building.

Questions

1 Give four reasons why we need food.

2 Give one example of a food which contains:

a lactose,
b sucrose,
c cellulose,
d sunflower oil,
e liquid protein.

3 What effect does cooking have on each of the following:

a potatoes,
b eggs,
c cabbage,
d beef?

Explain your answers.

4 Explain each of the following statements.

a Children suffering from shortage of protein do not grow as quickly as they should.
b Eggs are better for body-building than bread.

5 Look at picture 7 then answer these questions.

a Which food contains most protein?
b Which food contains most carbohydrate?
c Which food contains most water?
d Which food would you recommend for a person who wants to avoid heart disease?
e Which plant food would be best for a vegetarian, and why?

6 Each of the foods in the left-hand list is closely related to one of the words in the right-hand list. Write them down in the correct pairs.

wholemeal bread	protein
sugar	insulation
butter	artificial meat
eggs	roughage
soya beans	energy

7 The amount of protein present in a particular food, and how good that protein is for body-building, are two quite different things.

Explain what this statement means.

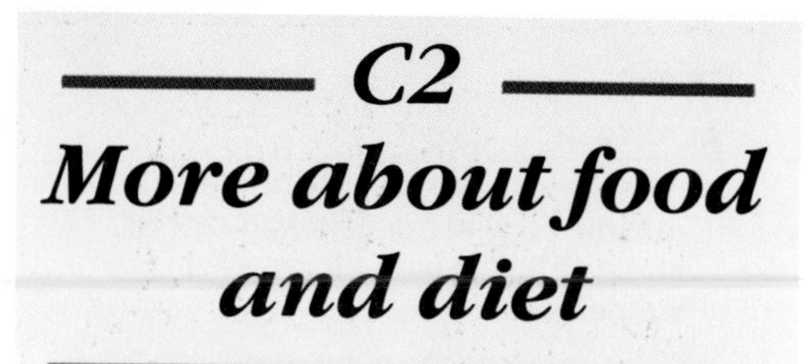

C2 More about food and diet

This topic is mainly about substances that we need in only small amounts.

Mineral salts

Mineral salts contain certain chemical elements. All these elements have particular jobs to do in the body. Here are some of the more important ones:

■ Sodium

We take in sodium when we eat salt, for common salt is sodium chloride. Salt is present in most foods, though of course some are saltier than others.

Our blood must contain the right amount of salt. It helps our nerves to transmit messages and our muscles to contract. If you run short of it, you get cramp. We lose salt when we sweat. Miners, and other people who work in hot places, eat salt tablets to make up for the salt they lose in sweating.

It is important not to eat too much salt. It can cause high blood pressure, and may be linked with heart disease.

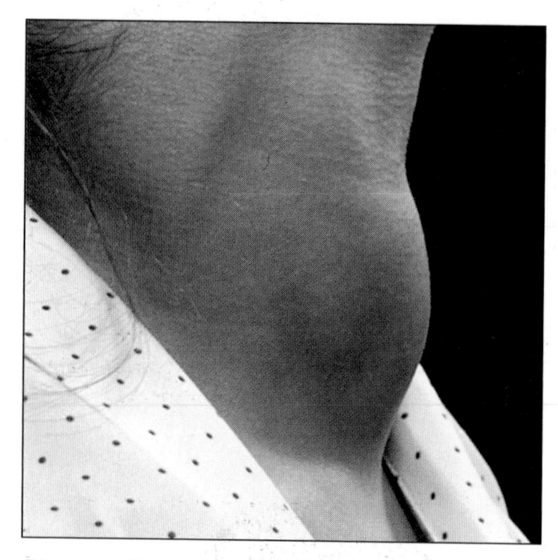

Picture 1 This person is suffering from goitre, caused by lack of iodine in the diet.

■ Calcium

When a baby is born, its bones are soft. For the bones to harden, they must take up calcium compounds. These compounds are calcium phosphate and calcium carbonate. The process is called **calcification**. A similar process makes the teeth hard.

Calcium occurs particularly in milk, cheese and fish. If a child does not get enough calcium, its bones remain soft and become deformed. This is called **rickets**. Calcium is also needed for making muscles contract, and it helps blood to clot when you cut yourself.

■ Iron

Iron is needed for our blood. It is present in haemoglobin, the red pigment which carries oxygen.

Iron is particularly plentiful in liver and kidneys. Small amounts occur in most drinking water, and we get quite a lot of it from metal utensils used in cooking. The amount of iron in a piece of beef can be doubled by mincing it in an iron mincer.

Shortage of iron results in the blood containing too little haemoglobin. This is a type of **anaemia**. The oxygen-carrying ability of the blood is reduced, resulting in tiredness and lack of energy. People who are anaemic are often recommended to take iron tablets.

Picture 2 Portrait of a man drawn by the Italian artist Leonardo da Vinci in the fifteenth century. Notice his swollen neck. He may have lived in an area where there was no iodine in the water.

■ Iodine

Some elements are needed in only tiny quantities. They are known as **trace elements** (**micronutrients**). One such element is **iodine**.

Iodine is present in most drinking water, and in sea foods. We need it for making a hormone called **thyroxine** (see page 201). This is produced by the **thyroid gland**, situated close to the 'Adam's apple' in the neck.

Thyroxine speeds up chemical reactions in the body, making us more active. If we do not get enough iodine, the thyroid gland cannot produce thyroxine. As a result, the gland enlarges, causing the neck to swell. This condition is called **goitre** (picture 1).

There are some places where the drinking water lacks iodine. One such place is Derbyshire in the middle of England. In the old days it was common for people in that area to have enlarged thyroid glands, so the condition was called 'Derbyshire neck'. Nowadays, iodine is added to salt, so the condition no longer occurs. Before reading on, look at picture 2 and see what you make of it.

Vitamins

In the early 1900s an English scientist, Frederick Gowland Hopkins, discovered something interesting. He fed some rats on a special food mixture containing purified carbohydrates, fats, proteins and minerals – all the substances known to be necessary for healthy life. After a few weeks the rats were dead.

However, a second group of rats was given exactly the same food mixture, plus a very small amount of milk. They flourished. Apparently the milk contained something extra which the rats needed. We now know that this extra 'something' was **vitamins** (picture 3).

Vitamins are organic substances needed in the diet. Each has a specific job to do. If any of them is missing from the diet, we become ill and may die. They are needed in only small amounts and are therefore called micronutrients.

Vitamins are known by letters: A, B, C, etc. This way of naming them was introduced before their chemical structure was known. It is still used, though we can now give them proper chemical names as well.

For vitamins to do their jobs they must be in solution. Some of them dissolve in water, others in fat. This is one of many reasons why we need water and fat in our diet.

Now let's look at some of the more important vitamins in detail.

Picture 3 The two rats at the top were fed on a full diet including vitamins. The two rats at the bottom were given a full diet minus vitamins.

■ Vitamin A (*fat soluble*)

Vitamin A (retinol) is important for our eyes. It protects their surface, and helps us to see in dim light.

The best source of this vitamin is fish liver oil. We can also get it by eating carrots. The orange substance in carrots (called **carotene**) is turned into vitamin A inside our bodies.

Shortage of vitamin A makes it hard to see in dim light. This is known as **night blindness**. Severe lack of it causes the cornea to become thick and dry, a condition called **xerophthalmia**. In extreme cases this can lead to total blindness.

■ The B vitamins (*water soluble*)

This group of vitamins helps our cells to transfer energy in respiration.

The first B vitamin to be discovered was **nicotinic acid**, or **niacin** for short. There is plenty of it in liver, meat and fish. Lack of it results in a disease called **pellagra** (picture 4). How the vitamin was discovered is described on page 117.

One of the most important B vitamins is **vitamin B1 (thiamine)**. It occurs in yeast and cereals. Lack of it causes a disease called **beri-beri**. This is an African word meaning 'I cannot'. The muscles become very weak, and in the end the person becomes paralysed and dies.

There is a lot of vitamin B1 in rice; it is in the husk, the tough coat surrounding the grain. When rice is prepared, the husk is usually stripped off and the grain is polished. This removes the vitamin. Beri-beri is therefore common in parts of the world where people live on polished rice.

Another important B vitamin is **vitamin B2 (riboflavin)**. It is found particularly in leafy vegetables, eggs and fish. Lack of it causes sores round the mouth, and poor growth.

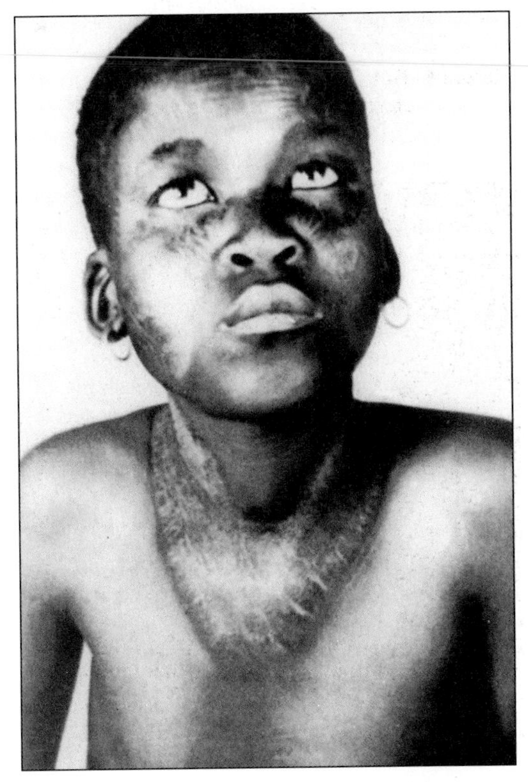

Picture 4 This boy is suffering from pellagra, caused by lack of the vitamin niacin (nicotinic acid). Notice the marks on his neck, shaped like a necklace. This is a characteristic feature of the disease.

■ Vitamin C (*water soluble*)

In the 1740s Admiral Anson led a fleet into the Pacific to fight the Spanish. During the voyage, 626 of his 961 men died of a disease called **scurvy**. In this disease, bleeding occurs in various parts of the body, particularly the gums (picture 5 on next page).

Scurvy is caused by lack of **vitamin C (ascorbic acid)**. This vitamin keeps the delicate lining of the mouth cavity and other body surfaces in a healthy state. It is abundant in green vegetables such as spinach, and citrus fruits such as lemons and limes. If people eat this kind of food, they will not get scurvy. In the 1800s British naval ships always carried a supply of lemons or limes to combat scurvy. This is why British sailors were called 'limeys'.

One snag about vitamin C is that it is destroyed by heating. As a result, a lot of it can be lost during cooking, and while the food is being kept hot afterwards. In restaurants and canteens, where food is kept hot for a long time, over ninety per cent of the vitamin may be lost.

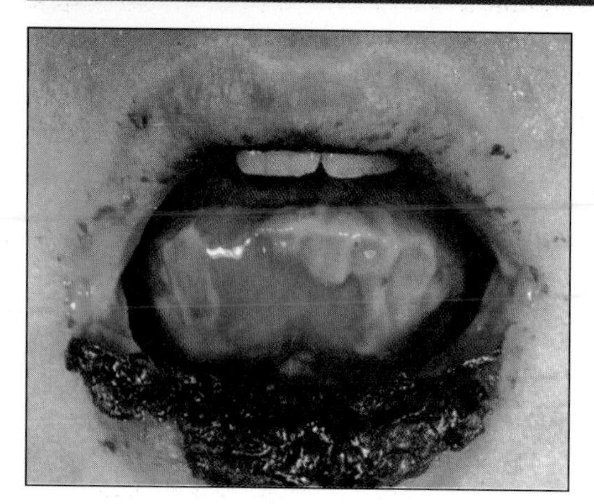

Picture 5 Lack of vitamin C causes scurvy, which is characterised by bleeding gums. Vitamin C helps our cells to stick together. When we don't get enough of it, the cells come apart.

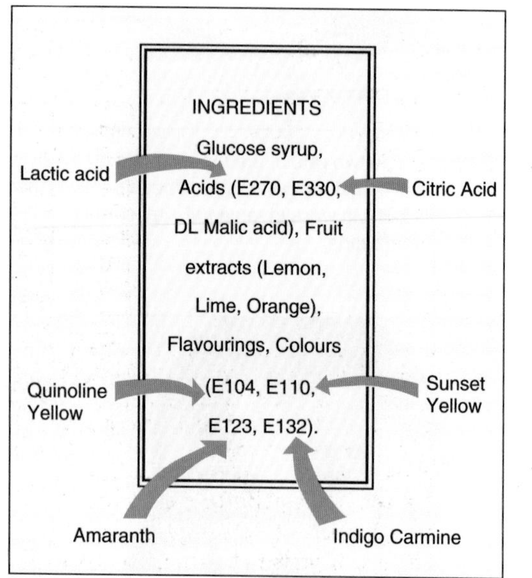

Picture 6 The additives in a packet of boiled sweets. Lactic acid and citric acid are both naturally occurring substances which are produced commercially by fermentation with microbes. They help to preserve food. In addition, citric acid enhances flavour and provides Vitamin C which is an anti-oxidant. Quinoline Yellow, Sunset Yellow, Amaranth and Indigo Carmine are all synthetic dyes which colour food in various ways. Certain people are allergic to these dyes, and some dyes may trigger hyperactivity in children.

■ **Vitamin D** (*fat soluble*)
We saw earlier that as a child grows, its bones become hard by taking up calcium salts. For this to happen **vitamin D** (**calciferol**) is needed. A child who does not get enough vitamin D will develop rickets, the same disease that occurs if there is not enough calcium in the diet.

Vitamin D occurs in fish liver oil. A certain amount of vitamin D can be made by the body itself. It is made in the skin provided sunlight is present. In a sunny climate an adult can get all the necessary vitamin D this way.

Vitamin deficiency in the modern world

Years ago vitamin deficiency diseases were very common, even in countries like Britain. Nowadays they are common only in poorer parts of the world. But even in rich countries they occur from time to time. Those at risk include pregnant women, old people living alone, and people who prefer not to eat certain kinds of food. It is a good idea for such people to take **vitamin tablets**.

Can you have too much of a vitamin? The answer is yes in some cases. However, it is unlikely to result from our food. It is more likely to result from taking too many vitamin tablets. For example, large doses of vitamin D can cause tissues other than the bones to become calcified – the liver for instance.

Vegetarians

Vegetarians eat plant food, but not meat. Their diet includes animal products such as milk and eggs, but not the animals themselves. Strict vegetarians, known as **vegans**, don't even eat animal products.

A vegetarian diet that includes dairy products can provide everything needed for a healthy life. Plants can provide lots of fibre, vitamins and polyunsaturated fat, so a vegetarian diet is good in that respect.

Vegans have to make sure that their food gives them all the substances they need. In general their diet needs to be bulky and varied. This will ensure that they get enough carbohydrate and protein, together with the full range of vitamins and essential amino acids.

Food additives

Nowadays all sorts of food additives, natural and synthetic, are added to food by the manufacturers. Some sweeten, flavour or colour the food. Others (**preservatives**) increase its shelf life by checking the growth of microbes. **Emulsifiers** and **stabilisers** give the food the right consistency, perhaps thickening it or making it into a jelly. **Anti-oxidants** prevent oxidation of substances in the food when exposed to air; oxidation makes fats and oils go rancid, and fruit such as apples and bananas turn brown when cut.

Some people are allergic to certain additives, and in rare cases children may become over-active because of them. Such additives include the colourings tartrazine and sunset yellow. Another additive which may cause problems is monosodium glutamate. This occurs naturally in a Japanese seaweed, but is made commercially from sugar beet and wheat gluten. It is used to flavour meat products. It gives some people headaches and nausea. More serious are artificial sweeteners called cyclamates. They can damage our chromosomes, and have now been banned.

Food additives have had bad publicity, and more and more foods are appearing in the shops that are additive-free. However, it is only fair to point out that some people are allergic to *natural* foods too (see page 161). If the public wants foods that look and taste nice, and can be kept for a reasonable time, then additives have to be used. It is the manufacturers' responsibility to carry out all the necessary safety tests beforehand.

All packaged foods and drinks in European Union countries are required to display a full list of additives, either by name or 'E number'.This enables shoppers to know exactly what they are buying (picture 6).

The man who ate faeces

Some advances in science have been made by heroic people who have done experiments on themselves. Such was the case with Joseph Goldberger.

Goldberger was an American doctor. In 1916 he came across some convicts who had a strange disease. They had skin rashes, swollen tongues, upset stomachs and headaches. Some of them were mentally ill.

Now this disease might have been caused by germs, but Goldberger suspected it was due to their poor diet.

He tested this idea on himself. He took some blood from one of the convicts, and injected it into his arm. He ate the powdered skin rash of one of the others, and swallowed some faeces from another.

Goldberger reasoned that if the disease was caused by a germ, he should catch it. On the other hand, if it was caused by poor diet, he should remain healthy. Goldberger did not get the disease. He tried the experiment on his wife, and she didn't get the disease either.

We now know that the convicts were suffering from lack of the vitamin niacin. There is plenty of this vitamin in liver and fish, just the kind of food which the unfortunate convicts were not getting. The disease resulting from its absence is called pellagra.

1 Does the fact the Goldberger did not get the disease *prove* that it was caused by a poor diet? Explain your answer.

2 Suggest one way in which Goldberger might have confirmed his conclusions?

3 Was Goldberger heroic or daft? Opinions differ. What do you think?

Re-enacting a scientific discovery

This is a group activity, requiring at least five people. Read how Goldberger discovered the cause of pellagra. Plan, write and perform a short play entitled 'How Goldberger discovered the cause of pellagra'. You may use your imagination as much as you like in writing the dialogue, but don't alter the facts of the story.

Questions

1 Each of the diseases in the left-hand column is caused by lack of one or more of the substances in the right-hand column. Which causes which?

night blindness	iron
rickets	vitamin A
anaemia	calcium
goitre	vitamin D
xerophthalmia	iodine

2 Explain the reasons for each of the following statements.

a Carrots are good for you.
b A person who has been sun-bathing all day eats a salt tablet.
c Old people who live alone tend to get scurvy towards the end of the winter.

3 This question is about the diet of a growing child.

a Suggest *four* particularly important nutrients which should be included in the diet. In each case give an example of a food which contains a lot of the nutrient, and explain why the nutrient is so important.
b Suggest *two* nutrients which should be left out of the diet. In each case give an example of a food which contains a lot of the nutrient, and explain why it should be avoided.

4 Describe an experiment which could be done to find out if the husk surrounding the rice seed contains a substance which prevents beri-beri.

5 Scientists carried out an experiment to find the effect of cooking a finely shredded cabbage on the amount of vitamin C in it. They put the cabbage in boiling water and boiled it for 10 minutes. They found the vitamin C content at intervals, expressing it as a percentage of the amount in the uncooked cabbage.

Here are the results.

Time after putting the cabbage in the water	Vitamin C content
0 min	100%
1/2 min	66%
1 min	55%
4 min	49%
7 min	43%
10 min	37%

a Plot these results on graph paper.
b Suggest reasons why the vitamin C content of the cabbage falls.
c What experiments could you do to test your suggestions?
d What advice would you give to a chef about cooking vegetables?

6 Make a list of all the advantages of being a vegetarian that you can think of. Include advantages to society as well as to the individual. What precautions should you take if you are a vegan, and why?

Activity

Testing food for vitamin C

1 Obtain a lemon, and squeeze some of its juice into a beaker.

2 Pipette one drop of blue DCPIP solution onto a white tile. (DCPIP is short for 2,6-dichlorophenol indophenol.)

3 With a pipette or syringe add lemon juice to the DCPIP solution, drop by drop, and stir with a needle.

Count how many drops of lemon are needed to make the DCPIP solution turn colourless.

The disappearance of the blue colour tells us that vitamin C is present in the lemon juice.

4 Use this test to compare the vitamin C content of different foods. In each case get some juice out of the food. Then find out how many drops of the juice are needed to decolourise one drop of DCPIP solution.

Do you think this is an accurate way of comparing the vitamin C content of different foods?

How could you make the experiment more accurate?

Why can't this test be done with blackcurrant juice?

5 Boil some lemon juice in a test tube. *Wear eye protection*. Then test it for vitamin C with DCPIP solution.

EYE PROTECTION MUST BE WORN

What effect does boiling have on vitamin C?

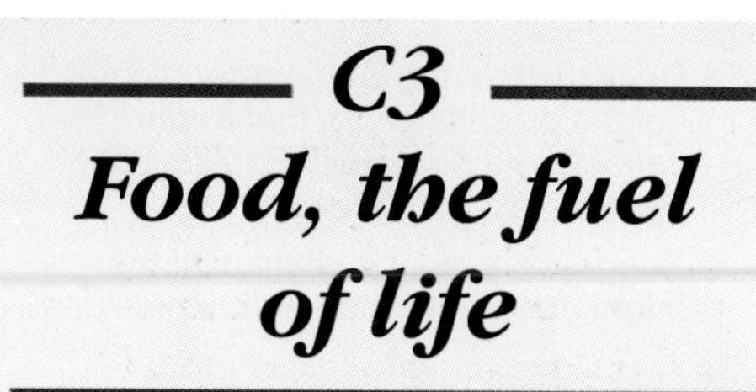

C3 Food, the fuel of life

We need energy for almost everything we do. We get it from our food which serves as a fuel.

Table 1 How much energy is there in various everyday foods? You can find out by looking at this list.

	kJ per gram
margarine	32.2
butter	31.2
peanuts/groundnuts	24.5
chocolate (milk)	24.2
cake (plain)	18.0
sugar (white)	16.5
sausages (pork)	15.5
cornflakes	15.3
rice	15.0
bread (white)	10.6
chips	9.9
chicken (roast)	7.7
eggs (fresh)	6.6
potatoes (boiled)	3.3
milk	2.7
beer (bottled)	1.2
cabbage (boiled)	0.34

Table 2 Approximate amounts of energy required daily by different types of people.

	kJ per day
Newborn baby	2 000
child 1 year	3 000
Child 2–3	6 000
Child 5–7	7 500
Girl 12–15	9 500
Boy 12–15	12 000
Office worker	11 000
Factory worker	12 500
Heavy manual worker	15 000
Pregnant woman	10 000
Woman breast-feeding	11 000

Food as a fuel

We can show that food is a fuel by burning it. When the food is burned, energy is transferred to the surroundings, warming them.

Now for practical details. We weigh a sample of food. Then we put the sample of food under a measured quantity of water. We set fire to the food and let it heat up the water. The rise in temperature of the water tells us how much energy has been transferred from the food. To do this accurately we use an apparatus called a **food calorimeter** (picture 1). In activity A you can use a simple but less accurate method based on the same principle.

The energy is measured in a unit called the **kilojoule (kJ)**. The unit of energy used to be the Calorie. This is still used sometimes, and you hear people talking about losing calories or eating low-calorie foods. However, this unit has now been officially replaced by the kilojoule.

The three main substances found in food are carbohydrate, fat and protein. We can estimate the amount of energy obtainable from each of these substances. We can then compare their energy values. Here they are:

- Carbohydrate: 1 gram gives 17 kJ
- Fat: 1 gram gives 39 kJ
- Protein: 1 gram gives 18 kJ

Notice that fat gives about twice as much energy as either carbohydrate or protein.

How much energy is given by different foods?

Table 1 tells you how much energy is given by some everyday foods. The amount of energy given by a particular food depends on the substances it contains. For example, margarine and butter consist almost entirely of fat, so they give a lot of energy. On the other hand, a cabbage is ninety per cent water, so it gives very little energy.

Another thing that determines how much energy a particular food gives is how it is cooked. For example, potatoes fried in fat give three times as much energy as potatoes boiled in water.

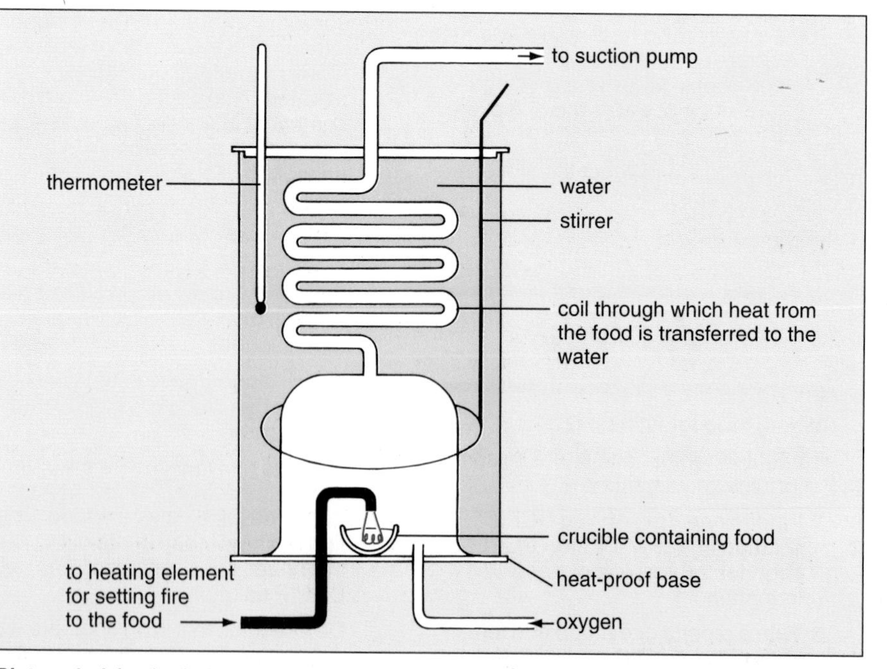

Picture 1 A food calorimeter. It is used to find how much energy there is in a sample of food.

How much energy do we need?

Imagine you are lying in bed doing nothing. Even in this inactive state you need energy to breathe, make your heart beat, and drive all those countless chemical reactions which keep you alive. The rate at which these 'ticking over' processes take place is called your **basal metabolic rate**.

How much energy do you need to maintain your basal metabolic rate? It is difficult to say, because it varies from one individual to another. Very roughly, the amount needed is 7000 kJ per day. This is about the same amount of energy needed to boil water for 100 cups of tea.

Few of us spend our days lying in bed – most of us do something. Table 2 tells you roughly how much energy is needed by different people in the course of a normal day. You will see that it depends on the person's age, sex and occupation. A person who spends most of the time sitting in an office needs far less energy than a very active person.

Picture 2 A person takes in energy from food, and gives out energy in muscular exercise, etc. Excess energy is stored in the body as fat.

What happens if we eat too much?

Suppose you eat more food than is needed for supplying you with energy. What happens to the food left over? Most of it is turned into fat and stored. The result is that your body weight increases. (Strictly speaking we should call this the body *mass*. However, the word 'weight' is normally used in this context so we shall use it here.) If your body weight increases, you run the risk of becoming fat – or **obese**, to use the proper word.

We can think of a person as taking in energy (**energy intake**) and giving out energy (**energy output**) (picture 2). Obesity is caused by a person's energy intake being greater than the energy output. The most fattening foods are therefore those that provide most energy, such as cakes, sweets and butter.

In our bodies, fat is stored in **fat cells** under the skin (see page 186). An overweight person has too much fat in these cells, and the total number of fat cells is too high.

For everyone there is a 'correct' weight. This will depend on the person's sex, age and height.

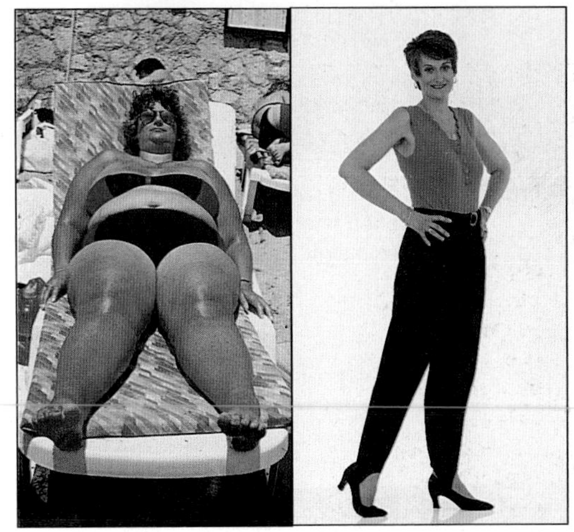

Picture 3 Going on a diet can be a good way of losing weight. This lady reduced her weight dramatically in a few months.

How can we lose weight?

The only way to lose weight is to make our energy intake less than the output. This can be done in two ways:

- By taking more exercise: this will increase the energy output.
- By eating less energy food; this will decrease the energy intake.

The first method is certainly helpful, but it is no use taking exercise if you don't keep a check on your diet as well. For example, yesterday I played tennis for half an hour. During the game, I lost about 700 kJ of energy. Afterwards, I felt thirsty and had a glass of beer. The result was that I put back all the energy I'd just lost!

The second method is very good if carried out properly. A person on a well planned weight-reducing diet can lose about 1 kg of body mass per week. Such diets contain relatively little high-energy food and a lot of low-energy food. The result of going on a diet of this sort is shown in picture 3.

The best results can be obtained by combining both methods, i.e. by going on a diet *and* taking more exercise. Regular sessions of steady exercise are better than occasional bouts of very strenuous exercise. The ultimate aim should be to achieve a balance between energy intake and energy output. Then your weight should stay more or less constant.

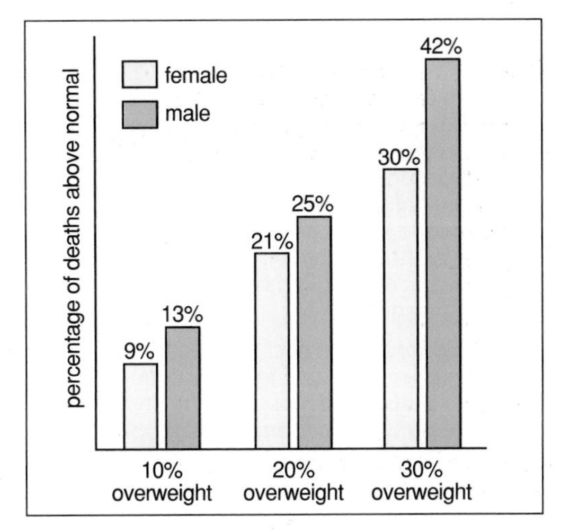

Picture 4 This bar chart shows the relationship between people's body weight and the death rate in the United States.

Why lose weight?

Look at the bar chart in picture 4. It shows the relationship between people's body weight and the death rate. We may draw this simple conclusion from the

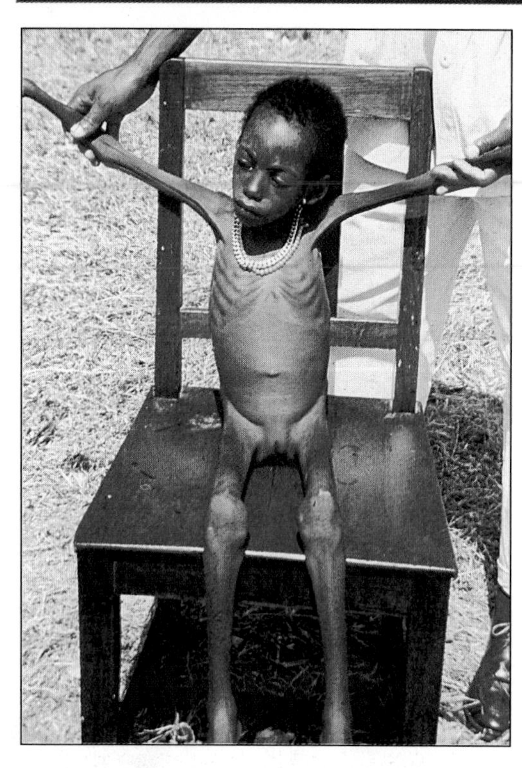

Picture 5 This child is suffering from general starvation (marasmus). Her thinness is caused mainly by lack of energy food.

chart: there are more deaths amongst people who are overweight than amongst people of normal weight. In other words, overweight people don't live as long, on average, as people who are the normal weight.

An overweight person has a greater chance of having a stroke or heart attack (see page 154). Other illnesses, too, are connected with being overweight. The risk of death is greater for men than for women, and it increases with the amount of excess weight.

What happens if we eat too little?

Suppose you eat no energy foods at all. What happens? At first you obtain energy from your fat stores. As a result you lose weight.

Eventually all your fat gets used up. In order to stay alive, the body starts getting energy from your tissue proteins, particularly your muscles. As a result, you waste away, becoming thin and weak. Death will occur after about two months. This has happened in concentration camps, and in famine areas.

Starvation

There are many countries in the world where people don't get enough to eat. They may not die from lack of food, but they become thin and weak and find it difficult to work. The wasting of the body from general starvation is called **marasmus** (picture 5).

A person who is starving may be short of all sorts of other nutrients besides energy food, and the effects may be complex. For example, a person who is not getting enough protein or vitamins may be just as lacking in energy as a person who is not getting enough energy food.

Where are people starving?

An adult person needs at least 9200 kJ of energy to get through the day. Anyone receiving less than this can be said to be starving.

Now look at picture 6. You can see that in the developing countries of Central and South America, Africa and the Far East, many people get less than this minimum amount of energy. On the other hand, in the developed countries, people get plenty of energy food – often far more than is good for them.

With modern methods of farming, countries like Britain produce more food than can be used. The surplus results in 'grain mountains' and 'butter mountains'. The tragedy is that this surplus doesn't get to the people who need it.

Anorexia and bulimia

Some people suffer from a psychological condition in which they eat very little. This is called **anorexia nervosa**. It sometimes happens to young people, particularly women, and it may stem from an obsessive fear of becoming fat. Such people often become thin and frail, and if nothing is done to remedy the condition, they may die.

The opposite also occurs. A person may go on eating and eating and eating. This is called **bulimia nervosa**. Such people may deliberately make themselves sick.

Anorexia and bulimia are both distressing conditions which may require the expert help of a doctor or psychiatrist.

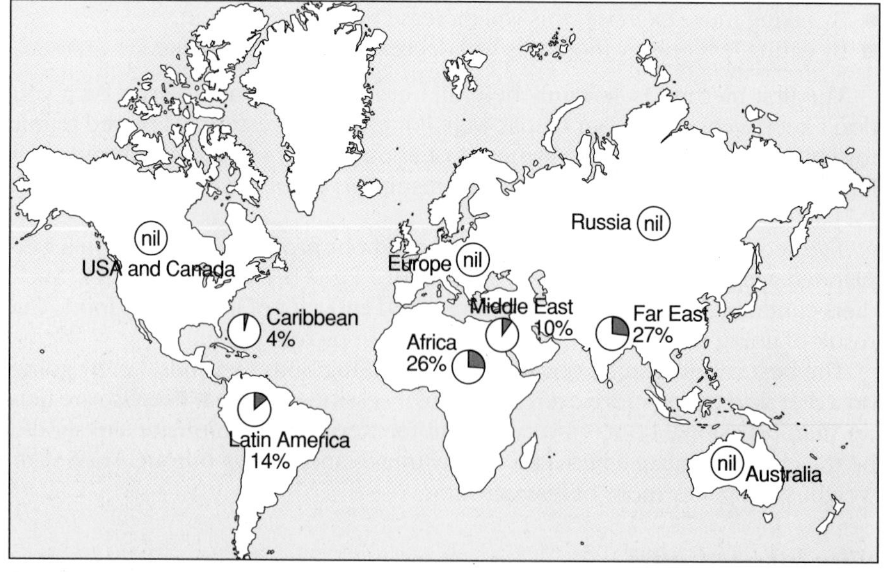

Picture 6 This picture shows the main parts of the world where people are starving. The red area in each pie chart shows the proportion of the adult population who receive less than 9200 kilojoules of energy per day from their food.

Activities

A A simple way to find out how much energy a piece of food contains

1 Measure out 20 cm^3 of water with a measuring cylinder, then transfer it to a large test tube.

2 Clamp the test tube to a stand, and put a thermometer in it, as shown in the illustration.

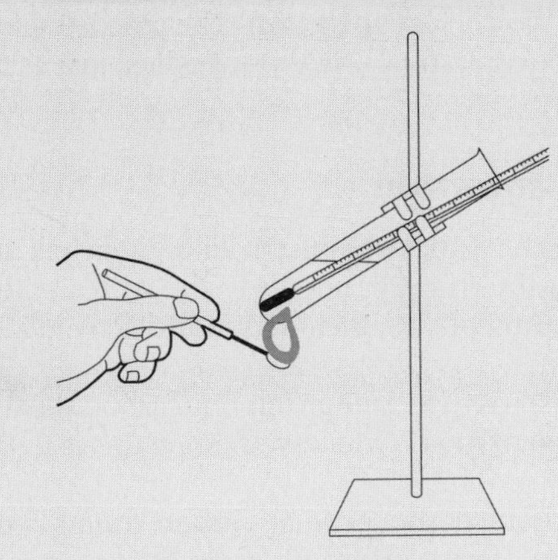

3 Record the temperature of the water.

4 Weigh a peanut (groundnut), then stick it onto the pointed end of a mounted needle. (Twist the needle – this avoids cracking the nut.)

5 *Wear eye protection.* Hold the nut in a bunsen flame until it starts to burn, then place it under the test tube as shown in the illustration.

EYE PROTECTION MUST BE WORN

6 When the nut stops burning, record the temperature of the water again.

By how much has the temperature of the water risen?

Energy released from the nut in joules (J) = mass of water in grams × rise in temperature in °C × 4.2.

Work out the amount of energy in joules released from the nut.

Convert the joules into kilojoules (kJ) by dividing by 1000.

Knowing the mass of the nut, work out how much energy in kilojoules is obtainable from one gram of peanut.

Compare your figure with others in your class and find the average.

How does the class average compare with the figure given in table 1?

Do you think this is an accurate way of finding out how much energy there is in a piece of food?

B Investigating the energy provided by different foods

1 Your teacher will give you a tin can, crucible, glass rod, thermometer, two tripods (one larger than the other), wire gauze and insulating material. Use them to make a simple calorimeter. Your calorimeter should be based on the one in picture 1, but simpler.

2 Use your calorimeter to find out how much energy different foods contain. First, plan your experiment, writing down any safety precautions which should be taken. Show your plan to your teacher. When your teacher has approved your plan, carry it out.

3 Write an account of your experiment, and give your results. Mention any difficulties you came up against. What are the sources of inaccuracy in your method. How might it be improved?

Questions

1 What mass of roasted peanuts (groundnuts) would the heavy manual workers referred to in table 2 have to eat in a day to just satisfy their energy needs?

2 The data summarised in picture 4 were compiled by an American life insurance company.

a Explain in full how you think the data were obtained.
b Why should a life insurance company want to compile such data?

3 Give examples of the sort of food you would recommend to

a someone who is going on a hiking holiday and
b someone who wishes to lose weight.

4 Study picture 1 carefully, then answer the following questions.

a What is the oxygen for?
b Why is it necessary to stir the water?
c The top, sides and bottom of the calorimeter should be insulated. Why?
d What does the suction pump achieve?

5 The following table shows the approximate amounts of energy used up in different activities by a normal person.

Activity	Energy used
sleeping	4.5 kJ per min
sitting	5.9 kJ per min
standing	7.1 kJ per min
washing/dressing	14.7 kJ per min
walking slowly	12.6 kJ per min
walking fairly fast	21.0 kJ per min
walking up stairs	37.8 kJ per min
carpentry	15.5 kJ per min
playing tennis	26.0 kJ per min
playing football	36.5 kJ per min
running	42.0 k J per min

a From these figures work out the approximate total amount of energy which you yourself use up in 24 hours. Show your working in full.
b Using table 1 draw up a menu for breakfast, lunch and supper which would give you just enough energy to satisfy your need. Give the amount of each food item which you would need.
c A person who ate the food listed in your menu might still be getting an inadequate diet. Why?

6 The table below shows the daily energy requirements of people of different ages.

Age (years)	Energy requirements (kJ per day)
1	3 000
2	6 000
6	7 500
12	10 000
15	12 000
18	13 000

a Plot these figures as a graph.
b How would you explain the shape of the graph?
c What assumptions are made in drawing up figures of this sort?
d What can you predict about a person's energy requirements after the age of 18?

7 If all the food produced in the world was evenly distributed, no one would starve. Discuss why this has never happened. Is there any sign that it might happen in the future?

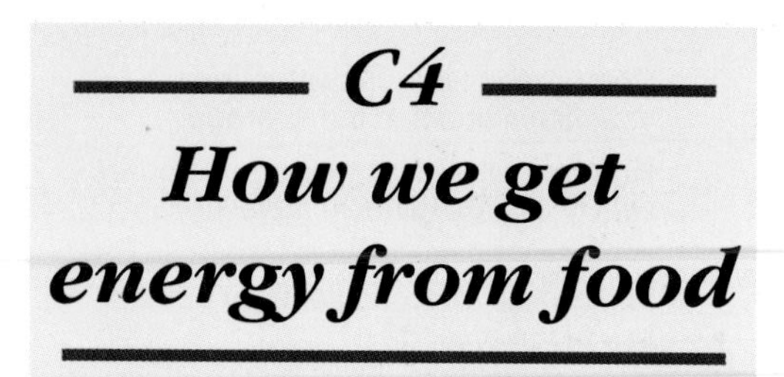

C4
How we get energy from food

How is energy transferred from a sprinter's food to her muscles?

What happens to the food?

First, think what happens when a fuel like petrol is burned. Oxygen is used up and carbon dioxide and water are produced. At the same time, energy is transferred to the surroundings, warming them.

The same kind of thing happens in our bodies. Substances, derived from our food, are oxidised to give carbon dioxide and water, and energy is transferred.

The main substance oxidised is glucose. We can summarise what happens as an equation:

$$\text{glucose} + \text{oxygen} \rightarrow \text{carbon dioxide} + \text{water} + \text{energy}$$

$$C_6H_{12}O_6 \quad 6O_2 \quad 6CO_2 \quad 6H_2O$$

This process takes place in our cells. We call it **respiration**. The glucose serves as a fuel. The oxidation of glucose in respiration drives our bodies, just as the burning of petrol drives a car.

Confirming the respiration equation

Various experiments can be done to show that organisms take in oxygen and give out carbon dioxide, and also that they transfer energy to the surroundings by warming them. Some of the experiments are included in the activities on page 124.

But how can we show that taking in oxygen and giving out carbon dioxide are connected with the breaking down of *food*? One way of doing this is to use **radioactive** tracers.

Scientists have made glucose in which the normal carbon atoms were replaced with the radioactive isotope of carbon. In other words, the carbon atoms in the glucose were 'labelled'. They then fed this labelled glucose to a mouse, and traced what happened to the radioactive carbon by means of a **Geiger counter**. The experiment is outlined in picture 2. The results show that the carbon in the carbon dioxide which the animal breathes out comes from the glucose which it has eaten.

Picture 1 A sprinter in action. How do his muscles get the energy they need?

What is energy needed for?

Here are the main things that organisms need energy for.

- All organisms need energy for growth, cell division, transporting chemicals, and just staying alive.
- Animals need energy for making muscles contract, sending messages along nerves and keeping warm.
- Plants need energy for taking up mineral salts from the soil and opening and closing their air pores (stomata).

Measuring the rate of respiration

You can measure an organism's rate of respiration by estimating how quickly it takes up oxygen. The apparatus is called a **respirometer**. You can use a respirometer in activity C.

One of the things you can do with a respirometer is to measure the rate of respiration at different temperatures. If you do this, you find that respiration speeds up as the temperature rises. In fact, a rise of 10°C roughly doubles the rate of respiration. This is true of chemical reactions in general, and it suggests that respiration is basically an ordinary chemical process.

If the temperature rises much above 40°C, respiration slows down and soon stops altogether. This is the temperature at which **enzymes** are destroyed (see page 24). It suggests that respiration is a chemical process which is catalysed by enzymes.

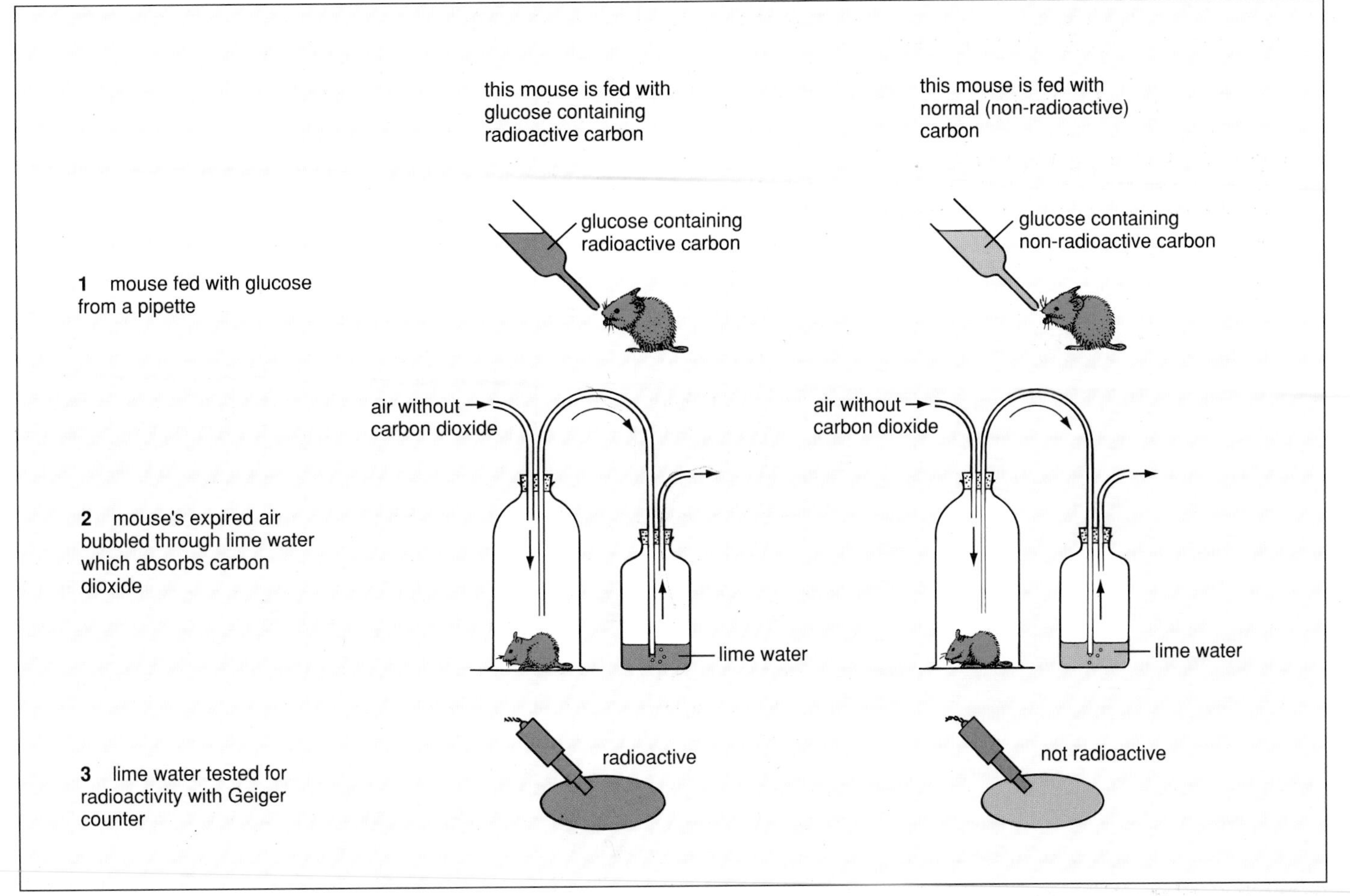

Picture 2 This experiment shows that the carbon dioxide which an animal breathes out comes from its food.

The chemistry of respiration

Scientists have shown that glucose is not oxidised in a single chemical reaction, as the equation on page 122 suggests. Instead, it is broken down in a series of small steps, each catalysed by a particular enzyme. The energy is transferred, not all at once, but bit by bit.

Why is this important? Think of it this way. A slice of apple pie contains as much energy as a stick of dynamite. If the energy was released in one go, as when dynamite explodes, your temperature would shoot up by at least 10°C and you would die.

ATP, the essential link

The energy released when glucose is oxidised is not used directly. It is linked to activities such as muscle contraction by another substance called **ATP**. ATP stands for **adenosine triphosphate**. When glucose is oxidised, the energy is transferred to ATP. When a muscle contracts, energy is transferred from the ATP to the muscle. This is illustrated in picture 3 which shows ATP as a kind of 'energy carrier', taking the energy from the glucose to the muscle.

If you put a drop of glucose solution on a muscle fibre, nothing happens. But if you put a drop of ATP solution on the muscle fibre, it contracts. This shows that glucose, by itself, cannot provide the energy needed for muscle contraction. It is the ATP, made as a result of oxidising glucose, that provides the energy.

ATP supplies energy in all living organisms. It is made in the **mitochondria** inside the cells (see page 13). This is why mitochondria have been described as the 'power-house of the cell'.

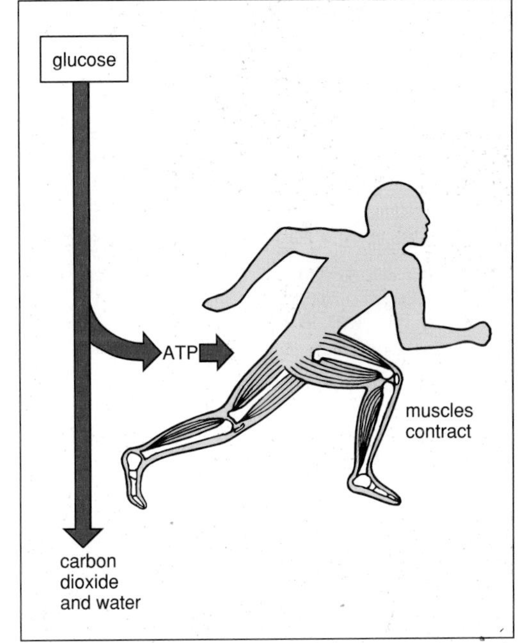

Picture 3 ATP (adenosine triphosphate) transfers energy from glucose to the muscles, enabling them to contract.

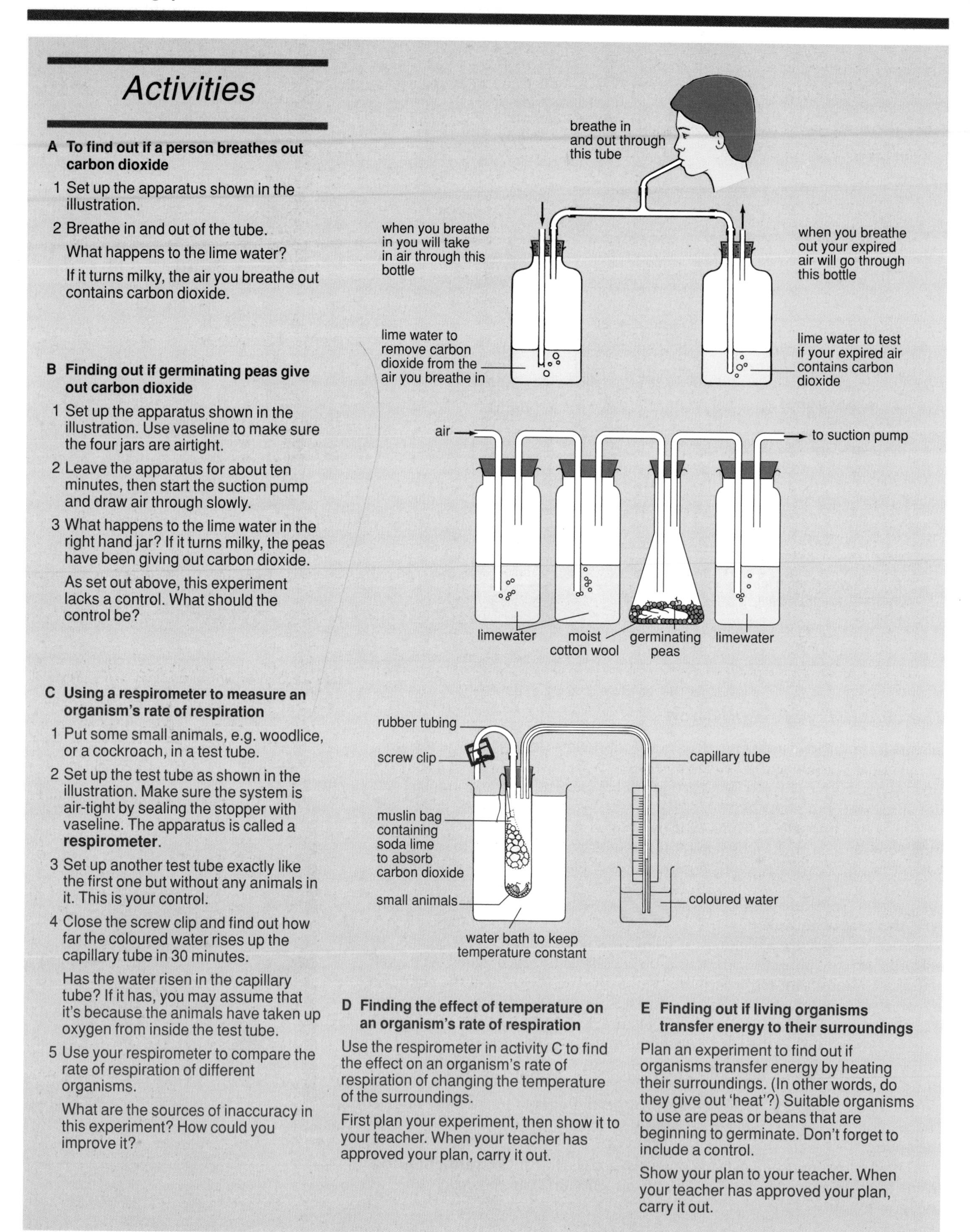

Activities

A To find out if a person breathes out carbon dioxide

1. Set up the apparatus shown in the illustration.
2. Breathe in and out of the tube.

   What happens to the lime water?

   If it turns milky, the air you breathe out contains carbon dioxide.

B Finding out if germinating peas give out carbon dioxide

1. Set up the apparatus shown in the illustration. Use vaseline to make sure the four jars are airtight.
2. Leave the apparatus for about ten minutes, then start the suction pump and draw air through slowly.
3. What happens to the lime water in the right hand jar? If it turns milky, the peas have been giving out carbon dioxide.

   As set out above, this experiment lacks a control. What should the control be?

C Using a respirometer to measure an organism's rate of respiration

1. Put some small animals, e.g. woodlice, or a cockroach, in a test tube.
2. Set up the test tube as shown in the illustration. Make sure the system is air-tight by sealing the stopper with vaseline. The apparatus is called a **respirometer**.
3. Set up another test tube exactly like the first one but without any animals in it. This is your control.
4. Close the screw clip and find out how far the coloured water rises up the capillary tube in 30 minutes.

   Has the water risen in the capillary tube? If it has, you may assume that it's because the animals have taken up oxygen from inside the test tube.
5. Use your respirometer to compare the rate of respiration of different organisms.

   What are the sources of inaccuracy in this experiment? How could you improve it?

D Finding the effect of temperature on an organism's rate of respiration

Use the respirometer in activity C to find the effect on an organism's rate of respiration of changing the temperature of the surroundings.

First plan your experiment, then show it to your teacher. When your teacher has approved your plan, carry it out.

E Finding out if living organisms transfer energy to their surroundings

Plan an experiment to find out if organisms transfer energy by heating their surroundings. (In other words, do they give out 'heat'?) Suitable organisms to use are peas or beans that are beginning to germinate. Don't forget to include a control.

Show your plan to your teacher. When your teacher has approved your plan, carry it out.

Questions

1 What would you conclude from each of these observations?

a A piece of food will only burn if oxygen is present.

b When food is burned a gas is given off which turns lime water milky.

2 Describe a simple experiment which could be done to find out if a piece of burning food produces water.

3 In order to show that the air we breathe out contains carbon dioxide, a teacher blows bubbles through a drinking straw into a glass of lime water.

Why is this not as good an experiment as the one given in activity A?

4 Your uncle did not do any science at school, and he does not believe that the air he breathes out contains carbon atoms from the food he eats. Write a short letter to convince him. Use picture 2 to help you, but don't include the picture with your letter!

5 Give two differences between the way energy is transferred from a piece of food when you burn it and when it is broken down inside our cells.

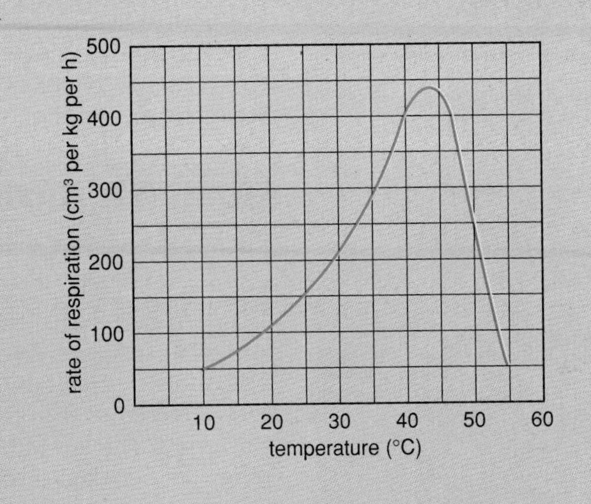

6 The graph above shows the effect of temperature on the rate of respiration of a small cold-blooded animal. The respiration rate is expressed as the volume of oxygen consumed per kg of body mass per hour.

a What do we call the apparatus that was used in this experiment?

b What conclusions would you draw from the graph about the general nature of respiration?

On being a good scientist

On page 123 it says that when respiration takes place, glucose is not oxidised in a single chemical reaction but in a series of small steps.

Many of the individual steps were discovered by the late Sir Hans Krebs who was a pupil of a great German scientist called Otto Warburg. Sir Hans Krebs was awarded a Nobel Prize in 1953 for his research on respiration.

Here he suggests some of the qualities which a successful scientist should have:

'Technical skills are, of course, prerequisites for successful research. What is critical in the use of skills is how to assess their potentialities and their limitations; how to improve, to rejuvenate, to supplement them. But perhaps the most important quality is humility, because from it flows a self-critical mind and the continuous effort to learn and to improve. If I try to summarise what I learned from Warburg. I would say it was asking the right kind of question, forging new tools for tackling the chosen problems, being ruthless in self-criticism, taking pains in verifying facts, and expressing results and ideas clearly and concisely.'

Asking the right kind of question in research means avoiding those which may give a quick result and concentrating on those which are really worth while tackling. Paul Weiss remarked: 'The primary aim of research must not be more facts and more facts, but more facts of strategic value'. Goethe expressed the same idea much earlier. 'Progress in research is much hindered because people concern themselves with that which is not worth knowing, and that which cannot be known'.

Medawar has stated very succinctly: 'If politics is the art of the possible, science is the art of the soluble'. How to select worthwhile soluble problems and how to create the tools required to achieve a solution is something that scientists learn from the great figures in science rather than from science books.

(Adapted from Hans Krebs. *Reminiscences and Reflections*, Oxford University Press, 1981)

1 Why is it important for scientists to show humility?

2 What are 'facts of strategic value'? Can you think of examples?

3 Suggest examples of things which, in your opinion, are not worth knowing or cannot be known.

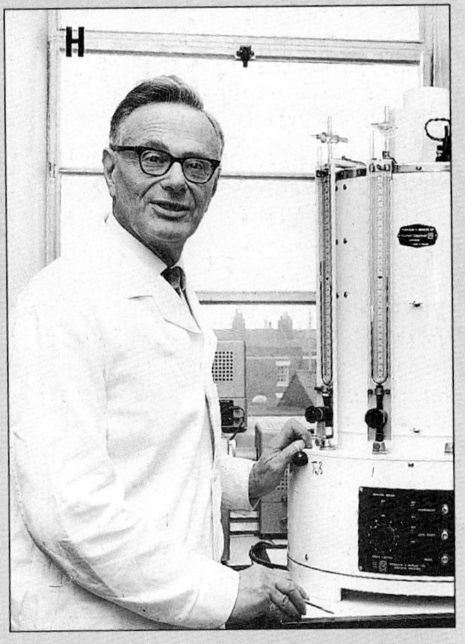

Picture 1 Hans Krebs in his laboratory at Oxford.

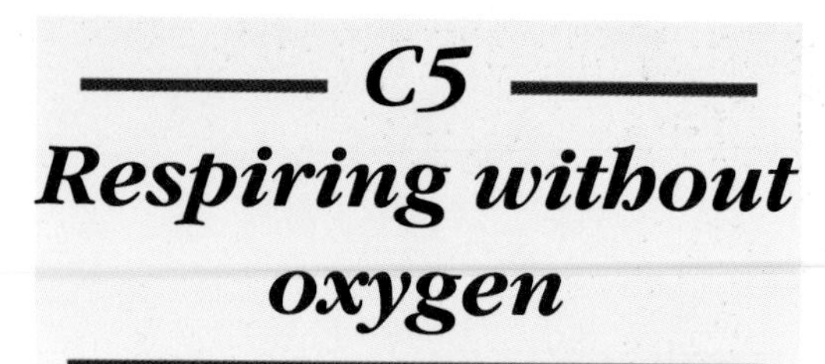

C5 Respiring without oxygen

Sometimes respiration can take place without oxygen. This is very important for us, as we shall see.

Picture 1 Yeast cells seen under the microscope, greatly magnified.

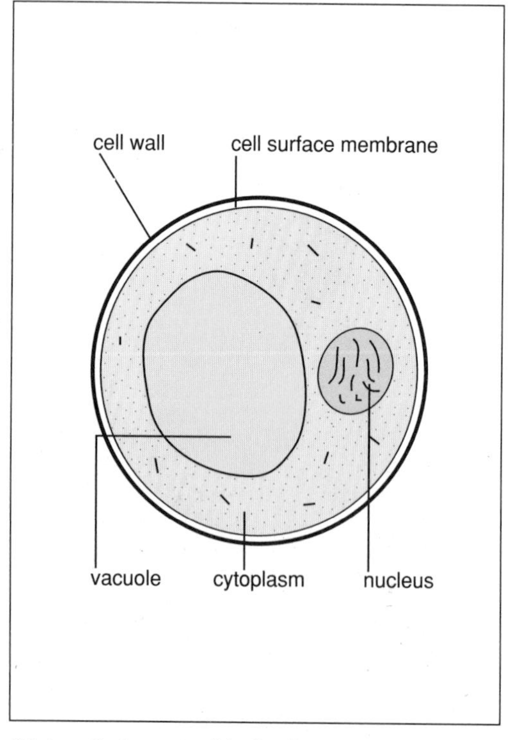

Picture 2 A yeast cell in detail.

Aerobic and anaerobic respiration

In the last topic we saw that organisms obtain energy by respiration. Glucose is broken down into carbon dioxide and water, and useful energy is transferred. Since oxygen is needed for this process, it is called **aerobic respiration** – 'respiration with air'.

Now what happens if oxygen is not available? Usually the organisms will die. However, in some cases glucose may still be broken down, and energy transferred. As this happens without the use of oxygen, it is called **anaerobic respiration** – 'respiration without air'.

Alcoholic fermentation

One way of respiring anaerobically is to convert glucose into alcohol:

$$\text{glucose} \rightarrow \text{alcohol} + \text{carbon dioxide} + \text{energy.}$$

This process is called **alcoholic fermentation**. As with aerobic respiration, carbon dioxide gas is given off.

Alcoholic fermentation gives a lot less energy than aerobic respiration. When a mole (180 g) of glucose is respired aerobically, 2880 kJ of energy are transferred. The equivalent figure for anaerobic respiration is 210 kJ. A lot of energy stays locked up in the alcohol. This can be shown by burning some alcohol; the energy is then transferred to the surroundings which become warmed.

So alcoholic fermentation is less efficient than aerobic respiration. However, it can supply certain organisms with enough energy to keep them going when oxygen is scarce.

Yeast, the great fermenter

One such organism is yeast (pictures 1 and 2). Yeast is a single-celled fungus which lives on the surface of fruit, feeding on sugar. It multiplies rapidly by **budding**: each cell pinches off new ones and a large number can be formed in a short time (page 248).

For centuries humans have used yeast for making bread and alcoholic drinks such as wine and beer.

Making alcoholic drinks

All you need for making alcohol is a sugar solution and yeast. But to make a pleasant alcoholic drink is not so simple, as any wine-maker will tell you.

Wine is usually made from grapes. The juice contains sugar, and wild yeast grows on the skin. The grapes are crushed and the juice is extracted. The yeast cells multiply, fermenting the sugar and turning it into alcohol (picture 3).

Beer is made from barley. The process is called **brewing**. The partly germinated barley grain contains malt sugar. The grain is mashed with water and the resulting liquid is boiled with hops to give it the right flavour (picture 4). Yeast is then added and fermentation soon gets underway (picture 5).

Wine and beer making are major industries. In Britain over 6000 million litres of beer are drunk each year. Yeast too is manufactured on a large scale – you can buy it from shops. For the pleasure, and problems, which people get from a night in the pub we have to thank this lowly organism.

Making bread

Imagine you are a baker. You mix some flour and water with a small amount of sugar and yeast. This makes dough. You then leave the dough in a warm place for an hour or so. The living yeast cells multiply and ferment the sugar, giving off carbon dioxide gas. The gas makes the dough expand (picture 6). The gas is prevented from escaping by the stickiness of the dough (see page 246). When you bake the dough in a hot oven, the heat kills the yeast, and the alcohol evaporates away.

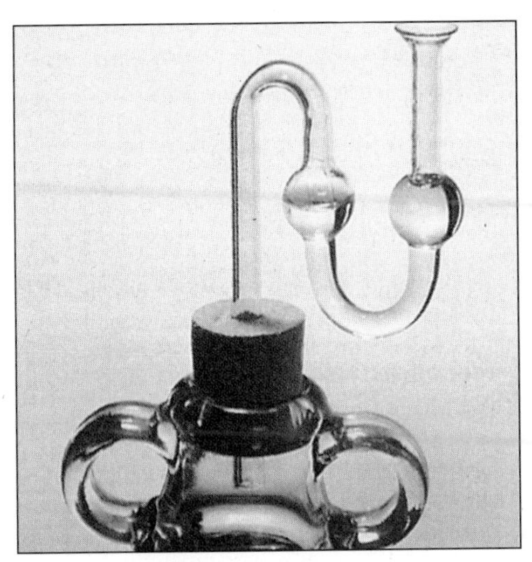

Picture 3 In making wine the jar is fitted with a special kind of valve which allows carbon dioxide to escape but prevents bacteria from getting in.

Picture 4 Hops are climbing plants which are grown in south-east England (Kent). The unpollinated female flowers are gathered and dried, and used in brewing beer.

Picture 5 Beer fermenting in a large tank in a brewery. The carbon dioxide gas given off by the yeast creates a froth on the surface. The men are holding this back so as to view the fermenting mixture underneath.

Can humans respire without oxygen?

During strenuous exercise we cannot breathe fast enough, and our circulatory system is not efficient enough, to get enough oxygen to our muscles to keep them going. So the muscles obtain extra energy by respiring anaerobically.

Unlike yeast, our muscles don't produce alcohol – the effect of that would be alarming to say the least. Instead they produce another substance called **lactic acid**:

$$\text{glucose} \rightarrow \text{lactic acid} + \text{energy}$$

Think of running a 100 metre sprint. During the race lactic acid builds up in your muscles. Lactic acid is a mild poison, and it makes the muscles ache. When the race is over it must be got rid of. It is broken down into carbon dioxide and water. Oxygen is needed for this – that's why we pant after a race.

The oxygen needed to get rid of the lactic acid is called the **oxygen debt**. If we build up an oxygen debt during a bout of exercise, we must pay it off immediately afterwards.

Because the muscles can work for a short time without oxygen, sprinters can hold their breath while running 100 metres.

In a long-distance race, lactic acid builds up to begin with, but soon the body adjusts and the acid is destroyed while you are actually running. When this point is reached we have got our 'second wind'.

Anaerobic respiration produces far less energy than aerobic, and it can't go on for very long. However, it can make the difference between life and death. An antelope fleeing from a cheetah may owe its life to the fact that for a short time its muscles can respire without oxygen.

Picture 6 Dough before and after rising. Carbon dioxide gas, given off by the yeast, makes the dough expand.

Anaerobic respiration in bacteria

Many kinds of bacteria respire anaerobically. They ferment sugar and other substances. The end products vary. Some produce alcohol, others lactic acid. Some of these bacteria are useful to us. For example, those that produce lactic acid are used for making butter, yoghurt and cheese. There is more about this on page 92.

Certain bacteria produce methane gas as the end product of anaerobic respiration. These bacteria thrive in situations where there is no oxygen, such as stagnant ponds, waterlogged soil, and sewage. They too are useful to us. The methane they produce can be used as a fuel (see page 94).

Industrial fermentation

Microbes such as bacteria and yeasts produce all sorts of things that are useful for humans. Penicillin, insulin and washing powder enzymes are just three examples.

To obtain these products in sufficient amounts, we need to grow microbes on a large scale. This is done by means of an **industrial fermenter**.

One type of fermenter is shown in picture 1. It consists of a large stainless steel vessel. The vessel is filled with a suitable **medium**. This consists of a food solution such as sugar, together with any other substances which the microbes need in order to multiply and grow quickly.

Certain bacteria need only carbohydrate and minerals; others need protein and vitamins as well. Most fungi need carbohydrate, minerals and vitamins, though yeast needs only sugar and minerals.

The right kind of microbes are then added to the medium, and left to multiply.

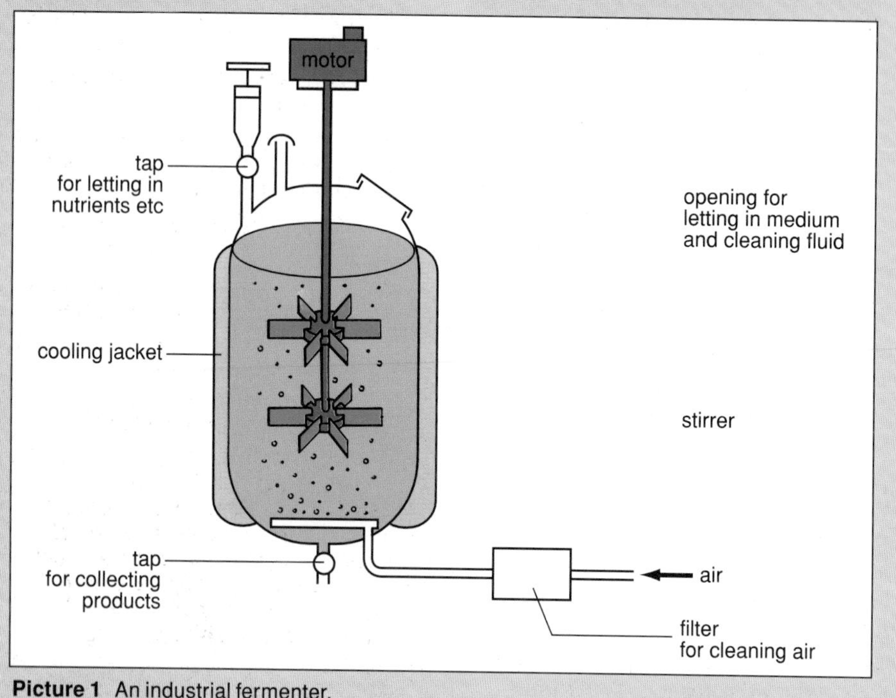

Picture 1 An industrial fermenter.

Paddle-like **stirrers** keep the contents of the vessel moving about. If the microbes are aerobic, they must be given a supply of oxygen. This is done by an **aerator** which bubbles air through the mixture.

Fermenting microbes give off a lot of energy which heats up the medium. This means that the fermentation vessel has to be cooled. Small fermenters can be cooled by a **cooling jacket** surrounding the vessel. Larger fermenters need a **cooling coil** inside.

The vessel, medium and air supply are all sterilised beforehand to prevent any possibility of unwanted microbes getting into the mixture.

When fermentation is complete, the contents of the vessel are collected from a tap at the bottom. The product is then separated from the rest of the mixture and purified.

1 Suggest reasons why it is necessary to stir the contents of the fermentation vessel.

2 Why is it important that unwanted microbes should be kept out of the fermentation vessel?

3 What would happen if you did not cool the fermenter?

4 Suppose you decided to use an industrial fermenter to make beer. What medium would you put in, and what would you add to it?

Batch culture and continuous culture

In the method of fermentation described above, you get the process going, wait for the microbes to grow and then collect the products. You then clean the fermenter and start again. This is called **batch culture** and it is the traditional method used in industrial fermentation.

A more modern method is to use a fermenter with an overflow. Once you have got the process going, the contents of the fermenter overflow and the product is extracted continuously. As fast as the culture overflows, more medium is added to keep pace with the loss. This is called **continuous culture**.

Continuous culture systems are fully automated and go on working day and night.

Continuous culture occurs in nature too. For example, cellulose-digesting microbes (mainly bacteria) live in a special part of the stomach, called the rumen, of cattle and sheep. The microbes enable the animal to digest the vast quantities of grass that it eats (see page 139).

1 What are the advantages of continuous culture over batch culture?

2 Most fermentation factories use batch culture and are reluctant to change to continuous culture. Why do you think this is?

3 Why is it correct to regard the rumen of a cow as a continuous culture rather than a batch culture? What would happen if it was a batch culture?

Producing penicillin, an example of batch culture

A starter culture of the fungus *Penicillium* is added to a liquid medium in a fermenter. The medium contains various sugars. The temperature is kept at a steady 24°C and the oxygen concentration is carefully controlled. The yield of penicillin depends on the oxygen concentration and the relative concentrations of the different sugars.

During the first 24 hours the fungus grows rapidly. After that the sugar content of the medium begins to fall, and the fungus starts producing penicillin. After about a week the concentration of penicillin in the medium reaches a maximum. The culture is then collected and filtered. The liquid medium passes through the filter and the fungus gets left behind. Penicillin is extracted from the medium.

Activities

A Finding out about the products of alcoholic fermentation

1 Put a 10 per cent solution of glucose into a large test tube to a depth of 2 cm.

2 Wear eye protection. Boil the glucose solution to expel any oxygen present.

EYE PROTECTION MUST BE WORN

3 Cool it, then add a little yeast.

4 Pour a thin layer of liquid paraffin on top of the glucose solution to stop oxygen getting to the yeast.

5 Set up the test tube as shown in the illustration.

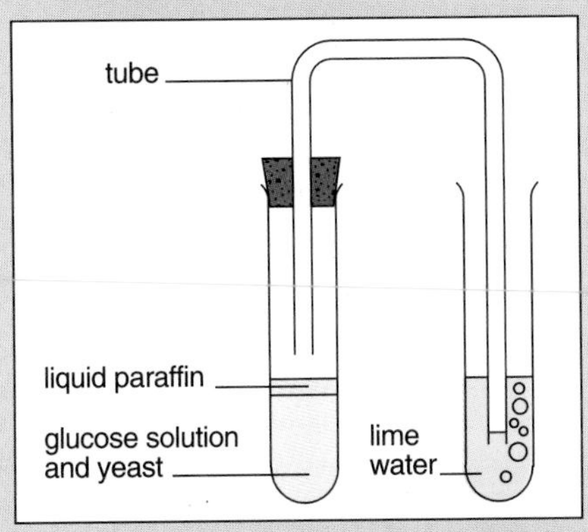

6 Set up a second test tube exactly like the first one but do not add any yeast to the glucose. This is your control.

7 Leave the test tubes in a warm place for at least an hour. Then examine them.

Has the lime water gone cloudy? If it has, carbon dioxide has been given off.

Sniff the contents of the test tubes. Does either smell of alcohol? If so, why?

Feel the two test tubes. Is one warmer than the other? If so, why?

What conclusions do you draw from this experiment?

B Does yeast make dough expand?

1 Make a small amount of dough as follows. To 50 g of flour add water a little at a time and mix with a spoon. Stop adding water when the dough feels rubbery. Don't let it get sticky.

2 Put a spatula-full of yeast in a test tube and shake it up with some warm water. Add about 10 g of sugar and shake again.

3 Divide the dough into two portions. To one portion add the yeast suspension and mix it in well with your hands. Do not put any yeast into the second portion.

4 Grease the inside of two beakers. Into one beaker put the dough which contains yeast.

5 Into the second beaker put the dough which does not contain yeast. This is your control.

6 Leave both beakers in a warm place for about an hour.

7 After an hour compare the appearance of the dough in the two beakers.

Has either risen? If so which one – and why?

Why was it necessary to put some sugar with the yeast before adding it to the dough?

C Does the temperature affect the expansion of dough?

Plan an experiment to find out if temperature affects the expansion of dough. When you have had your plan approved by your teacher, carry out the experiment.

Write up your results and draw conclusions. What advice would you give to a cook about making dough rise?

D How does lactic acid affect our muscles?

1 Raise one arm above your head.

2 Clench and unclench your fist twice a second for as long as you can.

Notice the feeling in your arm as you exercise your muscles.

3 When you can continue no longer, rest your arm on your lap and note what happens to the feeling.

How would you describe the feeling in your arm during and after the exercise?

Does your experience fit in with the idea that lactic acid builds up in the muscles during exercise and is removed afterwards?

E Can germinating seeds respire without oxygen?

Plan an experiment to find out if germinating seeds can respire without oxygen. Use pea or bean seeds which have been soaked in water and are just beginning to germinate (see page 244). You need to think about how you will show that respiration is happening.

After having your plan approved by your teacher, carry out the experiment and draw such conclusions as you can.

Questions

1 People who make wine at home sterilise their equipment before they start work. Why? Where is the best place to leave wine to ferment? Give reasons for your answer.

2 Why is yeast used for making bread? Do you think someone could get drunk by eating lots of bread? Explain your answer. What is unleavened bread, and what is it like to eat?

3 What sort of things make one type of wine different from another? Suggest as many possibilities as you can.

4 Mrs Matthews makes some marmalade and stores it in jars in a cupboard in the kitchen. Three months later she opens one of the jars and finds that the marmalade looks frothy and smells of alcohol. What do you think has happened and why? How might Mrs Matthews prevent this happening in the future?

5 The following organisms can all respire anaerobically: (a) whales, (b) a tapeworm in the human small intestine, (c) bacteria in the mud at the bottom of a lake, (d) the roots of rice plants.

Suggest why it is useful for each of these organisms to be able to respire anaerobically.

6 Yeast can respire without oxygen, but it also respires with oxygen. In fact yeast is poisoned if the alcohol concentration gets too high. Why is this information important for wine-makers and brewers?

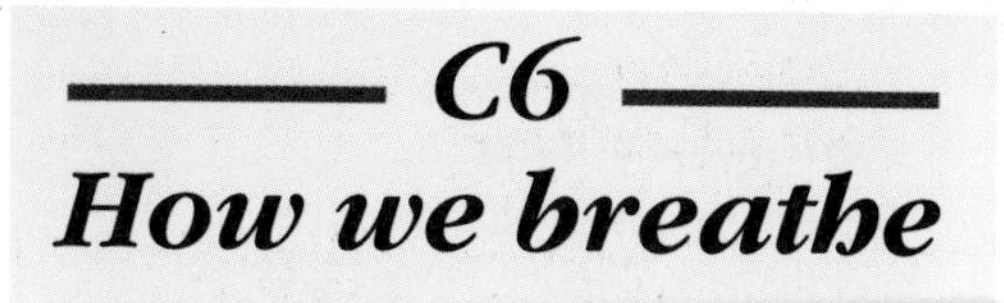

C6 How we breathe

Breathing is the process which moves air in and out of the body.

	Inhaled air	Exhaled air
Oxygen	20.93	16.4
Carbon dioxide	0.03	4.1
Nitrogen (and argon)	79.04	79.5

Picture 1 The air we breathe out contains less oxygen and more carbon dioxide than the air we breathe in.

Why is breathing important?

Breathing enables us to obtain oxygen and get rid of carbon dioxide. You can show this by comparing the air we breathe in (**inhaled air**) with the air we breathe out (**exhaled air**). Picture 1 shows such a comparison. You will see that the person's exhaled air contains less oxygen and more carbon dioxide than inhaled air.

When we breathe in, oxygen is taken out of the air and used for respiration (see page 122). Meanwhile carbon dioxide, produced by respiration, is added to the air which we breathe out. This exchange of oxygen and carbon dioxide is called **gaseous exchange**. It is a life-giving process in nearly all living things.

Our breathing system

When we breathe in, air is drawn into our **lungs**. These are the main organs of our **breathing system**.

The breathing system is shown in picture 2. There are two lungs, situated in the chest (**thorax**). The sides of the chest are bounded by the **ribs**. Between the ribs are **intercostal muscles**. Below the lungs is a sheet of muscle tissue, shaped like a dome, which separates the thorax from the abdomen. This is called the **diaphragm**.

The lungs are surrounded by two thin sheets of tissue with fluid in between. The fluid serves as a lubricant, allowing the membranes to slide over each other smoothly as we breathe in and out.

The rest of the breathing system consists of a series of cavities and tubes through which air flows to and from the lungs. Here are a few notes about the structures through which the air passes.

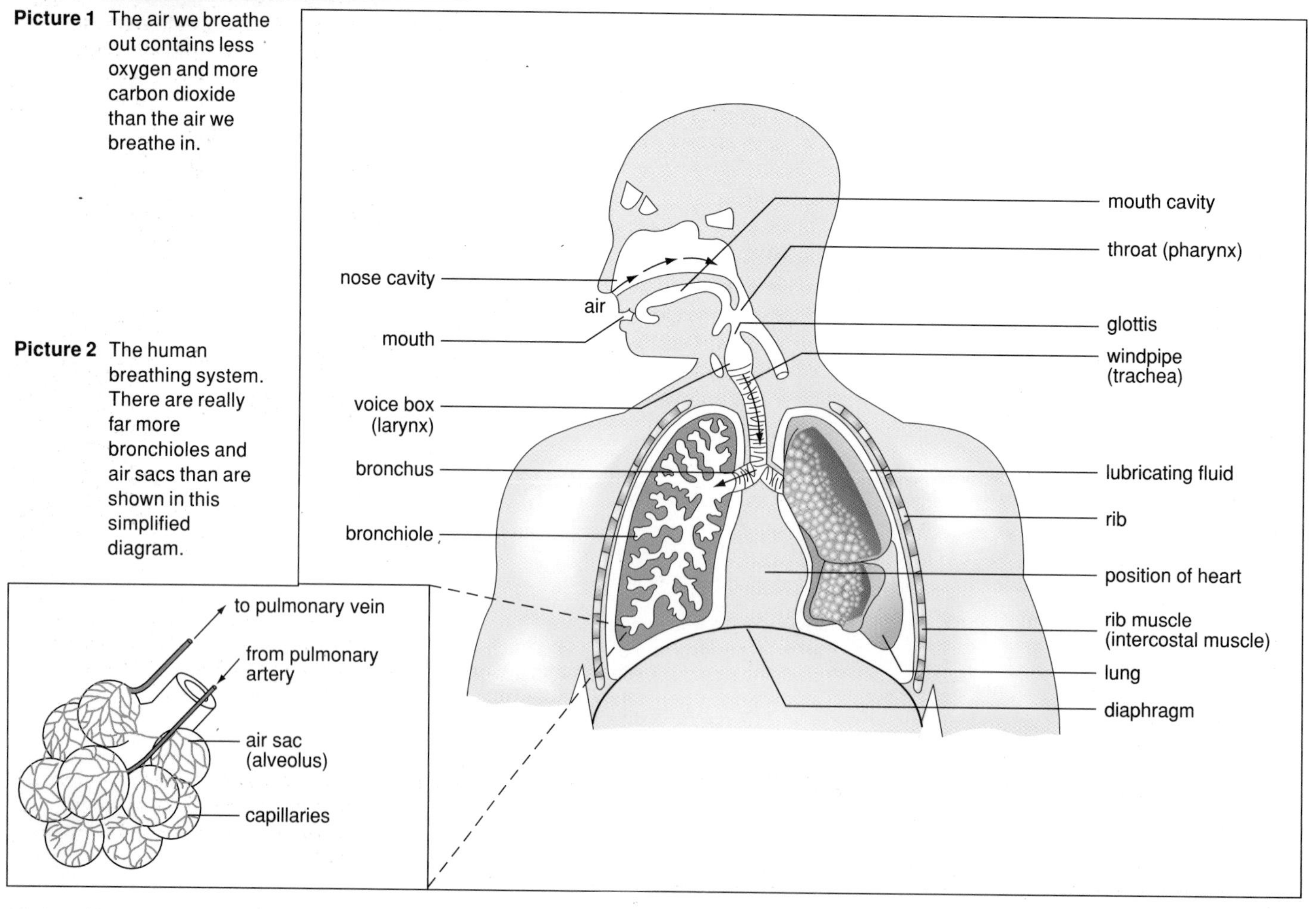

Picture 2 The human breathing system. There are really far more bronchioles and air sacs than are shown in this simplified diagram.

The nose

The cavity inside the nose is divided up by shelf-like partitions rather like the radiator of a car. As air passes through, it is warmed, moistened and cleaned. Dust and germs get caught up in mucus on the lining and are wafted towards the throat by thousands of tiny, beating 'hairs' called **cilia**. The cilia project from a special kind of epithelial tissue called **ciliated epithelium**. Dust and germs, thus trapped, are expelled from the nose and throat by **sneezing** and **coughing**. These are both reflexes which help to protect us from infection.

The throat (pharynx)

Here the airway crosses the throat. Air passes from the nose cavity across the throat and into a little hole called the **glottis** which leads to the **windpipe** (**trachea**). You cannot breathe and swallow at the same time. When you swallow, the breathing pathway is closed off so food does not go the wrong way (page 137). If a bit of food does get stuck in the airway, you can choke.

The windpipe (trachea)

Before air reaches the windpipe it has to go through the **voice box** (**larynx**). The windpipe itself runs from the voice box into the chest. For the air to flow freely, the windpipe must be open at all times. It is kept open by rings of cartilage in its wall (picture 3).

Dust and germs which escaped being caught in the nose and throat get trapped in mucus lining the windpipe. They are then wafted towards the throat by cilia. Every now and again you need to clear your throat to get rid of the mucus.

The bronchi and bronchioles

Inside the chest the windpipe splits into two short tubes called **bronchi** (singular: **bronchus**). These then divide like a tree into lots of small branches called **bronchioles** which get narrower and narrower towards their ends (picture 4).

The air sacs

Each bronchiole leads to a bunch of tiny air sacs called **alveoli** (singular: **alveolus**). The air sacs are surrounded by a network of blood capillaries, rather like a string bag (see inset in picture 2). Here, in the depth of the lungs, gaseous exchange takes place.

How does gaseous exchange take place?

The air sacs inside our lungs are in close contact with the blood capillaries, and the two are separated by a very thin membrane. Across this membrane gaseous exchange takes place. Oxygen diffuses from the air sacs into the blood, and carbon dioxide diffuses from the blood into the air sacs. The inner surface of the air sacs is covered by a thin layer of fluid in which the gases dissolve before they diffuse through (picture 5).

There are about 150 million air sacs in each lung, and together they cover a

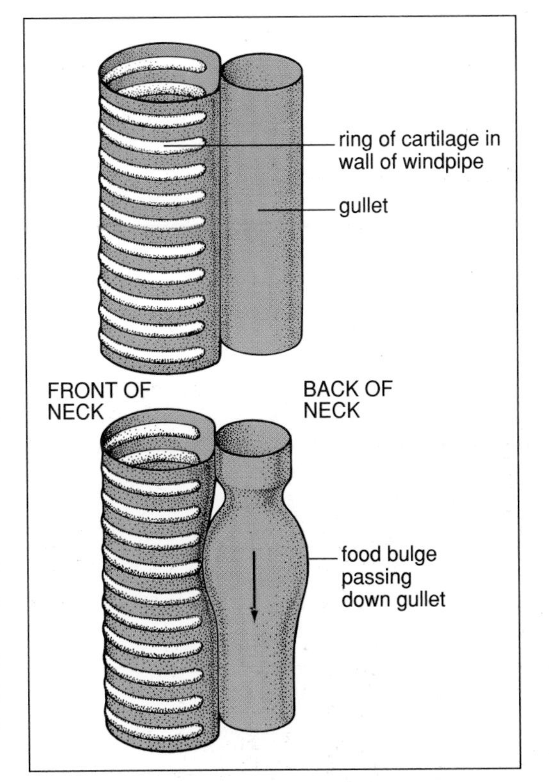

Picture 3 Rings of cartilage keep the windpipe permanently open and prevent its wall caving in. The rings are incomplete, like a pile of Cs. The open side of the C is next to the gullet; this allows the gullet to expand when food is passing down.

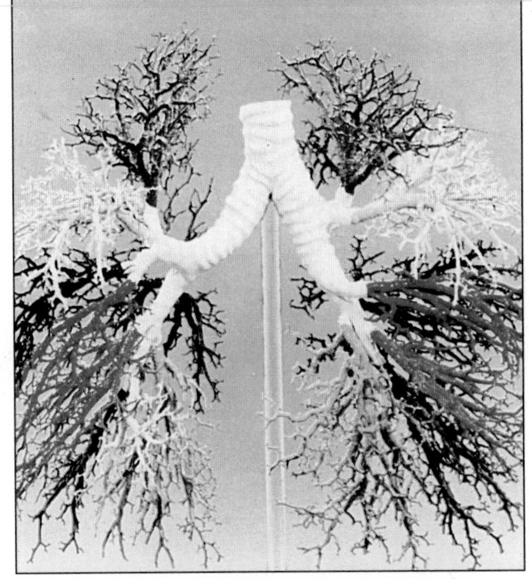

Picture 4 A museum exhibit of human lungs. The bronchial tree was filled with a solution which hardened. The rest of the lung tissue was then dissolved.

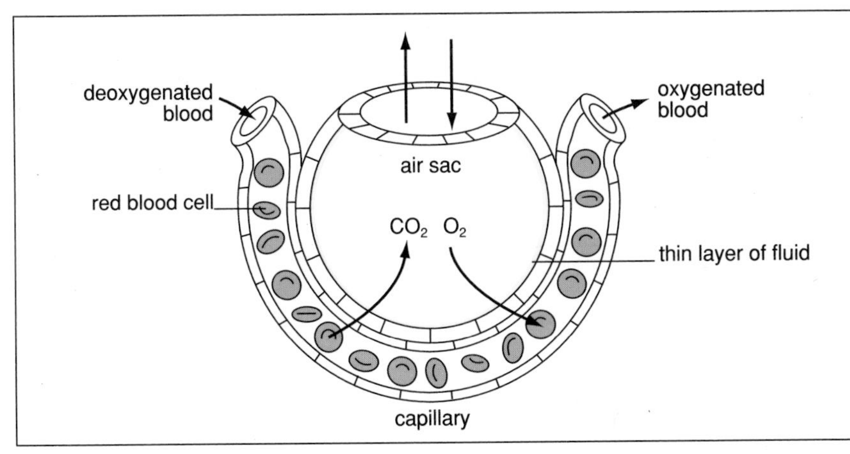

Picture 5 As blood flows past an air sac, it gives up carbon dioxide and picks up oxygen. These gases move in and out of the air sac by diffusion.

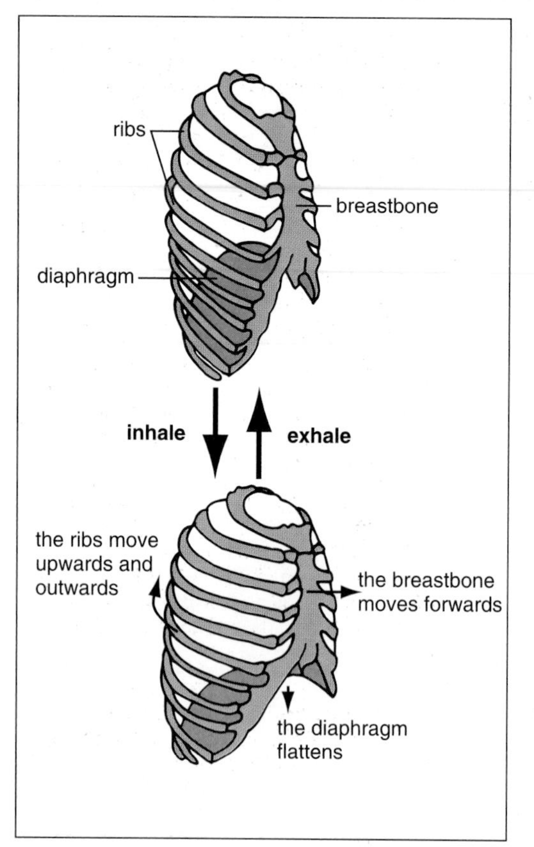

Picture 6 How the chest expands when we breathe in. Notice that its volume increases in all directions.

very large surface area. Someone has worked out that if you were to flatten them out like a sheet, they would cover an area the size of a tennis court! It is important that the lungs should have a large surface area, because it means that plenty of oxygen can be taken up by the blood when we breathe in.

The air sacs always contain some air, even when we breathe out as hard as we can. If there was no air inside the air sacs, their walls would cave in and stick together. The gaseous exchange surface would then be so reduced that we would suffocate.

How does air get in and out of the lungs?

Our breathing system works rather like a pair of bellows, sucking air in and then forcing it out. The details are shown in picture 6.

Inhaling is brought about by the chest expanding. The ribs, moved by their muscles, swing upwards and outwards and the breast bone moves forward. At the same time the diaphragm flattens. The result is that the volume of the chest increases. This lowers the pressure in the chest, so air passes into the lungs.

Exhaling is brought about by the reverse process. The ribs and breastbone return to their original positions, and the diaphragm bulges upwards again. The result is that the volume of the chest decreases. This raises the pressure in the chest, so air is forced out of the lungs.

How is breathing controlled?

We breathe without having to think about it. What, then, makes it happen? The answer is the brain. In the base of the brain there is a special group of nerve cells which controls breathing and makes it happen even when we are asleep.

The brain also makes us breathe faster and more deeply when, for example, we take exercise. This is important because during exercise our muscles work harder and so they need more oxygen. The muscles will also be producing more carbon dioxide. This must be removed quickly, otherwise it might build up and poison our tissues.

During exercise the control centre in the brain senses that there is too much carbon dioxide in the bloodstream. It then sends messages along the nerves to the chest, making the breathing muscles work faster.

Questions

1 Suggest why it is a good idea to:
- a breathe through your nose rather than through your mouth,
- b take deep breaths rather than shallow ones,
- c stop talking when you swallow,
- d blow your nose when necessary.

2 Explain each of the following in terms of breathing in and out: a yawn, a gasp, a cough, a sigh, a laugh.

3 Picture 1 on page 130 gives the percentage volumes of oxygen and carbon dioxide in the air inhaled and exhaled by a human.
- a What is the other main gas in inhaled air? Would you expect it to be in exhaled air as well?
- b Explain how the change in the composition of the air is brought about.

4 An experiment was carried out on a young man in which the volume of air taken in at each breath, and the number of breaths per minute, were measured at rest and after running. Here are the results:
- a What is the total volume of air breathed in per minute at rest and after running?
- b Twenty per cent of the air breathed in consisted of oxygen, but only sixteen per cent of the air breathed out consisted of oxygen. Assuming that these figures remain constant, work out the volume of oxygen entering the blood per minute at rest and after exercise.
- c Why does the amount of oxygen taken up into the blood increase after exercise?

	Volume of air per breath	Breaths per minute
at rest	450 cm^3	20
after running	1000 cm^3	38

Activities

A Examining lungs under the microscope

Usually when you look at things under the microscope, you are told what you are looking at. Not so with this investigation!

Your teacher will give you a slide on which a thin section of a lung has been mounted. The section has been stained with a dye to show up the various parts of the lung.

First decide what you would *expect* to see in your section. Use picture 5 on page 131 to help you. Then look at the section under the low power. Describe what you see, making sketches to illustrate. Can you say what the various parts are? Label them in your sketches even if you are not sure about them.

When you have finished your teacher will tell you which labels are correct.

In what way is the structure of the lung suited for gaseous exchange?

B How much air can you breathe in?

1 Fill a basin with water. Lay a bell jar on its side in the basin, so that the bell jar fills up with water. Then place the bell jar in an upright position on supports. Insert a bent tube under the rim of the bell jar as shown in the picture.

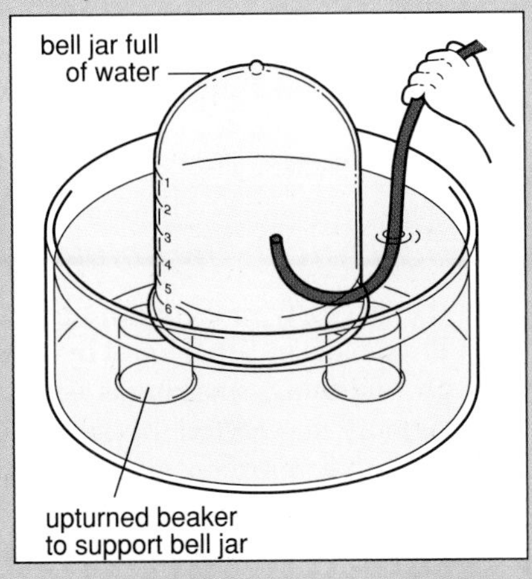

2 Take as deep a breath in as you can. Then put the end of the bent tube in your mouth and hold your nose. Now breathe out as much as you possibly can: your exhaled air will displace some of the water from the bell jar. As soon as you have finished breathing out, remove the tube from your mouth.

3 From the scale on the side of the bell jar, read off the volume of air in litres which you have breathed out. This is called your **vital capacity**. It is the maximum volume of air which you can take into your lungs.

4 Compare the vital capacities of the students in your class. Try to persuade your teacher to take part too. What sort of variation is there, and how would you explain it?

5 Use the apparatus to find the volume of air which you exhale in a single normal breath when you are at rest. This is called the **tidal volume**. Compare your tidal volume with your vital capacity. What conclusions can you draw?

Note: this experiment can also be carried out using a **gas meter** or **spirometer**. They should only be used under the strict supervision of your teacher.

Artificial respiration

If a person has an accident, the brain may stop working for a time and breathing stops. However, the person's life may be saved by **artificial respiration**. This must be carried out as soon as possible, otherwise the brain cells may run so short of oxygen that they die.

The best method of artificial respiration is **mouth-to-mouth resuscitation** or 'kiss of life'. With the person lying down, you keep breathing out into his or her mouth as shown in the pictures. There's enough oxygen in your exhaled air to keep life going, and hopefully the victim will soon start breathing again. Mouth-to-mouth resuscitation is something everyone should be prepared to do if necessary.

In a severe accident, such as a car crash, the brain may be so badly damaged that the person cannot start breathing again. It may then be necessary to attach the person to a **resuscitator**, a machine which forces air in and out of the lungs. An unconscious person can be kept alive for weeks or even months on a machine like this. Sometimes the brain recovers sufficiently for the person to start breathing again. If recovery does not take place, the family and doctors have to decide whether to keep the person alive on the machine or switch it off. Obviously this is an agonising decision to have to make.

1 Pinch the nostrils shut with the fingers of one hand, then tilt the head back and push the lower jaw forward so the chin juts out. This will force the tongue forward and open the air passages.

2 Take a deep breath, then open your mouth and seal your lips against the person's mouth. Breathe out firmly but gently into the person's mouth and so into the lungs.

3 Lift your mouth off, then turn your head so as to look at the person's chest. If you have been successful you will see that it has risen and is now falling as air comes out of the lungs.

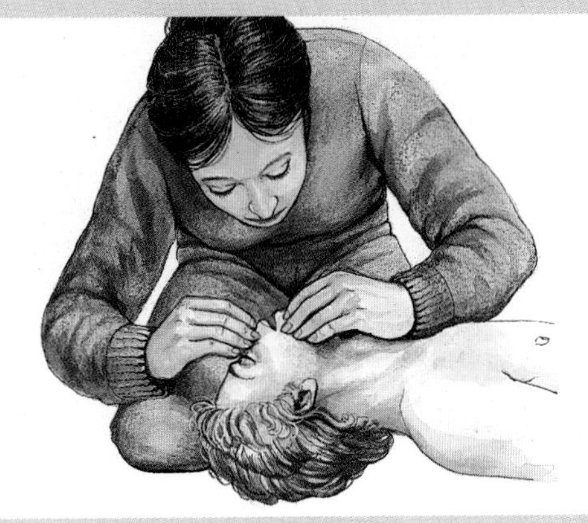

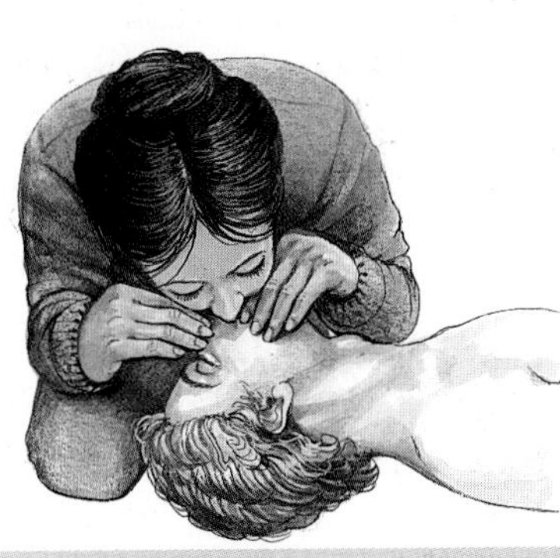

4 Repeat steps 2 and 3 at a steady rate. The person's colour should improve, and eventually breathing should start up again.

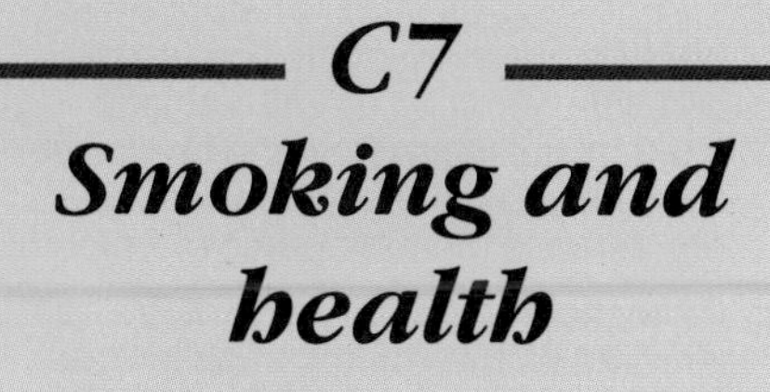

C7 Smoking and health

More people die as a result of smoking than from any other cause.

How do we know that smoking is harmful?

Most people who smoke inhale the smoke right down into their lungs. How do we know that this affects their health?

One way of finding out is to take two groups of similar people of the same age and sex. One group are smokers, and the other group are non-smokers. You then study each group over many years, and watch what happens to their health. If a certain disease is more common amongst the smokers than the non-smokers, we may suspect that the disease is linked with smoking. Investigations of this kind have to be repeated many times, and similar results obtained, before we can be sure that there is a connection between smoking and the disease.

Lung cancer, the smoker's disease

In the last forty years or so, surveys have been carried out by the Royal College of Physicians in Britain and by similar bodies in other countries. The results all show one thing: smoking is associated with **lung cancer** (picture 1). This conclusion is supported by laboratory experiments. These experiments have shown that tobacco smoke contains substances which can cause cancer. Substances which cause cancer are called **carcinogens**. The main carcinogens in tobacco smoke are a group of dark-coloured chemicals usually referred to as 'tar'.

Now here is an important point. Being a smoker doesn't mean that you're bound to get lung cancer; nor does being a non-smoker ensure that you won't get it. All we can say is that if you are a smoker you're more likely to get the disease. The odds are shown in picture 2.

Research suggests that smokers are less liable to get lung cancer if they:

- do not inhale,
- smoke a pipe rather than cigarettes,
- take fewer puffs per cigarette,
- smoke filter-tips,
- smoke 'low-tar' cigarettes.

But the best thing is never to start – or, if you have started, to give it up. If a heavy smoker stops smoking, the risk of getting lung cancer gradually falls until after a few years it is about the same as for a non-smoker.

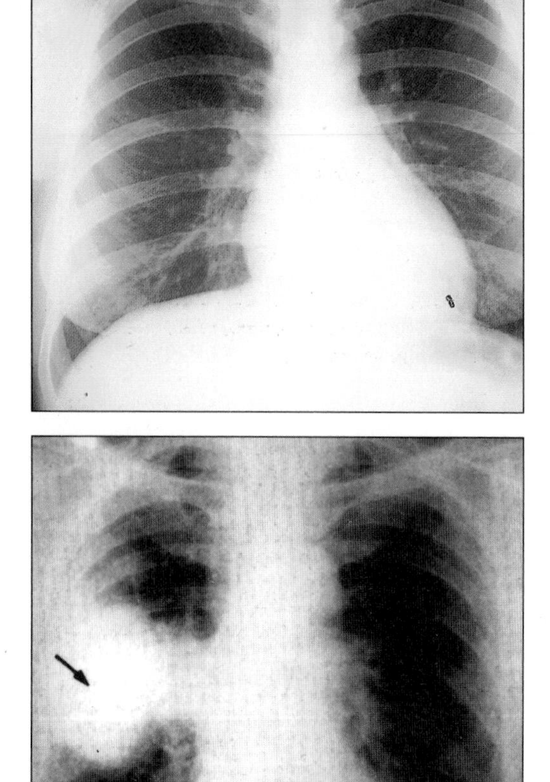

Picture 1 Two X-rays of the human chest. The top X-ray is normal: the lungs are in the dark areas enclosed by the ribs. The bottom X-ray shows cancer of the lung: the cancer consists of a growth (arrowed) which shows up as a white area in the X-ray.

Other effects of smoking

When a person smokes, tiny particles in the smoke get caught on the lining of the windpipe and bronchial tubes. Extra mucus is produced, and the cilia stop beating. Smoking one cigarette stops the cilia for about an hour. Instead of being wafted to the throat by the cilia, the mucus collects in the bronchial tubes and gives rise to a 'smoker's cough'. If the tubes become infected, the person may get **chronic bronchitis.** Bronchitis is inflammation (soreness) of the bronchial tubes, and 'chronic' means that it persists instead of getting better.

Further unpleasant effects may follow. Repeated coughing can cause the delicate walls of the air sacs in the lungs to break down. This reduces the surface area for gaseous exchange, so the person gets very short of breath. This is called **emphysema** (picture 3).

Although smoking mainly affects the lungs, it can also cause cancer of other organs such as the mouth, throat, gullet and bladder. It is also linked with stomach ulcers. A woman who smokes while pregnant may give birth to an under-sized baby, and could even have a miscarriage (page 228).

Smoking and heart disease

Scientists have shown that there are more cases of **heart disease** among smokers than among non-smokers. The effect of smoking on the heart is partly due to **nicotine**. This substance gets into the bloodstream when tobacco smoke is inhaled.

Nicotine is a habit-forming drug, which is one of the reasons why people get hooked on smoking and cannot give it up. Nicotine may make a smoker feel good, but like other drugs it is a poison. If all the nicotine was extracted from one cigarette and injected into your bloodstream in one go, it would kill you. Absorbed in small doses from smoking over many years, it raises the blood pressure and increases the amount of fatty substances in the blood. It is this that can lead to heart disease.

Tobacco smoke also contains **carbon monoxide**. This makes the blood less good at carrying oxygen (page 157). Carbon monoxide can also lead to heart disease. In fact scientists believe that it is the main component of cigarette smoke to cause heart disease.

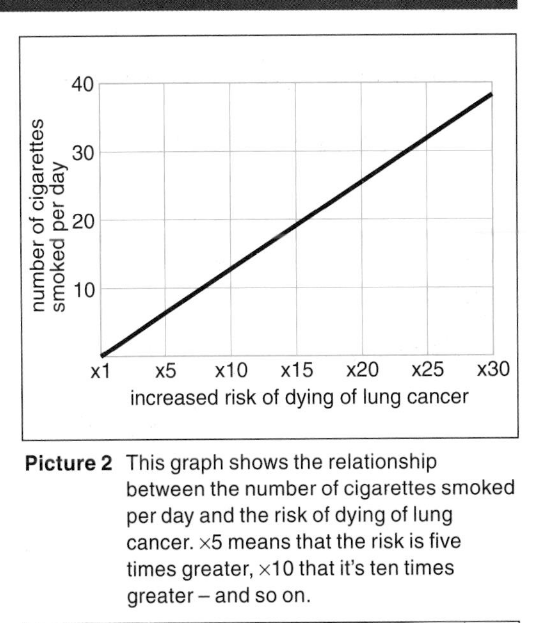

Picture 2 This graph shows the relationship between the number of cigarettes smoked per day and the risk of dying of lung cancer. ×5 means that the risk is five times greater, ×10 that it's ten times greater – and so on.

Smoking and society

In Britain between 60 and 70 thousand people die of lung cancer, chronic bronchitis and emphysema every year. This is over eight times as many people as are killed in road accidents.

The connection between these diseases and smoking is now well established. In many countries tobacco commercials on television have been banned, and cigarette packets carry health warnings (picture 4).

Smoking is prohibited in many public buildings and cinemas. The trouble is that governments get a lot of money from taxes on tobacco, and this has made some countries reluctant to campaign too hard against it.

Despite the warnings, many people still smoke. Amongst doctors, however, smoking has decreased. They understand the risks too well, and they know how unpleasant it is to die of lung cancer.

Of course smoking isn't the only cause of lung cancer. There is evidence that it can be brought on by motor vehicle exhaust, industrial smoke and dust, and radioactive materials. However, these causes are insignificant compared with smoking. It is claimed that if all smokers gave up smoking, deaths from lung cancer would fall to a tenth of what they are at the moment.

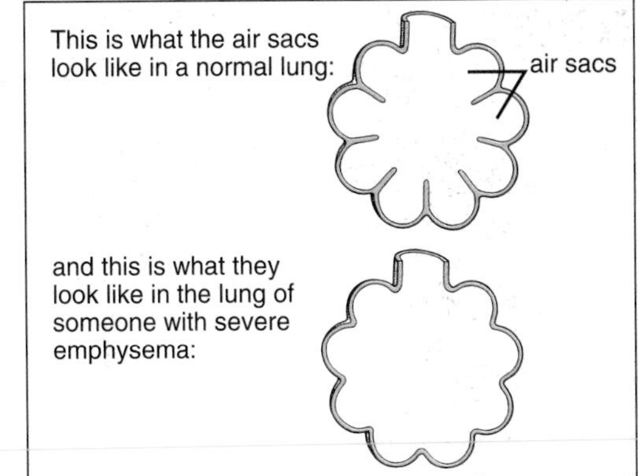

Picture 3 These diagrams show how emphysema affects the lungs.

EEC Council Directive (89/622/EEC)
SMOKING CAUSES HEART DISEASE

Picture 4 Cigarette packets carry health warnings like this one.

Questions

	Percentage of patients who smoked more than fifteen cigarettes per day	Percentage of patients who were non-smokers
Group A	25	0.5
Group B	13	4.5

1 Explain why each of these remarks is unscientific:

a 'My dad smokes like a chimney so he's bound to get cancer.'

b 'All that stuff about smoking and cancer is nonsense: my uncle died of lung cancer and he never touched a cigarette all his life.'

2 Two British scientists carried out a survey in a large hospital. They selected two groups of patients, both the same sex and roughly the same age. The patients in group A all had lung cancer, whereas those in group B had other diseases. The scientists then found out how many patients in each group smoked. The results are shown above.

a What do you think the scientists were trying to find out?

b Group B is called the control group. Why was it necessary to investigate this group of patients as well as group A?

c What conclusion would you draw?

d Suggest one other way the scientists might have carried out their survey.

3 It has been suggested that smoking and lung cancer appear to be connected, not because smoking causes cancer, but because people who need to smoke are the kind of people who get cancer. What sort of investigations would have to be carried out to test this suggestion?

C8
How we digest our food

What will happen to this hamburger after Rachel has put it in her mouth?

Picture 1 Taking the Big Bite, the start of digestion.

Some facts and figures about the gut

The human gut is about 9 metres long altogether, over four times a person's height. This is short compared with a cow, whose gut may exceed 50 metres.

The human small intestine is about 6 metres long, and the large intestine 1.5 metres. The small intestine is narrower than the large intestine, which is why it is described as 'small'.

The human stomach has a volume of 1 to 1.5 litres. However, the stomach wall is very elastic, allowing its volume to increase by three times or more during a large meal. Compare this with the rumen of a cow which has a volume of 20 litres.

A person produces 1 to 1.5 litres of saliva daily. This is small compared with a cow which can produce as much as 190 litres in a day.

The gut

The mouth leads into the **gut**. The gut is really a tube which runs from the mouth to the anus. It is between eight and nine metres long – that's about four times Rachel's height. Being so long, most of it is coiled up. This enables it to fit neatly into the **abdomen**.

The gut isn't a simple tube. It consists of a series of parts, each with a particular job to do. The parts of the gut are shown in picture 2.

What happens in the gut?

Let's look at the basic principles before we get down to details. A hamburger contains starch, fat and protein. The protein and fat are mainly in the meat, and the starch is in the bun.

After the hamburger has been taken into the mouth (**ingested**), it is chewed and swallowed. As the bits of hamburger pass along the gut, the starch, fat and protein are broken down into soluble substances. This process is called **digestion**. The soluble substances are then **absorbed** through the lining of the gut into the bloodstream. They are then carried round the body to where they are needed.

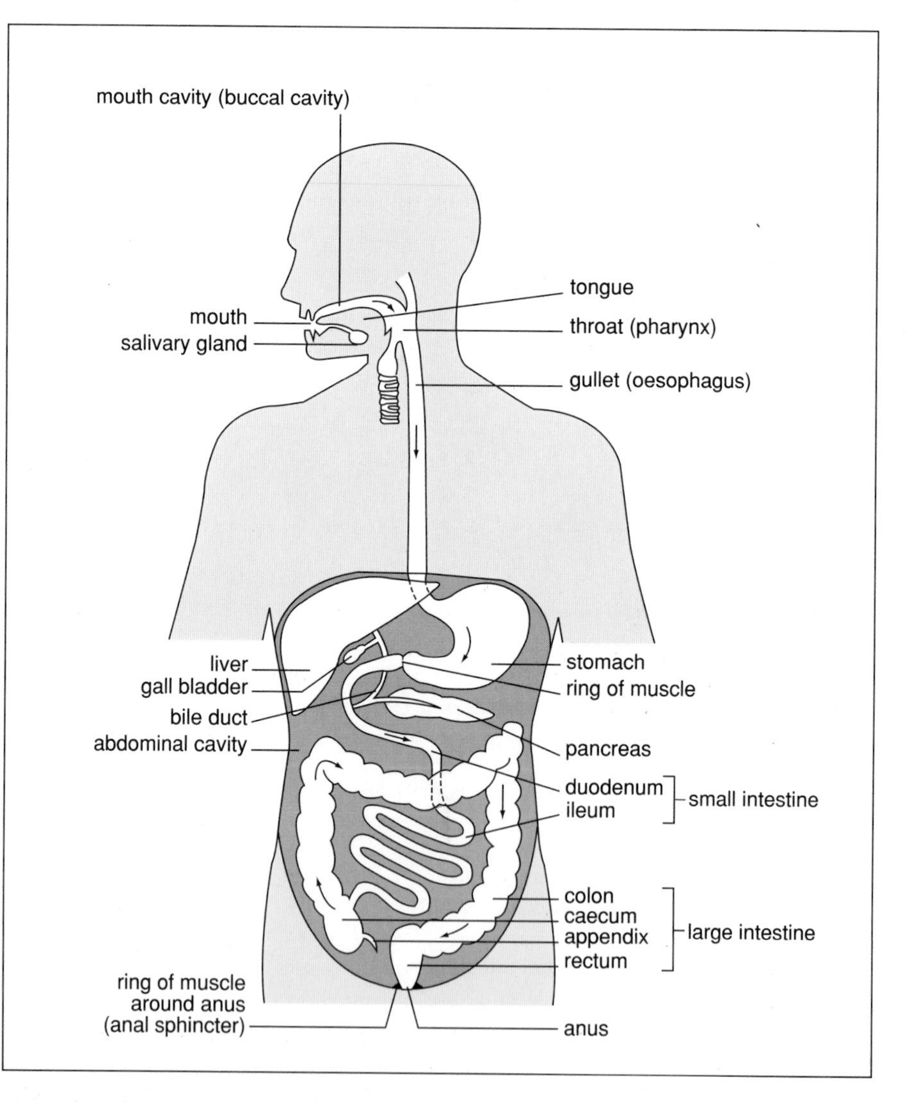

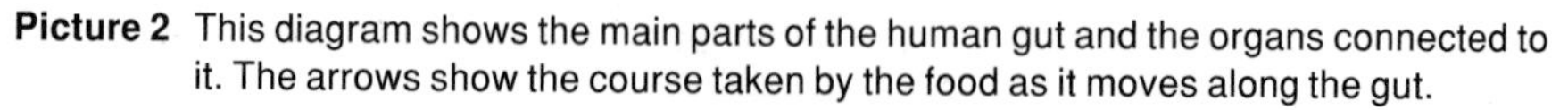

Picture 2 This diagram shows the main parts of the human gut and the organs connected to it. The arrows show the course taken by the food as it moves along the gut.

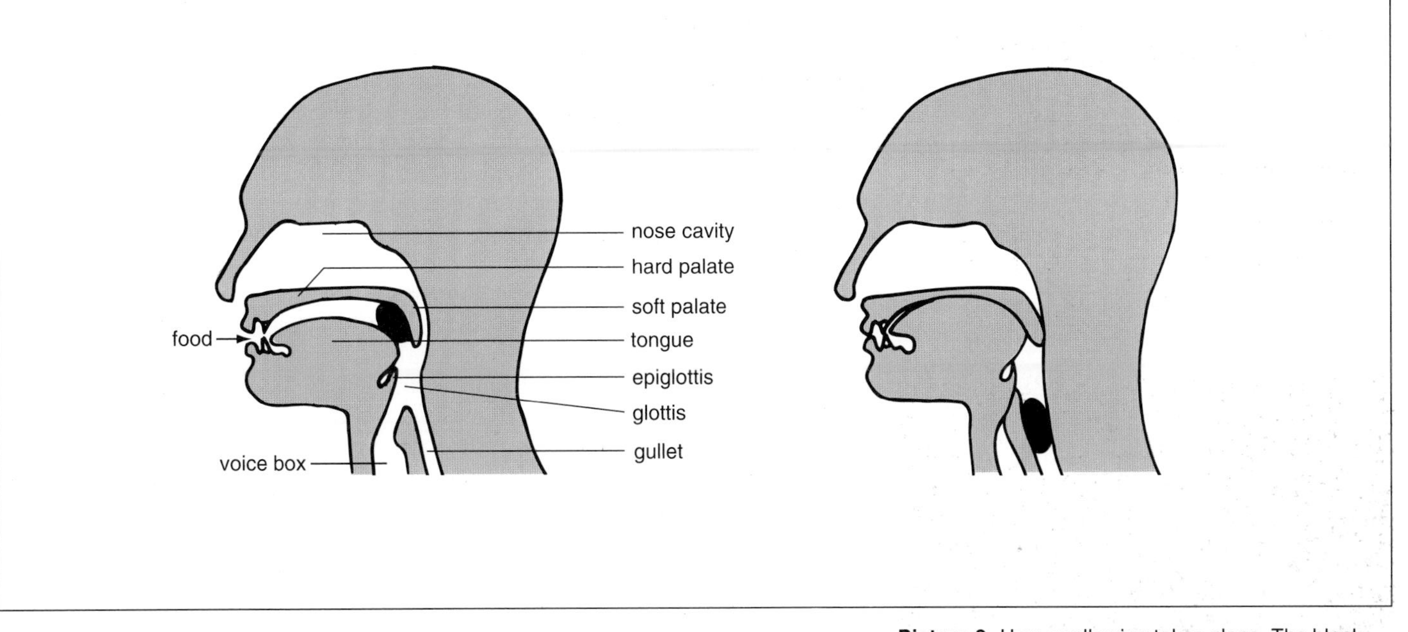

Picture 3 How swallowing takes place. The black blob is a lump of food. Swallowing is a reflex. When you swallow, the nose cavity and glottis are closed off, as shown in the right hand diagram, so food is prevented from going the wrong way. Notice how this closing off takes place. The soft palate moves back and presses against the back of the throat, and the front of the gullet moves up and presses against the epiglottis.

How does digestion take place?

Digestion is brought about by **digestive enzymes** (see page 23). There are several types, and each acts on a particular substance in the food. Thus **carbohydrases** act on carbohydrates such as starch, **lipases** on fat and **proteases** on proteins.

Digestion is helped by **chewing**. Chewing breaks the food into small pieces, increasing the surface area over which the enzymes can act.

Now let's look in more detail at what happens in each part of the gut. The enzymes are summarised in table 1 on the next page.

In the mouth cavity

When you take some food into your mouth, your mouth waters – it becomes full of **saliva**. Saliva is produced by **salivary glands** which are connected to the mouth cavity by ducts.

Actually your mouth starts watering before you eat. Salivating is a reflex which is triggered by the sight, smell and even the thought of food. (Reflexes are explained on page 194).

Saliva contains an enzyme called **amylase**. This acts on starch, breaking it down into malt sugar (maltose).

Saliva also contains a slimy substance called **mucus**. This serves as a lubricant, enabling the food to slip easily through the throat when it's swallowed.

Saliva also contains an enzyme called **lysozyme** which kills bacteria.

Through the throat and down the gullet

Swallowing is a reflex which happens without you having to think about it. When you swallow, the food is pushed down your throat by the back of your tongue. At the same time the glottis is closed off, so the food is prevented from going down the wrong way (picture 3).

If food does try to enter the glottis, you cough and this dislodges it. Coughing is a reflex, triggered by irritation of the sensitive epithelium in that region.

Once swallowed, the food passes down the gullet to the stomach. The gullet has muscle tissue in its wall. A ring of muscular contraction moves slowly downwards, pushing the food in front of it. This process is called **peristalsis** (picture 4). Mucus acts as a lubricant, enabling the food to slip down easily.

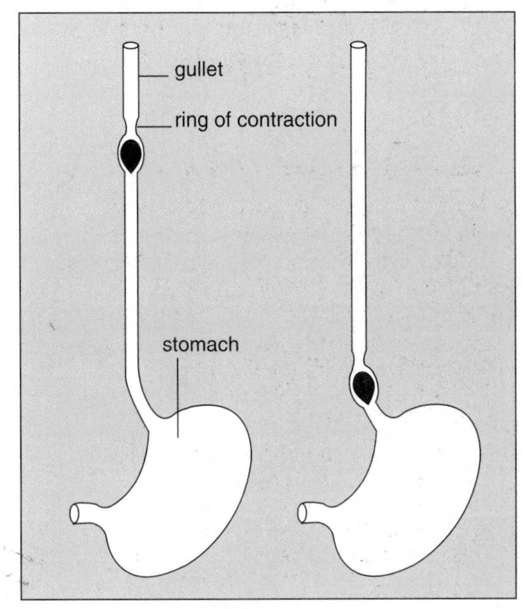

Picture 4 Food passes down the gullet by a ring of contraction which moves downwards sweeping the food before it. This is called peristalsis.

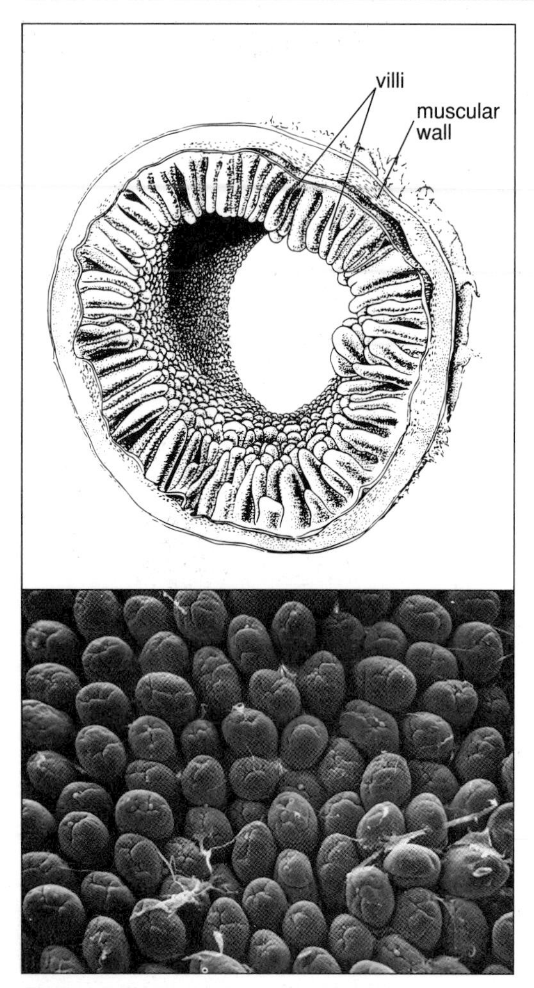

Picture 5 Looking into the small intestine. Notice the finger-like villi projecting into the cavity. The bottom picture, taken with a scanning electron microscope, shows the villi in surface view. There are 4–5 million villi in the human gut.

In the stomach

The wall of the stomach is thick and muscular, and it contains lots of tiny **gastric glands**, about 35 million altogether. The gastric glands produce a fluid called **gastric juice**. This contains an enzyme called **pepsin** which breaks down protein into simpler substances called polypeptides.

Pepsin works best in acid conditions. For this reason gastric juice contains **hydrochloric acid**. The acid also helps to kill germs that happen to get into the stomach. The wall of the stomach produces lots of mucus which protects the stomach lining from being damaged by the acid.

Food spends three or four hours in the stomach. Every now and again the stomach wall contracts, churning up the food and turning it into a mushy fluid.

Between the stomach and small intestine there is a ring of muscle. This is tightly shut most of the time, but every now and again it opens. At the same time a wave of contraction passes along the stomach and sweeps some of the food into the small intestine. If there is anything wrong with the food, violent contractions occur in the other direction. As a result the contents of the stomach are shot up the gullet and out through the mouth. This is **vomiting** and it is a useful way of getting rid of germs and poisons.

The stomach of calves contains an enzyme called **rennin**. Rennin solidifies milk protein, which is then broken down by pepsin. If it was not solidified first, milk would pass through the stomach so quickly that digestion would not have time to take place.

In the small intestine

Despite its name, the small intestine is the longest part of the gut – it may be over six metres in length.

The small intestine has enzymes which continue the work already started in the stomach. There is more amylase to break down starch, and an enzyme called **trypsin** acts on protein. Other enzymes then complete the digestion of the hamburger: **maltase** turns maltose into glucose, **lipase** turns fat into fatty acids and glycerol, and **peptidases** turn polypeptides into amino acids. And if Rachel had been daft enough to sprinkle sugar on her hamburger, this would have been broken down in her small intestine by an enzyme called **sucrase**.

All the new substances formed by the action of these enzymes are soluble and can be absorbed through the lining of the intestine into the bloodstream. More about this in a moment.

Some of these enzymes are produced by the wall of the small intestine itself. Others are produced by the **pancreas** from which they flow down a duct into the small intestine. The pancreas also produces sodium hydrogencarbonate which neutralises the acid from the stomach. This is necessary because the enzymes in the small intestine will only work properly in alkaline conditions.

Table 1 The main enzymes found in the human gut. Bile salts are also included although they are not enzymes and don't alter the food chemically. In the case of some substances, digestion starts in the small intestine and is finished off inside the surface layer of cells.

Where it comes from	Where it works	Name of enzyme	Food acted on	Substances produced
salivary glands	mouth cavity	amylase	starch	maltose
stomach wall	stomach	pepsin	protein	polypeptides
liver	small intestine	bile salts (not enzymes)	fat	fat droplets
pancreas	small intestine	amylase trypsin lipase	starch protein fat	maltose polypeptides fatty acids and glycerol (can be absorbed into blood)
wall of small intestine	small intestine	maltase sucrase peptidases	maltose sucrose polypeptides	glucose glucose and fructose amino acids (can be absorbed into blood)

The part played by the liver

The liver produces a fluid called **bile**. Bile is stored in the gall bladder and after a meal it is squirted, bit by bit, into the small intestine.

Bile contains substances called **bile salts**. These act on fat, breaking it up into very small droplets. The same thing happens when washing-up liquid comes into contact with fat – the process is called **emulsification**. Emulsification helps the digestion of fat by increasing the surface area on which the enzyme lipase can act.

The hamburger is now more or less completely digested and the soluble products are ready to be absorbed.

Absorption

The inner surface of the small intestine has many finger-like projections called **villi** (singular **villus**). The villi greatly increase the surface area for absorption (picture 5).

The wall of the small intestine has a very good blood supply (pictures 6 and 7). The villi contain blood and lymph capillaries into which the digested food substances are absorbed. Muscle tissue in the wall enables the intestine to contract, squeezing the contents this way and that, and strands of muscle in the villi enable them to wave about. As a result the contents of the small intestine are constantly brought into contact with the lining.

Recent research suggests that the final step in the digestion of carbohydrates and proteins takes place inside the cells lining the villi.

Every now and again a ring of contraction passes along the small intestine, sweeping the contents towards the large intestine.

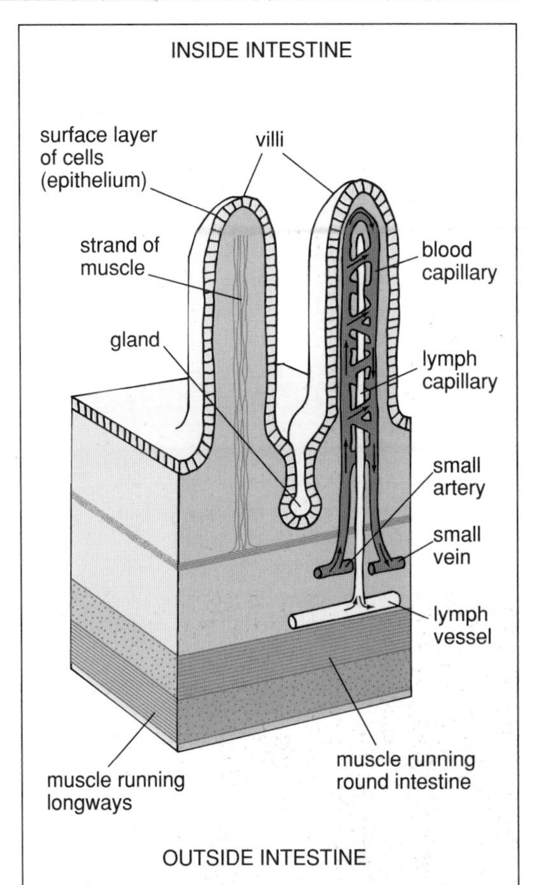

Picture 6 Diagram of the wall of the small intestine showing two villi in detail. Some of the fat we eat is absorbed into the lymph capillaries. Other digestive products are absorbed into the blood capillaries. The capillaries have been omitted from the left hand villus so as to show the muscle.

In the large intestine

The hamburger came with a piece of lettuce which contains cellulose. Humans don't have an enzyme for breaking down cellulose, so it can't be digested. Along with any other indigestible matter (**fibre**), it passes on to the colon, the first part of the large intestine.

As material moves along the colon, water is absorbed from it. As a result it becomes more and more solid. The solid matter moves on to the rectum where it is stored as **faeces**. The lining of the rectum produces mucus which acts as a lubricant and eases the passage of the faeces.

Eventually the faeces are **egested** through the anus. This is a reflex, triggered by the build-up of pressure in the rectum. The ring of muscle surrounding the anus opens up, and the wall of the rectum contracts forcing the faeces out. The polite word for this is **defaecation**.

Normally it takes between 24 and 48 hours from when the food is eaten to when the faeces are ready to be expelled through the anus. To keep up this sort of time scale, it is a good idea to eat plenty of fibre (see page 110). This stretches the wall of the large intestine, inducing it to contract. In this way the contents are kept on the move.

The caecum and appendix

The caecum and appendix are an offshoot from the large intestine, a kind of cul-de-sac or blind alley. They have no function in the human, but in grass-eating mammals such as rabbits they contain large numbers of helpful bacteria which can digest cellulose and break it down into glucose.

In humans the appendix sometimes gets infected with harmful bacteria. As the appendix sticks out from the intestine, the bacteria don't get flushed out by the normal passage of material along the gut. So they multiply there, and may cause severe inflammation leading to **appendicitis**. Normally appendicitis is cured by removing the appendix in an operation.

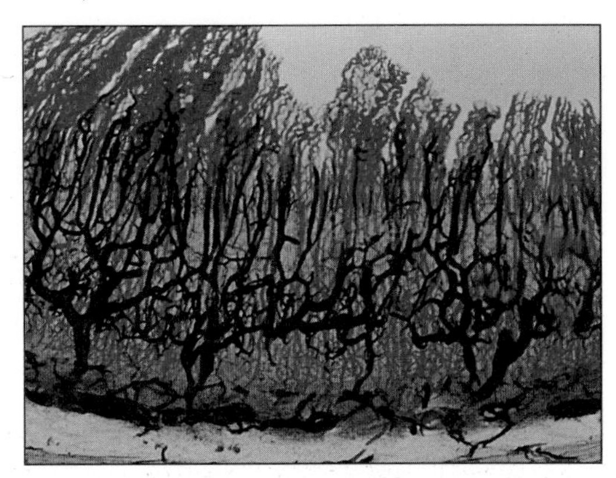

Picture 7 Part of the wall of the small intestine as seen under the microscope. The blood vessels have been injected with a dye so as to show them up. The arteries are red and the veins black.

Activities

A Does amylase break down starch?

1 Your teacher will give you a test tube containing a solution of amylase, the enzyme found in saliva.

2 Half fill another test tube with starch suspension (4 per cent).

3 With a pipette place fifteen separate drops of iodine solution, side by side, on a white tile.

4 With a glass rod lift a drop of the starch suspension from the test tube and mix it with the first drop of iodine on the white tile. A blue-black colour indicates starch. This will serve as your control.

5 Pour the amylase solution into the test tube of starch suspension, and shake quickly.

6 With the glass rod place a drop of the starch-amylase mixture in a drop of iodine on the white tile, and mix them together.

7 Repeat step (6) with the other drops of iodine at half-minute intervals. Note the colour given each time.

Explain the colour reactions as fully as you can.

Approximately how long does it take the amylase to break down the starch?

B Observing pepsin in action

1 Obtain two large test tubes and label them A and B.

2 Place a small piece of hard-boiled egg white (albumen) in each test tube.

3 Cover the egg white in tube A with a solution of acidified pepsin which your teacher will give you.

4 Cover the egg white in tube B with water. This will serve as a control.

5 Leave both tubes in a warm place, preferably an incubator at 30°C, for about 48 hours.

6 After 48 hours examine the contents of both tubes.

In which tube or tubes has the egg white changed?

What do you think has happened to it?

C At what pH does pepsin work best?

Plan an experiment to find out the pH at which pepsin works best. You can do it the same way as in activity B, or you can invent your own method.

After having your plan approved by your teacher, carry out the experiment and draw conclusions.

D Observing emulsification

1 Obtain three test tubes, and label them A, B and C.

2 Pour some corn oil into each test tube to a depth of about 3 cm.

3 To A add a few drops of water.

To B add a pinch of powdered bile salts.

To C add a few drops of washing-up liquid.

4 Shake the test tubes, then let them stand for a while.

What has happened to the oil in each test tube?

How do bile salts help digestion?

How does washing-up liquid help with washing up?

5 Half fill a test tube with water and add a small piece of solid fat. Shake well.

6 Half fill a second test tube with a solution of bile salts and add a small piece of fat. Shake well.

What happens to the fat in each test tube? Explain your observations.

Questions

1 Why is the human intestine coiled rather than straight?

2 Which part of the human gut:

a absorbs water from indigestible material,
b contains the enzyme pepsin,
c is normally acidic,
d receives bile from the bile duct?

3 Explain each of the following:

a If you chew a piece of bread for long enough, it eventually begins to taste sweet.
b When you swallow a piece of food, the food normally does not go up into the nose cavity.

4 There is a disease of cattle in which the villi in the small intestine are destroyed and the inner lining of the small intestine becomes smooth. As a result the animal gets weak and wastes away. Why do you think the disease has this effect?

5 It has been suggested that saliva produced during a meal digests starch faster than saliva produced before the meal. Describe an experiment which could be done to find out if this is true.

6 The diagram on the right shows an experiment which is intended to show what happens in the human gut.

After being set up, glucose, but not starch, passes out of the bag into the surrounding water.

a How could you show that glucose has leaked out, but starch has not?
b How would you explain this result?
c To what extent is this similar to what happens in the human gut?

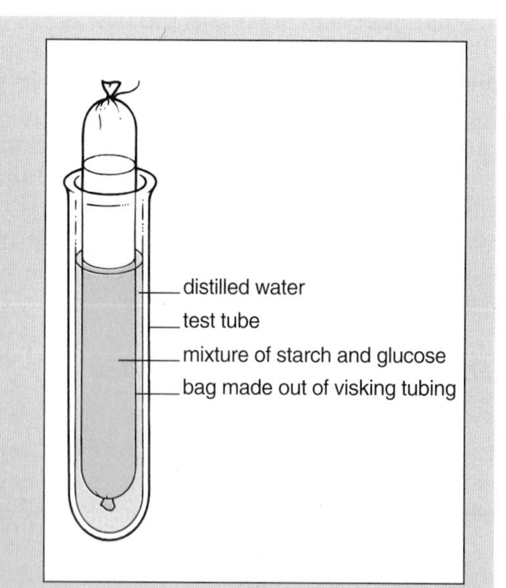

Constipation

The record time for constipation is claimed to be 102 days! Most people suffer from constipation at one time or another, but not for that long. It is caused by indigestible matter moving too slowly along the large intestine. The result is that more water is absorbed than usual, and the faeces become hard and dry.

Constipation can be caused by bad bowel habits. People usually feel the urge to defaecate after a meal, generally breakfast: this is a natural reflex arising from the stretching of the gut wall. If you keep suppressing this reflex, the faeces are held in the intestine and you may become constipated.

We can help to prevent constipation by eating foods which contain plenty of fibre, such as wholemeal bread and fresh fruit (see page 110). Fibre can't be digested so it adds to the bulk of material in the large intestine, stretching its wall. This stimulates the muscle tissue in the wall to contract, pushing the faeces along and keeping them moving.

1 Some people think that you should get into the habit of defaecating at a regular time each day. Others feel that you should defaecate only when you need to. What do you think?

2 People who have a mainly vegetarian diet defaecate more often than people who eat a lot of animal food. Why do you think this is?

3 A person who is constipated a lot is more likely to get cancer of the bowel than a person who defaecates regularly. Suggest reasons for this.

Seeing your own gut

Spread out your hands and place them on your abdomen. Immediately beneath your hands is your stomach and intestine, approximately eight metres of it. So near, and yet you can't see any of it. Is there any way a doctor can see it without cutting you open?

Yes there is, and it's done with **X-rays**. As you know X-rays are used to take pictures of our bones. A machine sends a beam of X-rays through the body; the X-rays do not pass easily through the solid bones. On the other side of the person there is a photographic film which, when exposed to the X-rays, goes dark everywhere except where the bones are.

Unlike bones, the gut lets X-rays through so it doesn't show up in X-ray pictures. How, then, can we see it? The answer is to get the person to drink a fluid which won't let X-rays through. The fluid fills the stomach and intestine, so when the person is X-rayed these parts of the gut show up clearly. The substance used is barium sulphate. It is taken as a thick suspension and is called a **barium meal** (picture 1).

A doctor may use this technique to find out if there's something wrong with a patient's gut. For example, a stomach ulcer will show up in an X-ray picture as an irregularity in the lining of the stomach. In hospitals X-ray pictures are taken by a specially trained person called a **radiographer**. Nowadays, continuous pictures are obtained on a television screen.

Another way of seeing the gut is to push a flexible tube. and a light, down the patient's throat into the stomach or small intestine. The doctor then looks through the tube at the lining of the gut (picture 2). This can help to confirm a diagnosis which has been made from X-ray pictures.

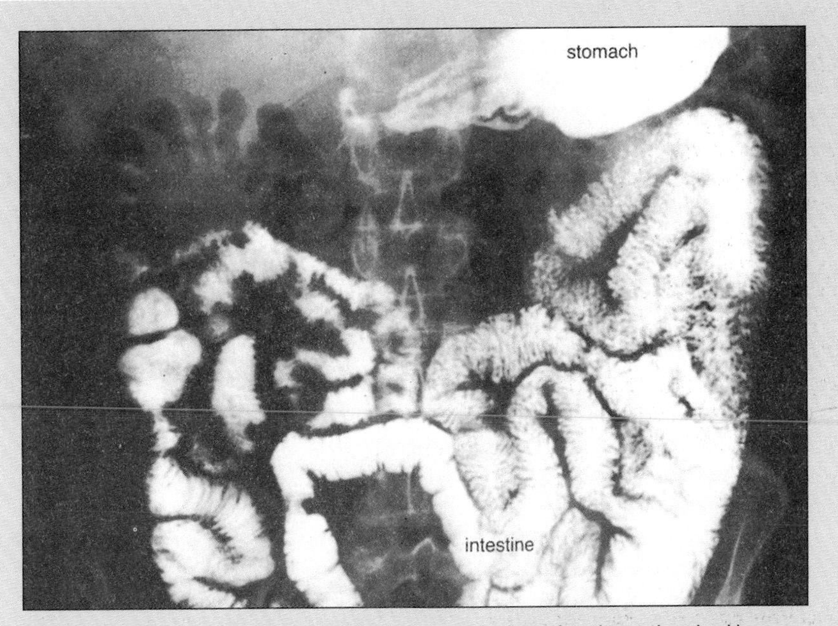

Picture 1 An X-ray photograph of the human gut taken after the patient had been given a barium meal.

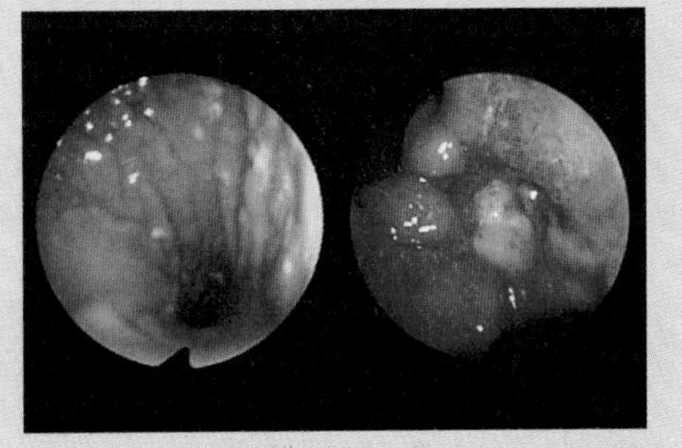

Picture 2 The lining of the stomach seen through a tube. Left: a healthy stomach. Right: a stomach ulcer.

1 Explain the meanings of the following words, used in the passage:

(a) small intestine, (b) suspension, (c) irregularity, (d) radiographer.

2 Patients who are to have a barium meal are usually asked not to eat anything for about six hours beforehand. Why?

3 In picture 1 there appears to be no connection between the stomach and intestine. Suggest two possible reasons for this.

4 The tube which is used for examining a patient's stomach enables the doctor to actually see the lining of the stomach. How do you think it works?

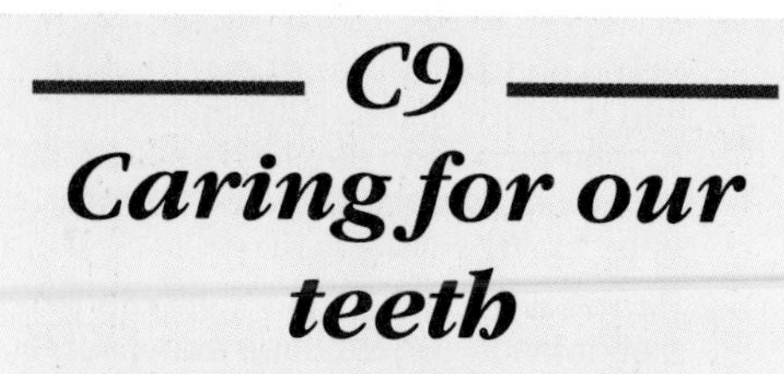

C9 Caring for our teeth

Teeth are one of our most useful possessions. Here we look at our teeth and see how we can best look after them.

A full set of teeth

An adult human has 32 teeth at the most. You can see them in pictures 1 and 2. The top part of each tooth (the part you can see) is called the **crown**. The bottom part, buried in the jaw bone, is called the **root**.

Run the tip of your tongue over your teeth and feel them. You will find that they are not all the same. The ones towards the front of your jaw have simple crowns with a straight or curved cutting edge. Those further back (**cheek teeth**) have broader crowns with several pointed **cusps**.

If you have seen teeth that have been taken out by a dentist, you will know that their roots differ too. The front teeth have roots with a single projection, whereas the cheek teeth generally have two or three projections (picture 3).

When do we get our teeth?

Babies are usually born without any teeth. During the first few years of life a set of **milk teeth** develops. The baby has fewer teeth than an adult because the jaw is too small.

Between the ages of six and twelve the milk teeth fall out, one by one, and are replaced by a set of **permanent teeth**. There are four cheek teeth on either side of each jaw. Further cheek teeth may be added behind the others after the age of seventeen: these are the famous **wisdom teeth**.

They bring the total number of teeth to 32. This is your final set – if any teeth are lost now, they will not be replaced.

If your jaw is small, a wisdom tooth may break through and push against the one in front, which may then push against the one in front of that. This can cause a lot of pain. Usually the dentist extracts the wisdom tooth and the one in front of it.

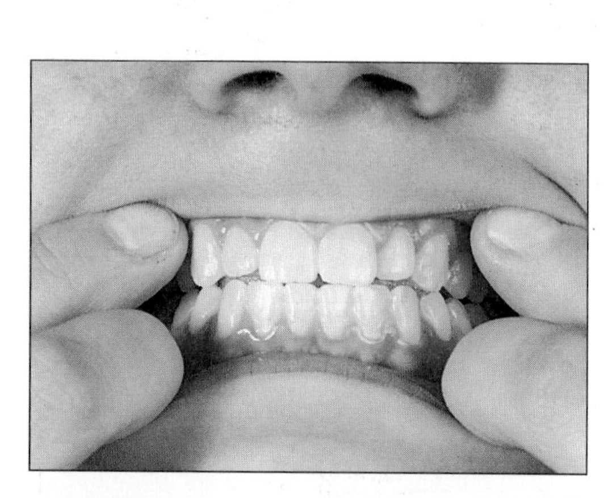

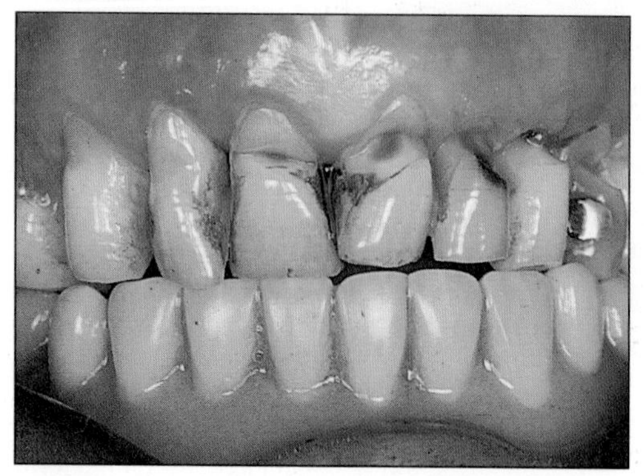

Picture 1 Well-kept and badly-kept teeth.

upper jaw
lower jaw
i i c pm pm m m m
i i c pm pm m m m

i incisor
c canine
pm premolar } cheek teeth
m molar

Picture 2 A complete set of human teeth, looking from the side.

Inside a tooth

Suppose you were to cut a thin section of one of your front teeth and look at it under the microscope. Picture 4 shows what you would see.

The crown is made up of three layers. On the outside is a thin layer of extremely hard **enamel**. Beneath this is a layer of hard **dentine**. In the centre is a soft area called the **pulp cavity** which contains small blood vessels and a nerve. Tiny channels containing extensions of living cells run out from the pulp cavity into the dentine. These make the dentine sensitive. The enamel and dentine are made hard by the presence of calcium phosphate, the same substance that makes bones hard.

The outside of the root is covered not by enamel but by another material called **cement**. Attached to the cement are tough fibres which run into the jaw bone. These fibres hold the tooth in its socket: they allow the tooth to move slightly, and cushion it from being jarred when it hits something hard.

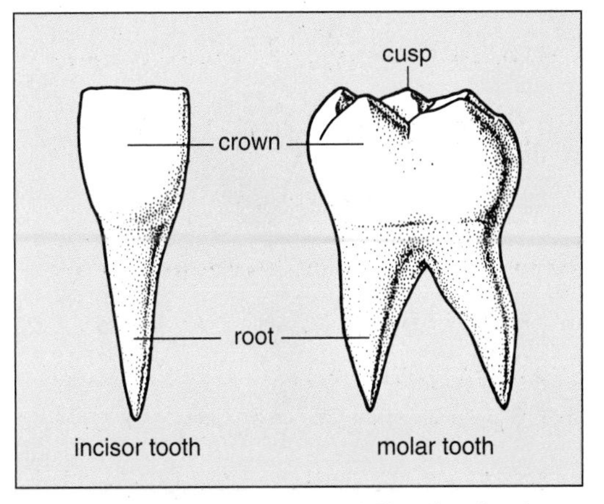

Picture 3 This is what teeth look like after they have been extracted from the jaw.

Tooth decay

Tooth decay is caused by bacteria in the mouth. These bacteria form an invisible layer called **plaque** on the surface of the teeth.

After a meal, the bacteria feed on any sugar present and turn it into acids. The acid eats into the teeth. Within an hour or so the acid is neutralised by the saliva – but it's too late, the rot has begun.

Decay usually starts between the teeth and in the crevices on the crowns. The acid eats through the enamel into the dentine, allowing bacteria to get into the pulp cavity (picture 5). In severe cases the bacteria may spread to the base of the tooth, causing an **abscess**.

Bacteria may also get between the tooth and the gum, causing the gum to bleed. And sometimes the fibres by which the root is attached to the bone get infected. Such teeth gradually become loose and eventually fall out.

In developed countries tooth decay is a serious problem. In Britain, for example, nearly 50 per cent of five year olds and over 90 per cent of 15 year olds suffer from it. Until recently nearly one in three people had lost all their teeth by the age of twenty and were wearing false teeth (dentures).

Fortunately the amount of tooth decay is now falling. This is partly due to better health education, and partly to the skills and technology of dentists.

Picture 4 This diagram shows the inside of a tooth.

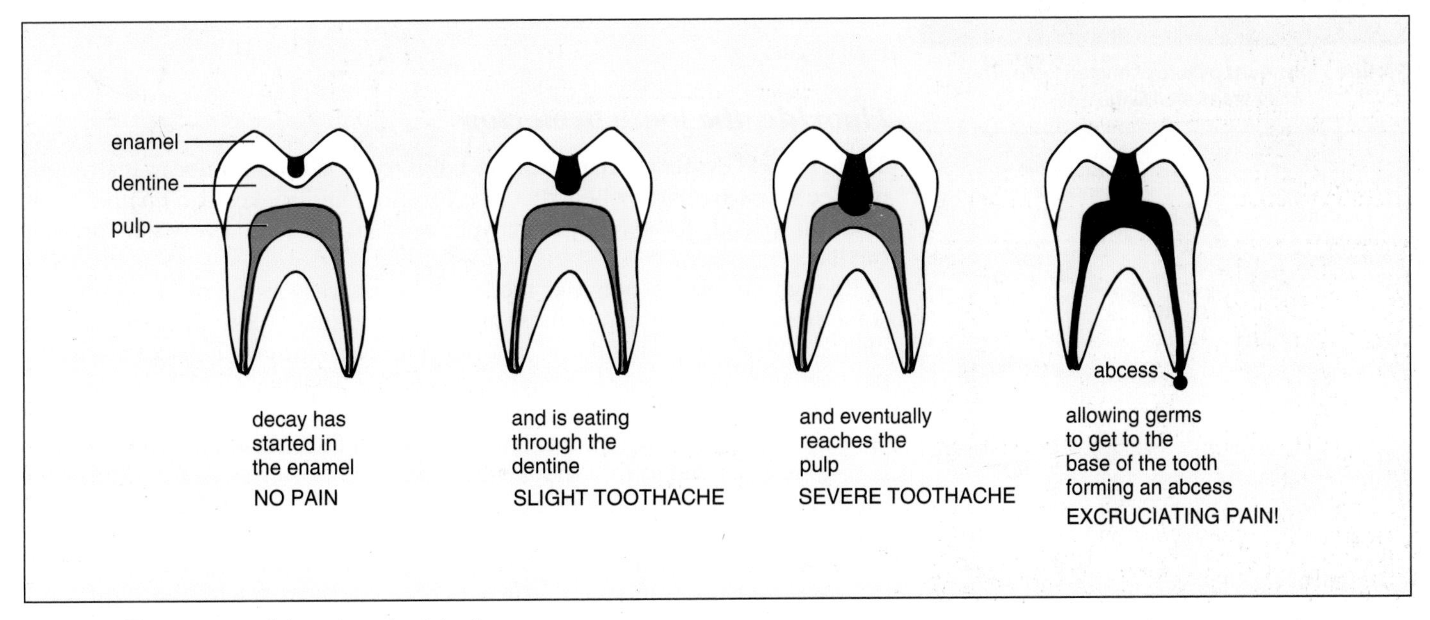

Picture 5 The progress of decay in a cheek tooth.

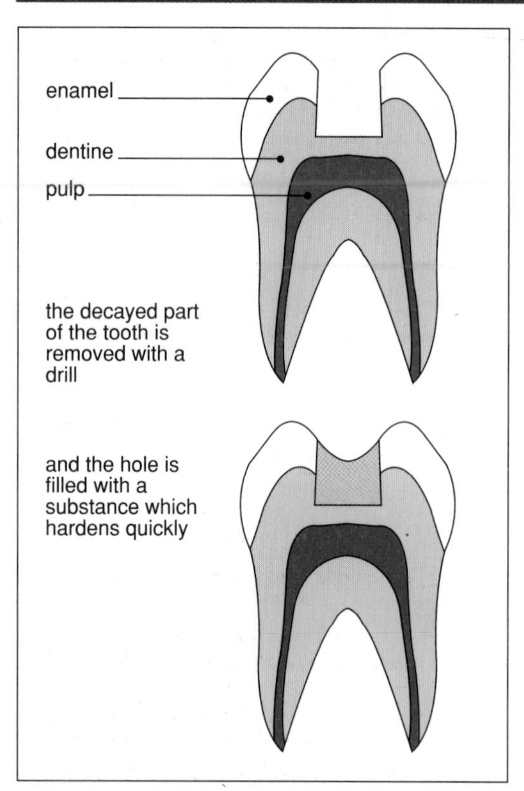

Picture 6 How a dentist fills a tooth.

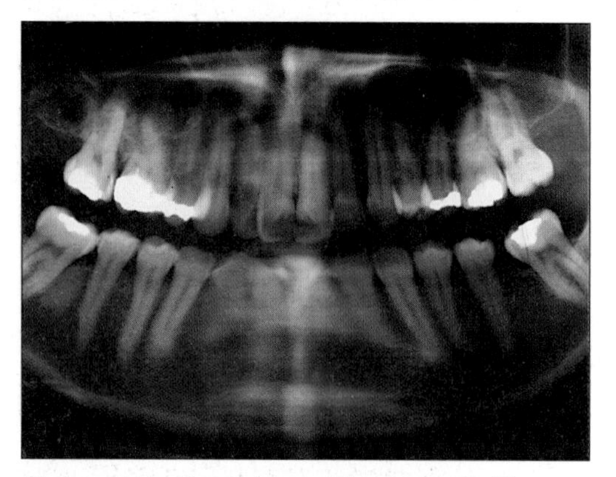

Picture 7 An X-ray picture of human teeth. The white areas are fillings.

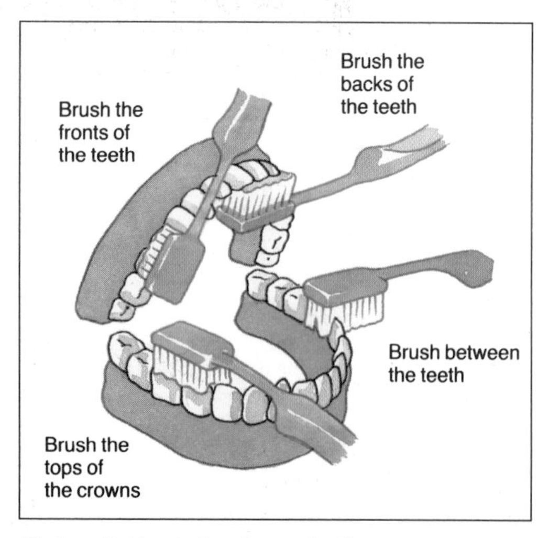

Picture 8 How to brush your teeth.

Dentist to the rescue

So long as the decay has not spread too far, the tooth can be repaired by a dentist. The dentist cuts away the decayed part of the tooth with a drill, and then fills the hole with a substance which hardens quickly (picture 6).

Back teeth are usually filled with a mixture of metals such as silver and tin. Front teeth are filled with a ceramic or plastic-like material, the same colour as the teeth.

If the decay has got into the pulp cavity, the dentist may not be able to save the tooth. It may then have to be pulled out, or the crown replaced by an artificial one. Before deciding what to do, the dentist takes X-ray pictures of the person's teeth to find out what state they are in (picture 7).

Looking after our teeth

Many studies have been carried out which show that tooth decay is caused by eating sugary foods, particularly sweets, and drinking sugary drinks. Scientists have found that the amount of tooth decay amongst people is directly related to the amount of sugar consumed.

The best way to prevent tooth decay, therefore, is to keep sugar out of your mouth. This means not eating sweets, and not drinking sugary drinks, particularly between meals. If you avoid sugar, the bacteria are starved of their food supply so plaque does not develop on your teeth.

To some extent tooth decay can be kept under control by cleaning your teeth with a **toothbrush**. This removes at least some of the plaque. It takes about 24 hours for plaque to be re-formed after it has been removed, so it's essential to brush the teeth at least once a day.

Plaque left on the teeth gradually hardens into a crusty material called **tartar**. You can't remove this yourself – it has to be scraped off by a dentist or dental hygienist.

For cleaning teeth dentists recommend using a toothbrush with a small head whose bristles can get between the teeth and into the crevices (picture 8). You can also pull a piece of thread (**dental floss**) backwards and forwards between the teeth. This helps to remove plaque from places which the toothbrush can't reach.

You should throw away your toothbrush and use a new one as soon as the bristles look bent. An old toothbrush with bent bristles is useless.

It is a good idea to visit the dentist every six months or so, even if you think there is nothing wrong with your teeth. Decay may have started without you realising it.

Fluoride, the tooth protector

There is good evidence that **fluoride** helps to prevent tooth decay. It strengthens teeth, particularly when they are forming, and makes the enamel more resistant to acid. To be really effective it needs to be taken every day throughout life.

Enough fluoride to help our teeth occurs naturally in the drinking water in some parts of the country. In these areas the amount of tooth decay in the population is about half what it is in non-fluoride areas. In certain places, for example Birmingham, fluoride has been added to the drinking water for many years and this has led to a great improvement in people's teeth. Only tiny amounts are needed: in Birmingham the concentration of fluoride in the drinking water is one part per million. In areas where there is not enough fluoride in the water, people can take fluoride tablets or mouth-rinses. Cleaning the teeth with a fluoride toothpaste is also helpful.

Some people feel that it is wrong for local authorities to add fluoride to drinking water, even though it may help people's teeth. What do you think?

Activities

A You and your teeth

The full number of teeth which an animal possesses is given by the **dental formula.**

The dental formula of an adult human is:

$$\mathbf{i}\,\frac{2}{2}\quad \mathbf{c}\,\frac{1}{1}\quad \mathbf{pm}\,\frac{2}{2}\quad \mathbf{m}\,\frac{3}{3}$$

i stands for incisors, **c** for canines, **pm** for premolars, and **m** for molars.

The top figures are the number of teeth on one side of the upper jaw. The bottom figures are the number of teeth on one side of the lower jaw. So an adult human has a total of 32 teeth.

Look at the inside of your mouth with a mirror, and try to answer these questions:

1 How many teeth have you got altogether?

2 Which teeth, if any, do you lack, and why?

3 What is your dental formula at the moment?

4 What was your dental formula likely to have been when you were five years old?

5 What has happened to your teeth since you were five?

B Examining individual teeth

1 Examine healthy human teeth in detail.

Using picture 2 to help you, decide whether each tooth is an incisor, canine or cheek tooth.

2 Draw each type of tooth, showing as many of the structures in picture 3 as you can see.

3 Examine decayed teeth which have been extracted by the dentist.

Compare them with healthy teeth. Whereabouts is the decay?

Why do you think the decay is situated where it is?

C Going to the dentist then and now

John Betjeman, who was Britain's Poet Laureate in the early 1980s, said that the only thing in Britain which had improved since he was a boy was going to the dentist.

Talk to elderly people and find out what going to the dentist was like when they were young.

Make a list of the ways dental treatment has improved in the last 50 years.

D To see the plaque on your teeth

This can be done using plaque-staining tablets available from a pharmacy.

1 Chew a tablet and spread it over your teeth with your tongue. Then spit it out. (These tablets are not meant to be swallowed, but it will do you no harm if you do.)

2 Rinse your mouth with water.

3 Look at your teeth in a mirror. Any plaque will be stained pink. Where is the plaque located?

4 Brush your teeth with toothpaste in the way you normally do, then rinse your mouth out with water. *Do not share your toothbrush with anyone.*

Do not share your toothbrush with anyone.

5 Look at your teeth in the mirror again.

Has all the plaque been removed? If not, where is it still left?

6 Brush your teeth again, using the method given in picture 8. Work the toothbrush this way and that, and try hard to remove all traces of plaque. Then rinse your mouth out with water.

7 Look at your teeth in the mirror again.

Has all the stained plaque been removed now?

8 If there is still some plaque between your teeth, try removing it with dental floss.

Does dental floss remove the plaque?

Questions

1 In about 100 words give advice to the general public on how to prevent tooth decay.

2 Look at picture 8 opposite. What are the advantages of using a toothbrush in the way suggested? Are there any possible disadvantages and how might they be overcome?

3 Two toothpaste manufacturers, Teethcleen and Dentagleem, compete with each other. Each claims that its toothpaste is the best one for preventing tooth decay. How could you, as a scientist, decide between them?

4 You decide to have a snack. To what extent is each of the following foods liable to cause tooth decay: an apple; a bar of chocolate; a slice of bread; a piece of cheese?

5 The graphs show the pH of plaque at different times of the day for two people, Pat and Ann. The shaded areas show the pH values at which decay occurs.

a For approximately how many hours, between 6 a.m. and 11 p.m., is tooth decay likely to occur in (i) Pat and (ii) Ann?

b Explain the pattern of the curves in each graph.

c Suggest *two* ways by which Ann might reduce the rate at which her teeth decay.

d The bacteria that form plaque respire anaerobically.
(i) What does *anaerobic* mean?
(ii) How do these bacteria bring about tooth decay? (Hint: what happens when bacteria respire anaerobically?).

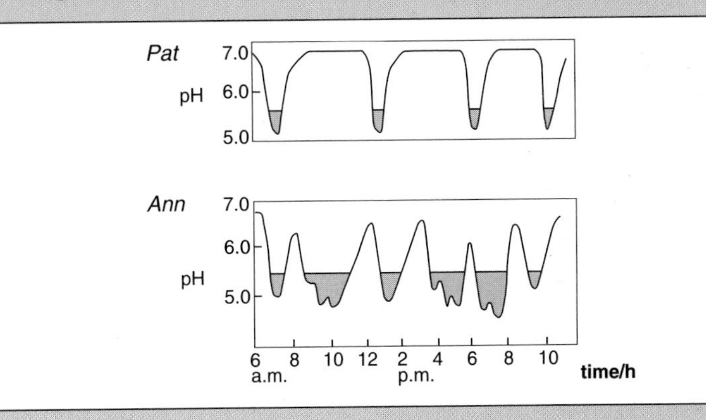

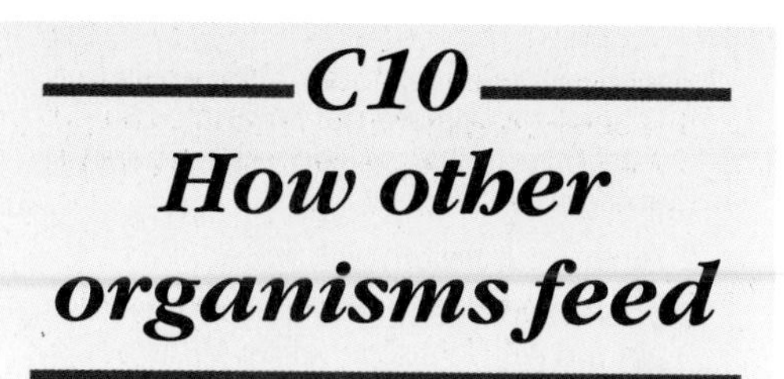

C10 How other organisms feed

Here we compare the way we feed with the methods used by certain other organisms.

Picture 1 A young male lion displaying its teeth.

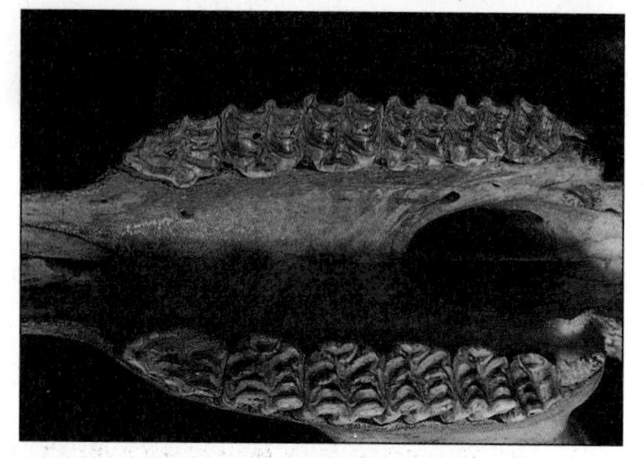

Picture 2 The cheek teeth (molars and premolars) of a horse showing enamel ridges on the surface of the crowns.

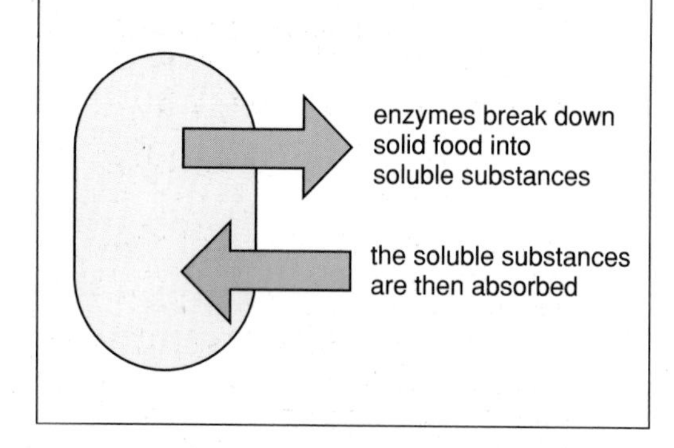

Picture 3 Fungi and many types of bacteria digest their food outside the body, and then absorb the soluble products of digestion.

Carnivores

Carnivores eat other animals, and this presents problems. Take the lion for example. It has to chase, catch, kill and then eat its prey. The lion has special features that help it to do this. In particular, its canine teeth are large and dagger-like (picture 1). What does it use them for, and what other features help the lion to feed?

Herbivores

Now consider a herbivore such as the horse. It doesn't have to chase its food, but the food (grass) is difficult to digest. One reason for this is that it contains **cellulose**. Animals don't have an enzyme for breaking down cellulose; the only organisms that possess this enzyme are certain microbes, notably some species of bacteria.

Herbivorous mammals such as the horse have special features which help them to digest cellulose:

- **Their cheek teeth have enamel ridges on the surface**. When the animal is chewing, the ridges of the upper and lower teeth rub together and grind up the food (picture 2).
- **They have an extra-long small intestine**. Plant food takes a long time to digest. Having a long intestine gives the food plenty of time to be digested before it reaches the end. A cow's small intestine may be 50 metres long.
- **The gut contains cellulose-digesting microbes**. The microbes turn the cellulose into soluble sugar. Some of this is used by the microbes for themselves, the rest is used by the herbivore. In cattle and sheep these microbes are in a special part of the stomach called the rumen.

How do fungi and bacteria feed?

Humans take in solid food, chew it, and then break it down into soluble substances by means of digestive enzymes inside the gut. Most animals feed this way.

However, fungi and most bacteria tackle their food quite differently. They have no gut and no teeth. They break down the food into soluble substances *outside* their bodies, and then absorb the soluble substances through their cell walls (picture 3).

These organisms get their food from the bodies of living or dead animals and plants.

- Those that feed on living organisms are **parasites**. They can cause serious diseases of humans, domestic animals and crop plants. An example is the potato blight fungus (picture 4).
- Those that feed on dead material are called **saprobionts**. They help to bring about decay (see page 52). Some of them cause food spoilage. An example is the penicillin fungus which you can read about on the next page.

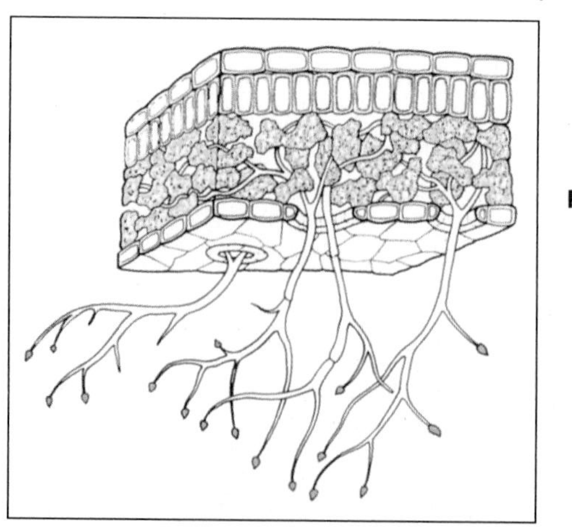

Picture 4 The potato blight fungus *Phytophthora infestans* in a leaf. The hyphae form a network between the cells of the leaf. Short side branches grow into the cells and digest the contents. In this picture aerial hyphae have grown out of the underside of the leaf and are forming spore bodies for infecting other plants.

Activities

A Observing animals feeding

Watch these animals feeding:

- sheep, cow or horse,
- dog or cat,
- rabbit, hamster or guinea pig.

Observe the movement of the jaws, and make notes on what you see. How do the jaw movements, tongue, and the structure of the teeth enable each animal to eat its particular kind of food?

If available, examine the skulls of the animals, and study the teeth in detail.

B Fungi and their method of feeding

Put a piece of moist white bread in a dish, and cover it with a lid.

Fix the lid to the dish with sticky tape.

BIOHAZARD

Examine it at intervals during the next two weeks for the growth of fungi. The main fungus which is likely to develop is *Mucor* (pin mould). Notice that the fungus consists of a network (mycelium) of fine threads (hyphae) on the surface of the bread.

What substance in the bread is the fungus probably feeding on? How could you test your suggestion? How does the fungus manage to feed on it?

You may see black spore-containing bodies at the tops of vertical hyphae (hence the name 'pin mould'). Try to find out how the spores are released. What is their function?

C To find out if pin mould needs moisture to grow

Plan an experiment to test the suggestion that moisture is needed for pin mould to grow on a piece of bread. Make sure you set up a control, and that you keep all conditions the same except for the one whose effect you are investigating.

Show your plan to your teacher. When your teacher has approved your plan, carry it out.

If time permits, you could investigate other conditions that are needed for pin mould to grow, e.g. temperature and oxygen.

The penicillin fungus

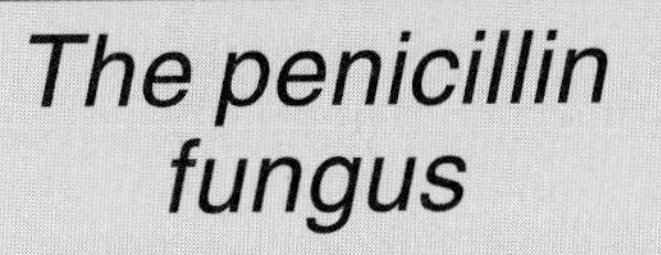

The penicillin fungus, *Penicillium*, feeds on sugary substances on the surface of fruit and other foods. Its life cycle is shown in the picture.

The fungus starts off as a spore. If the spore finds itself on a suitable food source, a thread-like **hypha** grows out. The hypha consists of a chain of cells which grows over the surface of the food, branching this way and that. Eventually the food becomes covered by a tangled mass of hyphae called a **mycelium**. This is visible as a greenish mould on the food.

The hyphae produce digestive enzymes which break down the food into soluble substances. These are then absorbed by the fungus. The hyphae don't penetrate far into the food; they stay near the surface because they need oxygen for respiration.

Eventually the food will run out, so the fungus must move to another source of food. Hyphae grow upwards, and form branches. Spores are pinched off from the tips of the branches. The spores are released and carried away by air currents or on people's fingers. If a spore lands on a suitable surface, a new mould develops and the cycle is repeated.

Picture 1 The structure and life cycle of the penicillin fungus, *Penicillium*.

Growing mushrooms

Mushrooms are fungi. The mycelium of a field mushroom is in the soil, and the mushroom itself is its spore-producing body. The mushroom is the only part of the fungus above the ground.

Mushrooms are a popular food, and they have to be mass produced. They do not require light and are grown in places where it is cool and moist such as cellars, sheds, caves and disused mines.

The food on which the fungus develops is usually woody material such as the stalks of wheat and other cereal plants, composted with horse manure. Farm stubble, the stalks of cereal crops left behind after harvesting, is used for growing mushrooms.

Farm stubble used to be burned in the fields after harvesting. However, it is much better to use it for growing mushrooms.

1 Why is it better to use farm stubble for growing mushrooms than to burn it?

2 Why is the stubble composted with horse manure before it is used for growing mushrooms? Do you think it could be used without composting it first?

Questions

1 What sort of teeth would you expect each of these mammals to have:

a one that feeds on mussels and breaks them open,

b one that feeds on fish and swallows them whole?

2 You may have noticed that when jam goes mouldy the mould is only at the surface. Suggest a reason for this? How could your suggestion be tested?

3 How do *plants* feed? Fungi used to be regarded as belonging to the plant kingdom. Now they are given a kingdom of their own. Why is it better *not* to regard them as plants?

C11 The heart and circulation

Blood constantly flows round and round the body. This is called the circulation.

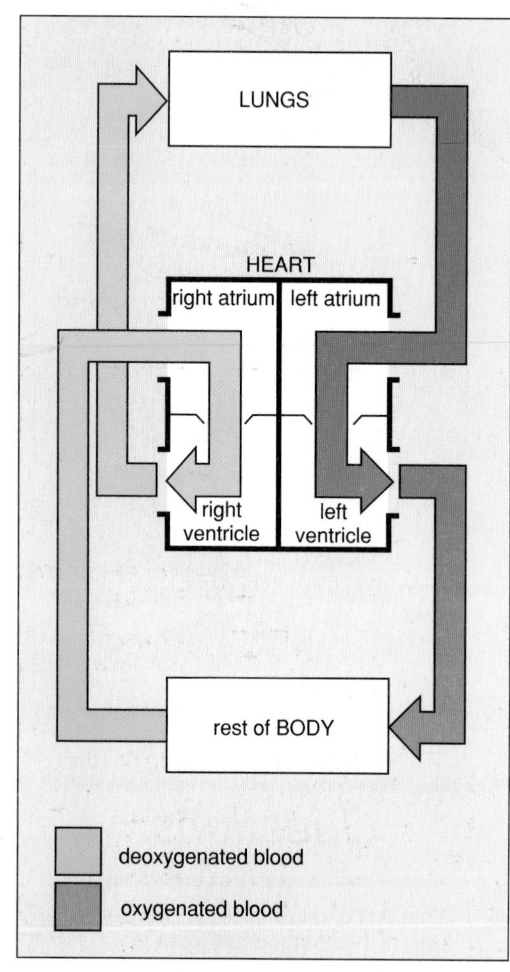

Picture 2 A simple view of the human circulation. It is usual to show pictures of human anatomy from the belly (ventral) side, so in this diagram the right side of the heart is on your left, and the left side is on your right.

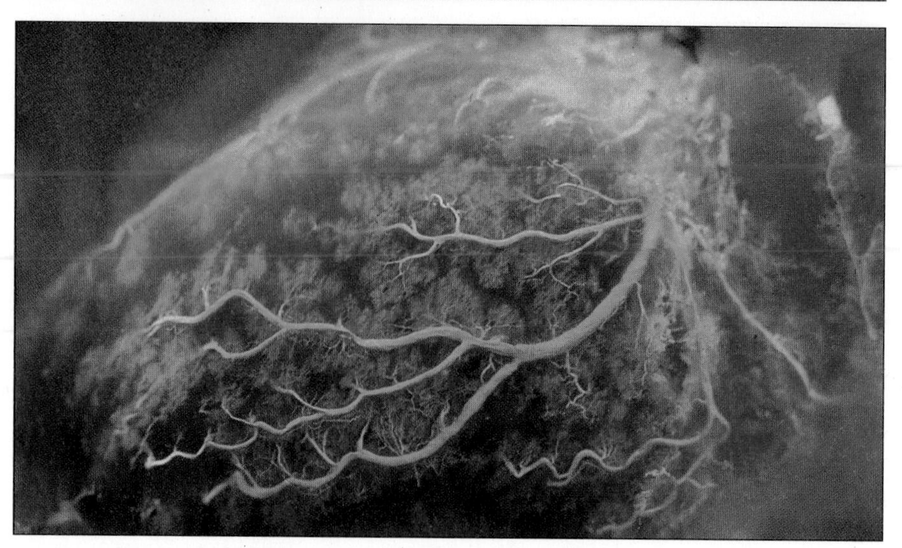

Picture 1 The heart, the body's life-giving pump. The branched tubes on the surface are the coronary arteries which carry oxygen to the heart muscle.

The general plan of the circulation

The structures which blood flows through as it goes round the body make up the **circulatory system**. The main organ in the circulatory system is the **heart**, which is situated between the lungs in the chest (picture 1). The heart's job is to pump the blood round the body. More about that presently.

The rest of the circulatory system consists of tubes called **blood vessels**. These are of two types: **arteries** carry blood away from the heart to the various organs, and **veins** carry blood back from the organs to the heart. Within each organ the arteries and veins are connected by numerous very narrow blood vessels called **capillaries**.

As blood flows along the capillaries, oxygen and other useful substances diffuse out to the surrounding cells, and unwanted substances diffuse in the other direction. In this way the capillaries keep our tissues in a healthy state.

The capillaries are extremely numerous and every organ contains thousands of them. No cell is more than a twentieth of a millimetre from the nearest one. If a person's capillaries were laid end to end, they would stretch round the world two and a half times!

We really have two circulations

Look at picture 2. The heart is divided by a partition into two halves, left and right. Blood is pumped from the right side of the heart to the lungs where it takes up oxygen. The **oxygenated blood** then passes back to the left side of the heart, which pumps it to the rest of the body. The oxygen is then taken up by the various organs, and the **deoxygenated blood** returns to the right side of the heart.

So there are really two circulations, one serving the lungs and the other serving the rest of the body. Putting it a different way: blood passes through the heart twice for every complete circuit of the body.

When blood arrives at the heart from the veins, it goes first into a chamber called the **atrium** (plural: **atria**). This is the Latin word for an entrance hall. From here the blood flows through an opening into another chamber called the **ventricle**. This has a very thick, muscular wall for pumping the blood into the arteries. So the heart has four chambers altogether: left and right atria, and left and right ventricles.

The human heart and circulation are shown in detail in pictures 3–6. Follow the route by which blood flows through the heart and round the body.

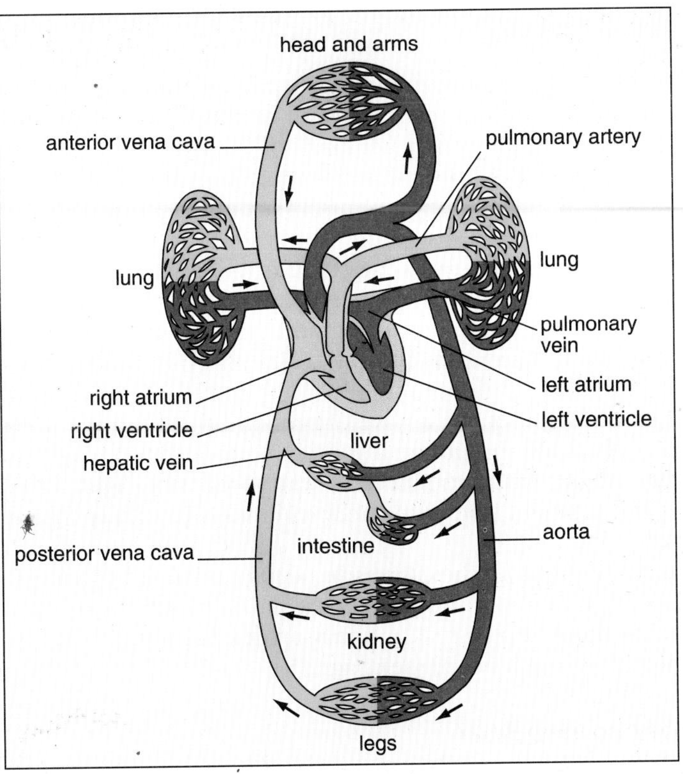

Picture 3 General plan of the human circulatory system. Oxygenated blood, red; deoxygenated blood, blue.

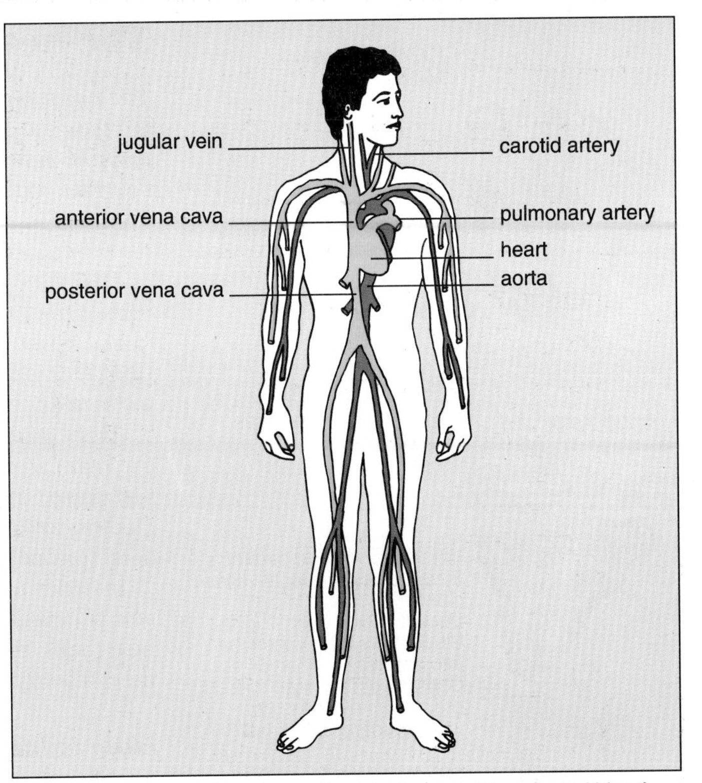

Picture 4 The main blood vessels of the human. Oxygenated blood, red; deoxygenated blood, blue.

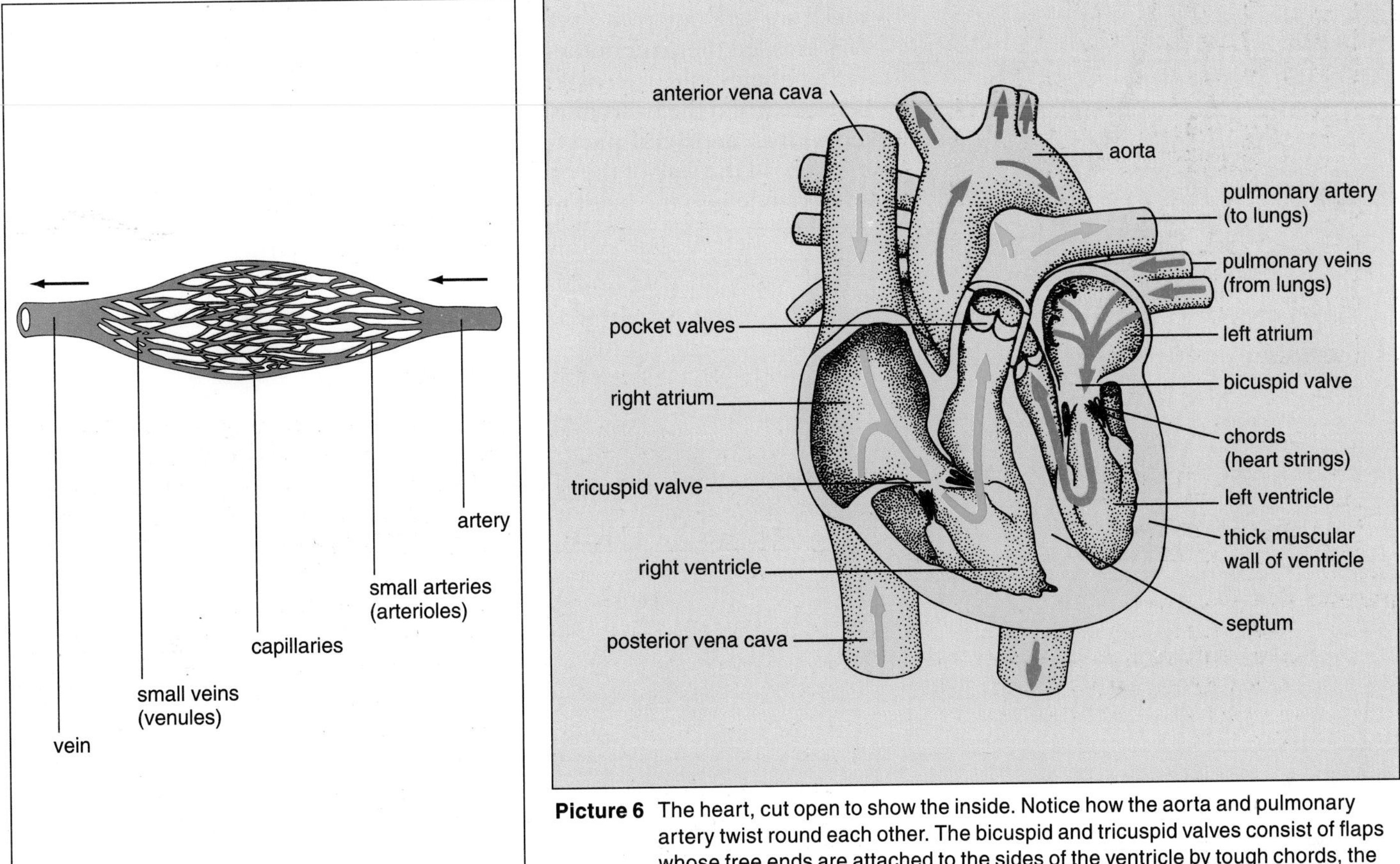

Picture 5 A capillary system. The arrows show the direction of blood flow.

Picture 6 The heart, cut open to show the inside. Notice how the aorta and pulmonary artery twist round each other. The bicuspid and tricuspid valves consist of flaps whose free ends are attached to the sides of the ventricle by tough chords, the 'heart strings'. The bicuspid valve has two flaps, the tricuspid valve has three. The pocket valves at the entrance to the aorta and pulmonary arteries are similar to the valves in the veins (see picture 10).

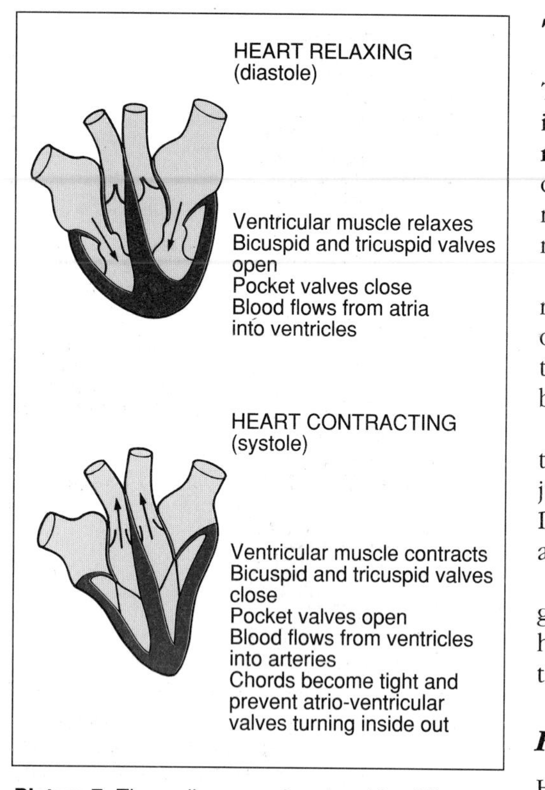

Picture 7 These diagrams show how blood flows through the heart. The valves stop the blood flowing backwards.

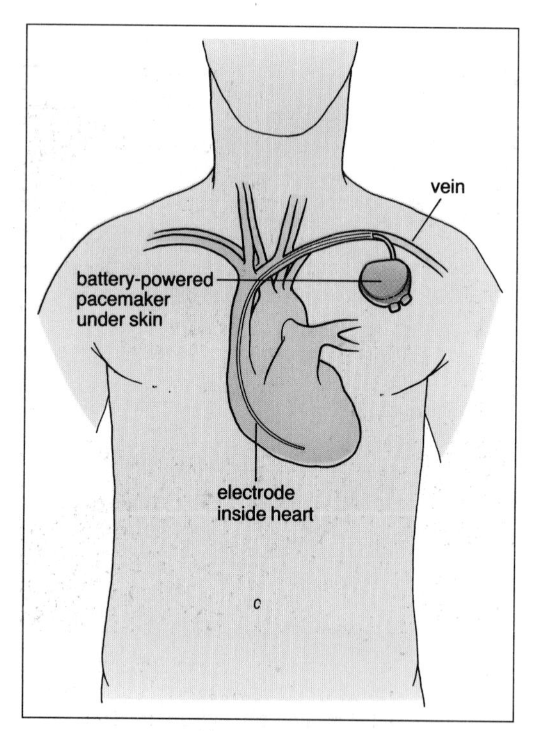

Picture 8 An electronic pacemaker. In this particular type an electrode is inserted into the heart through a vein, so there is no need for the surgeon to cut the chest open. Electrical pulses pass from the pacemaker into the heart muscle via the electrode.

The heart as a pump

The heart beats about 70 times a minute, that's over 100 000 times a day. This is made possible by the muscle tissue in the wall of the ventricles. **Heart muscle** (**cardiac muscle**) differs from other kinds of muscle in being able to contract repeatedly without getting tired. Try clenching your fist 70 times per minute and your hand muscles will soon give up. Heart muscle, however, has no difficulty working at this rate.

After the heart has contracted, it relaxes back to its original position. When it relaxes, blood flows into it from the veins. When it contracts, blood is pumped out of it into the arteries. So blood flows through the heart in only one direction. This is made possible by flap-like **valves** which prevent the blood flowing backwards (picture 7).

Every time the heart beats it sets up a wave of pressure which travels along the main arteries. This is called the **pulse**. If you put your finger on your skin just above the artery in your wrist, you can feel your pulse as a slight throb. Doctors and nurses feel a patient's pulse to check whether the heart is beating at its normal rate.

In order to keep contracting, the heart muscle needs a good supply of oxygen. It gets this from a system of **coronary arteries** which branch out over the heart wall. You can see them in picture 1. A **heart attack** is caused by one of the coronary arteries becoming blocked (see page 154).

How is the heart controlled?

Here is a remarkable fact: if the heart was removed from the body, it would go on beating on its own. In other words the mechanism which makes the heart beat is in the heart itself. The heart is made to beat by tiny electrical pulses which are sent out from a patch of special tissue in the wall of the right atrium. This is called the **pacemaker**.

Sometimes the pacemaker fails to work properly, or a block develops between it and the heart muscle. If this happens, the person may need to be fitted with an **artificial pacemaker**. This is an electronic device, placed under the skin on the wall of the chest (picture 8). It sends electrical stimuli through an electrode into the heart muscle, making it contract.

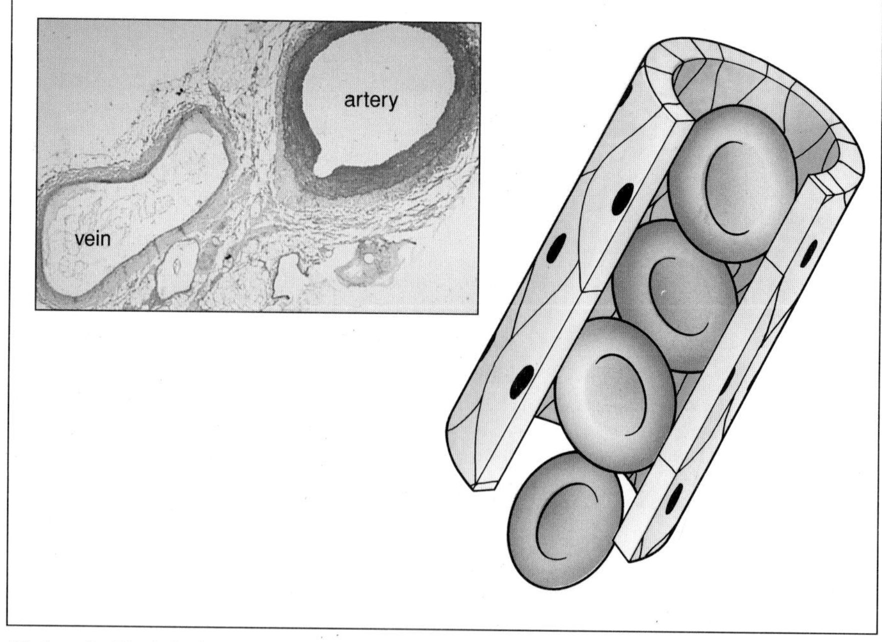

Picture 9 The left picture is a cross-section of an artery and a vein. The artery has a thicker wall than the vein. The right picture shows a capillary with four red blood cells inside it.

The blood vessels

Picture 9 shows an artery and a vein as they appear in a thin section under the microscope. The **arteries** have tough elastic walls containing muscle tissue. The high pressure from the beating of the heart forces the blood along quickly, much as water is forced along a narrow hosepipe.

By the time the blood reaches the **veins** the pressure pushing it along is much less. Also a lot of the blood is now moving against gravity as it flows upwards from the legs. This makes it difficult for the blood to get back to the heart. However, the veins are relatively wide and let the blood flow along easily. Also they contain **valves** which prevent the blood slipping back (picture 10).

Nevertheless the bloodflow through the veins may become sluggish. The pressure of blood may stretch the walls of the veins, making them flabby like thin bags. These are called **varicose veins** and they often develop in the legs, particularly in older people. Movement and exercise help to prevent varicose veins: contraction of the leg muscles squeezes the blood along and helps to keep it moving.

The **capillaries** are just wide enough to let the red blood cells pass along in single file. Their walls are very thin, consisting of just one layer of flattened cells. This enables oxygen and other substances to diffuse through easily.

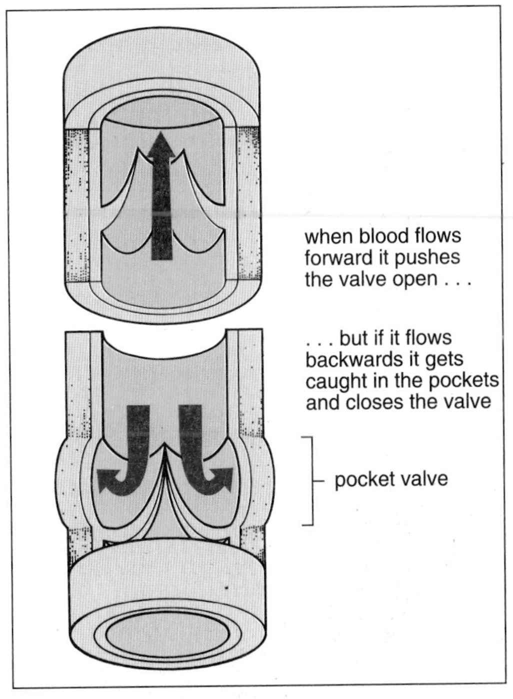

Picture 10 Pocket valves inside the veins stop blood flowing backwards. The valves at the entrance of the aorta and pulmonary artery work in the same way.

Blood pressure

The pumping of the heart, combined with the narrowness of the smaller blood vessels, produces a considerable pressure in the arteries. This is what we mean by **blood pressure**. It is important that our blood pressure should be reasonably high because it keeps the blood on the move.

Our blood pressure varies according to what we are doing. In general anything which makes the heart beat faster, or the arteries get narrower, will increase the blood pressure. For example, anger, excitement and exercise all have this effect.

Some people's blood pressure is too high all the time. This puts an extra strain on the heart. The pressure may even burst a blood vessel, particularly in old people whose vessels have become weak with age. If this happens in the brain, the cells in the region of the burst are killed, resulting in a **stroke**. A stroke may leave a person partly paralysed and unable to speak properly. A severe stroke can be fatal. A stroke can also be caused by a blood vessel in the brain becoming blocked by a blood-clot.

What causes high blood pressure? We don't know for certain, but it seems to be connected with stress and tension, over-eating, smoking and drinking too much alcohol. If a person feels tired and run down, one of the first things the doctor does is to measure the blood pressure (picture 11).

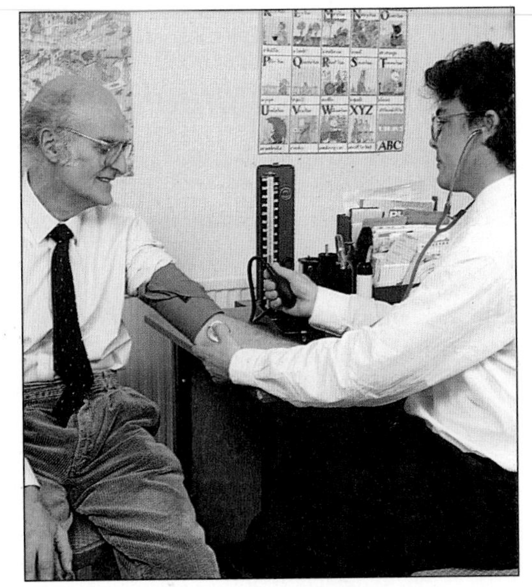

Picture 11 A doctor taking a patient's blood pressure. The pressure is registered by a U-tube containing mercury. The pressure is highest when the heart contracts, and lowest when the heart relaxes. Both pressures are measured and expressed as a fraction with the highest figure on top. A healthy person's blood pressure should be around 120/70 millimetres of mercury.

The circulation during exercise

Suppose you run a race. During the race your heart beats faster and your blood pressure increases. The arteries serving the muscles widen, whereas those serving less active organs such as the gut get narrower. The result is that blood is diverted to the structures that need it most, particularly the leg muscles.

All this begins to happen before you start the race. It is part of a complicated reflex which prepares the body for an emergency (see page 200).

Activities

A Looking at the heart

1 Look at the heart of a mammal obtained from the butcher. The heart has been cut open, so you can look inside.

2 Decide which side of the heart is which. In a human heart the more rounded side faces your front (ventral).

3 Identify the two atria, and the ventricles.

How do they differ in size and shape?

4 Feel the atria and ventricles with your fingers.

How do they differ in the way they feel?

Explain the reason for the difference.

5 Look at the large blood vessels attached to the heart. (Only their stumps will be left.)

Can you recognise the vessels shown in picture 6?

Which ones are arteries and which ones are veins?

How do the arteries and veins differ from each other?

6 Observe the narrow blood vessels on the surface of the ventricles, and notice where they come from. These are the coronary vessels.

What is their function?

What would happen if one of them became blocked?

7 Look at the cut which has been made in the wall of the ventricles.

What is the wall made of?

Has one of the ventricles got a thicker wall than the other?

Why do you think they differ in this way?

8 Look inside one of the ventricles.

Which structures shown in picture 6 can you see?

In particular notice the valves.

What are their functions?

9 Do simple experiments on the valves to find out how they work. For example, attach a rubber tube to a tap, and run water into the aorta to see what the valve does.

In what way is the heart suited to its job of pumping blood round the body?

When you have finished your investigations, wash your hands with soap and water.

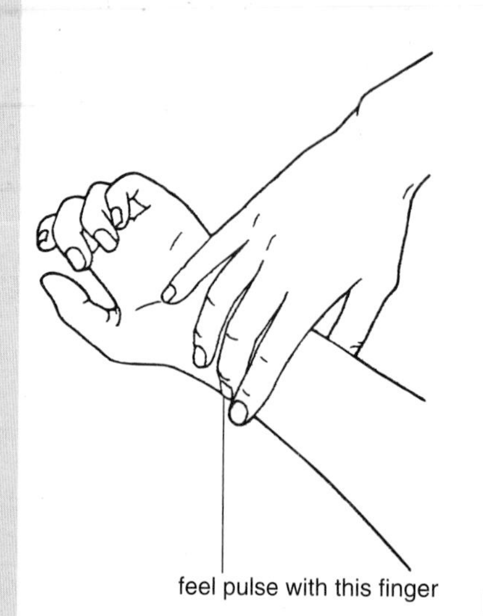

B Finding how fast your heart is beating

You can find how fast your heart is beating, that is your **heart rate**, by feeling your pulse as shown above.

1 Sit down comfortably in a chair with the palm of your hand facing upwards.

2 Gently place the middle finger of your other hand on your wrist as shown in the illustration. Can you feel your pulse as a repeated throb? If necessary change the position of your finger, until you can feel your pulse really well.

3 Count the number of heart beats in one minute.

4 Repeat step 3 four times.

Write down the number of beats per minute each time.

Work out your average heart rate. As this is your heart rate when sitting down, it is called the **resting heart rate**.

5 Stand up for one minute.

6 Still standing, take your pulse another five times.

Work out your average heart rate in beats per minute. This is called your **standing heart rate**.

How do your resting and standing heart rates differ?

Why do you think they are different?

The effect of exercise on the heart rate is explored on page 164.

The heart-lung machine

Some people are born with a heart defect, or they may develop a heart defect in later life. The most common heart defects are faulty valves or having a hole in the septum between the two sides of the heart.

A heart defect may be put right by an operation. In **open heart surgery**, the job of the heart and lungs is taken over by a **heart-lung machine**. The blood bypasses the heart and lungs, so the surgeon can carry out the operation without the heart beating all the time.

Picture 1 shows one type of heart-lung machine, greatly simplified. Blood is taken out of the person's circulation by tubes inserted into the right atrium. The blood then flows to an **oxygenator**. After being oxygenated, the blood is pumped back into the circulation by a tube inserted into the aorta. If the operation involves the aorta, the tube is inserted into the main artery of the leg.

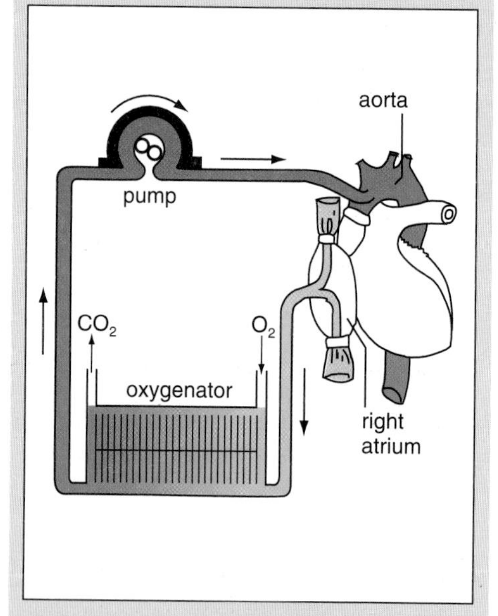

Picture 1 A heart-lung machine.

Various types of oxygenator are used. Some bubble oxygen through the blood, in others the blood flows over thin metal plates where it is exposed to oxygen, and others consist of thin membranes through which oxygen diffuses into the blood.

Of the three kinds of oxygenator described, which one is most like the lung itself? Explain your answer.

Questions

1 A person's blood pressure can be recorded continuously by means of an electronic pressure gauge placed inside one of the arteries.

Here is a recording obtained in this way:

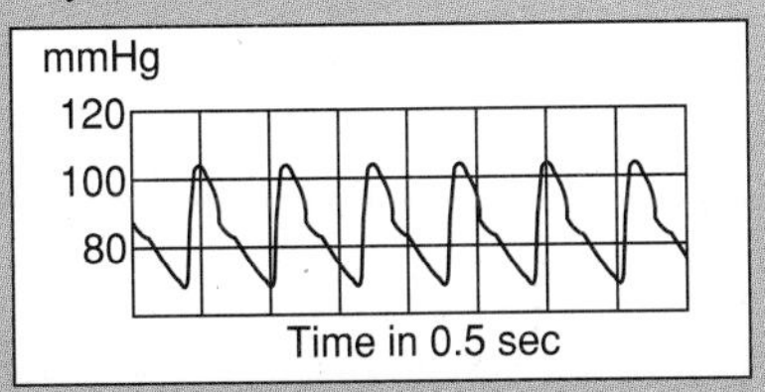

a Why do you think the pressure goes up and down all the time?
b What might make the frequency of the waves increase?

2 Suggest a reason for each of the following:

a the right atrium is larger than the left atrium,
b the left ventricle has a thicker, more muscular wall than the right ventricle,
c arteries have more muscle in their walls than veins,
d capillaries have very thin walls,
e veins contain valves.

3 Devise an experiment which you could do to test the suggestion that veins have more elastic walls than arteries.

4 The average speed of the blood in the arteries is 45 cm per second, but the average speed in the capillaries is only 0.5 mm per second.

a Give the speed in the capillaries as a percentage of the speed in the arteries.
b What do you think causes the difference?
c Why is it desirable for blood to flow through the capillaries slowly?

5 The chart below shows the pulse rate of a patient measured at four hourly intervals every day.

a Can you detect a regular pattern in the way the pulse rate changes? If so, describe the pattern.
b Do you have any criticism of the way the pulse rate is graphed in the chart?
c What were the highest and lowest pulse rates and when were they recorded?
d Give possible reasons why the pulse rate reached these particular values.

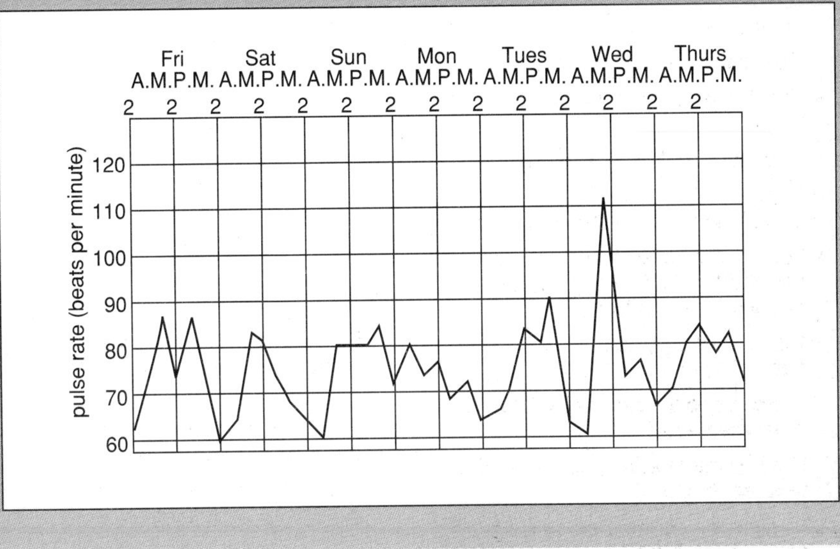

The lymphatic system

If you are unwell, the doctor will probably feel the 'glands' in your neck. If they are swollen, it's a sign that you may have an infection. What are these 'glands', and what are they doing? To answer this we must go back to the circulatory system.

As blood flows along the capillaries, a certain amount of fluid leaks out of the capillaries into the surrounding tissues. This is called **tissue fluid**. It is formed by a kind of filtration: the blood cells and proteins stay behind in the capillaries, but the other components of the blood can pass through the capillary walls.

The tissue fluid seeps round amongst the cells. As more is formed, the surplus is drained into narrow channels called **lymph capillaries**. Once inside these capillaries, the fluid is known as **lymph**. It is a colourless liquid.

There are lymph capillaries all over the body. Together they make up the **lymphatic system**. The lymph capillaries eventually lead to the veins, so lymph gets back into the bloodstream. Valves help to keep it flowing in the right direction.

There are little swellings in the course of the lymph capillaries. They are called **lymph nodes**. Each one is like a sponge, full of spaces, and the lymph has to pass through these spaces on its way back to the bloodstream. As it trickles through, any germs that happen to be present are attacked by germ-fighting cells – the same kind that are found in the blood (see page 157).

The lymph nodes are the 'glands' which the doctor feels when you are ill. Suppose you have a mouth infection. The germs get trapped in the lymph nodes in your neck. Here the germ-fighting cells do their best to kill them and prevent them spreading to other parts of the body. As the battle rages, the nodes swell up and become tender and painful. That's what the doctor is testing when he feels your neck. The lymph nodes aren't really glands, but that's what doctors usually call them.

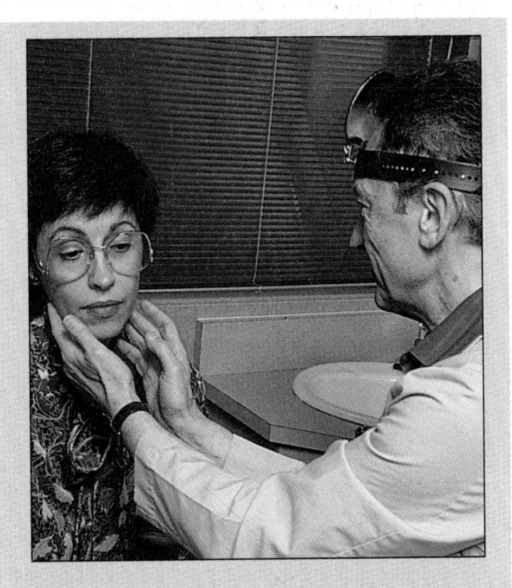

Picture 1 The doctor is feeling this patient's neck for signs of infection.

The main lymph nodes are in the neck, armpits and groin. Lymph tissue is also found in the **tonsils** and **adenoids** in the throat region. Many a battle is fought in these parts of the body.

(*Continued on next page*)

1 Bob cuts his foot and the cut goes septic. Within a short time his groin hurts whenever he touches it. Explain the reason.

2 Why is it particularly useful to have lymph tissue in the throat region?

3 What useful functions do you think tissue fluid performs?

4 Elderly people sometimes get swollen legs and ankles. What do you think causes this?

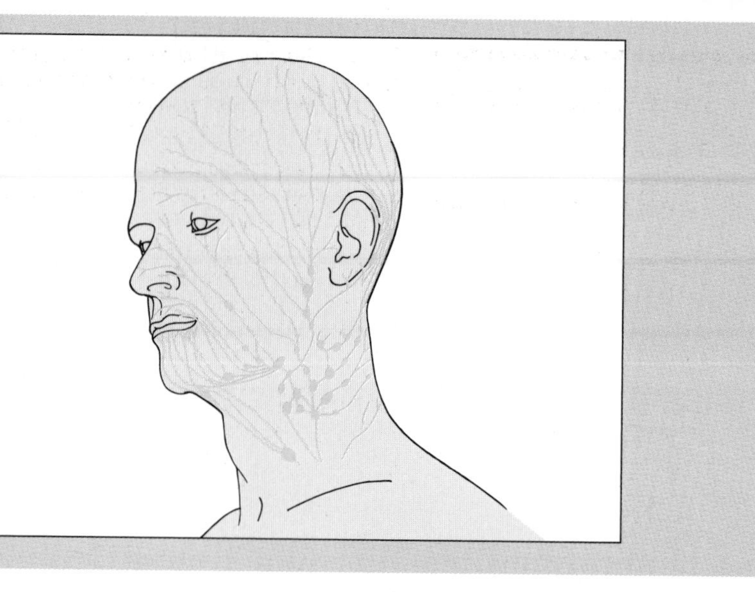

Picture 2 The lymphatic system in the head and neck.

What caused Jim's heart attack?

It all started many years before, when Jim's arteries began to harden. The culprit was the fat-like substance **cholesterol** (page 111). Cholesterol was laid down in the walls of some of Jim's arteries, making them narrower and slowing the flow of blood through them.

Jim didn't realise this was happening. Why should he? He had been goalkeeper for the local football team for many years, and although he had a desk job now, he felt fit. He had no idea that blood was likely to clot inside one of his hardened arteries.

One day it happened. Jim was watching his team play a match. He suddenly felt stabbing pains in his chest, then he passed out. A **clot** had formed in one of his coronary arteries. The artery was blocked, and the heart muscle served by it was starved of oxygen and stopped beating. He had had a heart attack. Fortunately an ambulance was at hand: the driver massaged his chest to keep the blood flowing. It turned out that only a small part of Jim's heart was damaged, so he recovered. If a larger part had been affected, he would have died.

How could Jim have avoided having a heart attack? Well, he should never have got hardening of the arteries. This would have meant keeping down the amount of cholesterol in his blood. He might have done this by eating unsaturated rather than saturated fat, and by not smoking. He knew these were linked with heart disease. But he loved his food and smoked twenty cigarettes a day.

1 What is saturated fat, and what sort of foods contain it? (Use the index if necessary.)

2 What advice would you give Jim on how to avoid having another heart attack?

3 In some cases hardening of the arteries causes pain. In what way can the pain be helpful?

4 Some people eat a lot of saturated fat and live to an old age without ever having a heart attack. What conclusions do you draw from this?

Picture 1 A person who has had a heart attack can sometimes be saved by cardiac massage.

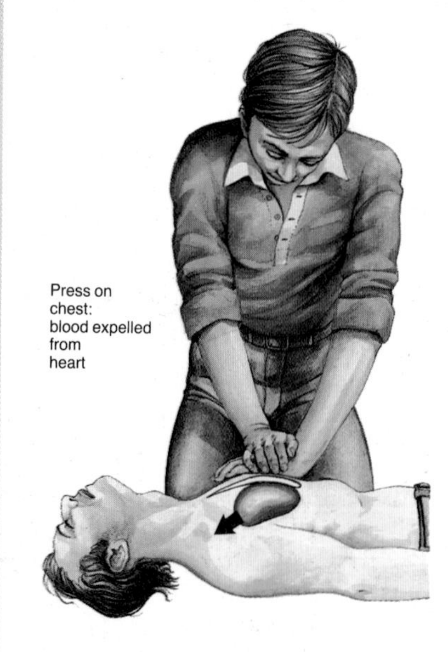

How the circulation was discovered

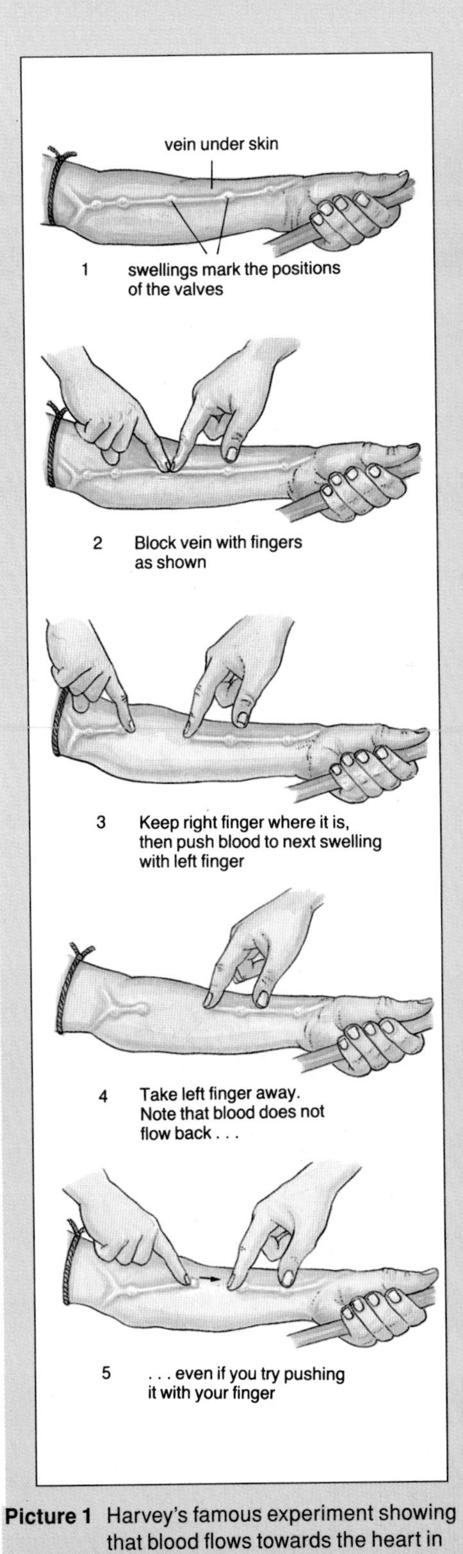

Picture 1 Harvey's famous experiment showing that blood flows towards the heart in the veins.

Some of the best experiments are very simple. A simple, but clever, experiment led William Harvey to discover the circulation of the blood in the early 1600s.

Harvey was Charles I's doctor. At that time scientists thought that blood seeped out of the heart to the various parts of the body, and then back again in the same vessels – rather like the tide flowing in and out of an estuary. This 'ebb and flow' theory was put forward by a Greek physician called Galen in the second century AD and it was still believed at the time of Harvey.

But Harvey had a different idea. He thought that the blood circulated round the body, flowing away from the heart in the arteries, and back to the heart in the veins.

Harvey's experiment is illustrated in picture 1. Study it carefully. The experiment shows that blood flows in only one direction in the arm vein – towards the heart. The valves stop the blood flowing in the other direction.

Picture 2 shows Harvey demonstrating this experiment to some young doctors in London. Try the experiment yourself, on a friend. But be careful: tying a band round the arm can be dangerous, so do the experiment only when your teacher is present.

This is only one of many experiments which Harvey did. He also dissected animals, studied the heart and blood vessels, and made calculations on the flow of blood. All his observations supported the idea that blood is pumped by the heart into the arteries and returns in the veins.

However, he never discovered the connection between the arteries and veins, namely the capillaries. He predicted that such a connection must exist, but he never found it.

Harvey has been called the father of modern medicine. He saw the human body as a machine, obeying the laws of physics and chemistry in the same way that non-living things do. At that time most people believed that the human body was created by God and obeyed special laws. The heart was the seat of the soul, not just a pump. Some of Harvey's patients were so upset by his ideas that they went to other doctors. Even his fellow scientists were critical, particularly as the final piece of evidence – the capillaries – was missing.

The capillaries were discovered seven years after Harvey's death by an Italian scientist, Marcello Malpighi. Malpighi used a microscope to examine the webbed foot of a frog. There before his eyes were the capillaries, with the blood flowing through them. Although the microscope was invented during Harvey's lifetime, he never used it in his work.

1. How did Malpighi's approach differ from Harvey's? Which approach provided the best evidence that the blood circulates, or is it impossible to say?
2. Harvey made a prediction which was tested by Malpighi. What was the prediction, and what piece of technology enabled Malpighi to test it?
3. How do you think people's lives were changed by Harvey's discoveries?

Picture 2 William Harvey showing his experiment to a group of young doctors at the Royal College of Physicians.

C12
Blood, the living fluid

Blood is much more complicated than you might think.

Picture 1 Blood highly magnified.

What does blood consist of?

An average sized person has about five litres of blood, that's nearly a bucket full. To the unaided eye blood looks like a simple fluid. But if you look at a sample of it under the microscope, you can see that there is more to it than that. It is a special type of tissue consisting of millions of cells in a fluid.

The cells are of two kinds: **red blood cells**, and **white blood cells**. The fluid part of the blood is called **plasma** (picture 2).

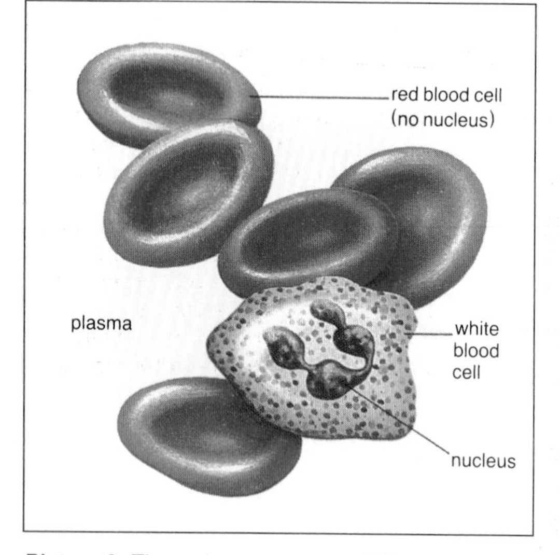

Picture 2 The main components of blood. There are several types of white blood cell of which only one is shown here. A cubic millimetre of blood (that's one small drop) contains about five million red cells and seven thousand white cells.

What does our blood do?

Our blood does three main things for us:

- It transports substances, including oxygen, within the body.
- It helps to defend us against disease.
- It keeps conditions right for the working of our cells.

Most of the substances that the blood transports are carried in solution in the plasma. They include soluble food substances (e.g. glucose), excretory products (e.g. urea) and hormones (e.g. insulin).

Its other jobs, carrying oxygen and fighting disease, we shall now look at in detail.

How does blood carry oxygen?

Oxygen is carried by our red blood cells. A single drop of blood contains millions of these cells. The red blood cell has a distinctive shape: it's like a disc which has been pressed in on each side, like the wheel of a car (picture 3). This gives it a large surface area for taking up oxygen.

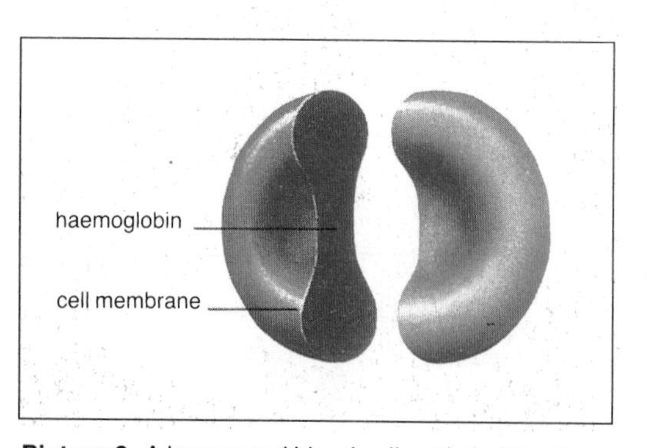

Picture 3 A human red blood cell cut in half to show its inside.

The red blood cell does not have a nucleus. The inside is filled with a red pigment called **haemoglobin** – this is what makes blood look red. Haemoglobin is a special protein and contains iron. The iron enables the haemoglobin to carry oxygen. That is why we need iron in our food.

How does haemoglobin work? It combines with oxygen to form a substance called **oxyhaemoglobin**. In this form the oxygen is carried by the blood from the lungs to the tissues. When it gets to the tissues, the oxyhaemoglobin lets go of the oxygen and is turned back into haemoglobin.

The blood also carries carbon dioxide from the tissues to the lungs. The carbon dioxide is picked up by the red blood cells, but is then carried mainly in the plasma until it gets to the lungs. The carriage of oxygen and carbon dioxide is summarised in picture 4.

The remarkable thing about haemoglobin is how readily it takes up oxygen in the lungs. It's as if it has a special liking for oxygen.

Haemoglobin has an even greater liking for another substance: carbon monoxide. This is present in motor vehicle exhaust (see page 40). It combines with haemoglobin about 300 times more readily than oxygen does. If we breathe it in, less oxygen can combine with the blood, so the tissues are starved of oxygen.

Breathing carbon monoxide gas can kill you in a few minutes – that's why car exhaust is poisonous, even though less than five per cent of it is carbon monoxide. Small amounts of carbon monoxide are also present in cigarette smoke – not enough to kill you, but enough to make you feel faint if you're not used to smoking.

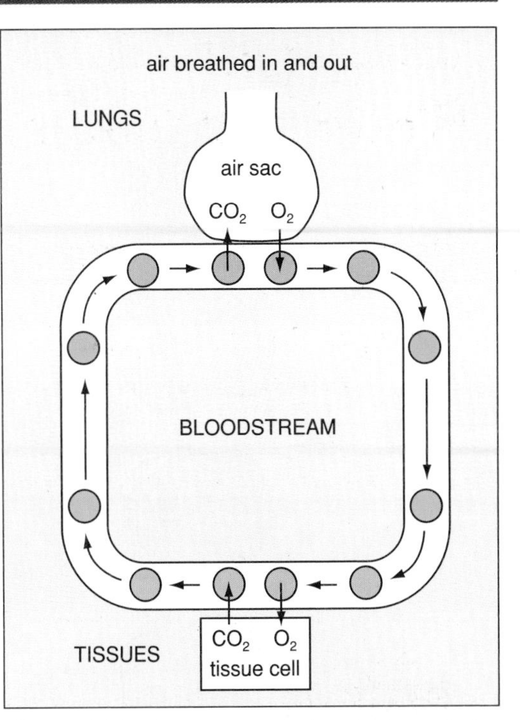

Picture 4 Blood carries oxygen from the lungs to the tissues, and carbon dioxide from the tissues to the lungs. Oxygen is carried from the lungs to the tissues by haemoglobin inside the red blood cells. In the tissues carbon dioxide enters the blood via the red blood cells, but most of it is then carried in the plasma as hydrogencarbonate ions. In the lungs carbon dioxide leaves the blood via the red blood cells.

Non-stop production

Red blood cells live for only about four months. They are then destroyed in the liver, rather like used cars are destroyed when they're finished with. To keep up the right number in our bloodstream, new ones must constantly be produced. They are made in the **bone marrow**, a soft tissue inside certain bones (see page 215). About two million are manufactured every second!

A person who doesn't have enough red blood cells or haemoglobin suffers from **anaemia**. It can make you feel very tired. We can become anaemic by not having enough iron in our food, or by losing a lot of blood.

What happens if we lose a lot of blood?

People sometimes lose a lot of blood, haemophiliacs for example and people injured in car accidents. If more than a couple of litres are lost, the person's life is in danger for two reasons:

- The blood pressure falls, and this slows down the flow of blood round the body.
- The number of red blood cells falls, so the oxygen-carrying ability of the blood is reduced.

All sorts of consequences follow, but the main one is that not enough oxygen gets to the brain. The result is that the person goes unconscious and may die. However, life may be saved by a **blood transfusion**.

Blood transfusions

In a blood transfusion the person is given blood which has been donated by a **blood donor** (picture 5). If the person is very short of blood cells, a blood cell transfusion is given. Otherwise plasma alone will do. This restores the blood pressure, so the blood flows round the body at its normal speed. Over the next few weeks the patient makes new red blood cells to replace the ones which have been lost.

Before carrying out a transfusion with red blood cells the doctors make sure that the donor's blood and the patient's blood belong to the same **blood group**. Otherwise the donor's red blood cells may clump together, blocking the patient's blood vessels and causing death. Blood groups are explained on page 161.

Picture 5 This man is being given a blood transfusion.

How does blood fight disease?

Fighting disease is the job of the white blood cells. They are fewer in number than the red blood cells. They do not contain haemoglobin, and they have a nucleus. Their job is to attack and destroy germs which get into the body. This is called our **immune system**. Like the red blood cells, the white blood cells are manufactured in the bone marrow.

There are two main kinds of white blood cell: **phagocytes** and **lymphocytes**. They attack germs in different ways. Let's look at them in turn.

Phagocytes

Phagocytes destroy germs by eating them (see page 14). They are remarkable cells; they can change shape and wriggle about from place to place. They can

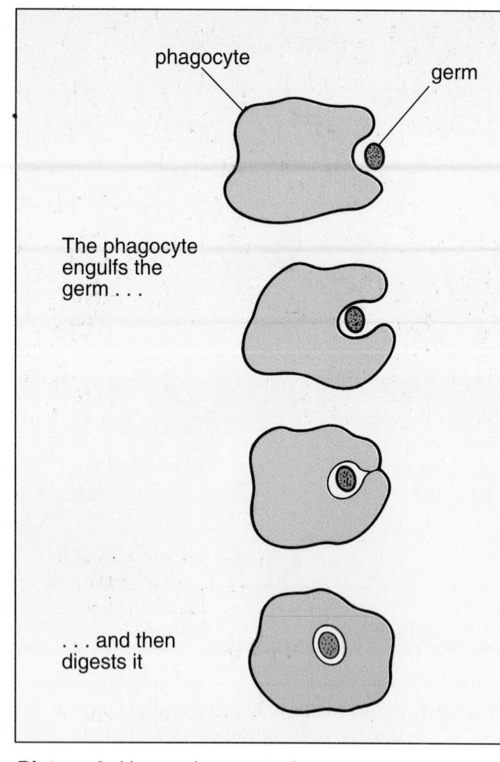

Picture 6 How a phagocyte destroys a germ. Phagocytes move by changing their shape. This is called amoeboid movement (see page 221).

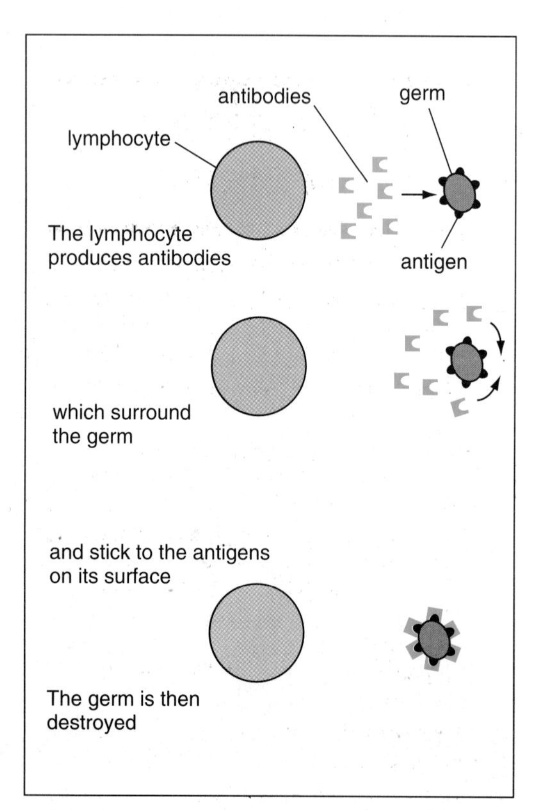

Picture 7 How a lymphocyte destroys a germ.

even worm their way through the walls of the capillaries into the tissues. When a phagocyte comes into contact with a germ, it engulfs it and takes it into the cell. The germ is then killed and digested (picture 6).

Phagocytes are like an army. They patrol the bloodstream and tissues waiting for germs to arrive. If you get an infection, in, say, your big toe, phagocytes move to the infected area and attack the germs. It's like a battle, and the place where it happens may become red, swollen and painful. This is called **inflammation** and it may result in the formation of **pus**.

Lymphocytes

If a particular type of germ gets into your bloodstream, it is detected by lymphocytes. These cells then destroy the germs.

This is how they do it. Germs have on their surface certain proteins which we call **antigens**. When a germ gets into your bloodstream, its antigens cause certain lymphocytes to produce matching proteins called **antibodies**. The antibodies then combine with the antigens, destroying the germ in the process (picture 7).

Antibodies destroy germs in different ways. Some make the germs burst. Others make it easier for phagocytes to eat them. And there are some which make the germs clump together, after which they are eaten by phagocytes.

Some germs do their damage by releasing poisonous substances (**toxins**). In these cases the antibodies react with the poison, making it harmless. Such antibodies are called **antitoxins**.

After they have been produced in the bone marrow, our lymphocytes move to special structures in various parts of the body called **lymph nodes**. Lymph nodes are explained on page 153.

Killers, helpers and suppressors

The lymphocytes that produce antibodies are called **B lymphocytes**. In addition we have other kinds of lymphocytes called **T lymphocytes** which do not produce antibodies.

Our T lymphocytes include **killer cells**. They destroy cells in our body that have been attacked by viruses, so the viruses are eliminated before they have a chance to multiply. The killer cells also attack cells in an organ that has been put into the body in a transplant operation. This is why transplants are sometimes rejected (see pages 21 and 193).

Some of our T lymphocytes are **helper cells**. They help the other cells of our immune system to do their jobs. For example, they help the phagocytes to destroy germs, and they help the B lymphocytes to produce antibodies quickly.

Other kinds of T lymphocyte *inhibit* the phagocytes and antibody-producing cells. These are called **suppressor cells**. They act as brakes in the immune system, preventing it from over-reacting. If we get an infection, the helper and suppressor cells cooperate with each other to give just the right response.

Our immune system is a very complex, finely tuned system. All the time it is quietly working away, defending us from all sorts of diseases and infections. Anything which interferes with it can put us in great danger. HIV, the virus that causes AIDS, damages the immune system by attacking the helper cells. This lowers the performance of other cells in the immune system. People with AIDS are therefore liable to get diseases which normally they would be protected against (see page 238).

Blood clotting

If you cut yourself, you usually bleed for a short time, but soon the blood thickens and the bleeding stops. The thickening of the blood is called **clotting**. Clotting is important because it plugs up wounds, preventing blood being lost and stopping germs getting in. It is also the first step in the healing process in which the damaged tissues join up again.

How does clotting take place? Damage to a blood vessel causes tiny cell fragments in the blood, called **platelets**, to start a chain reaction. The final part of the chain is shown in picture 8. A soluble protein called **prothrombin** turns into an enzyme called **thrombin**. The thrombin then causes another soluble protein called **fibrinogen** to turn into solid threads of **fibrin**. This is the clot.

Many different substances are needed for blood to clot. We are born with some of them, others we get from our food. Occasionally a person is unable to make one of the substances. For example, lack of a protein known as **factor VIII** results in a disease called **haemophilia**. The blood takes a vary long time to clot, so a lot of blood may be lost from even a small cut. This is an inherited disease which can be passed from parents to their children (page 272).

Blood which is kept in hospitals for medical use must not be allowed to clot. This is prevented by adding an **anticoagulant** such as sodium citrate. Or you can let the blood clot and then take out the clot. If you remove the clot from blood plasma, you are left with a pale yellow fluid called **serum**.

It is obviously a good thing for our blood to clot when we cut ourselves. However, it would be disastrous if this happened inside our blood vessels. To prevent this, the vessels contain natural **anticoagulants**. Even so, clotting sometimes occurs in an artery, causing a heart attack or stroke. Cholesterol is laid down in the wall of the artery, hardening it. Eventually the wall gets damaged and this triggers the clotting process. This is called a **thrombosis**.

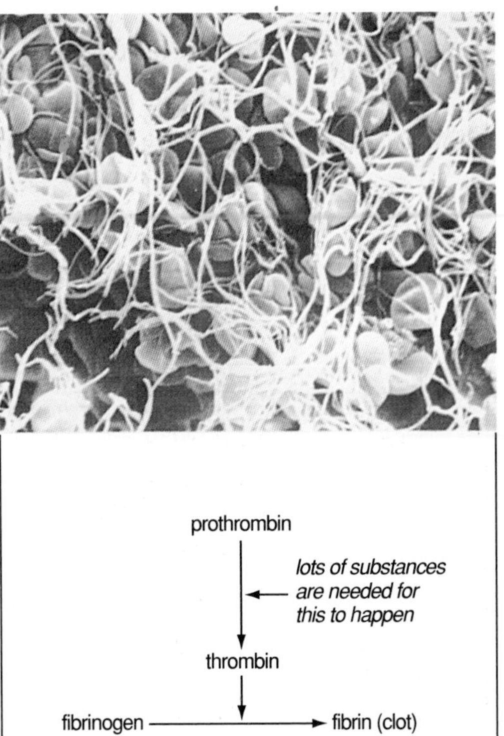

Picture 8 When blood clots a meshwork of fibres (called fibrin) is formed. The picture was obtained with an electron microscope. The things that look like deflated footballs are red blood cells. The main chemical reactions involved in the clotting process are shown below the picture.

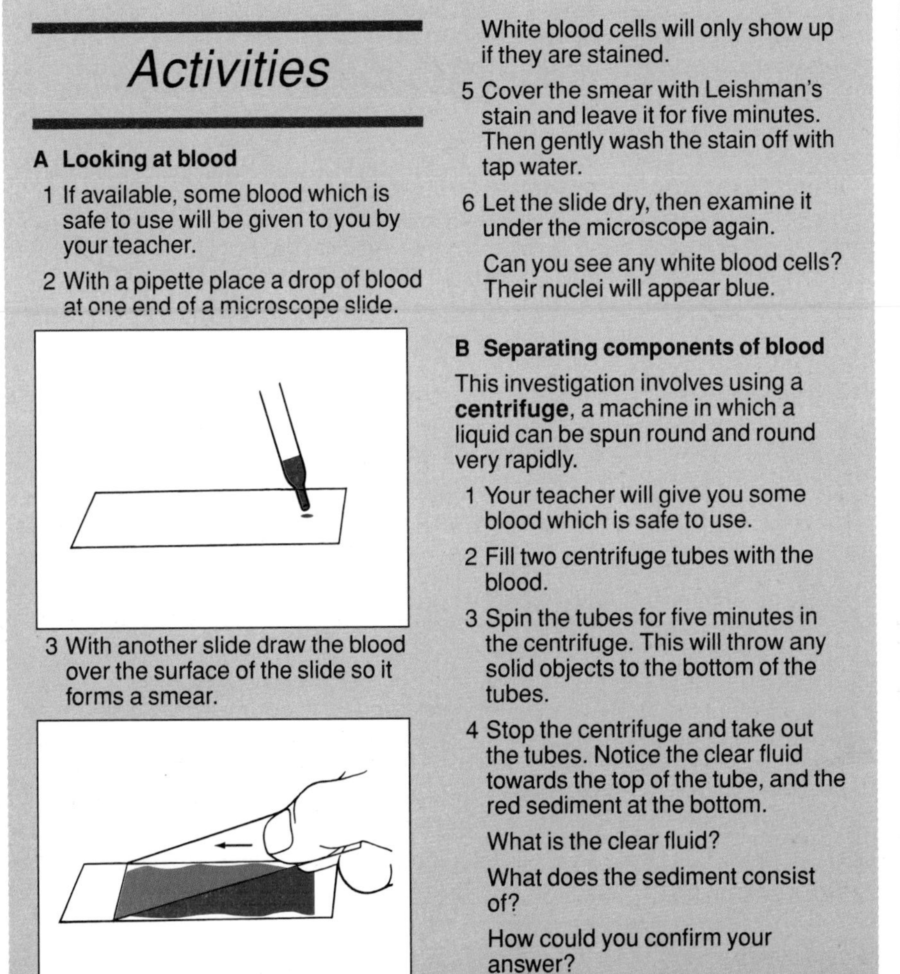

Activities

A Looking at blood

1 If available, some blood which is safe to use will be given to you by your teacher.

2 With a pipette place a drop of blood at one end of a microscope slide.

3 With another slide draw the blood over the surface of the slide so it forms a smear.

4 Let the blood smear dry, then examine it under the microscope: low power first, then high power.

Can you see red blood cells?

White blood cells will only show up if they are stained.

5 Cover the smear with Leishman's stain and leave it for five minutes. Then gently wash the stain off with tap water.

6 Let the slide dry, then examine it under the microscope again.

Can you see any white blood cells? Their nuclei will appear blue.

B Separating components of blood

This investigation involves using a **centrifuge**, a machine in which a liquid can be spun round and round very rapidly.

1 Your teacher will give you some blood which is safe to use.

2 Fill two centrifuge tubes with the blood.

3 Spin the tubes for five minutes in the centrifuge. This will throw any solid objects to the bottom of the tubes.

4 Stop the centrifuge and take out the tubes. Notice the clear fluid towards the top of the tube, and the red sediment at the bottom.

What is the clear fluid?

What does the sediment consist of?

How could you confirm your answer?

First aid for bleeding

1 Get the person to lie down and relax.

2 Press the edges of the wound together with thumb and finger, or press down on the wound with the palm of your hand.

3 Raise the site of the wound above the level of the heart.

4 Place a thick pad (e.g. a folded handkerchief) on the wound, continuing to press all the time.

5 Bandage the pad very firmly with, e.g. a tie or stocking. If blood oozes through the bandage, add a further pad and bandage.

6 If a lot of blood has been lost, raise the legs above the level of the head and trunk.

Give a reason for each of the steps in this procedure.

Questions

1 Write down *three* ways in which red and white blood cells differ in their appearance.

What job does each do?

2 Why is it dangerous to breathe in motor car exhaust? Explain your answer.

3 There are approximately five million red blood cells in a cubic millimetre of human blood, and the total volume of blood in the whole body is about five litres. Each red blood cell has a surface area of about 120 square micrometres.

a How many red blood cells are there in the entire bloodstream?
b What will be the total surface area of all the red blood cells? Give your answer in square metres.
c What is the significance of these measurements?

4 Suggest two reasons why it is dangerous to lose more than two litres of blood.

5 A scientist investigated the number of red blood cells possessed by people living at sea level and in a mountainous region at a height of 5860 metres. Here are her results:

sea level	5.0 million per mm^3
5860 metres	7.4 million per mm^3

Why do you think they differ?

6 Explain the reasons for each of the following:

a Not more than half a litre of blood is normally taken from a blood donor.
b After giving blood, the donor is advised to sit down quietly for about half an hour.
c A little sodium citrate is usually added to blood which has been given by a donor.
d Complete blood is only kept for about a month after it has been obtained from a blood donor, but plasma may be kept much longer.

Summary of our defences against disease

Innate immunity.

Defences which are present by the time we are born, against disease in general, e.g. the skin for keeping out invading germs, cilia for sweeping germs out of the windpipe, acid for killing germs in the stomach, and phagocytes for eating germs in the bloodstream and tissues.

Acquired immunity

Defences which we develop during our lives against specific diseases:

- **Active acquired immunity** is when we make antibodies in response to receiving particular antigens. We may receive the antigens naturally by being exposed to germs, or artificially by being immunised by a doctor. This type of immunity may last a lifetime.
- **Passive acquired immunity** is when we receive ready-made antibodies either naturally from mother across the placenta or via her milk, or artificially from a doctor by injection. This type of immunity does not last long.

Why we get some diseases only once

Once you have had a disease like measles, you are protected from getting it again. You are now **immune** to that disease.

We can explain immunity like this. The first attack causes you to produce antibodies against the antigens on the germs that cause the disease. When you get better, the antibodies disappear. However, special **memory cells** remain in the body long after the attack is over. If there is a second attack, the memory cells recognise the germ's antigens at once and quickly develop into appropriate antibody-producing cells. The antibodies are produced so quickly and in such large numbers that the germs are destroyed before they have a chance to make you ill.

Having a particular disease will protect you against that disease in the future, but it won't protect you against other diseases. This is because the antibodies which you produce against, say, measles will act only against the measles germs – they won't act against any other kind of germ. To use the jargon, antibodies are *specific* to particular antigens.

Sometimes people get a mild attack of a disease when they are young – so mild that they may not even notice it. However, it causes them to develop the appropriate memory cells, so they are protected against that disease when they grow up. The same thing is achieved artificially when doctors immunise us against a disease by means of a **vaccine** (see page 101).

Some germs constantly change by a process called **mutation** (see page 284). The result is that new strains of the germ keep arising. This is the case with the common cold and flu viruses. The antibodies needed to fight one strain are different from those needed to fight another. That's why we get colds and flu *more* than once.

The constant changing of germs into new strains makes it difficult to develop vaccines against them. By the time a vaccine has been developed against one strain, the germ may have changed into another strain. This is one reason why it is so difficult to develop vaccines against flu and HIV.

1 Make a list of the diseases which you think you are immune to. Why do you think you are immune to these diseases?

2 Many people go to the doctor for a 'flu jab', innoculation with a vaccine against flu. To be effective, what would the vaccine have to consist of?

3 When you are immunised against diseases such as typhoid and cholera, further doses of the vaccine ('boosters') need to be given from time to time. Suggest a reason for this.

What are blood groups?

The ABO system

Everyone's blood belongs to one of four groups called **A**, **B**, **AB** and **O**. The letters refer to particular protein antigens which may be present on the surface of the red blood cells:

Group A blood has type A antigens on the red blood cells.

Group B blood has type B antigens on the red blood cells.

Group AB blood has both types of antigen on the red blood cells.

Group O blood has neither type of antigen on the red blood cells.

In the plasma there are antibodies. If these antibodies were to combine with the antigens on the red blood cells, they would cause havoc.

However, a person never possesses blood in which the antigens and corresponding antibodies occur together. Nature simply doesn't allow it. For example, group A blood has type A antigens on the red blood cells but anti-B antibodies in the plasma; and group B blood has type B antigens on the red blood cells but anti-A antibodies in the plasma.

If the corresponding antibodies were present in the plasma, they would combine with the antigens and cause the red blood cells to clump together. It's as if the red blood cells were treated as germs.

In a blood transfusion the patient must be given blood from a donor with the right blood group. The donor's red blood cells will clump together if the patient's plasma contains the corresponding antibodies (see picture 1). This is what doctors have to avoid. The best way is to use blood which belongs to the same group as the patient's.

Picture 1 This diagram shows what may happen if blood of different blood groups is mixed.

The rhesus system

There are other blood groups besides the ABO system. For example, some people are described as **Rhesus positive**. They have a certain type of antigen on their red blood cells. People who don't have these antigens are called **Rhesus negative**.

Now suppose a Rhesus negative person is given Rhesus positive blood in a transfusion. In this case the Rhesus negative person makes antibodies which combine with the antigens on the donor's red blood cells. This can cause the donor's red blood cells to clump together and burst. Here again, the red blood cells are treated as germs.

Blood groups and transfusions

Before blood transfusions are carried out, doctors always make sure that the patient's blood is compatible with the donor's blood. Their blood groups are found by carrying out a simple test on small drops of blood.

We inherit our blood groups from our parents (page 272). The percentages of people belonging to the different blood groups in Britain are as follows:

ABO system		Rhesus system	
O	47%	Rh+	85%
A	41%	Rh–	15%
B	9%		
AB	3%		

These figures tell us which blood groups are most likely to be needed for transfusions.

Blood which has been given by donors is stored in blood banks. The blood is kept in bottles which are labelled with the blood groups.

1 People belonging to blood group O have anti-A and anti-B substances in their plasma.
   a What sort of antigens, if any, do they have on their red blood cells?
   b What sort of antibodies and antigens are present in the blood of a person belonging to blood group AB?

2 Four young people donate blood in a blood donation centre. A medical technician finds their blood groups by mixing drops of their blood with different kinds of serum.

   John's blood goes lumpy with anti-A serum but not with anti-B.

   David's blood goes lumpy with anti-B serum but not with anti-A.

   Anna's blood goes lumpy with both kinds of serum.

   Susan's blood does not go lumpy with either kind of serum.

   Which blood group does each person belong to?

Allergies

Sometime we produce antibodies against harmless substances such as certain foods, pollen or even clothing. This can cause unpleasant effects, as anyone who has suffered from hay fever knows. Hay fever is caused by pollen, and it particularly affects the nose and eyes. People differ in the way they react to food substances. Some people are sensitive to lettuces – something in the lettuce leaf makes you feel really ill.

These sort of reactions are called **allergies**. Sometimes it is difficult to know exactly what is causing a particular allergic reaction, and various tests have to be carried out. Once the cause has been discovered, steps can be taken to prevent it or at least make it less severe.

A well known symptom of allergic reactions is the release from certain cells in the body of a substance called **histamine**. One way of treating allergies therefore is to give the person an **anti-histamine drug**.

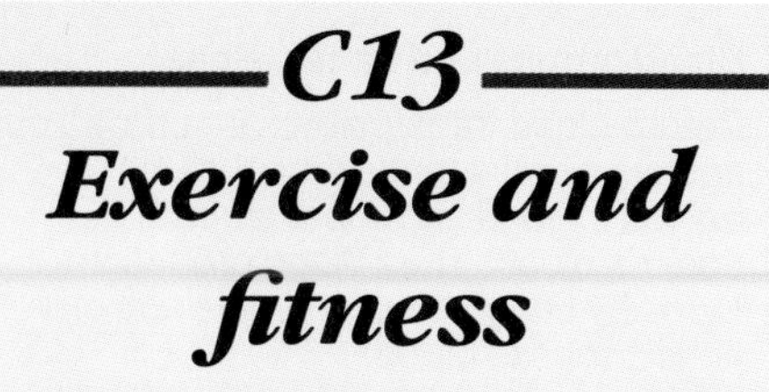

C13 Exercise and fitness

Fitness can be achieved by leading a healthy life and taking regular exercise.

Picture 1 Exercise is an important aspect of being fit.

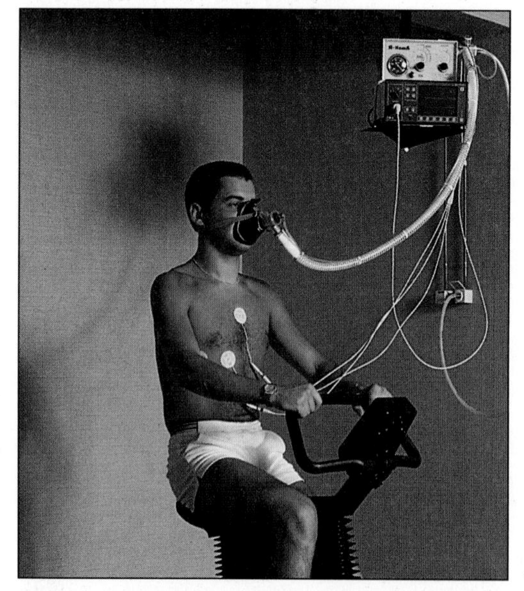

Picture 2 This person is having his oxygen consumption measured while he pedals an exercise cycle.

Why be fit?

Here are six reasons why it is a good idea to lead a healthy life and take regular exercise:

- It keeps our body mass down.
- It increases our resistance to disease.
- It helps to prevent circulatory problems such as heart disease and varicose veins.
- It helps us to cope with everyday tasks such as climbing stairs and walking uphill.
- It makes us *feel* better, physically and mentally.
- It helps us to withstand stress.

What does being fit mean?

In a general sense, being fit means having a well tuned and efficient body. In practice this means having the following attributes:

■ Strong muscles

You don't need huge bulging muscles to be fit. What you do need are muscles that are taut and ready for action. Tautness of the muscles is called **muscle tone**. Muscle tone is important for keeping our posture and for movement. You can maintain good muscle tone by exercising the muscles regularly. This makes the individual muscle fibres contract more efficiently.

Prolonged strenuous exercise also increases the total number of muscle fibres in a muscle, so the muscle gets larger and its mass increases. Weight lifters take this to an extreme.

■ Being able to endure prolonged physical activity

How far can you walk, or run, before you get tired? Would you be able to complete a marathon? Your ability to endure such physical activity depends on how fit you are. Of course your attitude is also important. You must *want* to succeed, and must be determined not to give up.

■ A flexible body

By flexible we mean being able to bend in various directions. It is caused by the elasticity of the ligaments and the amount of movement that the joints allow. Regular exercise helps to keep the ligaments elastic and the joints mobile. Tight muscles can limit your flexibility too, which is why athletes do stretching exercises.

■ The right body mass

The danger of being overweight is discussed on page 119. Having a body mass that is correct for your sex, age and height is an important aspect of being fit. An overweight person has too much fatty tissue. A person can acquire the correct body mass by combining regular exercise with sensible eating and drinking.

■ An efficient breathing system

When you take exercise your muscles need more oxygen (picture 2). So you breathe faster and more deeply. Exercise improves the efficiency of the breathing muscles, just as it improves the efficiency of the leg muscles. A fit person breathes more slowly and deeply than an unfit person, and less energy has to be spent on breathing movements.

■ An efficient circulatory system

During exercise your heart beats faster. Regular exercise is good for you because it exercises the heart muscle, making it stronger. Also, when the exercise is over, the heart rate returns to its normal resting rate more quickly if you are fit.

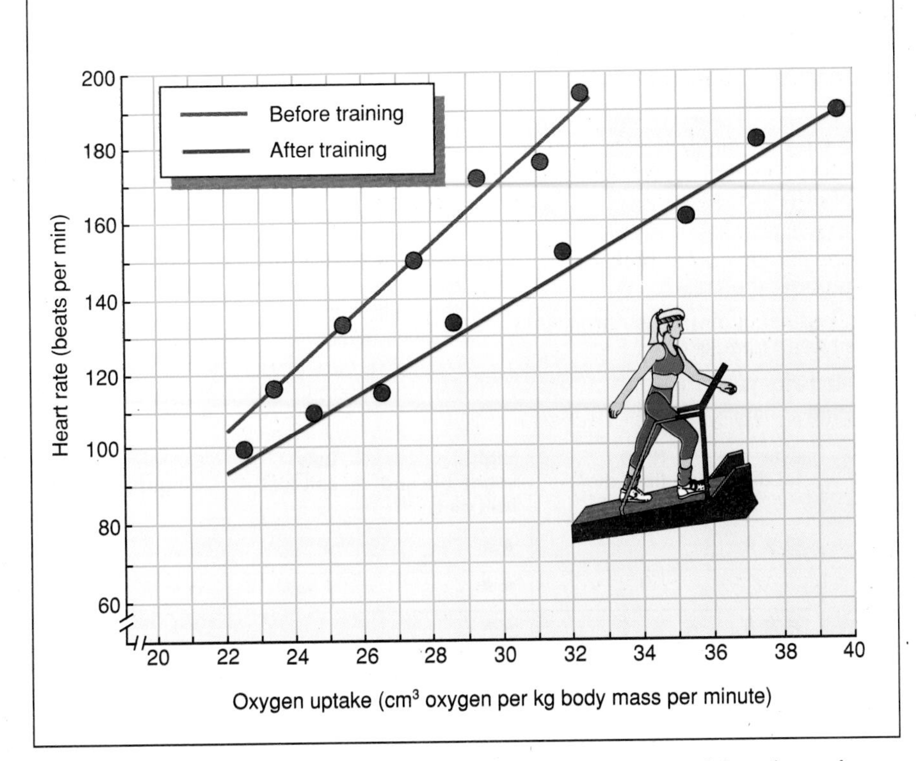

Picture 3 This graph shows the relationship between the heart rate and the volume of oxygen taken up during exercise on a treadmill.

Regular exercise also improves the coronary blood supply, so more oxygen can be delivered to the heart muscle.

The volume of blood pumped out of the heart every time it beats is called the **stroke volume**. The stroke volume is greater in fit people than in unfit people. This means that a fit person's heart does not need to beat at such a rate (picture 3). A normal person's resting pulse rate is about 70 to 80 beats per minute; an athlete's pulse rate may be as low as 40 beats per minute.

Regular exercise also improves the circulation to the working muscles, so oxygen is delivered to them more quickly. The muscle fibres become more efficient at taking up oxygen, and when the muscles respire anaerobically, lactic acid is removed more quickly, so you are less likely to get muscle pains (see page 127).

How fit are you?

Finding out how fit you are involves testing some or all of the attributes listed above. In very general terms, you are likely to be unfit if:

- your body mass is more than 2.3 kg above what it should be for your age, height and sex,
- a fold of skin, grasped between your thumb and forefinger just above the navel, is more than approximately 1.5 cm (half an inch),
- you cannot lift half your own body mass above your head,
- you cannot do six pressups, or three two-handed pullups, one after the other,
- you cannot touch your toes without bending your legs,
- your resting pulse rate is more than 85 beats per minute if you are female, and 80 beats per minute if you are male.

The trouble with these tests is that each assesses only one aspect of fitness. Together, they give you a rough idea of whether you are fit or not, but they do not tell you *how* fit you are. What we need are tests which measure quantitatively as many aspects of fitness as possible.

Measuring fitness

A simple way of measuring fitness is to time how long it takes you to run one mile on the flat at a steady pace. You then read off from a table how fit you are for your age and sex. A young woman who can run a mile in less than seven minutes is very fit. For a young man the time should be more like six minutes.

Another way of measuring fitness is to find out how quickly the pulse rate returns to normal after a standard bout of exercise. When you take exercise, your pulse rate increases. When you stop the exercise and rest, your pulse gradually returns to its resting rate. The quicker your pulse returns to the resting rate, the fitter you are.

You can measure fitness over a longer period of time by working through a series of tests. The first test is to walk one mile in 20 minutes, the second test is to walk one mile in 18 minutes, the fourth test is to walk two miles in 40 minutes – and the hundredth test is to run ten miles in 60 minutes. You work through the tests, and see how many you can manage. From a table you then read off how fit you are for your age.

This method can be part of a training programme in which you monitor your fitness as you go along.

Keeping healthy and avoiding illness

Much more is involved in keeping healthy than just taking exercise. The table shows ways of avoiding some major illnesses. Of course, following the advice in the table does not mean that you will never get any of these illnesses. It simply means that you are less likely to get them.

Some interesting anomalies have been discovered. For example, fish oil and olive oil help to prevent heart attacks. So does alcohol – in moderation!

1 What sort of investigations would have had to be carried out to obtain the information in the table?

2 Which ingredient of wine is *not* likely to prevent heart attacks?

Give a reason for your answer.

How to avoid some major illnesses.
★★★ very effective,
★★ fairly effective,
★ slightly effective.
The table is based on information from the American Medical Foundation.

	No smoking	Little or no alcohol	Low fat diet	High fibre diet	High vegetable and fruit diet	Low salt diet	Exercise and weight control
Heart attack	★★★		★★★		★★	★	★★★
Stroke	★				★★	★★★	★★
Diabetes (adult)			★★★	★	★★		★★
Cancer: lung	★★★		★		★		
liver		★★★			★★		
colon			★★★	★★★	★★★		★
breast			★★★	★	★★		★

Activities

A Measuring your fitness

WARNING! Do not attempt this activity if you have a health problem or are recovering from an illness.

CAUTION

We shall use a simple test which is used in the army. It involves stepping up onto a stool or chair, 43 cm high, 30 times per minute. Each step (up and then down) should take 2 seconds.

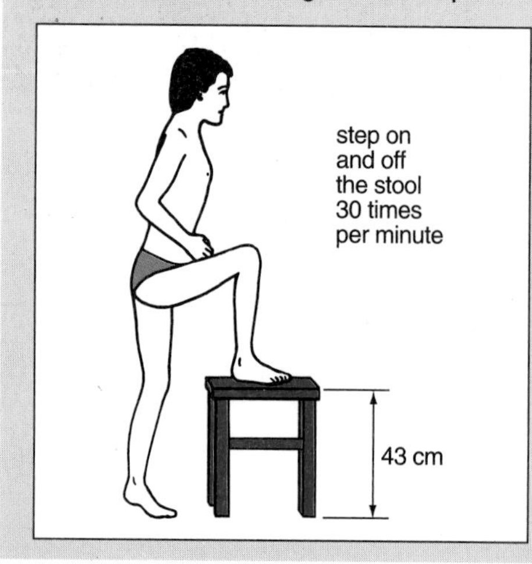

Work in pairs: one of you (the subject) should take the exercise, and the other one (the tester) should administer the test. The tester should count the seconds to keep the subject stepping at the right rate. Try it for a short time to get used to the rhythm. Also practise taking the subject's pulse (see page 152).

Now proceed as follows:

1 Step up and down for 5 minutes.
2 Sit down and rest for 1 minute.
3 Take the pulse for 30 seconds = **A**.
4 Rest for 30 seconds.
5 Take the pulse for 30 seconds = **B**.
6 Rest for 30 seconds.
7 Take the pulse for 30 seconds = **C**.
Add together **A** + **B** + **C**.

Assess your fitness from this scale:

	Male	Female
Very fit	175 or less	190 or less
Fairly fit	200 approx	220 approx
Rather unfit	215 approx	235 approx
Very unfit	230 or more	250 or more

These figures, suggested by former Olympic athlete and coach, Bruce Tulloh, are suitable for teenagers.

Which attributes of your body are being measured by this test? Do you think it is a valid test of fitness? What criticisms have you of it? Could it be used to find the best athlete in your school?

B Devising your own fitness test

Devise a fitness test, with a scoring system, which assesses as many physical attributes of the human body as possible, e.g. muscular strength, efficiency of the heart, flexibility and so on.

The test can be divided into a number of parts, each assessing a particular attribute, but a person on whom the test is carried out should finish up with a single score that reflects his or her overall fitness.

Discuss your test with your teacher, then try it out on students in your class.

Questions

1 Of the three methods of measuring fitness outlined on page 163, which one is the most valid, and why?

2 Mr Jones sits at a computer all day and watches television all evening. The only exercise he gets is walking to and from his car.

What dangers are there in his lifestyle?

3 Read about how humans respire without oxygen on page 000. Then use information from the present Topic to explain why a fit person pants less after a race than an unfit person.

4 The graph shows the heart rates of two students before, during and after a three-minute bout of exercise.

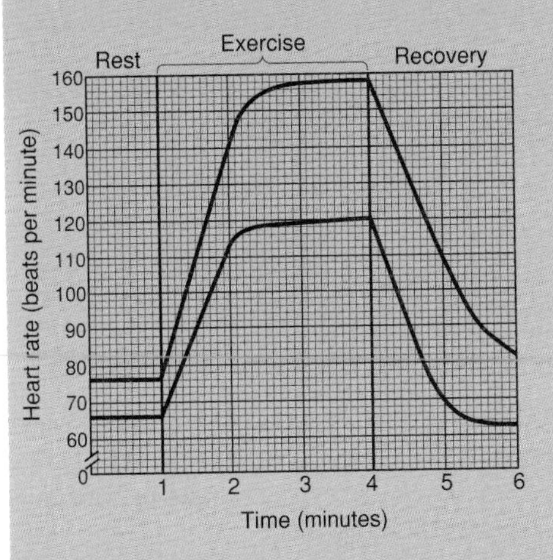

a Comment on the possible life styles of the two students.
b What can you say about the working of the students' hearts?
c Suppose you wanted to carry out the same investigation on a student in your class. Describe in detail how you would do it.

5 The table below shows the world records achieved by women and men in eleven track events (data from the 1995 Guinness Book of Records).

Distance	Men	Women
	min:sec	min:sec
100 m	9.86	10.49
200 m	19.72	21.34
400 m	43.29	47.60
800 m	1:41.73	1:53.28
1000 m	2:12.18	2:30.6
1500 m	3:28.82	3:50.46
1 mile	3:44.39	4:15.61
2000 m	4:50.81	5:28.69
3000 m	7:28.96	8:06.11
5000 m	12:56.96	14:37.33
10 000 m	26:58.38	29:31.78

a Calculate the average speed of the athletes in each event.
b Plot a graph of the data, putting the curves for women and men on the same grid. Plot distance on the horizontal axis and the average speeds on the vertical axis.
c What does the graph show?
d Suggest reasons why athletes are unable to maintain the same speed over 800 metres as they can over 400 metres.
e Comment on the difference in the average speeds for the 100 metres and 200 metres.
f Suggest reasons why the world records for women are below those for men. Is the difference the same for all the events?

6 One hundred years ago the world record for one mile was 4.5 minutes. In 1954 Dr Roger Bannister ran the mile in just under 4 minutes. The current record is 3 minutes 44.39 seconds.

a Suggest why athletes continue to improve on world records.
b Do you think a time will come when further improvements cannot be made? What do you think will set the limit?

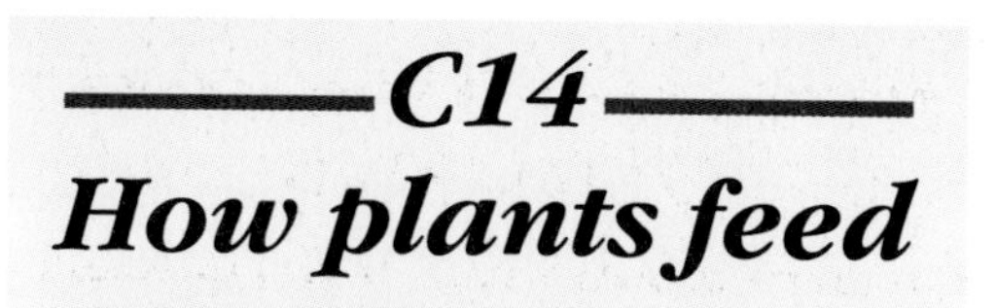

C14 How plants feed

You and I feed by taking in organic food. Plants feed quite differently, as we shall see.

Picture 1 This plant is full of food substances. How did they get there?

Picture 2 Mount Everest, the highest mountain in the world. Its peak is well over 8000 metres above sea level. Imagine a pile of sugar that high!

Photosynthesis

Look at the plant in picture 1. It has no mouth and no gut, and yet it is full of food such as sugar. Where does it get its food from? There are only two possible places: the soil and the air. But there's no sugar in soil or air, only simple substances like carbon dioxide and water. What the plant does is to absorb these simple substances from its surroundings and build them up into sugar and other complex substances.

Making complex substances from simpler ones is called *synthesis*, and plants need light for doing this. So the process by which plants make food is called **photosynthesis**.

What happens in photosynthesis?

The simple substances which the plant uses for photosynthesis are carbon dioxide and water. A land plant like the one in the picture gets **carbon dioxide** from the air, and **water** from the soil. These two substances are turned into **sugar**, and **oxygen gas** is given off as a by-product. The reaction requires energy, and this normally comes from **sunlight**. The green substance **chlorophyll** enables the plant to transfer energy from sunlight to sugar. It is therefore an essential helper in the process.

Photosynthesis is not a single reaction, but takes place in a series of small steps. However, it is usually summed up by this simple equation:

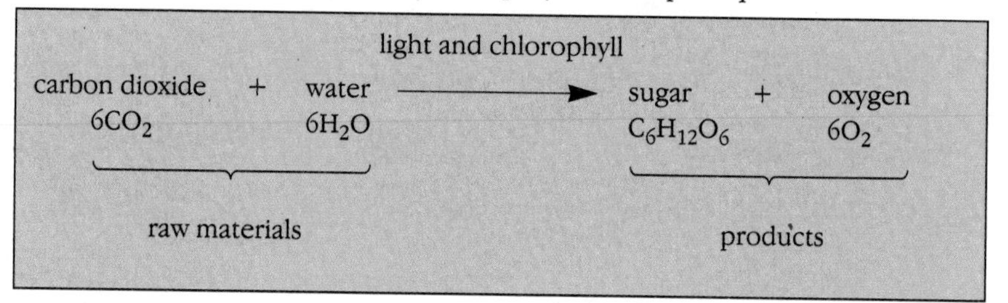

$$\underbrace{\underset{6CO_2}{\text{carbon dioxide}} + \underset{6H_2O}{\text{water}}}_{\text{raw materials}} \xrightarrow{\text{light and chlorophyll}} \underbrace{\underset{C_6H_{12}O_6}{\text{sugar}} + \underset{6O_2}{\text{oxygen}}}_{\text{products}}$$

Some of the sugar formed by photosynthesis is turned into **starch** for storage. Light is needed for making sugar, but not for turning the sugar into starch.

Chemical tests can be carried out to see if plants contain sugar or starch. If they do, it shows that photosynthesis has been taking place.

Why is photosynthesis important?

Animals can't make organic food substances for themselves. The only way an animal can get these substances is by eating plants, or by eating animals which have eaten the plants (or by eating animals which have eaten the animals which have eaten the plants!)

So animals depend on plants for food. When you eat a beefburger, you are able to do so only because the cow ate grass. We can sum this up by saying that plants produce food which can then be consumed by animals. This is the basis of **food chains**. You will find more about this on pages 44–46.

To give you some idea of the importance of photosynthesis, here are some figures. A hectare of maize – that's a field about the size of two football pitches – can produce more than 20 000 kg of sugar in a year. This is enough to sweeten over one million cups of tea.

Here is another way of looking at it: if all the sugar made by the world's plants in three years was piled up, it would form a heap the size of Mount Everest!

And there's another reason why photosynthesis is important. The oxygen in our atmosphere has been put there by plants over millions of years as a result of photosynthesis. This is the oxygen on which we depend for respiration. Without it animals could not exist.

Activities

A Testing a plant for sugar

Try this test on an onion bulb.

1 Put a few pieces of onion into a mortar and add a pinch of sand.

2 Grind up the pieces of onion with a pestle and cover them with water.

3 Filter the contents of the mortar into a test tube to a depth of about one centimetre.

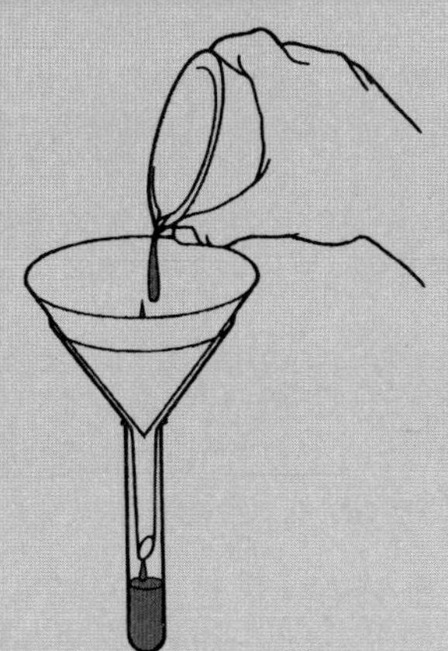

4 Pour the same amount of Benedict's solution into the test tube. Wear eye protection. Stand the test tube in a beaker of boiling water until its contents boil.

EYE PROTECTION MUST BE WORN

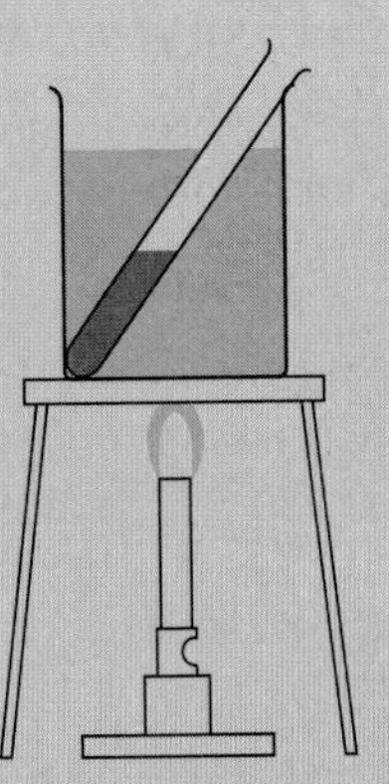

5 Repeat step 4 on some water in a test tube to serve as a control.

What happens to the solution in the test tubes?

A green, brown or red colour means there's sugar present.

Is there any sugar in the onion?

B Testing a leaf for starch

Try this test on a geranium leaf. Wear eye protection.

EYE PROTECTION MUST BE WORN

1 Dip your leaf into a beaker of boiling water for about ten seconds. This will kill it and make it soft.

2 Turn out the bunsen.

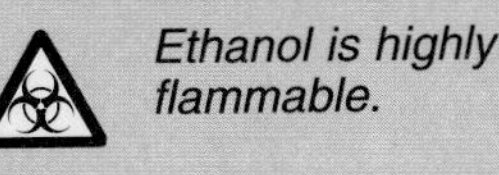

Ethanol is highly flammable.

FLAMMABLE

Put the leaf into a test tube of ethanol. Stand the test tube in the beaker of hot water for about ten minutes. The ethanol will decolorise the leaf.

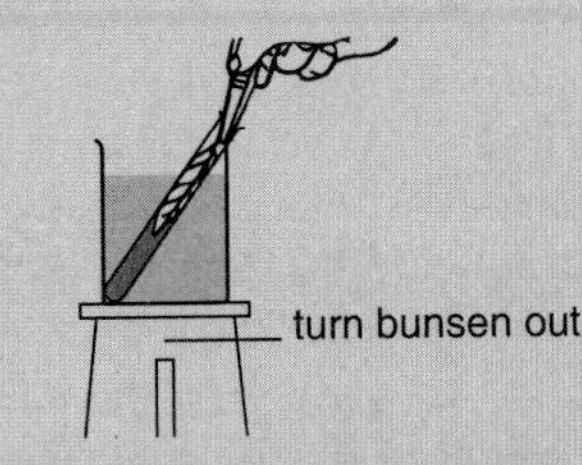

3 Wave the leaf to and fro in the beaker of water. This will remove the ethanol and soften the leaf.

4 Put the leaf in a petri dish and cover it with dilute iodine solution.

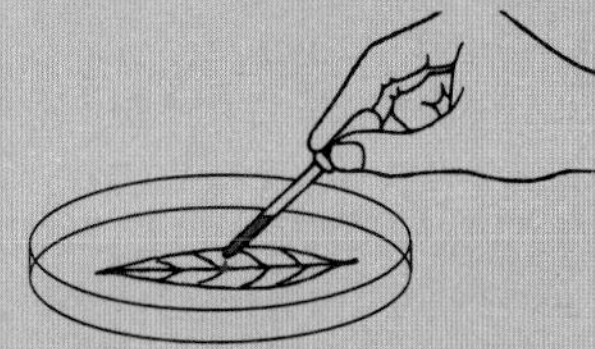

A blue-black colour shows that starch is present. Is there any starch in the leaf?

Questions

1 Plants are generally rooted to the ground and do not move around. How does this fit in with their method of feeding?

2 'When you eat a beefburger, you are able to do so only because the cow ate grass.' Explain this statement.

3 Suppose there was a great disaster and all plant life was suddenly wiped out. What effect might this have on humans, and why?

4 For many years scientists have tried to create artificial systems which will carry out photosynthesis, but the attempts have not been very successful.

a Why do you think this is so difficult?
b Research into this has been financed by the American space agency NASA. Why do you suppose NASA should be interested in it?

5 Someone has worked out that the total amount of organic matter made by all the world's plants in a year is 150 thousand million tonnes. But the total amount of food consumed by the world's human population is only 1/200th of this. If plants make more food than humans need, why are people starving?

6 One of the first experiments which led to the discovery of photosynthesis was done by a Dutchman called Van Helmont in 1692.

Van Helmont weighed a young willow tree. Then he filled a pot with soil, and weighed that. He then planted the tree in the soil.

He carefully covered the soil and then left the tree to grow, giving it nothing but water. After five years he weighed the tree and the soil, again. He found that the tree had gained 74 g in mass, but the soil had lost only 56 g. The tree seemed to have gained 18 g from somewhere other than the soil.

a Why did van Helmont cover the soil, and what may he have used for this?
b How had the tree gained 18 g?
c If you were to repeat Van Helmont's experiment today, what precautions would you take to make sure that your results were accurate?

7 In this topic you are told that plants make sugar. However, it is possible that they might absorb sugar from the soil in which they are growing.

Describe an experiment to find out if this might be true.

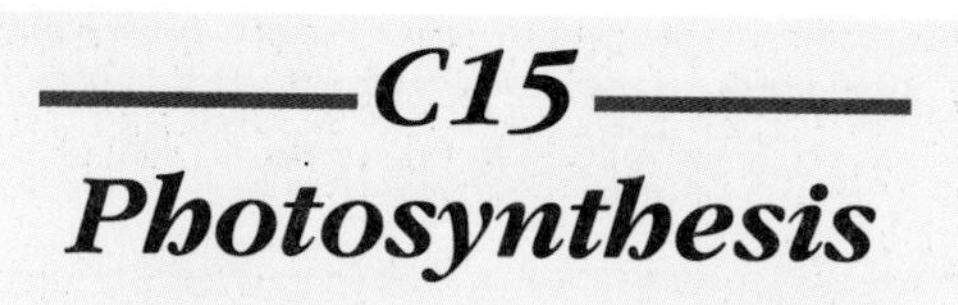

C15 Photosynthesis

Here we look in detail at photosynthesis and how it can be investigated.

What do plants need for photosynthesis?

Look at the equation for photosynthesis on page 166. From it we can say that for photosynthesis a plant needs **light**, **carbon dioxide**, **chlorophyll** and **water**.

We can do experiments to show that these things really are needed for photosynthesis. The experiments can be done on potted plants such as geraniums. We take the presence of starch in the leaves as an indication that photosynthesis has been taking place.

The principle behind the experiments is quite simple. First we make the plant use up all its stores of starch. This is called **de-starching**. It can be done by putting the plant in the dark for a few days. To make sure that the plant has been completely de-starched, we do an iodine test on one of its leaves.

We then give the plant everything it needs for photosynthesis except the one thing we want to investigate. After several days, we again do an iodine test on one of the leaves to see if the plant has been making starch. If it has not made any starch, we conclude that the missing factor is needed for photosynthesis.

As in other biological experiments, we must have a **control** with which to compare the result. The control plant is given everything it needs, including the factor being investigated.

With these principles in mind, let's look at the experiments in detail. You can try some of them for yourself if you do the activities on page 171.

Do plants need light for photosynthesis?

We put one de-starched plant in the dark. We put a second de-starched plant in the light, to serve as a control. After a few days we test each plant for starch. We find that only the plant in the light has made starch. This suggests that light is needed for photosynthesis.

Here's another method which is fun to do. We take a de-starched plant and cover part of one leaf with a piece of foil or black paper which does not let light through. We then leave the plant in the light. After a day or two we test the leaf with iodine. As you might expect, starch is absent from the covered part of the leaf. In fact the characteristic blue-black colour develops only where the leaf was uncovered. This is called a **starch print** and it is a striking way of showing that light is needed for photosynthesis (picture 1).

Picture 1 A starch print on a geranium leaf. The top picture shows the mask on the leaf. The bottom picture shows the print.

Do plants need carbon dioxide for photosynthesis?

We set up two de-starched plants side by side. One of them is given air from which all the carbon dioxide has been removed. The other one is given air containing plenty of carbon dioxide – this is the control. The details of the experiment are shown in picture 2.

After several days a leaf from each plant is tested with iodine to see if it has made any starch. It turns out that the plant without carbon dioxide has not made starch. This suggests that carbon dioxide is needed for photosynthesis.

Do plants need chlorophyll for photosynthesis?

The ideal way to investigate this would be to take all the chlorophyll out of a leaf, and see if this stops it making starch. However, it is impossible to remove the chlorophyll without killing the leaf.

So what can we do? Luckily nature comes to our aid. It so happens that the leaves of some plants (including certain varieties of geranium) are green in some places but white or yellow in others. Chlorophyll is present in the green areas, but absent from the non-green areas. Such leaves are described as **variegated** (picture 3).

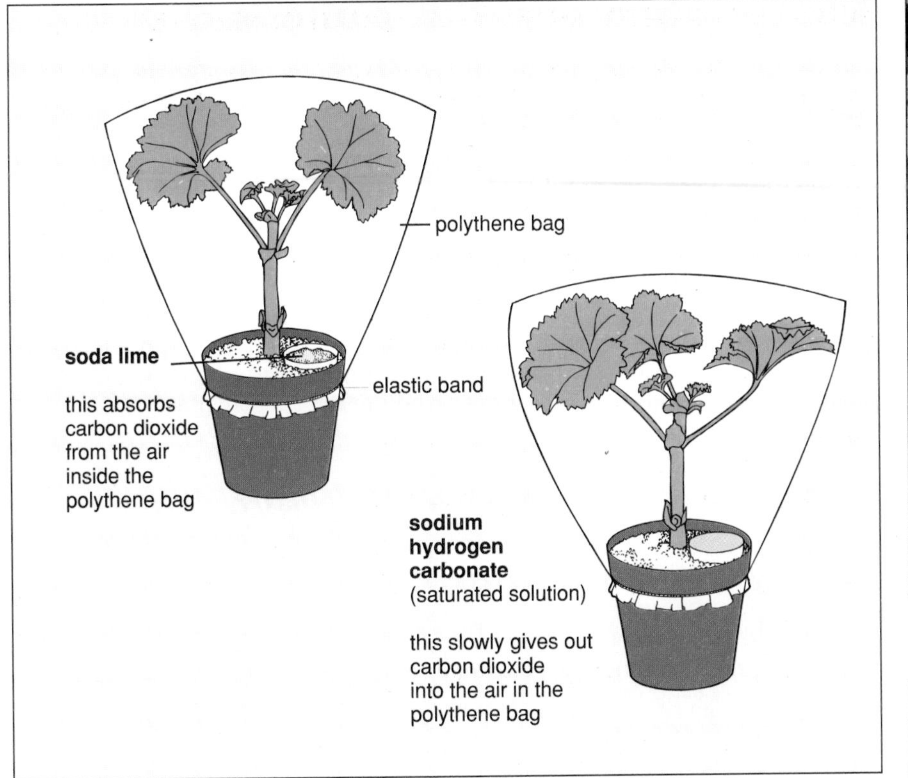

Picture 2 An experiment to find out if plants need carbon dioxide for photosynthesis. The top plant was given air with no carbon dioxide. The bottom plant was given air with carbon dioxide.

Picture 3 A plant with variegated leaves.

To find out if chlorophyll is needed for photosynthesis, all we have to do is to de-starch a variegated plant and then put it in the light.

After a few days we do the iodine test on one of its leaves. The blue-black colour develops only in the parts of the leaf that were green. This suggests that photosynthesis takes place only where chlorophyll is present.

Do plants need water for photosynthesis?

There is no *simple* experiment that can be done to answer this question. You can't do it by taking all the water out of a plant, because that would kill the plant. The only way is to find out what happens to the water which a plant takes up into its cells. Scientists have done this by giving plants a special form of water called **heavy water** and then tracing what happens to it. The results show that water is definitely needed for photosynthesis.

Can you say what the water is needed for? (Hint: look at the equation on page 166).

What does photosynthesis produce?

We have seen that plants make substances such as sugar and starch. However, they will only do this in the light. This strongly suggests that these substances are produced by photosynthesis.

Plants also produce oxygen. This can be shown quite easily by using a water plant such as Canadian pondweed. Canadian pondweed gives off bubbles of gas when put in the light. Picture 4 shows an experiment which can be done with this plant. The bubbles of gas are collected in an upturned test tube, and then tested for oxygen.

We now know that all green plants give off oxygen. They will only do so in the light, and this suggests that oxygen is a product of photosynthesis.

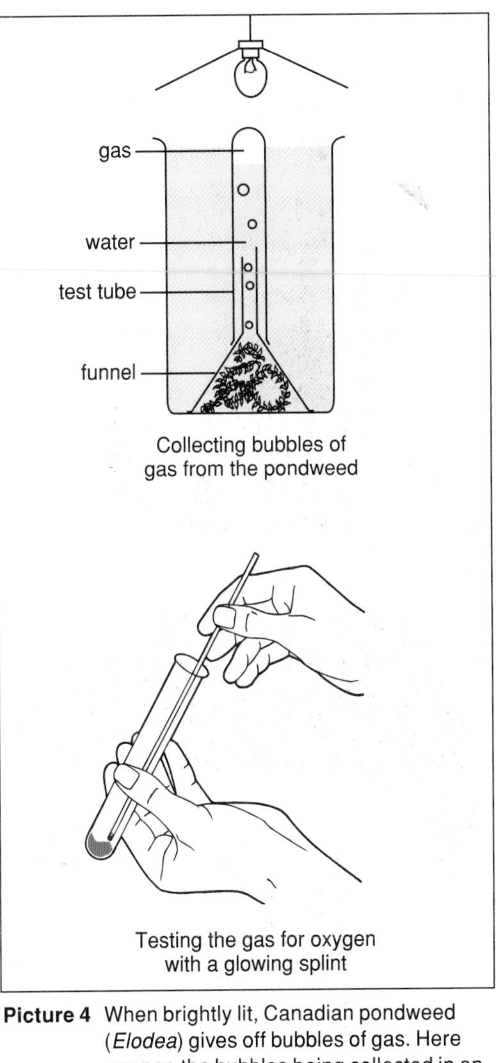

Picture 4 When brightly lit, Canadian pondweed (*Elodea*) gives off bubbles of gas. Here you see the bubbles being collected in an upturned test tube, and then tested for oxygen with a glowing splint.

Picture 5 A scientist, wearing protective gloves, is giving a simple plant-like organism carbon dioxide containing radioactive carbon atoms.

The chemistry of photosynthesis

In recent years scientists have discovered a lot about photosynthesis, thanks to the use of **isotopes**. Radioactive carbon has been particularly useful.

In one experiment, plant-like organisms were given carbon dioxide whose normal carbon had been replaced by radioactive carbon (picture 5). In this way the scientists were able to 'label' the carbon in the carbon dioxide and follow what happened to it.

In the light, the organisms took up the carbon dioxide in the usual way. The scientists then extracted the contents of the cells and tested them for radioactivity.

The results are summarised in picture 6. In the presence of light, the carbon in the carbon dioxide got into the sugar and other carbohydrates which the organisms made. This shows that the carbon in the carbohydrates comes from carbon dioxide.

What happens to the sugar made by photosynthesis?

The scientists who did the experiments with radioactive carbon dioxide detected radioactivity in the sugar quite quickly. Later on, other substances inside the plant became radioactive. We now know that plants make sugars first and then convert them into these other substances.

One of these substances is **cellulose** which is needed for cell walls. **Amino acids** are also formed from the products of photosynthesis. They are the building blocks of **proteins** which are needed for growth and as enzymes.

To make proteins a plant needs nitrogen and sulphur as well as carbon, hydrogen and oxygen. Land plants obtain these extra elements from the soil in the form of **mineral salts** (see page 75). Water plants get them from the surrounding water.

Some of the sugar made by photosynthesis is used for respiration or converted into starch for storage. Certain plants form other storage substances such as fats and oils. These, and all other substances which plants contain, come from photosynthesis (picture 7).

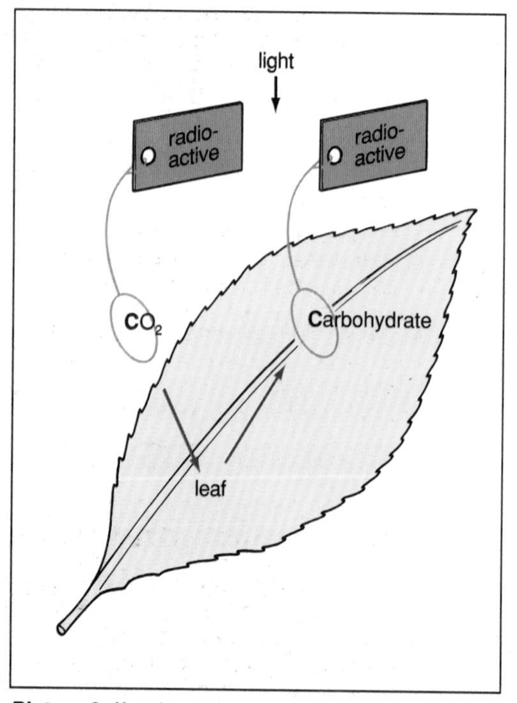

Picture 6 If a plant is given carbon dioxide whose carbon is radioactive, the radioactive carbon gets into the carbohydrate which the plant makes.

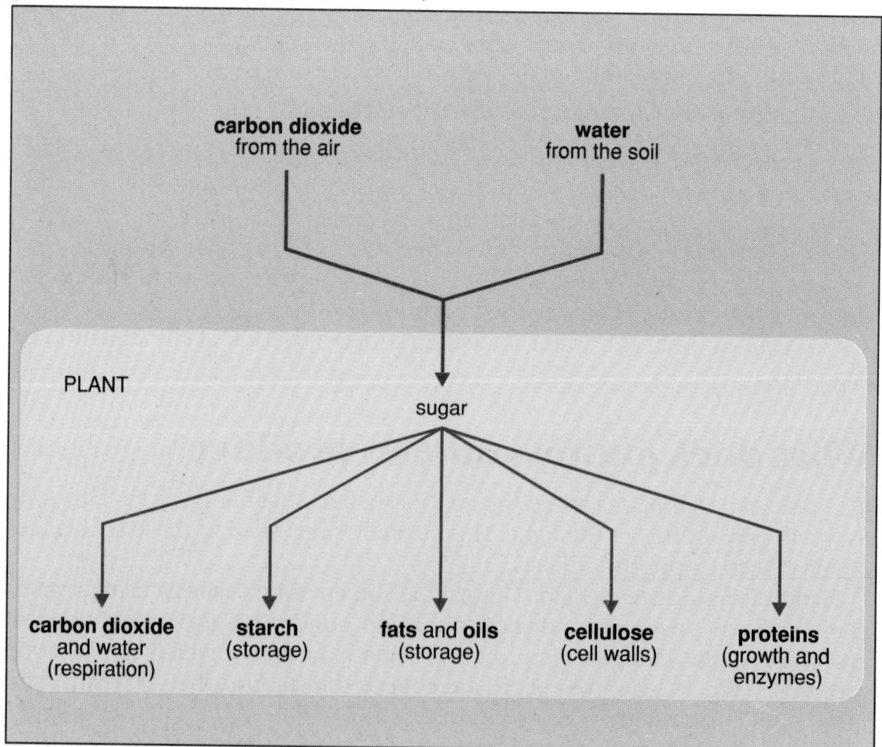

Picture 7 The diagram shows what can happen to the sugar which a plant makes by photosynthesis.

Activities

A Is light needed for photosynthesis?

First method

1 You will need two potted plants which have been de-starched.

2 Place one of them in the dark, and the other in the light. The plant in the light is your control plant.

3 After several days take a leaf (or part of a leaf) from each plant. Test them for starch (see page 167). Don't forget which leaf is which!

Has either plant formed starch? Is light needed for photosynthesis?

Second method

1 You will need a potted plant which has been de-starched.

2 Attach a strip of black paper or foil to the upper and lower sides of a leaf, as shown in the illustration.

3 Put the plant in a well lit place.

4 After several days detach the leaf and test it for starch (see page 167). Make a drawing of the leaf to show your result.

This is called a **starch print**. What conclusion do you draw?

B Is carbon dioxide needed for photosynthesis?

1 You will need two potted geranium plants which have been de-starched.

2 Set up the two plants as shown in picture 2 on page 169.

3 Place both plants side by side in a well lit place.

4 After several days take a leaf (or part of a leaf) from each plant. Test them for starch.

Which plant contains starch? Is carbon dioxide needed for photosynthesis?

C Do plants need chlorophyll for photosynthesis?

1 You will need a potted plant with variegated leaves, e.g. geranium. The plant should have been put in the light for several days.

2 Detach a leaf and draw its upper side, making a clear distinction between the green and non-green areas.

3 Now carry out a starch test on the whole leaf (see page 167).

Which parts of the leaf turn black when treated with iodine?

Indicate your answer in your drawing by writing B in the black areas.

Where is the control in this experiment?

Is chlorophyll needed for photosynthesis?

D Another way of finding out if carbon dioxide is needed for photosynthesis

1 A possible way of finding out if carbon dioxide is needed for photosynthesis is illustrated below. Study the picture, then answer the questions underneath.

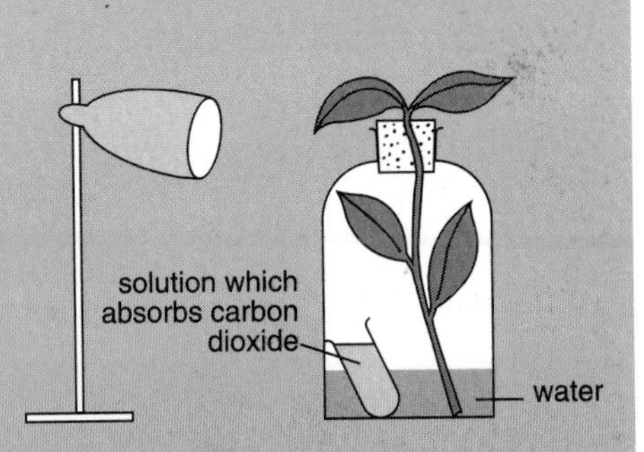

a What should be done to the plant beforehand, and why?
b Where is the control in this experiment?
c At the end of the experiment, how would you find out if the plant has been photosynthesising?
d Is this a satisfactory experiment? Give reasons for your opinion.

Now carry out the experiment yourself to see if it will work. If you wish to modify it in any way, do so, but discuss your plans with your teacher first.

To think about

In all these experiments we have assumed that the presence of starch indicates that photosynthesis has been taking place. Is this assumption justified? In what circumstances might photosynthesis take place without the formation of starch? How could you find out if your answer is true?

Questions

1 If you were making a starch print on a leaf, how could you print your initials in a light colour on a dark background?

2 Starch prints are made to find out if a plant needs light for photosynthesis. However, some people think that a starch print is not a good scientific way of doing this. What do you think?

3 When finding out if a particular condition is needed for photosynthesis the plants should be de-starched first.

a How are they de-starched?
b Why is this necessary?
c How could you make sure they have been completely de-starched before you begin the experiment?

4 Elizabeth wants to find out if a potted plant needs carbon dioxide in order to make starch. She is not satisfied with the method given in picture 2 on page 169 so she tries a different way. She selects two leaves on the plant and, without cutting them off, she encloses each one in a small polythene bag. In one bag she puts some soda lime, and in the other bag she puts some saturated sodium hydrogencarbonate solution.

Make a diagram of the set-up. Do you think Elizabeth's method is as good as the one in picture 2. Give reasons for your answer.

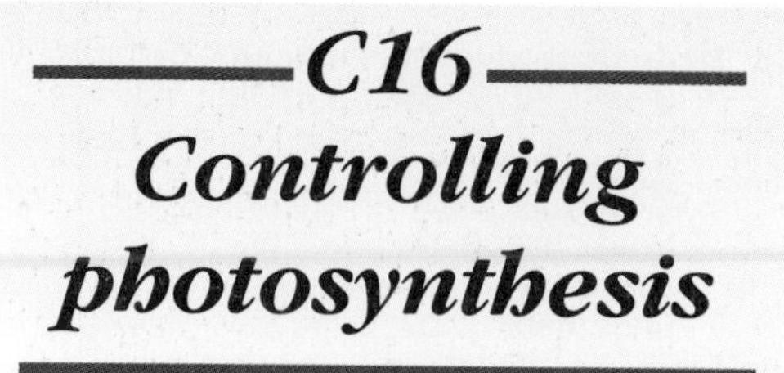

C16 Controlling photosynthesis

Photosynthesis can occur rapidly or slowly, depending on circumstances.

Picture 1 The lights in this commercial greenhouse enable the plants to photosynthesise even when it is dark or gloomy outside.

Picture 2 Primroses can photosynthesise efficiently in shady places such as woods.

Picture 3 The lettuces in the right hand box were grown in an atmosphere containing extra carbon dioxide. The lettuces in the left hand box were grown in a normal atmosphere. The lettuces that were given extra carbon dioxide are larger than the other ones, and fewer of them can be fitted into the box.

Getting a good yield

Think of a wheat field. The faster the wheat plants photosynthesise, the more food they produce. In other words, the yield is greater.

Four main things affect the rate of photosynthesis and therefore the yield. They are: **light**, **carbon dioxide**, **temperature** and **water**. Let's look at each in turn.

Light

Up to a point, the brighter the light, the faster is the rate of photosynthesis. On a sunny day plants photosynthesise faster than on a dull day, and plants growing in an open field photosynthesise faster than plants growing in the shade.

This is important to gardeners and farmers. If you want your crops to do well, you must grow them in a place that gets plenty of sunlight. Sometimes artificial lights are shone on greenhouse plants to increase productivity (picture 1). However, certain wild plants are adapted to living in shady places such as a wood. They can photosynthesise even in dim light (picture 2).

As with many other things in life, it is possible to have too much of a good thing: in very bright sunshine photosynthesis actually slows down. This is because bright sunshine contains a lot of ultraviolet light which can damage plants.

Which is the best type of light for photosynthesis?

Ordinary white light, such as sunlight, is made up of different colours or wavelengths. These make up the **spectrum**.

Not all the colours of the spectrum are used by plants for photosynthesis, only **blue** and **red**. In fact chlorophyll only absorbs these two colours. Other colours, particularly green, pass straight through chlorophyll or are reflected from it. The reason why leaves look green is that green light is reflected from the chlorophyll inside them. Green is the colour they do *not* use.

A plant which is deprived of red or blue light cannot photosynthesise properly and it will not make much food. Sunlight provides these colours in the right proportions. People who use artificial light in greenhouses must make sure that it contains these colours. Light manufacturers produce special lamps for this purpose.

Carbon dioxide

The more carbon dioxide there is in the air surrounding a plant, the faster the plant photosynthesises. The amount of carbon dioxide in the atmosphere is about 0.03 per cent, and it doesn't vary much. Even so, there are slight differences from place to place which may affect the rate of photosynthesis. For example, the concentration of carbon dioxide close to the ground in a dense forest is higher than in an open field. Why do you think this is?

Extra carbon dioxide is sometimes pumped into greenhouses, or produced by a 'burner', so as to increase the rate of photosynthesis. Picture 3 shows how helpful this can be.

Picture 4 Britain is too cold for melons to be grown out of doors. However, the warm conditions inside a greenhouse enable these excellent melons to be produced.

Temperature

Up to a point, the warmer it is, the faster is the rate of photosynthesis. Normally an increase in temperature of 10°C roughly doubles the rate. This is true of chemical reactions in general, including photosynthesis.

In the natural world there are tremendous variations in temperature, both from place to place and at different times of the year. One of the reasons why plants do so well in greenhouses is because of the warmth (picture 4).

But there is a limit. If the temperature gets much above 40°C, photosynthesis slows down rapidly and then stops altogether. This is because heating destroys the enzymes which are responsible for making the chemical reactions work.

Picture 5 The conditions in this jungle in Malaysia are just right for a high rate of photosynthesis.

Water

A plant which is beginning to wilt through lack of water may photosynthesise at only half the normal rate. This is not because it hasn't got enough water for photosynthesis – even a wilted plant has enough water for that. It's because lack of water produces other effects, and these then slow down photosynthesis. Can you think what these other effects might be?

Which places are best for photosynthesis?

One of the best natural places for photosynthesis is the tropical rain forest. Long hours of sunshine, warmth and a high rainfall ensure a high rate of photosynthesis and good growth. The tropical rain forest has been described as a vegetative frenzy! (picture 5).

Crop plants grown in the tropics make particularly large amounts of food. This is true of sugar cane, for example, which gives a higher yield of organic matter than any other crop plant (picture 6).

Sugar cane needs a hot, moist climate with temperatures averaging around 25°C and an annual rainfall of about 150 cm. It gets this in places such as the Caribbean and South East Asia. When it is grown in drier regions like North America and southern Africa, water must be supplied by irrigation.

In parts of the world where conditions are less favourable for photosynthesis, plants may be grown in special air conditioned greenhouses. Here all the conditions affecting the rate of photosynthesis are carefully controlled. There is more about this on page 175.

Which parts of the world do you think are *worst* for photosynthesis, and why? What sort of plants can grow in such places, and how do they survive?

Picture 6 Sugar cane growing in Malaysia. Sugar cane can photosynthesise more efficiently than any other crop plant. It provides most of the world's sugar.

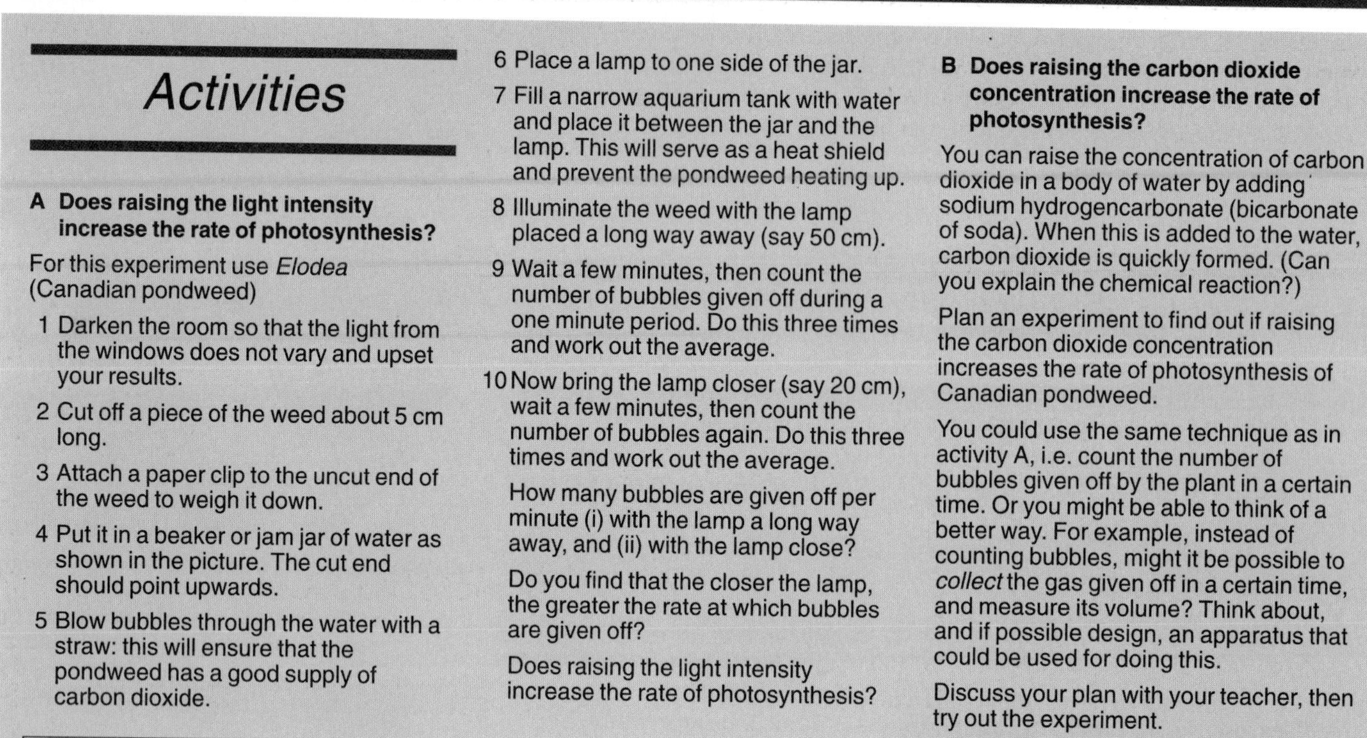

Activities

A Does raising the light intensity increase the rate of photosynthesis?

For this experiment use *Elodea* (Canadian pondweed)

1 Darken the room so that the light from the windows does not vary and upset your results.

2 Cut off a piece of the weed about 5 cm long.

3 Attach a paper clip to the uncut end of the weed to weigh it down.

4 Put it in a beaker or jam jar of water as shown in the picture. The cut end should point upwards.

5 Blow bubbles through the water with a straw: this will ensure that the pondweed has a good supply of carbon dioxide.

6 Place a lamp to one side of the jar.

7 Fill a narrow aquarium tank with water and place it between the jar and the lamp. This will serve as a heat shield and prevent the pondweed heating up.

8 Illuminate the weed with the lamp placed a long way away (say 50 cm).

9 Wait a few minutes, then count the number of bubbles given off during a one minute period. Do this three times and work out the average.

10 Now bring the lamp closer (say 20 cm), wait a few minutes, then count the number of bubbles again. Do this three times and work out the average.

How many bubbles are given off per minute (i) with the lamp a long way away, and (ii) with the lamp close?

Do you find that the closer the lamp, the greater the rate at which bubbles are given off?

Does raising the light intensity increase the rate of photosynthesis?

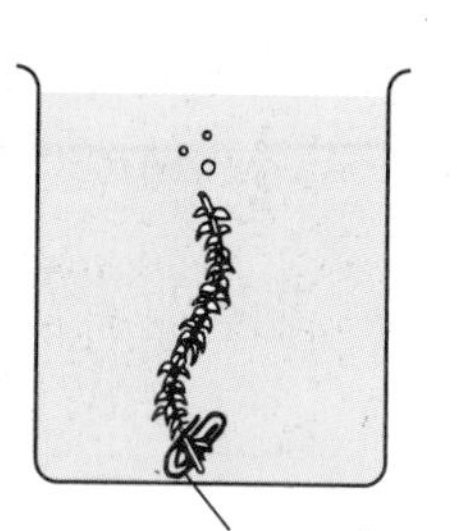

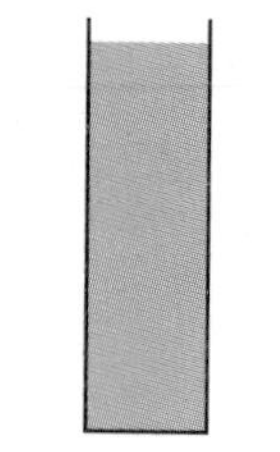

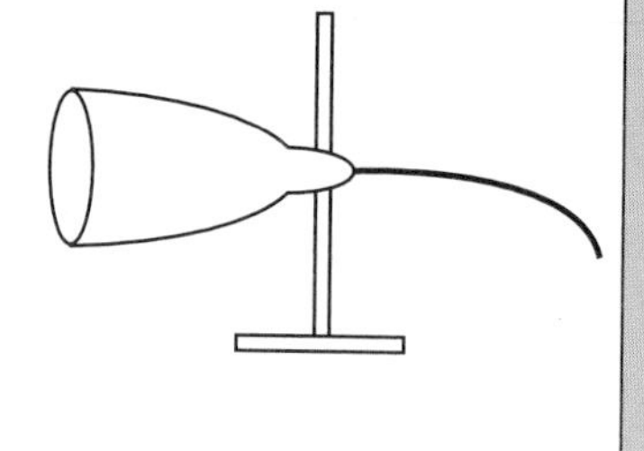

B Does raising the carbon dioxide concentration increase the rate of photosynthesis?

You can raise the concentration of carbon dioxide in a body of water by adding sodium hydrogencarbonate (bicarbonate of soda). When this is added to the water, carbon dioxide is quickly formed. (Can you explain the chemical reaction?)

Plan an experiment to find out if raising the carbon dioxide concentration increases the rate of photosynthesis of Canadian pondweed.

You could use the same technique as in activity A, i.e. count the number of bubbles given off by the plant in a certain time. Or you might be able to think of a better way. For example, instead of counting bubbles, might it be possible to *collect* the gas given off in a certain time, and measure its volume? Think about, and if possible design, an apparatus that could be used for doing this.

Discuss your plan with your teacher, then try out the experiment.

To think about

Think of the plants in a typical pond: water lilies, pondweed and so on. These plants get their carbon dioxide from the surrounding water. The carbon dioxide is in solution, and it diffuses into the plants' cells. However, carbon dioxide diffuses more slowly in water than in air. How might this affect the plants in a pond, and what could be done about it?

Questions

1 Mr Smith plants his onions in a shady place whereas Mrs Jones plants hers in the sun. Whose onions would you expect to do best, and why?

2 Someone observed that wheat grows taller, and gives a higher yield of grain, close to a certain coal-burning factory than further away.

a Suggest a reason for this.
b What investigations would you carry out to find if your suggestion is right?

3 Describe an experiment which you could do to find out which colours of the spectrum are used by a potted plant for photosynthesis.

4 The following figures give the total annual amounts of organic matter produced per hectare by plants in different parts of the world:

Sugar, cane, Java	87 tonnes
Tropical rain forest	59 tonnes
Pine forest, England	16 tonnes

Account for the differences. Approximately how much organic matter would you expect wheat to produce in England? Give a reason for your answer.

5 A scientist grew some cereal plants in a field. During the course of one day he took several plants every four hours and measured the amount of sugar in the leaves. The sugar concentrations, expressed as a percentage of the dry mass of the leaves, are given in the table at the top of the next column.

Time of day	Sugar concentration
4 am	0.45
8 am	0.60
12 noon	1.75
4 pm	2.00
8 pm	1.4
12 midnight	0.5
4 am	0.45

a Plot the data on graph paper, putting sugar concentration on the vertical axis.
b What is the probable concentration of sugar in the leaves at 10 am and 2 am?
c At what time of the day is sugar probably at a maximum in the leaf, and why?
d Explain why the sugar concentration changes over the 24 hour period.

Getting the best out of a greenhouse

Here are some of the things you need to bear in mind if you have a greenhouse:

- It must not get too hot, and the light must not be too bright. Too much heating and light can make plants produce less food, not more.
- The air must be fairly moist (humid). If it is too dry, a lot of water may evaporate from the plants and they may wilt. But if it is too moist, fungal pests may flourish.
- The plants must have a good supply of all the chemicals they need, including carbon dioxide, water and mineral salts.

Now look at the picture. This shows how a modern greenhouse meets these requirements. Of course, you want the running costs to be as low as possible. This means turning off energy-consuming devices such as the heater and lights when they are not needed. In large greenhouses used by market gardeners, the lights, heating, carbon dioxide and humidity are all regulated automatically.

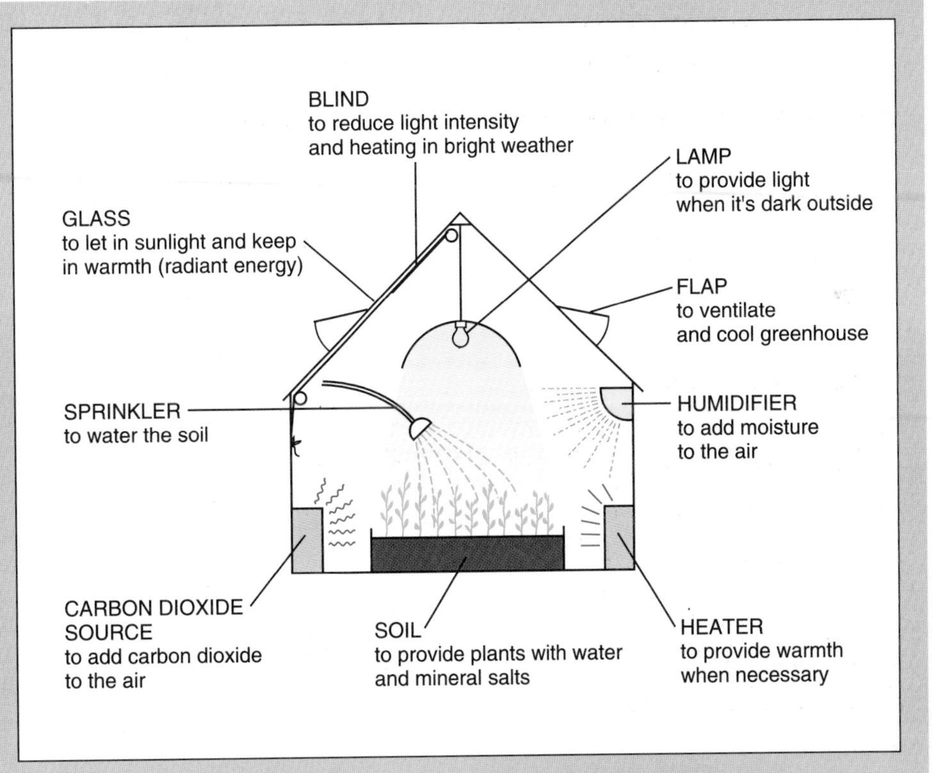

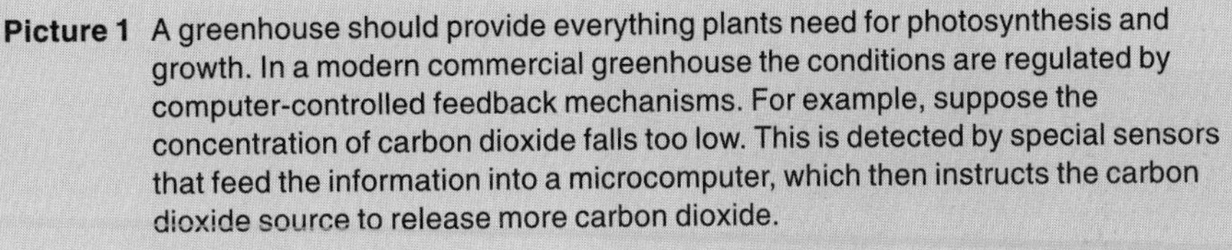

Picture 1 A greenhouse should provide everything plants need for photosynthesis and growth. In a modern commercial greenhouse the conditions are regulated by computer-controlled feedback mechanisms. For example, suppose the concentration of carbon dioxide falls too low. This is detected by special sensors that feed the information into a microcomputer, which then instructs the carbon dioxide source to release more carbon dioxide.

When several things control the same process

Look at the graph below. It shows the results of an experiment to find the effect of raising the light intensity on the rate of photosynthesis – like the activity on page 174.

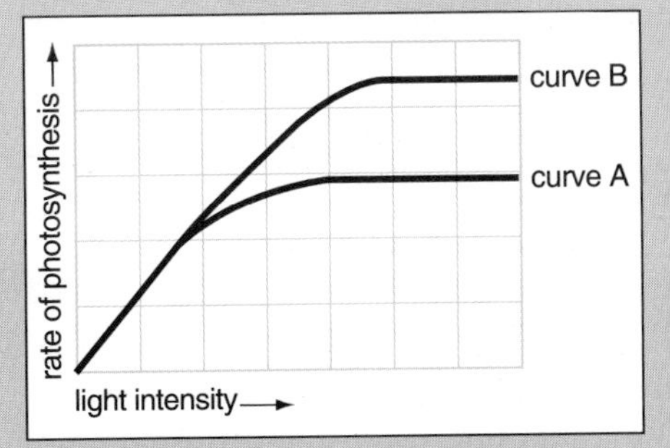

First look at curve A. As the light intensity is gradually raised, the curve rises, i.e. the rate of photosynthesis increases.

However, there comes a point when the curve flattens out – the rate of photosynthesis does not increase any more, however much you raise the light intensity.

Why do you think the rate of photosynthesis stops increasing at this point? Well, it could be that photosynthesis is going at its maximum possible speed. But it could be that something other than light is preventing the process going any faster.

What might this 'something' be? One possibility is carbon dioxide. How could we find out if it is carbon dioxide? One way would be to raise the concentration of carbon dioxide in the atmosphere surrounding the plant and repeat the experiment.

The result of doing this is shown in curve B. This time a much higher rate of photosynthesis is achieved. This tells us that carbon dioxide must have been controlling the rate of photosynthesis when the curve flattened out in the first experiment. We say that carbon dioxide was the **limiting factor** controlling the process – it was setting the pace.

What do we learn from this experiment? Well, here's one thing: in a greenhouse there is no point in giving plants extra light if the concentration of carbon dioxide is too low. It simply won't make any difference. This is an example of how an experiment done by scientists in a laboratory can be useful to people who grow plants for sale.

1. The three main conditions which affect the rate of photosynthesis are light, temperature and carbon dioxide. Which condition is likely to limit the rate of photosynthesis of wheat plants in a field in Britain on a cloudless day at:

   a noon in mid-summer,
   b dusk in mid-summer,
   c noon in mid-winter,
   d dusk in mid-winter?

   What effect do you think clouds have on the rate of photosynthesis?

2. A market gardener who is thinking of installing artificial lights in a greenhouse might benefit from reading the passage above. Why?

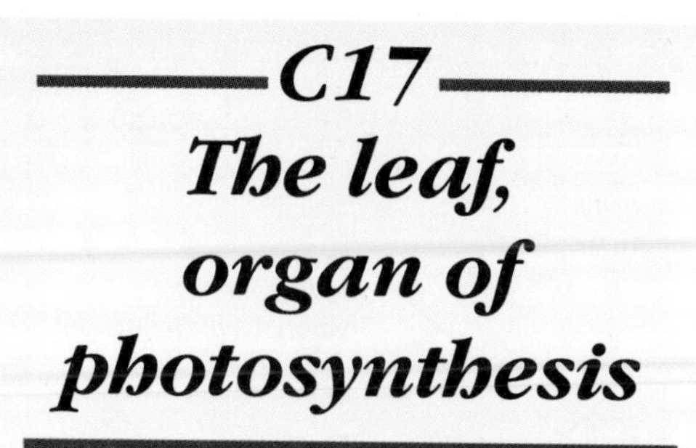

C17 The leaf, organ of photosynthesis

This topic is about leaves, and how their structure suits them for carrying out photosynthesis

The outside of the leaf

Look at picture 1. In this photograph you can see the main features of a typical leaf. Picture 2 will help you to identify its parts.

Each leaf is attached to the stem or branch by a **leaf stalk**. This leads to the **veins** in the leaf. The leaves of this particular plant have a **main vein** running down the middle, with **side veins** branching out on either side. The veins carry substances to and from the leaf. They also strengthen it.

You may bave noticed that leaves are often shiny, particularly on the upper side. This is because they are covered by a layer of waxy material. This is called the **cuticle** and, if thick enough, it is waterproof. It protects the leaf from losing too much water in hot, dry weather.

Immediately under the cuticle is a layer of cells, the **epidermis**, which forms the 'skin' of the leaf. The epidermis may be pierced by lots of tiny holes called **stomata** (singular **stoma**) (picture 3). The stomata are mainly on the lower side of the leaf. They allow gases to diffuse in and out of the leaf, and water vapour to escape. Each stoma is flanked by a pair of **guard cells** which can open and close rather like doors. They close in hot, dry weather to prevent too much water evaporating from the leaves.

Leaves are generally flat, sometimes large and often numerous. The result is that they have a large surface area for absorbing carbon dioxide and light. The veins help to support the leaf and hold it out flat, so that it can catch the maximum amount of light (picture 4). In many plants the leaves are positioned in such a way that they don't shade each other (picture 5).

The inside of the leaf

You can see the inside of a leaf by looking at thin slices (sections) of it under the microscope. The sections can be cut in various planes so as to give a three-dimensional picture of the leaf (picture 6).

Between the upper and lower epidermis are lots of cells which together make up the **mesophyll**. These cells contain **chloroplasts**, and this is where photosynthesis takes place – it is the 'business part' of the leaf.

The mesophyll towards the upper side of the leaf consists of cells shaped like bricks and arranged neatly side by side. They are called **palisade cells**. The other mesophyll cells are rounded and more irregular in their arrangement.

Between the mesophyll cells are **air spaces** into which the stomata open. When photosynthesis is taking place, carbon dioxide diffuses through the open stomata into the air spaces. It then diffuses into the cells.

Photosynthesis takes place mainly in the palisade cells. They contain most of the chloroplasts, and they are near the surface of the leaf that gets most light.

Picture 1 Leaves are the main place where photosynthesis takes place.

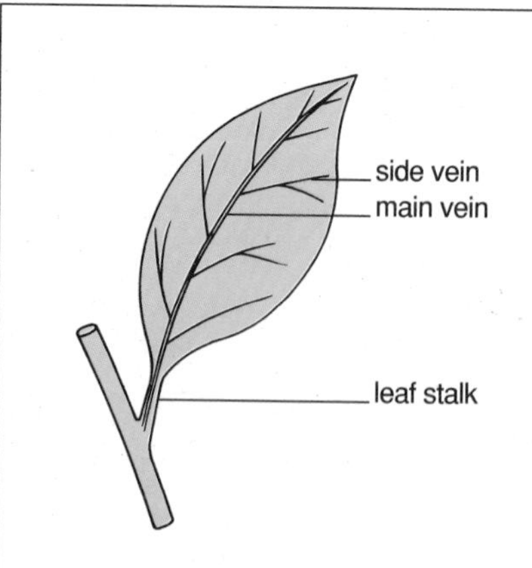

Picture 2 The parts of a typical leaf.

Picture 3 Photograph of the epidermis on the lower side of a leaf, showing the stomata. The photograph was taken down a microscope.

Picture 4 Notice the network of branching veins on the lower side of this bramble leaf.

Picture 5 In many plants the leaves are arranged so as not to shade each other. Two arrangements are shown here.

The chloroplasts are often clustered towards the tops of the cells, in the best position for catching light.

You can see a vein in picture 6. This is made up of two parts: the xylem towards the top, and the phloem below. The **xylem** brings water and mineral salts to the leaf. The **phloem** takes soluble sugar and other products of photosynthesis away from the leaf. We shall have more to say about these **vascular tissues** in the next topic.

Picture 6 A small part of a leaf, greatly enlarged, to show the structures inside it. The green dots in the cells are chloroplasts.

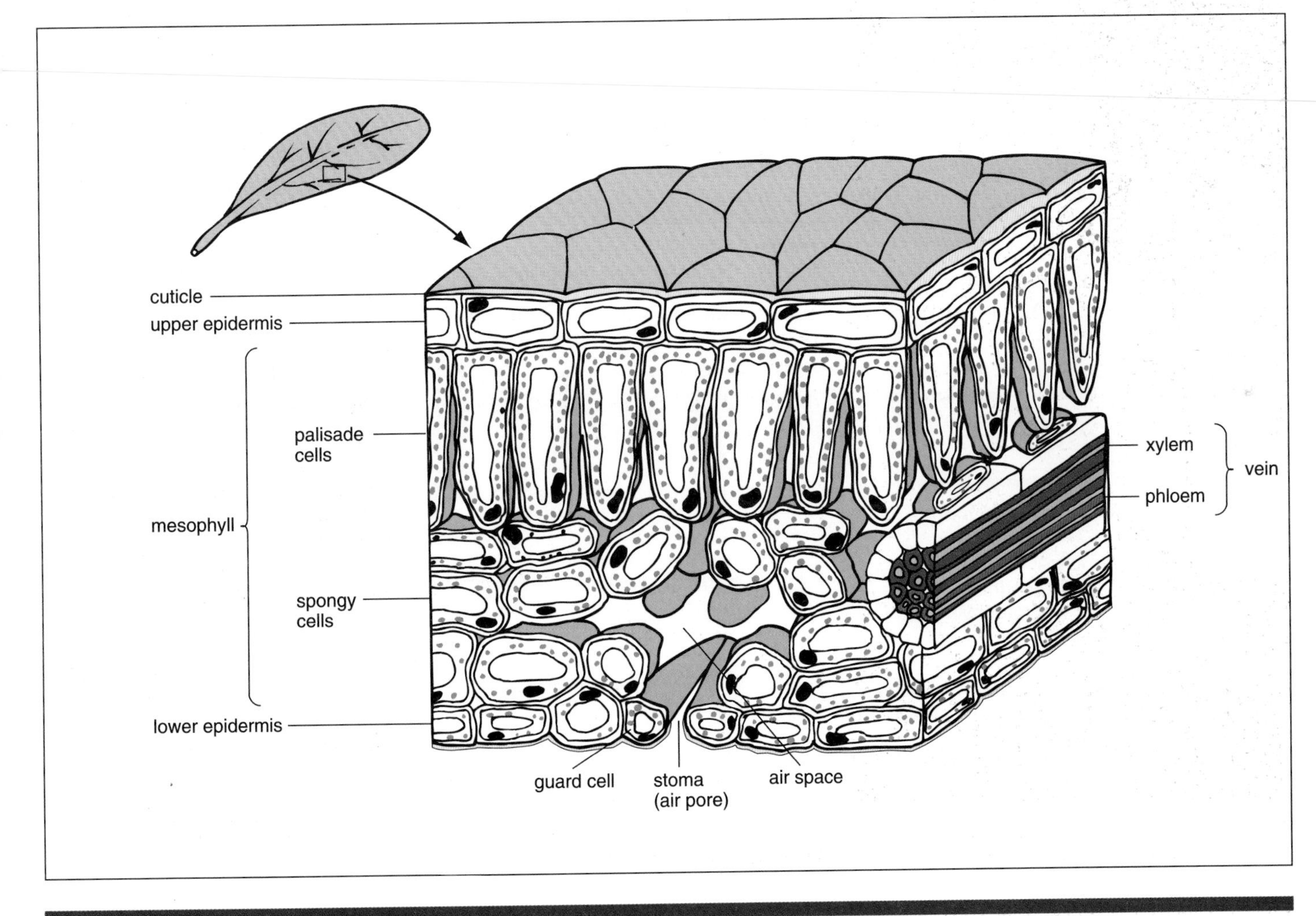

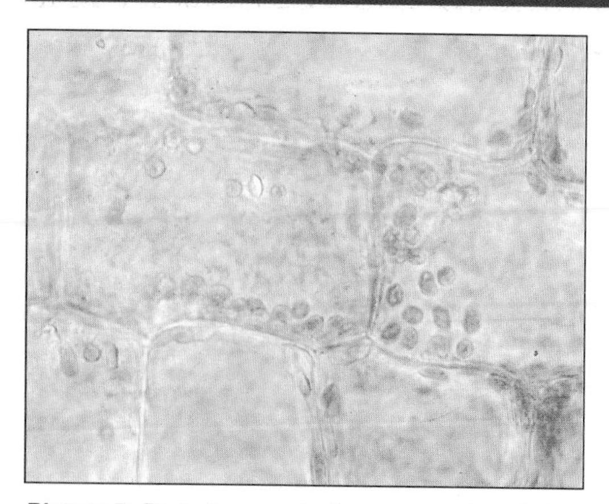

Picture 7 Part of a moss leaf as seen under a light microscope. Notice the chloroplasts in the cells. They are magnified about 1000 times.

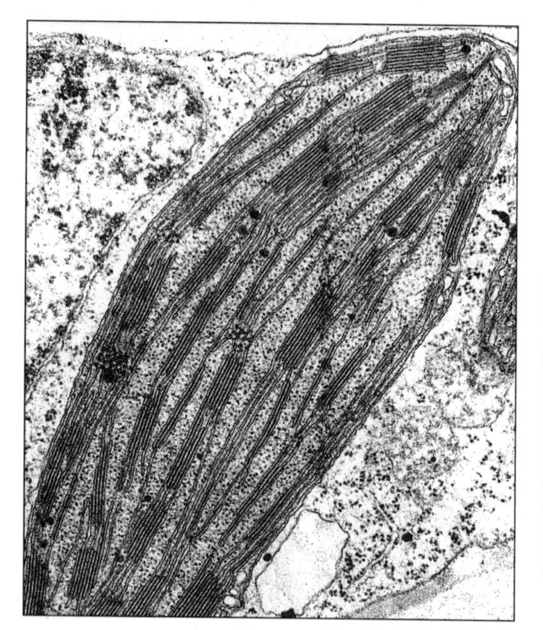

Picture 8 A single chloroplast seen in section in the electron microscope. It is magnified about 30 000 times.

Inside the chloroplast

Picture 7 shows part of a moss leaf under the light microscope. Notice the chloroplasts in the cells. Although they show up clearly, you cannot see anything inside them.

But if you look at a chloroplast in the electron microscope, you can see much more detail. Picture 8 shows a single chloroplast as it appears in the electron microscope. It is magnified about 30 000 times. If a whole moss leaf was magnified to this extent, it would be the size of a tree!

The electron microscope shows that the chloroplast is filled with rows of thin **membranes**. The way they are arranged is shown in picture 9. By careful experiments scientists have shown that millions of chlorophyll molecules are laid out on these membranes, rather like tiles on a roof.

A large surface area for photosynthesis

Think of a plant such as an oak tree. An oak tree has a large number of leaves. Inside each leaf are lots of cells, and inside each cell are lots of chloroplasts; inside each chloroplast are lots of membranes, and on each membrane are lots of chlorophyll molecules. So there's a vast number of chlorophyll molecules in an oak tree, and they cover a huge surface area. No wonder plants produce so much food!

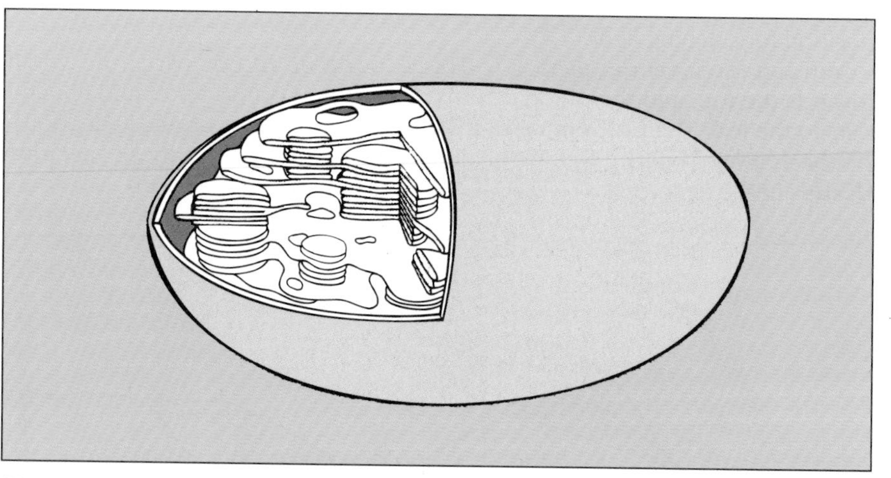

Picture 9 This picture shows the three-dimensional structure of a chloroplast. The inside is full of flattened membranes.

Questions

1 Each word in column A matches one or more words in column B.

A	B
air space	carbon dioxide
chloroplast	chlorophyll
stoma	light
xylem	water

a Against each word in column A write down the matching word or words from column B.

b What do the words in column A have in common?

c What do the words in column B have in common?

2 Why are the leaves of most plants (a) numerous and (b) thin?

3 Suggest a reason why

a there are normally more stomata on the lower side of a leaf than on the upper side,

b the cuticle is usually thicker on the upper side of a leaf than on the lower side.

4 Make a list of all the internal features of a green leaf which help it to carry out photosynthesis efficiently.

5 The photograph on the right shows the end of a branch of a chestnut tree.

a In what way might the positioning of the leaves help the tree with photosynthesis?

b How do the tree's trunk and branches help the leaves to photosynthesise?

Activities

A Looking at leaves

1 Examine a leaf, which your teacher will give you.

Which of the structures shown in picture 2 can you see?

2 How does the colour of the upper side of a leaf differ from the lower side? Why the difference?

3 Look at leaves from different trees. Can you explain their different shapes?

4 Examine the veins of different leaves. What sort of pattern do they form?

5 Tear leaves in two. Does this tell you anything about the function of the veins?

B Examining the surface of a leaf

1 Your teacher will give you a leaf.

2 With a paintbrush apply a thin layer of clear nail varnish to a small area on the lower surface of the leaf.

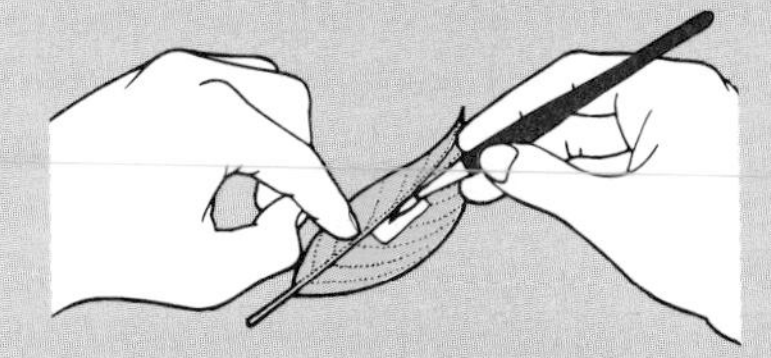

3 When the nail varnish is dry, peel it off carefully with a pair of forceps. The nail varnish will have made an impression of the leaf surface.

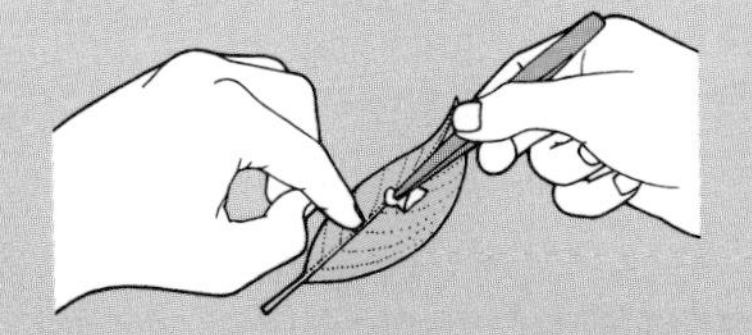

4 Put the nail varnish in a drop of water on a slide, and cover it with a coverslip.

5 Examine it under the low power of the microscope.

Can you see the stomata?

Approximately how many stomata are visible in the field of view?

6 Repeat the experiment on the upper surface of the leaf.

How many stomata are visible in the field of view this time?

Which side of the leaf has the greater number of stomata?

Why do you think the two sides of the leaf differ in this respect?

C Looking inside the leaf

You will need either a prepared slide, or a thin section of a leaf which you can mount on a slide yourself. If you are mounting a section, proceed as follows:

1 Put a drop of water on a microscope slide.

2 Carefully transfer the leaf section to the drop of water on the slide.

3 Cover the section with a coverslip.

4 Examine it under the microscope.

Can you see the structures shown in picture 6? (You may be looking at a different kind of leaf from the one shown in the picture, so watch out for differences.)

D Looking at chloroplasts in a moss leaf

1 With a pair of tweezers, carefully remove one leaf from a moss plant.

2 Put the leaf in a drop of water on a slide, and cover it with a coverslip.

3 Examine the leaf under a microscope.

Can you see the chloroplasts?

What colour are they?

What gives them their colour?

4 Lift up the coverslip and put a drop of dilute iodine solution on the leaf.

5 Put the coverslip back.

6 Examine the leaf under the microscope again.

A black colour indicates starch. Is there any starch in the cells? If there is, where is it?

What conclusions would you draw from this experiment about the function of the chloroplasts?

E Measuring the rate at which leaves lose water

Cobalt chloride paper is used to detect water. It is blue when dry, but turns pink when moist.

Plan an experiment, using cobalt chloride paper, to compare the rates of water-loss from:

1 the upper and lower sides of the leaves of a particular type of plant,

2 the lower sides of the leaves of different types of plants.

Get your plan approved by your teacher before you start. Draw as many conclusions as you can from your results.

How stomata open and close

Stomata allow carbon dioxide and oxygen to diffuse in and out of leaves. They are also the main route by which water vapour escapes from the plant. In hot, dry weather there is a risk that the plant may run short of water. For this reason it is important that the stomata should be able to open or close according to the weather conditions.

How do the stomata open and close? Look at the picture and notice the sausage-shaped **guard cells** lining the stoma. The top diagram shows the stoma when it is almost closed. This is what happens when the stoma opens: the guard cells take up water by osmosis from the neighbouring epidermal cells; as a result the guard cells swell up and bend, so the gap between them widens. This is shown in the bottom diagram.

Why do the guard cells bend when they swell up? They do so because the inner wall (the wall lining the gap) is thicker and less stretchable than the outer wall. The same thing happens if you blow up a sausage-shaped balloon which has got a piece of sticking plaster stuck along the side.

The stoma closes by the reverse process. Water passes out of the guard cells. As a result the guard cells straighten and the gap between them gets narrower.

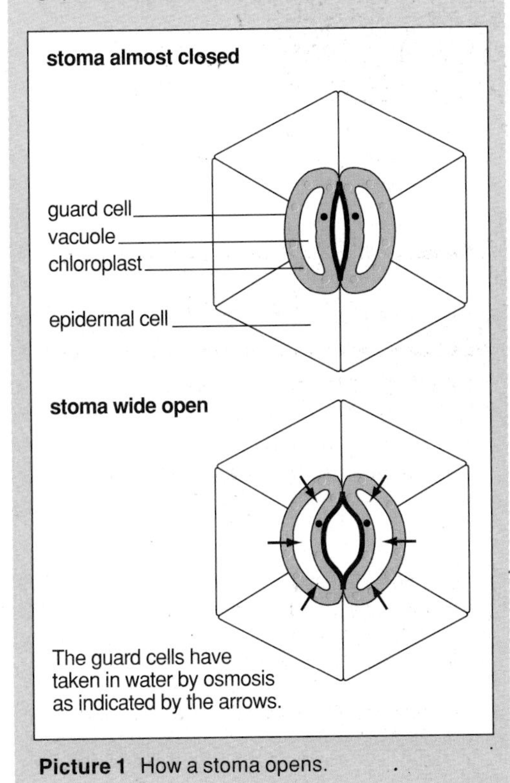

Picture 1 How a stoma opens.

C18
To and from the leaf

Plants have a transport system for taking things to and from the leaf.

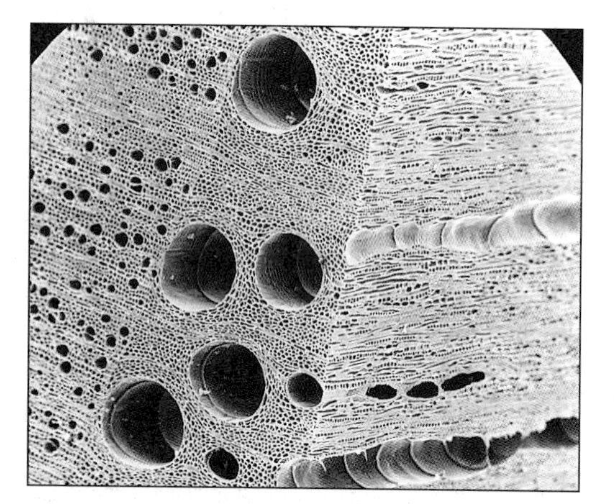

Picture 1 Looking into the xylem tubes. This picture was obtained using a scanning electron microscope. Notice the rings of thickening in the walls of the tubes. They prevent the walls caving in.

How does water get to the leaves?

Water exaporates from the leaves and other parts of the plant above the ground. The evaporation of water from the plant is called **transpiration**. It occurs mainly through the stomata. However, as quickly as water is lost from the leaves, more water enters the roots and flows up the stem.

The water flows up the plant in narrow pipes, rather like capillary tubes (picture 1). These tubes make up the **xylem**. The walls of the tubes are made of **lignin**, which is hard and waterproof. In trees and shrubs the xylem tubes are one of the main components of **wood**. They do not contain living material and are therefore dead.

Water rises up the stem mainly by being 'pulled' from above. The 'pull' is created by the evaporation of water from the leaves. If you stop this by, for example, cutting off the leaves or blocking the stomata, the flow of water up the stem slows down or stops.

The drier the atmosphere, the greater is the rate of transpiration. In very dry weather, water may evaporate from the leaves faster than it is replaced from the soil. The plant then suffers from water shortage.

Normally a plant's cells are full of water, and all the cells are pressing against each other within the epidermal covering. This helps to support the plant and holds the leaves out flat.

If a plant runs short of water, the cells lose water and go flabby. The plant then droops. This is called **wilting**. Plants living in dry places have all sorts of adaptations for preventing this (see page 60).

Roots, the link with the soil

The roots anchor a plant to the soil. They also provide a link with the soil water. From the soil water the plant gets not only water itself, but also mineral salts. The ends of the roots, except for the tip, are covered with delicate **root hairs** (picture 2). They increase the surface area for absorption. Water is absorbed by diffusion and osmosis (see page 14).

Certain substances can be absorbed by the roots even when they are more dilute in the soil than inside the root. This happens with nitrate and magnesium ions, for example. These ions are absorbed by the roots against a concentration gradient, and this requires energy from respiration. It is called **active transport** and is explained on page 14.

Because roots need to respire, the soil must contain plenty of air. This means that it must be well drained. If it is waterlogged, the uptake of salts by the roots is slowed down and growth may be poor.

Transporting food substances

Picture 2 The root hairs of wheat seedlings.

Sugar and other soluble food substances are made in the leaves. These substances are then transported to other parts of the plant where they are needed or can be stored. They move along special tubes in the **phloem**. Unlike xylem, phloem is a living tissue. The movement of food substances in the phloem is called **translocation**, and it requires energy from respiration.

The importance of the phloem in transporting food can be seen in trees. In a tree trunk, the phloem is in the soft inner part of the bark. If a ring of bark is cut from a tree trunk, food substances cannot get down to the roots. As a result, the roots starve and the tree dies. Grey squirrels damage trees by gnawing their bark and damaging the phloem, and in Africa elephants sometimes kill trees by stripping off the bark with their tusks.

Insects such as greenflies get their food from the phloem of plants. The insect sticks the tip of its needle-like proboscis into one of the tubes and feeds on the sugars which are flowing along it. The insect does not even have to suck. The pressure in the phloem tubes is sufficient to drive the sugar solution up the proboscis and into the insect's gut.

Activities

A Following the passage of water through a plant

1 Your teacher will give you a plant with its leaves and roots intact.

2 Wash the soil off the roots.

3 Stand the plant in a jar of water containing a coloured dye such as aqueous eosin or red ink.

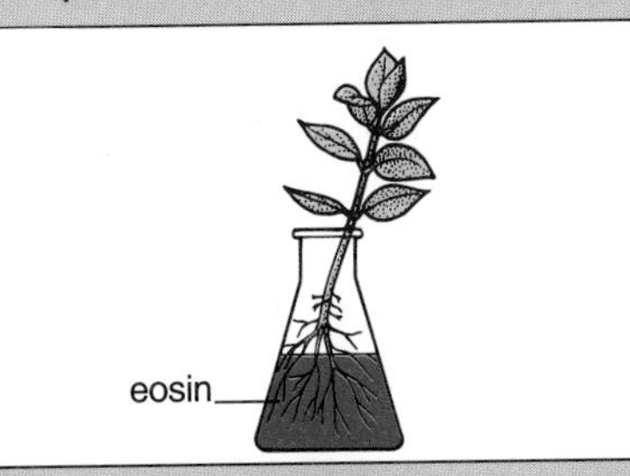

4 After about 4 hours, cut the stem in two with a sharp knife. Be careful not to cut yourself.

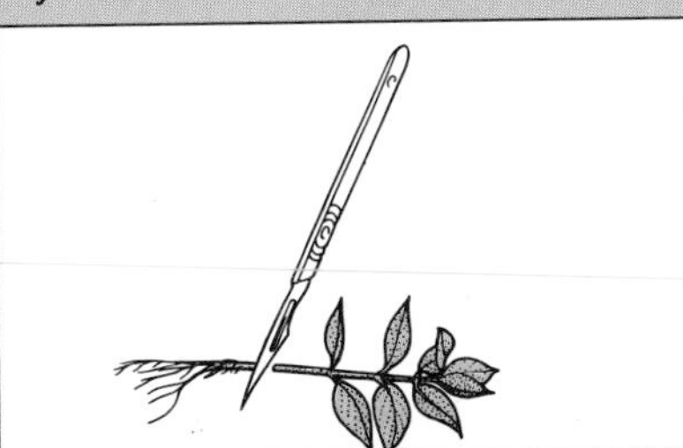

Whereabouts is the dye?

Compare the appearance of the stem with that of a plant which has been standing in water.

Explain your observations.

5 Cut the stem longways so as to find out more about where the dye is within the stem.

6 Stand a balsam plant ('Busy Lizzie') in a jar of dye, and leave it for 24 hours. This plant has a transparent stem, and you will be able to see where the dye is inside it.

7 Your teacher will give you a plant with white flowers. Stand it in a jar of dye for several days.

Does the dye eventually reach the flowers?

B What affects the rate of water uptake by a plant?

Use a potometer to find out which conditions, if any, alter the rate of water uptake by a leafy shoot.

The potometer is explained in the panel on the right.

Here are some of the conditions which you might investigate:

1 Cutting off some or all of the leaves.

2 Blocking the stomata by smearing vaseline over the leaves.

3 Placing the plant in a current of air from a hair drier.

4 Placing the plant in a humid atmosphere by covering it with a polythene bag.

First decide which conditions you are going to investigate.

Plan your experiment carefully, and have your plan approved by your teacher before you start.

Setting up a potometer

A potometer is used to measure the rate at which a leafy shoot takes up water. Set it up like this:

1 With the cut end of the shoot under water, attach it to the capillary tube by a short length of rubber tubing.

2 Clamp the capillary tube to a stand, with the bottom end in a beaker of water as shown in the picture. Make sure no air gets into the system.

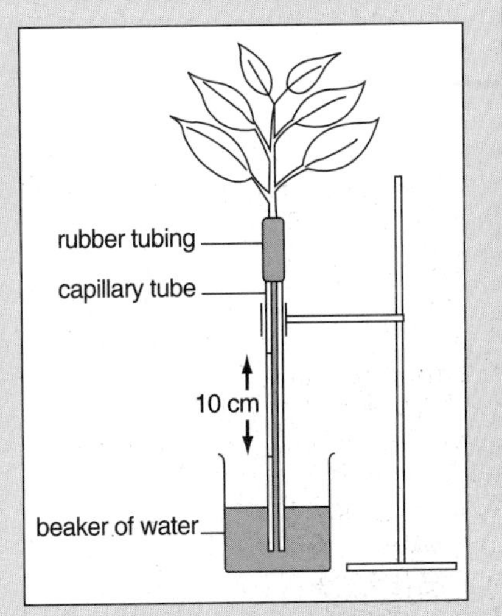

3 Make two marks on the capillary tube 10 cm apart.

4 Lift the capillary tube out of the beaker, touch the end of it with blotting paper, and then put it back. An air bubble will have been introduced into the capillary tube.

5 Time how long it takes for the air bubble to travel from the first to the second mark on the capillary tube.

6 When the air bubble has passed the second mark, push it back out of the capillary tube into the beaker of water by squeezing the rubber tubing. Then suck in water.

Questions

1 Suggest a reason for the following:

a It is better to water plants in the evening than in the middle of the day.
b Before transplanting a plant it is a good idea to remove some of its leaves.
c Water moves up a stem more quickly on a dry day than on a humid day.
d In very humid conditions water may drip from leaves.
e When a greenfly feeds on a plant it sticks its proboscis into the phloem.

2 If a tree is felled, a watery liquid may ooze out of the stump for a while. Comment on this observation.

3 The following features are found in different species of plants that live in hot, dry places. In each case explain how the feature helps the plant to cope with a shortage of water in its environment.

a Thick cuticle covering the leaves.
b Small leaves.
c Stomata sunk down into pits in the epidermis.
d Very deep roots.
e Roots just beneath the surface of the soil.

4 A scientist investigated the uptake of mineral salts by the roots of young barley plants. This is what she found:

a Salts were taken up even when they were more dilute in the soil water than inside the root.
b The rate of uptake was increased by raising the temperature, so long as it did not exceed 40°C.
c Uptake stopped if the roots were treated with a poison that stopped respiration.
d Uptake was much slower if the soil was waterlogged.

What conclusions can be drawn from these findings? How might they help farmers?

The vascular tissue of a plant

Together, the xylem and phloem make up a plant's **vascular tissue**. Picture 1 shows how this tissue is arranged in a typical non-woody flowering plant such as a buttercup or sunflower. The xylem carries water and mineral salts upwards. The phloem carries soluble food substances downwards.

In shrubs and trees the vascular tissue is arranged as in picture 2. The xylem forms the wood in the trunk and branches. The part of the wood in the centre (**heartwood**) is extemely dense and hard; its job is to support the plant. The part further out (**sapwood**) is less dense and therefore softer; it too supports the plant, but it also carries water and mineral salts upwards.

The phloem is the soft inner part of the bark. It carries soluble food substances from leaves to roots.

The hard outer part of the bark is made of **cork** which is a dead tissue. Cork is impervious to gases, but here and there the corky cells are more loosely packed, forming pimple-like **lenticels**. These serve as 'breathing pores'. Through the lenticels oxygen diffuses to the living tissues beneath, and carbon dioxide diffuses out.

1 Which sort of wood is more suitable for timber, heartwood or sapwood? Give a reason for your choice.

2 How could a five-year-old child kill a tree with a penknife? Explain your answer.

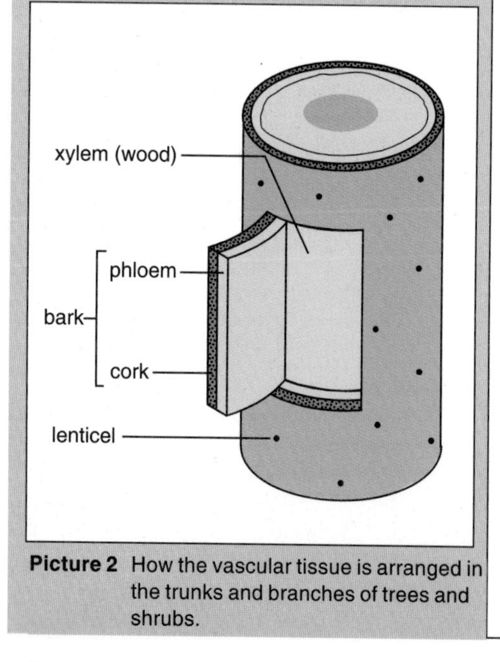

Picture 2 How the vascular tissue is arranged in the trunks and branches of trees and shrubs.

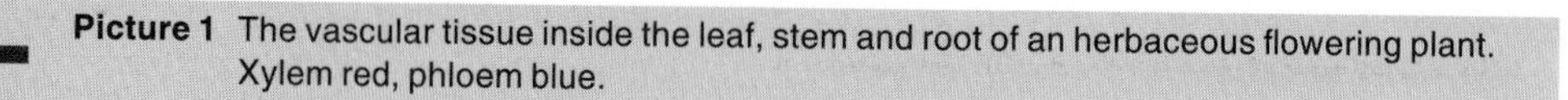

Picture 1 The vascular tissue inside the leaf, stem and root of an herbaceous flowering plant. Xylem red, phloem blue.

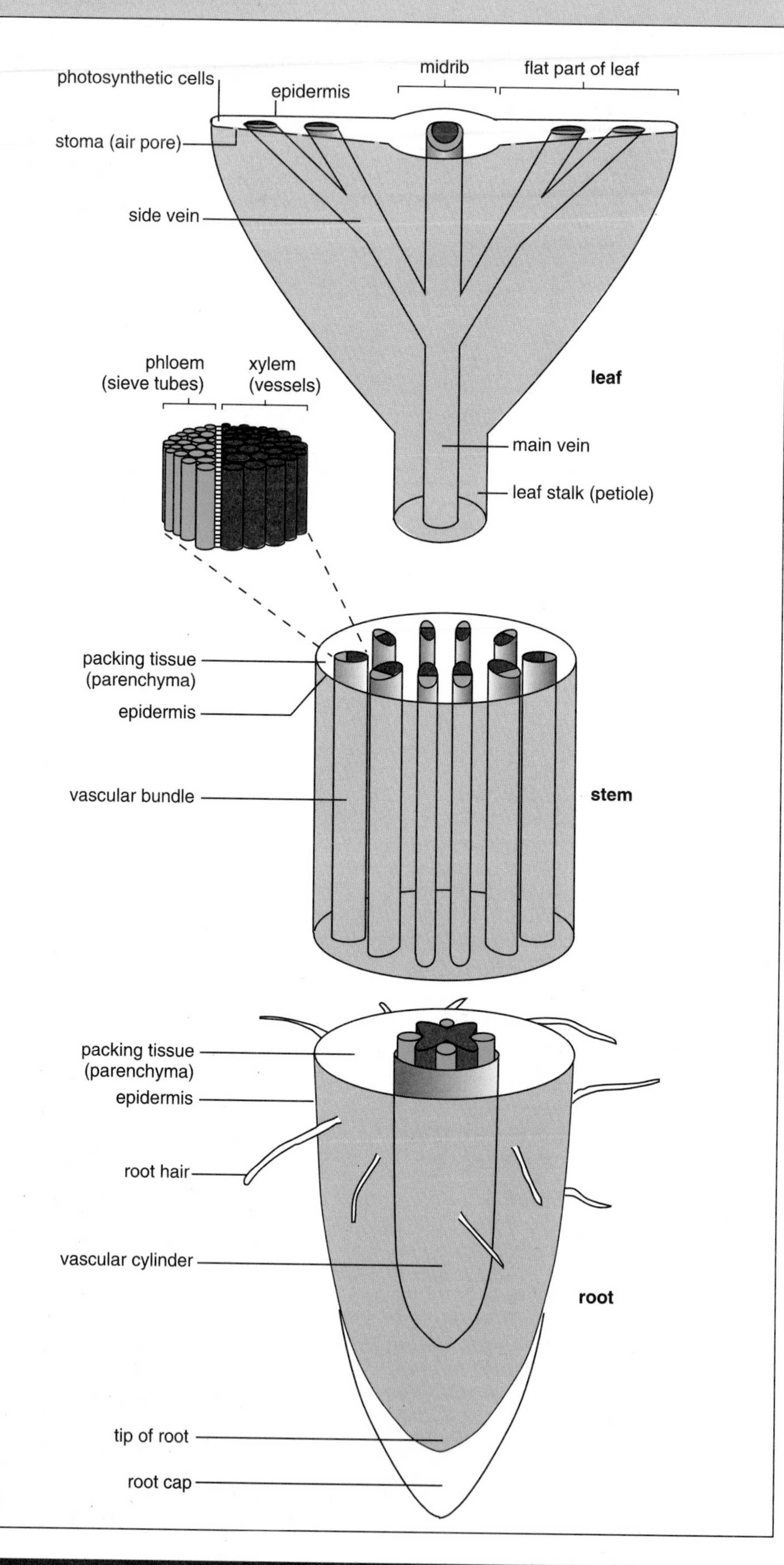

The flow of water through a plant

Look at the root hair in picture 1. It is a long extension of a single cell which sticks out from the outer layer of the root. It lies between the soil particles, which are normally surrounded by water.

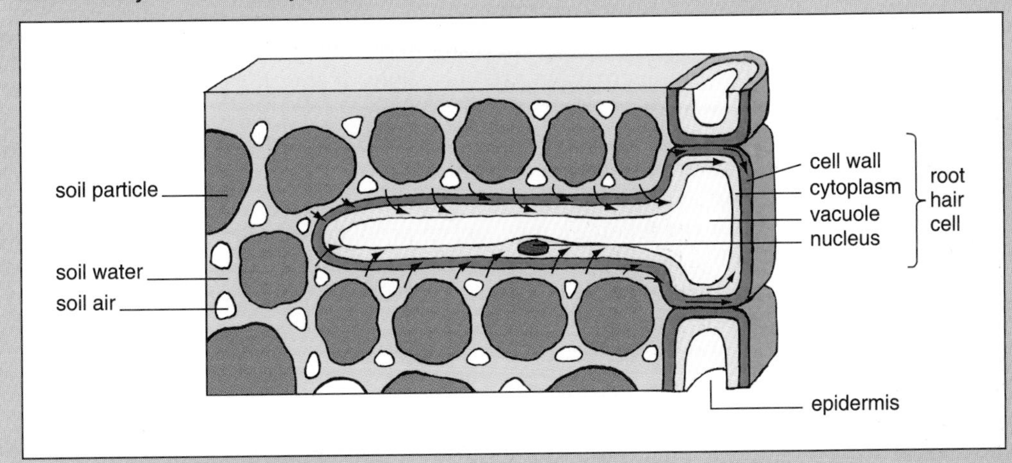

Picture 1 A root hair. The arrows indicate the absorption of water into the root hair cell.

The cell surface membrane of the root hair is partially permeable (see page 15). The concentration of water molecules inside the root hair is normally lower than in the soil water. The result is that water flows into the root hair by osmosis.

The water then diffuses from the root hair to the next layer of cells in the root, and then to the next layer – and so on. Eventually the water reaches the xylem tubes in the centre of the root.

Not all the water enters the root hair by osmosis. Much of the water simply diffuses into the cellulose wall of the root hair. The cell wall is like a sponge, and is normally saturated with water. The water then flows along the walls of the root cells until it reaches the centre of the root.

Once the water enters the xylem tubes, it flows from the roots to the stem and so to the leaves. The columns of water in the xylem tubes are prevented from breaking by strong cohesive forces that exist between the water molecules.

Water passes out of the xylem tubes in the leaves. It then flows from one leaf cell to the next mainly via the cell walls. The water then evaporates from the cell walls into the air spaces within the leaf (picture 2).

Finally, the water vapour diffuses through the open stomata to the outside of the leaf. This is **transpiration**.

The greater the rate of transpiration, the faster water flows from roots to leaves – provided there is enough water in the soil. The rate of transpiration depends on a number of things. In general, it is speeded up by high temperature, a dry atmosphere (low humidity) and air currents.

The flow of water through the plant is called the **transpiration stream**. Mineral salts absorbed by the root hairs are swept along in the transpiration stream. This is how essential elements needed by the plant reach the leaves.

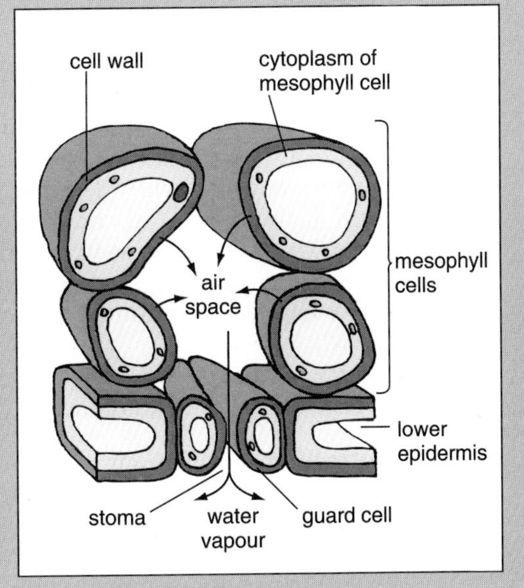

Picture 2 The internal structure of the lower part of a leaf, just inside a stoma. The arrows indicate the evaporation of water from the cells and the diffusion of water vapour to the outside.

1 Suggest two ways in which transpiration is useful to a plant.

2 In what circumstances would you expect transpiration to be very slow?

Think of as many circumstances as you can.

3 A giant sequoia tree may be over 80 metres tall. Are you surprised that water in a tree can reach such a great height? Scientists are! What is surprising about it?

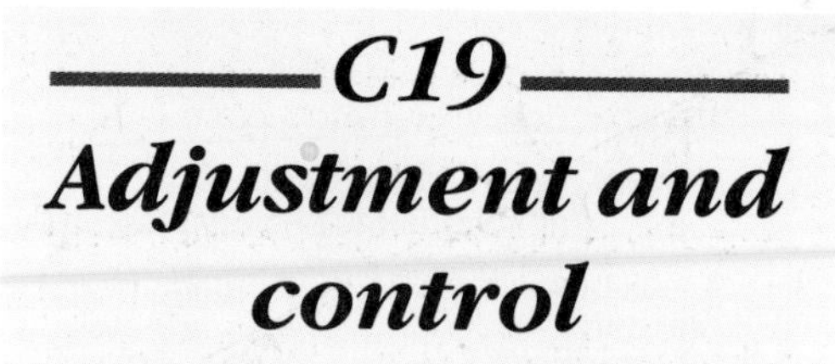

Conditions inside the body must always be steady and should not vary much.

A self-adjusting system

Think of an oven. If an oven is to cook things properly, it must stay at about the same temperature all the time. Imagine what a cake would be like if the oven suddenly got too hot – or too cold!

The oven is kept at a steady temperature by a **thermostat**. You set the thermostat at the desired temperature, and it keeps the oven very close to that temperature all the time.

Picture 2 shows how the thermostat works. If the oven gets too hot, the thermostat senses this and switches the heater off. As a result the temperature of the oven falls. As the oven gets cooler, the thermostat senses this and switches the heater on again. The result is that the temperature of the oven stays more or less constant. Not *absolutely* constant: it goes up and down all the time, but the fluctuations are only slight. If you think about it, there must be slight fluctuations – the mechanism depends on them.

The thermostat is an example of a **feedback system**. Information about the temperature of the oven is *fed back* to the thermostat, telling it to switch the heater on or off.

Notice that the feedback mechanism produces an effect which is the opposite of what has been happening before: if the oven is too hot, it is told to cool down; if it is too cool, it is told to heat up. This type of feedback is called **negative feedback**.

Picture 1 Once it has been set, the temperature of this oven will stay more or less the same all the time.

Keeping things constant

Feedback helps to keep conditions inside the body constant. This is important because if conditions were to vary greatly, our cells would not work properly and we would die. Keeping conditions constant is called **homeostasis**. This is a Greek word which literally means 'staying the same'.

Here are a few of the things that must be kept constant:

- the concentration of salts in the blood,
- the concentration of sugar in the blood,
- the acidity (pH) of the blood,
- the body temperature.

Let's look at two of these and see what happens.

Controlling the body temperature

A mammal's body is like a thermostatically controlled oven. Suppose the body temperature rises slightly. This is sensed by receptor cells in a special control centre in the brain. The control centre then sends out messages along the nerves. These messages switch on the various mechanisms that cool the body – sweating and so on (see page 186). When the body temperature falls, the reverse happens: the control centre switches on the mechanisms that warm the body.

So the control centre in the brain works like a thermostat, keeping the body temperature more or less constant (picture 3).

Controlling blood sugar

There is a certain concentration of sugar in the blood which is just right for the body. Now suppose you have a meal with lots of carbohydrate: cake, jam and so on. Shortly after the meal your blood sugar concentration will rise. However, your tissues, particularly the liver and muscles, get rid of the extra sugar. One way they do this is to turn the sugar into **glycogen** which is stored in the cells out of harm's way (page 22).

How do the tissues know that they should get rid of the sugar? The answer is that they are told to do so by another organ – the **pancreas**. Special cells in the pancreas sense that there is too much sugar in the blood. These cells then release the hormone **insulin** into the bloodstream. The insulin is carried in the bloodstream to the tissues which respond by turning the surplus sugar into glycogen.

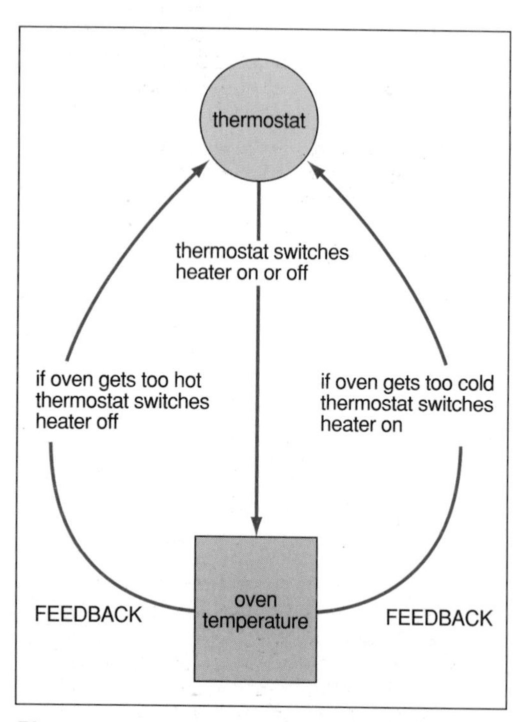

Picture 2 This diagram shows how the temperature of an oven is controlled by its thermostat.

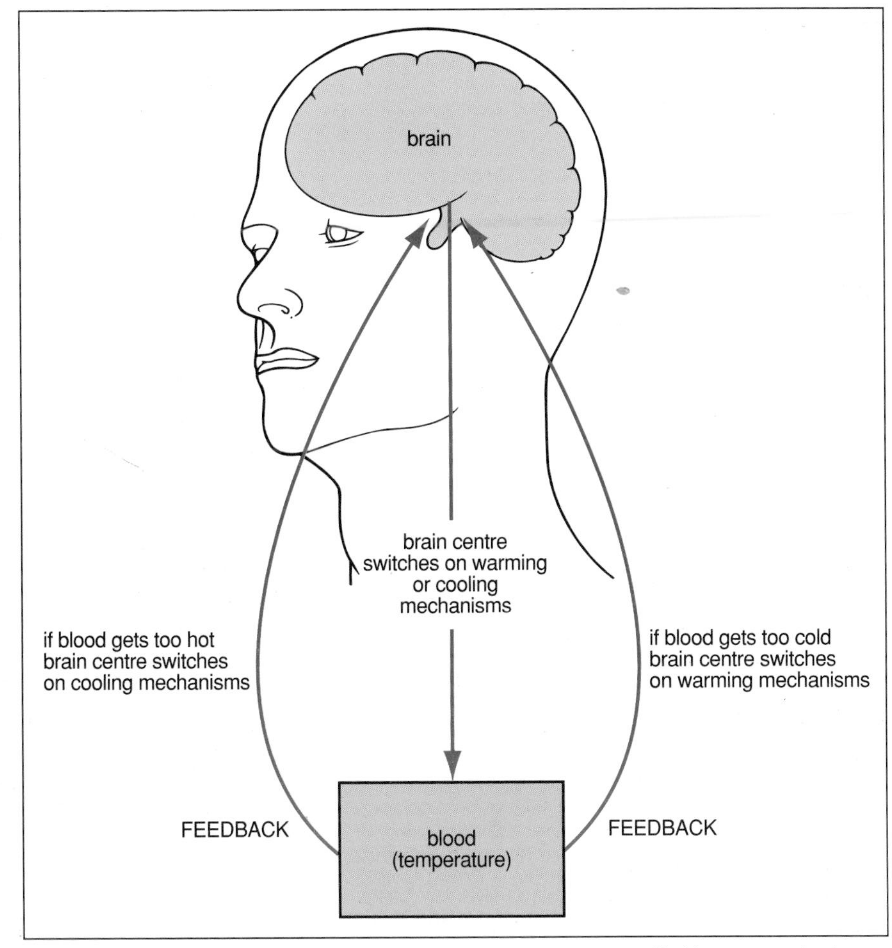

Picture 3 The body's cooling and warming mechanisms are controlled by a centre in the brain.

When the concentration of blood sugar gets too low, the reverse happens: the pancreas stops producing insulin, and the tissues turn some of their glycogen into sugar. The sugar is released into the bloodstream, so the blood sugar concentration rises again. This response is enhanced by another hormone called **glucagon**. This too is secreted by the pancreas but its effects oppose insulin's.

Insulin is very important in our lives. People who do not produce enough of it suffer from diabetes (see page 193).

Homeostasis – a world process

Homeostasis enables things to correct themselves when changes take place. In other words, it enables processes to be *self-adjusting*. We find this not only in biology but in other walks of life such as business and industry. For example, modern commercial greenhouses are equipped with feedback mechanisms which control the temperature and concentration of carbon dioxide (page 175).

Many machines work in the same kind of way. This is all part of **control technology**, which is becoming more and more important in business and industry. The same principles apply to economics. Manufacturers who produce more cars than people want will respond by producing fewer; and an hotel which charges so much that people won't come and stay will lower its prices.

Some people have suggested that the whole world is a huge self-adjusting system. They claim that environmental changes like the greenhouse effect will eventually be corrected by natural processes. However, if it is true, the time scale will be so long that most of us won't be around to see if they are right.

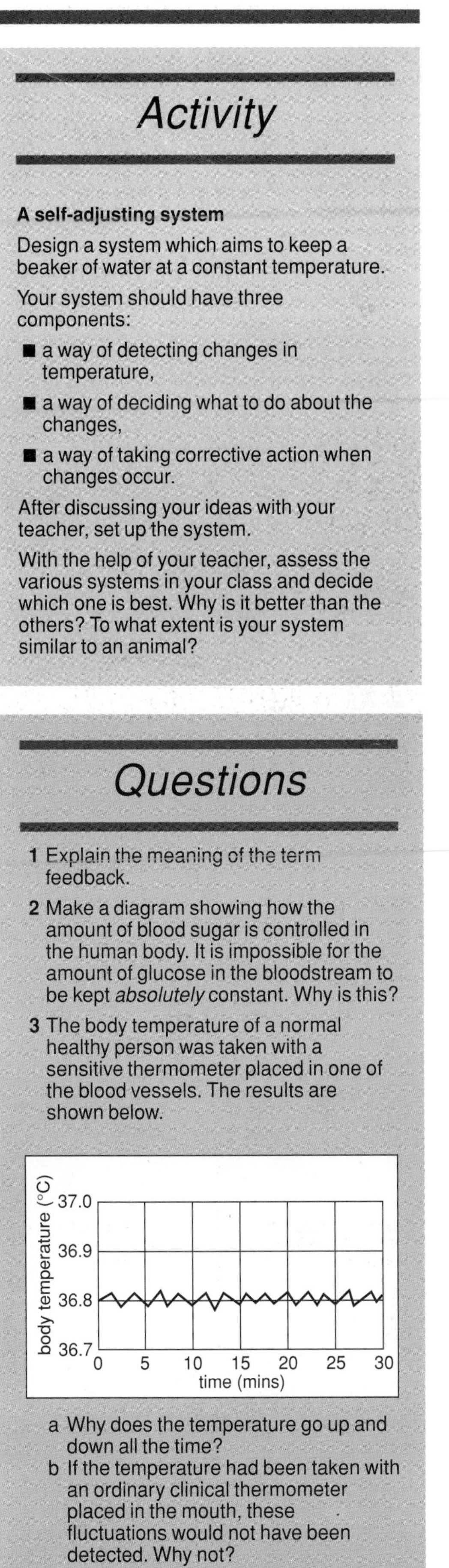

Activity

A self-adjusting system

Design a system which aims to keep a beaker of water at a constant temperature.

Your system should have three components:

- a way of detecting changes in temperature,
- a way of deciding what to do about the changes,
- a way of taking corrective action when changes occur.

After discussing your ideas with your teacher, set up the system.

With the help of your teacher, assess the various systems in your class and decide which one is best. Why is it better than the others? To what extent is your system similar to an animal?

Questions

1 Explain the meaning of the term feedback.

2 Make a diagram showing how the amount of blood sugar is controlled in the human body. It is impossible for the amount of glucose in the bloodstream to be kept *absolutely* constant. Why is this?

3 The body temperature of a normal healthy person was taken with a sensitive thermometer placed in one of the blood vessels. The results are shown below.

a Why does the temperature go up and down all the time?

b If the temperature had been taken with an ordinary clinical thermometer placed in the mouth, these fluctuations would not have been detected. Why not?

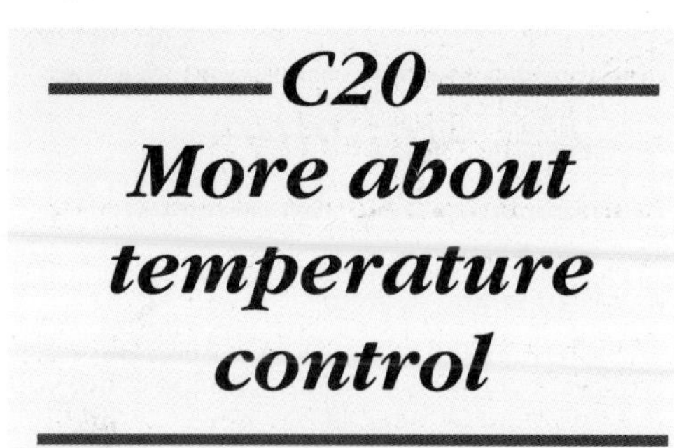

C20 More about temperature control

Here we look at how the body keeps its temperature constant.

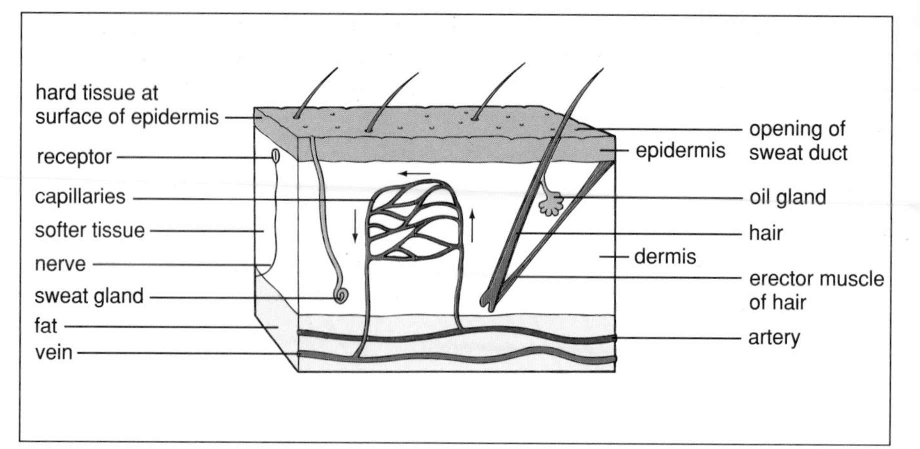

Picture 1 The main structures in the skin of a mammal.

Picture 2 The ruffled up hair of this cat helps to keep it warm.

The skin

Picture 1 shows the skin of a mammal greatly magnified. All the structures in the picture help the mammal to keep its body temperature constant. One of the most important parts of the skin is the **fat**. Fat is a good insulator and helps to prevent energy being lost from the body.

But the skin does more than just keep the body warm. It *adjusts* the energy loss from the body, thereby preventing the body getting too hot or too cold. This is how it happens:

Hairs are raised or lowered

Look at the cat in picture 2. Its **hairs** are ruffled up. When the hairs stand up like this, air is trapped between them. This layer of air insulates the body and prevents energy being lost. In hot weather the hairs lie down flat. Air is no longer held between them, so energy is lost and the body is kept cool.

The hairs are made to stand up by contraction of the **erector muscles**. Of course this response isn't much use to humans because we don't have much hair. But the hair muscles still contract in cold weather, giving rise to 'goose pimples'.

Blood flow through the skin is altered

In cold weather the blood vessels which take blood to the surface of the skin get narrower, so less blood flows through them. The result is that blood is kept away from the surface of the skin, and less energy is lost from it.

In hot weather the reverse happens. The vessels taking blood to the surface of the skin get wider, so more blood flows through them. Energy is lost from the blood as it flows near the surface.

Sweating and panting

In hot weather our **sweat glands** work overtime and produce lots of **sweat**. When the sweat evaporates, it cools the skin and the blood flowing in it.

Some animals keep cool by **panting**. When a dog pants, water evaporates from its mouth and tongue, and this cools the blood (picture 3).

Picture 3 A Bernese Mountain dog panting to keep cool after a sudden bout of exercise. Swiss basket weavers used to harness these dogs to carts and get them to haul their baskets to market.

Warm- and cold-blooded animals

Animals which can keep their body temperature constant, even when the surrounding temperature varies, are described as **warm-blooded**. They include all mammals and birds. All other animals are described as **cold-blooded.** Their body temperature is the same as their surroundings. If the surrounding temperature rises, so does the body temperature.

The main way a cold-blooded animal can control its body temperature is by making sure that it is always in a place where the surrounding temperature is suitable. A lizard, for example, will bask in the sun to keep warm, or go under a rock to keep cool. Of course warm-blooded animals do this sort of thing too, but for most cold-blooded animals it's the *only* way of controlling the body temperature.

Energy warms the body

Energy is constantly released by our metabolism, and this warms the body. In cold weather, we metabolise faster and produce extra energy. The liver is one of the main organs where this happens. Also we **shiver**. When we shiver our muscles contract. This too releases energy which helps to warm us.

Behaviour

All the responses mentioned so far are automatic – they happen without our thinking about them. In addition we can control our body temperature by our **behaviour**. For example, in cold weather we put on more clothes or stamp our feet. In hot weather we sit under a fan, or go for a swim.

In the skin there are **temperature receptors** which are sensitive to changes in temperature. These receptors send messages to the brain, enabling us to *feel* whether our surroundings are hot or cold. We can then take the appropriate action.

There's a limit

Despite these mechanisms, things sometimes go wrong. For example, if the body temperature falls much below 35°C, the control centre in the brain stops working and we can no longer control our body temperature. The metabolic rate falls and the body temperature gets lower and lower. This is called **hypothermia**. Eventually we go into a coma, and if no action is taken we will die.

Hypothermia is often brought on by damp clothes and a cold wind. Hikers and pot-holers are particularly at risk. So are babies and old people (picture 4). It is worse if you have not eaten recently. The trouble is that once hypothermia sets in, the brain stops working properly and you don't realise there is anything wrong – so you do nothing about it. This can even happen to trained soldiers.

Picture 4 Potholers and babies are both liable to hypothermia if precautions are not taken.

Activities

A How does insulation affect loss of energy?

Your teacher will give you two cans with lids, cotton wool, elastic bands, hot water, a thermometer and graph paper. Use them to test the hypothesis that an insulated object loses energy more slowly than an uninsulated object.

Plan your experiment first, and discuss it with your teacher before you start.

How is this investigation relevant to biology? What other investigations might you carry out on this topic?

B When do you sweat?

Cobalt chloride paper is blue when dry but turns pink when moist. Use cobalt chloride paper to find out in which conditions you sweat the most.

Plan your experiment first, and have it approved by your teacher before you start.

Questions

1 How do clothes help to keep us warm?

2 Explain the following:
  - a You feel cooler on a hot dry day than on a hot humid day.
  - b White people go pale in cold weather and pink in hot weather.

3 What advice would you give to an elderly person on how to avoid hypothermia? What can neighbours do to help elderly people in this respect?

4 The graph below shows the air temperature, and the body temperatures, of Jake and his pet lizard during 24 hours in the desert.

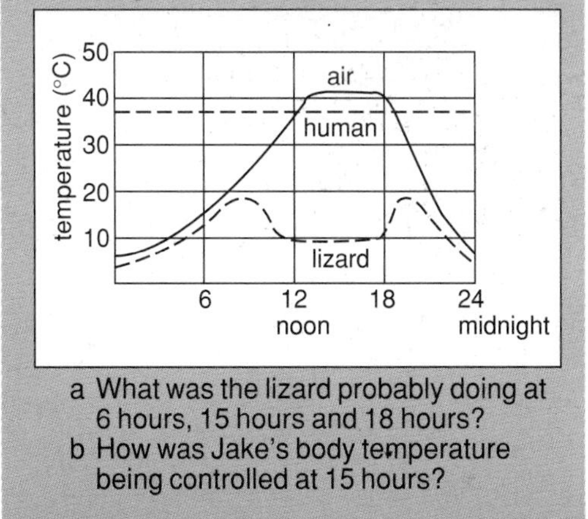

  - a What was the lizard probably doing at 6 hours, 15 hours and 18 hours?
  - b How was Jake's body temperature being controlled at 15 hours?

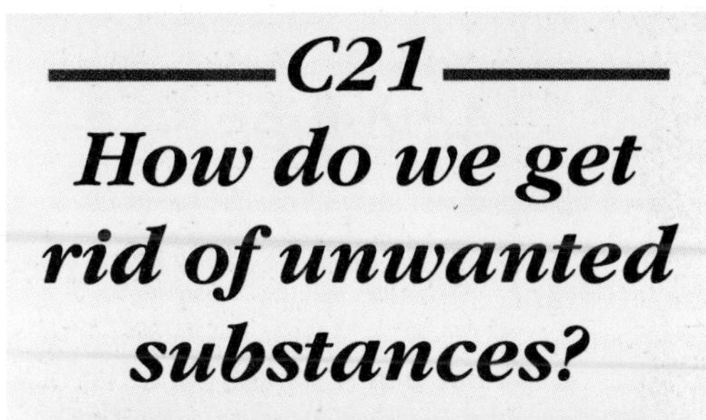

C21 How do we get rid of unwanted substances?

When you go to the toilet you are getting rid of unwanted substances.

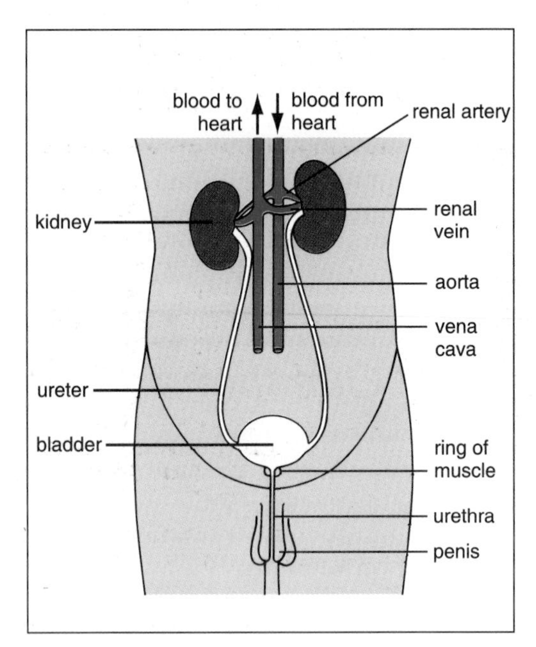

Picture 1 The urinary system of the human male. The arrows indicate the direction of the flow of blood. The ring of muscle where the bladder joins the urethra is shown here in sectional view. It encircles the urethra like a sheath. When contracted tightly it prevents the urine escaping from the bladder.

What do we get rid of?

Two obvious things that we have to get rid of are carbon dioxide and indigestible waste. We have already seen how they are dealt with. Here we shall see how two particular substances are got rid of from our blood. The two substances are:

- **Urea.** This is formed in the liver from protein which the body does not need. It contains nitrogen and is poisonous. Getting rid of it helps to keep the blood 'clean' and is called **excretion**.
- **Water**. We generally take in more water than we need, and we have to get rid of the surplus. This helps to keep the blood at the right concentration. It is called **osmoregulation**.

The part of the body responsible for getting rid of urea and water is called the **urinary system**.

Ins and outs of the urinary system

The main organs of our urinary system are the **kidneys**. We have two of them, one on each side the body. If you put your hands on your hips, your kidneys are just under your thumbs.

The urinary system is shown in picture 1. A narrow tube called the **ureter** runs from each kidney to the **bladder**. The bladder is a bag situated towards the bottom of the abdomen. Its wall is elastic and has muscle tissue in it. Leading from the bladder is a tube called the **urethra**. The urethra opens just above the vaginal opening in the female. In the male it runs down the penis and opens at its tip.

The kidneys produce a watery fluid called **urine** which contains substances that the body does not want. The urine trickles continuously down the ureters to the bladder which gradually expands like a balloon as more and more urine collects inside it. Every now and again the bladder is emptied. This is called **urination**.

How is the bladder emptied?

We all know the uncomfortable, almost painful, feeling when we want to urinate. This is the body's way of telling us that the bladder is full and needs emptying.

If you look at picture 1 you will see that the top of the urethra is surrounded by a ring of muscle. Most of the time this ring of muscle is tightly closed, so urine is kept in the bladder. When the bladder is emptied the ring of muscle opens, and at the same time the muscle tissue in the bladder wall contracts. The result is that the urine is forced out.

How is urea formed and why is it poisonous?

Much of the surplus food that we eat is stored by the liver. However, the liver can't store proteins or the amino acids of which they are composed. Instead the liver destroys them. Amino acids contain nitrogen, and the first thing that happens is that the nitrogen part of the amino acid molecule is broken off and turned into ammonia. This is called **deamination**. Ammonia is very poisonous so the liver quickly converts it into urea which is less poisonous. The urea is then carried by the blood to the kidneys which get rid of it in the urine.

Ammonia is not the only poisonous substance that the liver deals with. Many other poisons (including drugs) are either destroyed or made harmless by the liver. We call this process **detoxification**. It is a very important function of the liver.

How is urine formed by the kidneys?

Look at table 1. You will see that some substances are more concentrated in the urine than in the blood. These are the substances which the body does not want. The main one is urea. The figures in the table show that there is about sixty times more urea in the urine than in the blood. The reason is that as blood passes through the kidneys, urea is taken out of it and passed into the urine.

But the kidneys do more than cleanse the blood of urea. They also regulate the amounts of various substances in the blood. One of these is water.

Suppose you drink a lot of water in a short space of time. The water is absorbed from your gut into your bloodstream, and it has the effect of diluting the blood. As the blood flows through the kidneys, the unwanted water passes out of it into the urine. This is why you produce a lot of very watery urine after you have been drinking a lot.

By regulating the amount of water that goes out with the urine, the kidneys keep the concentration of the blood more or less the same all the time.

Table 1 This table compares the relative amounts of five different substances in the blood and urine of the human. The amount of water and salt in the urine varies according to the needs of the body.

Substances	Quantity (parts per hundred)	
	Blood	Urine
Water	92	95
Proteins	7	0
Glucose	0.1	0
Salt (chloride)	0.37	0.6
Urea	0.03	2

Inside the kidneys

If you slice a kidney open and look inside, you can see that it is divided into two areas, a light outer area and a darker inner area. The inner area is connected to the ureter as shown in picture 2.

To see any detail of the kidney itself you need a microscope. The microscope shows that inside each kidney there are about a million little structures called **nephrons**. One is shown in picture 3. It consists of a narrow tube, called a **renal tubule**. Its inner end is connected to a cup-like **renal capsule**. The other end is connected to a **collecting duct** which leads to the ureter.

Detecting substances in blood and urine

How can we measure the concentration of substances in samples of blood and urine? Nowadays, a **biosensor** can be used.

A biosensor is a sensitive device that can detect the presence of certain substances even at very low concentrations. It consists of a probe containing immobilised enzymes or microbes. When the substance touches the probe, it reacts with the enzymes, and the reaction is turned into an electronic signal. The concentration of the substance can be measured from the number of reactions that occur.

Biosensors have many other uses in biology. Can you suggest some of them?

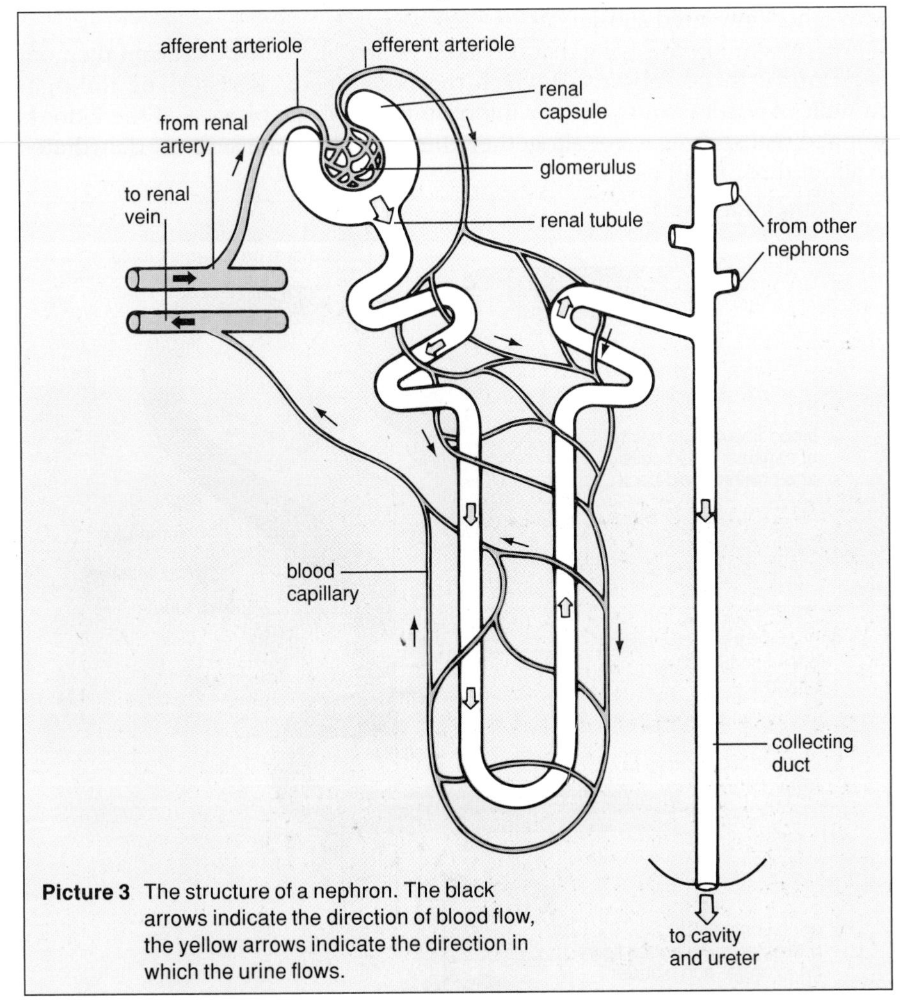

Picture 3 The structure of a nephron. The black arrows indicate the direction of blood flow, the yellow arrows indicate the direction in which the urine flows.

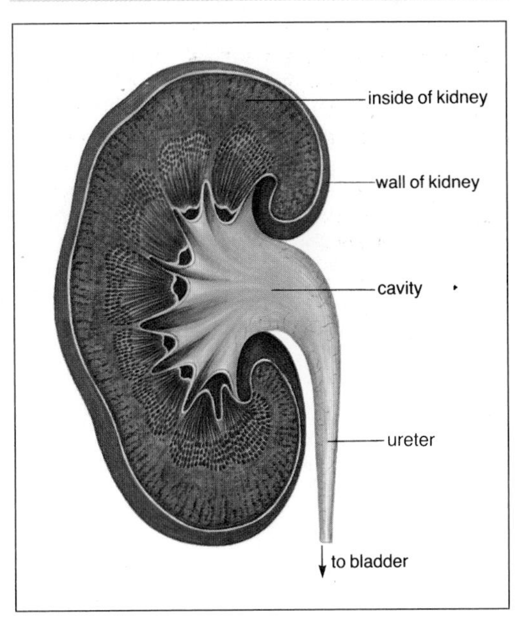

Picture 2 A kidney sliced horizontally to show the inside.

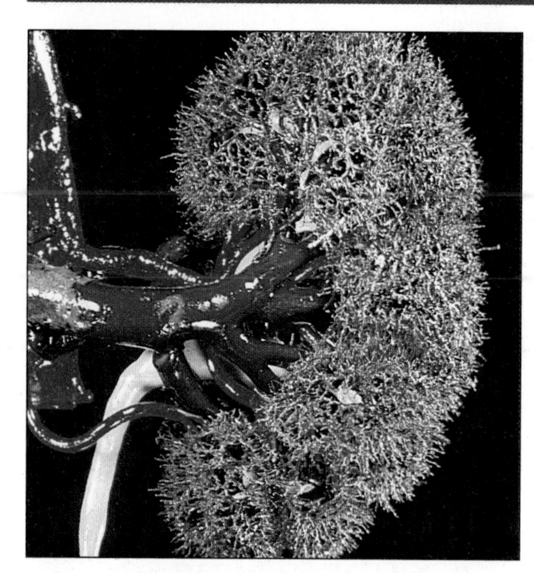

Picture 4 In this kidney all the tissues except the main blood vessels have been dissolved away. The yellow tube is the ureter which carries urine to the bladder. Arteries are red, veins blue.

The nephron's blood supply comes from a branch of the renal artery called the **afferent arteriole**. This enters the renal capsule where it splits up into a little bunch of capillaries called the **glomerulus**. The capillaries then join up again to form an **efferent arteriole** which comes out of the capsule and splits up into another set of capillaries. These capillaries are wrapped round the renal tubule – they then join up again to form a vessel which leads to the renal vein.

The two kidneys together contain about 16 km of tubules and 160 km of blood vessels (picture 4).

How do the kidneys work?

The kidneys work by each nephron 'cleaning' the blood that flows to it. Picture 5 shows what happens.

The blood which reaches the glomerulus is under high pressure. This is because the vessel which carries blood away from the glomerulus is narrower than the vessel which carries blood to it. The pressure forces the fluid part of the blood through the walls of the capillaries into the space inside the renal capsule. The fluid which goes through contains urea, glucose, water and salts. But the blood cells and proteins are too large to go through, so they stay in the capillaries. In this way the blood is filtered as it flows through the glomerulus. Because the things that are held back by the filter are very small, this process is called **ultra-filtration**.

The filtered fluid, or filtrate, then trickles along the renal tubule. As it does so, all the glucose, and most of the water and salts, are taken back into the capillaries wrapped round the renal tubule. The right amounts of water and salt are reabsorbed to give the blood its correct composition. In the meantime, urea passes on along the renal tubule and eventually reaches the bladder, together with other unwanted substances.

We can sum up by saying that the kidneys first filter the blood, and then put back into it those substances which the body needs. Reabsorbing the right amount of water is an extremely important part of the process: if the kidneys stopped reabsorbing water altogether, the body would be severely dehydrated in about three minutes.

Controlling water

How do a person's kidneys know how much water to reabsorb? Well, suppose your body is short of water. Under these circumstances the concentration of salts and other solutes in the blood will be too high.

This high concentration of solutes is detected by special cells in the brain which tell the pituitary gland to secrete a certain hormone into the blood. The hormone is carried to the kidneys and tells them to reabsorb extra water into the bloodstream instead of letting it out in the urine. You will also feel thirsty and drink some water. As a result, the solute concentration of the blood is brought down to its correct value.

The hormone is called **anti-diuretic hormone**. This is because it counteracts **diuresis**, which is the excessive flow of urine from the body.

1. When is your body likely to be short of water?
2. What happens when your body contains too *much* water? Include these words in your answer: *blood, brain, pituitary gland, anti-diuretic hormone, kidneys, urine.*

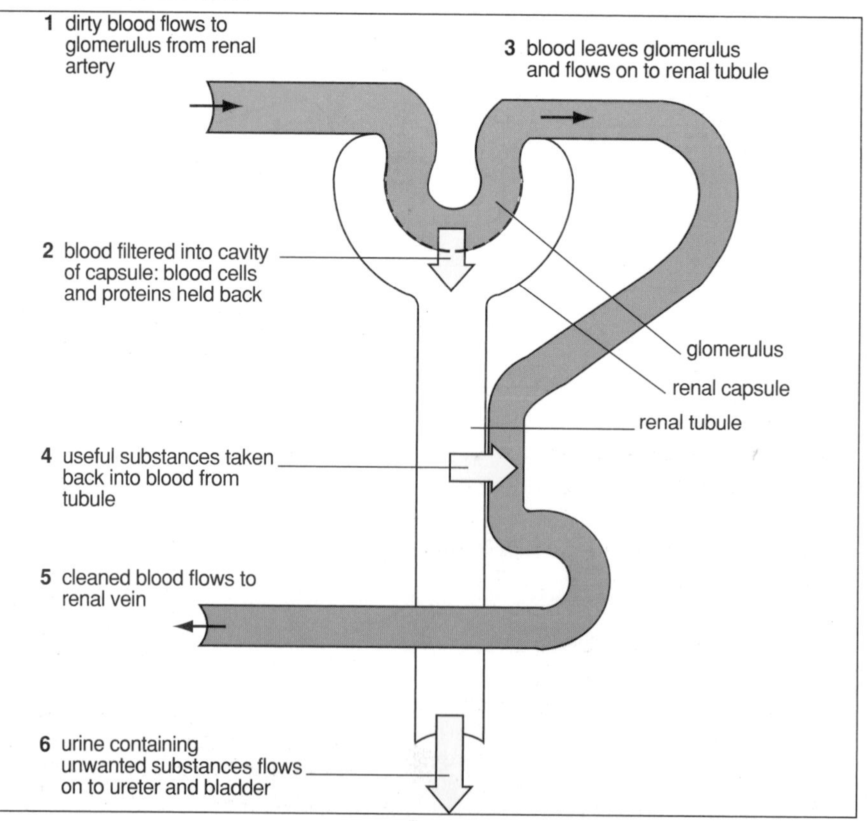

Picture 5 This diagram shows very simply how the nephron cleans the blood and makes urine.

What happens if the kidneys fail?

If one kidney fails it doesn't matter very much, because we can manage with the other one. But if both kidneys fail, the concentration of substances in the blood changes and the amount of urea goes up and up. The person becomes poisoned by the body's own waste products, and death occurs in about a week.

Once again technology comes to the rescue. A person with kidney failure can be attached to an **artificial kidney** (**kidney machine**). Blood from the patient flows through the machine where it is cleansed before returning to the patient.

Another way of dealing with kidney failure is to be given a **kidney transplant** (picture 6). This involves an operation in which the person's kidney is replaced by someone else's. Usually the new kidney comes from a close relative who is willing to donate a kidney, or from the victim of a fatal accident. There's more about the kidney machine and kidney transplants on pages 192 and 193.

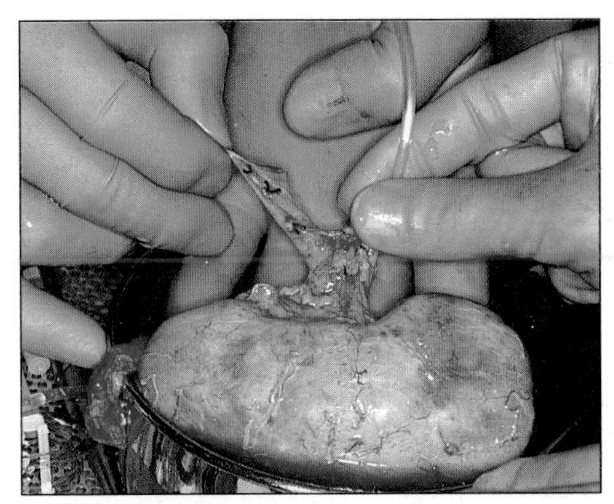

Picture 6 A kidney being prepared for a transplant operation.

Activities

A Looking at the kidney

1. Obtain a kidney of a mammal, such as a pig, from a butcher.
2. With a sharp knife slice the kidney across the middle as shown in the illustration.

⚠ DANGER *Be careful not to cut yourself.*

3. Which of the parts shown in picture 2 can you see?

   Whereabout does urine leave the kidney? Where does the urine go after it has left the kidney?

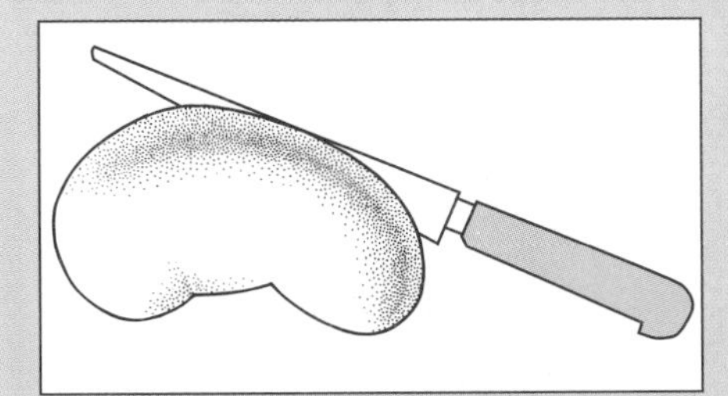

B Looking at the kidney under the microscope

The inside of the kidney consists of little else but thousands and thousands of nephrons.

What would you expect to see if you looked at a thin section of the kidney under the microscope?

Your teacher will give you a thin section of a kidney which has been specially stained to show up the structure.

Examine it carefully under the microscope and try to find these things:

- a renal capsule,
- a renal tubule,
- a blood vessel,
- a collecting duct.

What general conclusions about the kidney would you draw from your observations?

C To find the effect of drinking on the production of urine

Plan an experiment to investigate in detail the effect of drinking a large quantity of water on the rate at which urine is produced. First decide exactly what you want to find out, what measurements you should make and how you should make them.

You probably won't be able to carry out the experiment because of the possible risk of infection. However, your teacher may give you the results of an investigation which has already been carried out. Study the results and draw conclusions.

D Testing urine for glucose and protein

Your teacher will give you three samples of urine. One sample is normal, one contains glucose and one contains albumen (protein).

Test the urine samples for glucose and protein, by dipping paper test strips into them. Use **clinistix strips** to test for glucose, and **albustix strips** to test for albumen (protein).

What would the kidney be doing wrong for these substances to be present in a person's urine?

When you have finished, wash your hands with soap and water.

Questions

1. Explain the meaning of the terms excretion and osmoregulation.

   What job does the kidney do:

   a as an excretory organ, and
   b as an organ of osmoregulation?

2. Which of the substances listed in column A are found in each of the fluids listed in column B?

Column A	*Column B*
protein	blood entering kidney
glucose	blood leaving kidney
urea	fluid filtered into renal capsules
water	urine leaving kidney

3. What effect, if any, would you expect each of the following to have on the quantity and composition of the urine?

   a Eating a large quantity of salty food.
   b Having a bath.
   c Drinking a lot of beer.
   d Playing a hard game of squash.
   e Eating two bars of chocolate.

4. It has been suggested that in hot weather a person passes less urine than in cold weather.

   a Describe an experiment which could be done to find out if this is true.
   b If it is true, how would you explain it?

The artificial kidney

The artificial kidney, or kidney machine, can take over the job of the kidneys if they stop working.

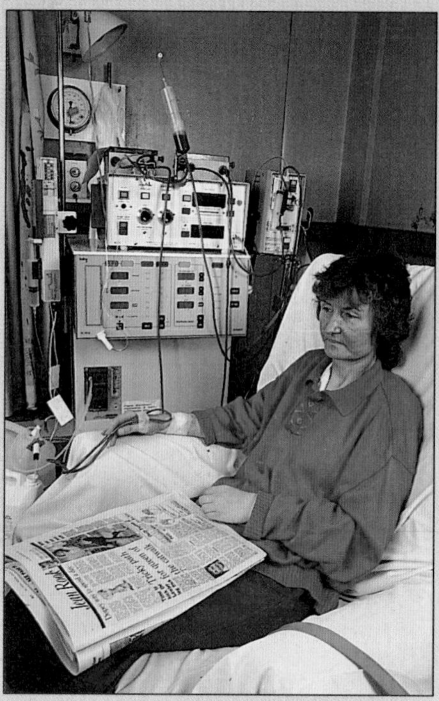

Picture 1 A person connected to an artificial kidney (kidney machine).

First a small operation is carried out in which an artery in the arm is joined directly to a vein. A tube is then connected to the vein. The blood flows along the tube to the machine.

The reason why the vein has to be joined to the artery is that the blood pressure in the vein would be too low to drive the blood to the machine. 'So why not connect the tube to the *artery*?' you ask. The answer is that the artery is narrower and more difficult to get at than the vein. The vein is just beneath the skin on the underside of the arm, and is easy to get at. Look at the underside of your own arm and you will see the veins through the skin.

Once inside the machine, the blood is pumped over the surface of a thin membrane made from a special polymer rather like cellophane. On the other side of the membrane is a watery solution containing the various substances that the body needs in the right concentrations. These substances include glucose and salts, but not urea.

The membrane will let through certain things, but not others. We call it a **dialysing membrane**, and the solution in contact with it is called the **dialysis solution**. As the blood flows over the membrane, urea and other unwanted substances pass through to the dialysis solution on the other side. Larger things in the blood, such as the cells and proteins, are held back. The dialysis solution is kept flowing all the time and is constantly replaced by fresh solution.

By the time the blood leaves the machine, the urea has gone and useful substances such as glucose have the same concentration as in the dialysis solution. The 'clean' blood is then returned to the patient by another tube inserted in the arm vein.

A person with total kidney failure needs to spend about five hours three times a week on a kidney machine, either in hospital or at home. Portable machines are now available, and people can learn how to attach themselves to them. Kidney machines are a wonderful example of how modern technology can save lives.

1. Draw a picture of what the dialysing membrane in a kidney machine might look like very highly magnified.

   Use your picture to explain how blood loses its urea as it flows through the kidney machine, and why blood cells don't pass through the membrane.
2. The dialysis solution contains glucose and salts at the same concentrations as they should be in the blood. Why is this important? Why does the solution have to be constantly replaced?
3. Kidney machines are extremely expensive and are used by relatively few people. Do you think money should be spent on expensive technology of this sort rather than on more basic things?
4. Unfortunately there are not enough kidney machines for everyone who needs them. What sort of things should be taken into account in deciding who should have them? Remember, without a kidney machine a person may die.

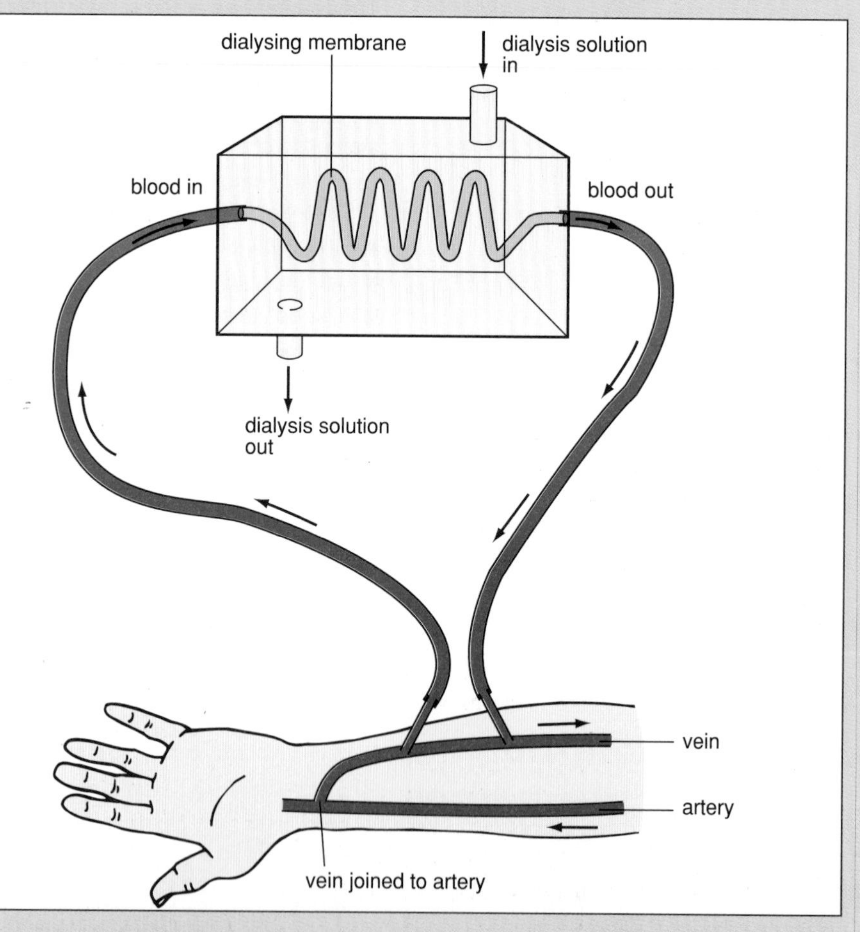

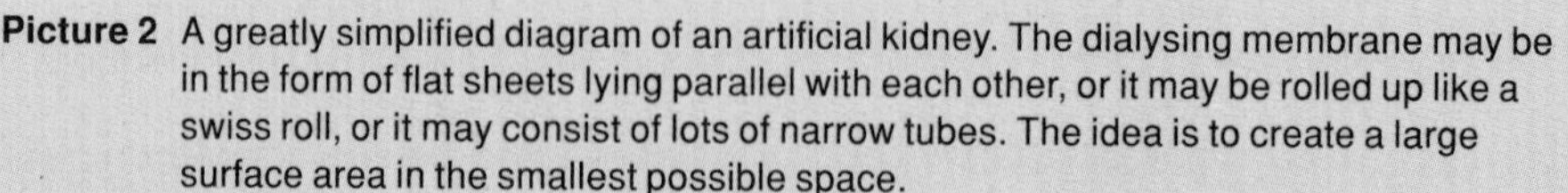

Picture 2 A greatly simplified diagram of an artificial kidney. The dialysing membrane may be in the form of flat sheets lying parallel with each other, or it may be rolled up like a swiss roll, or it may consist of lots of narrow tubes. The idea is to create a large surface area in the smallest possible space.

Having a kidney transplant

Giving someone a new kidney from a donor may sound a difficult thing to do, but a surgeon will tell you that the operation itself is relatively easy. The trouble is that the patient's immune system may treat the new kidney as an intruder and destroy it (see page 158). This is most likely to happen soon after the operation. To prevent this rejection, the patient is given drugs which suppress the immune system. Obviously this makes the body less efficient at fighting disease, so germs must be kept away from the patient until, hopefully, the kidney is accepted and the drugs can be reduced.

The more alike the donor and the patient are genetically, the less likely it is that the kidney will be rejected. So the doctors always find out how genetically similar the tissues are before the operation is carried out. This is done by a procedure called **tissue typing**. This applies not only to kidneys but to other transplanted organs too, such as hearts, lungs and livers.

There is more about organ transplants and spare part surgery on page 21.

1 What are the advantages of having a kidney transplant rather than a kidney machine. Are there any possible disadvantages?

2 Why is it important to keep germs away from a patient who has just had a transplant operation, and how do you think this is achieved?

3 A kidney transplant is most likely to be successful if the kidney comes from an identical twin. Why? If the person doesn't have an identical twin, who would be the next best donor?

4 A kidney donor can give a kidney at any time and stay alive. But a heart donor is always a person who has just died. Why the difference?

Diabetes

Some people have too much sugar in their blood. They are suffering from **diabetes** and are known as diabetics. The extra sugar in their blood makes them tired and thirsty. If nothing is done about it, the person loses weight and may eventually die.

The kidneys try to get rid of the extra sugar, so one of the signs of diabetes is that sugar is present in the urine. In the old days, doctors used to tell whether or not a patient had diabetes by tasting the urine to see if it was sweet. Nowadays, a simple test is used.

Diabetes is caused by the pancreas not producing enough insulin. The result is that the liver does not turn as much sugar into glycogen as it normally would. Diabetes may start early in life or it may develop in middle age. It cannot be cured, but it can be controlled by:

- eating carefully selected foods which do not cause a harmful increase in the concentration of sugar in the blood.
- taking tablets which lower the amount of sugar in the blood.
- taking a certain amount of insulin every day. This makes the tissues turn blood sugar into glycogen.

Unfortunately insulin cannot be taken by mouth. It's a protein and is broken down by digestive enzymes in the gut. So it must be injected through the skin with a hypodermic needle. Diabetics are taught to do this for themselves as shown in the picture.

The trouble is that it's difficult to get the dose exactly right. What sometimes happens is that diabetics give themselves too much insulin with the result that their blood sugar falls too low. This can produce all sorts of effects such as trembling, sweating and weakness. The diabetic learns to recognise these signs and, if they come on, eats a few lumps of sugar or glucose tablets to bring the blood sugar up to the right level.

With proper medical help, diabetics can learn to control their problem and to work, play games and lead a full and active life. Some leading sports figures are diabetics.

Exciting new developments in the treatment of diabetes include the use of a portable pump which is programmed to deliver exactly the right amount of insulin into the bloodstream continuously. Pancreas transplants are also possible, and research is being carried out on an artificial pancreas.

1 A person who is suspected of having diabetes produces a sample of urine which is tested for sugar.

a Describe a suitable test which could be carried out.

b What would be the cause of sugar being present in the urine?

2 Insulin is a protein. It cannot be taken by mouth because it would be broken down by digestive enzymes in the gut.

a Give the names of two enzymes which would attack the insulin.

b What would these enzymes break the insulin down into?

c How is insulin taken by a diabetic? Mention one danger of taking insulin this way.

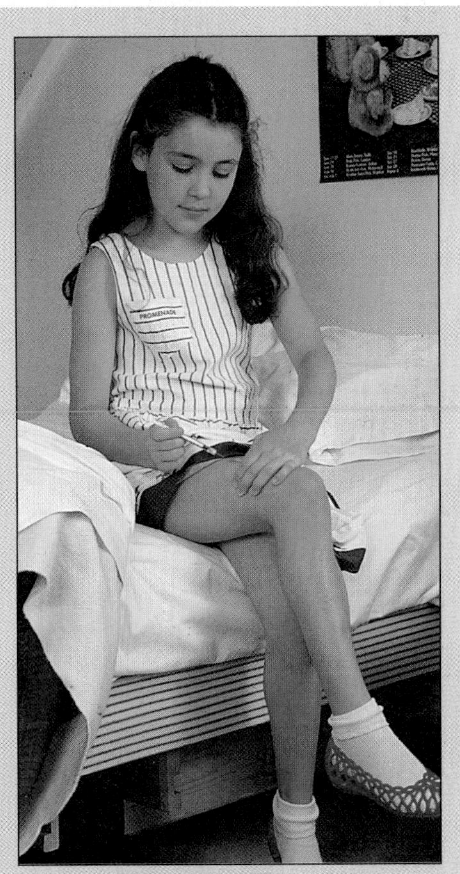

Picture 1 This girl has diabetes. Here you see her injecting herself with insulin.

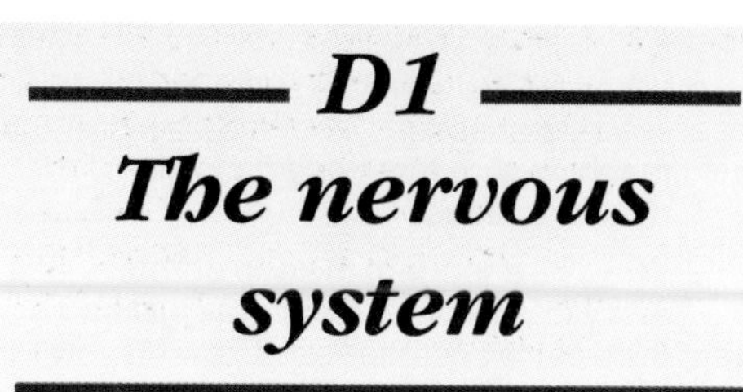

D1
The nervous system

At this moment you are reading this sentence. You can't do this without your nervous system.

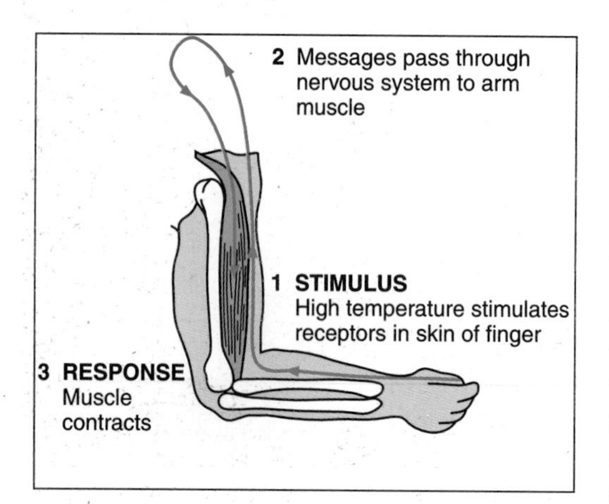

Picture 1 What happens when you pull your hand away from a hot object.

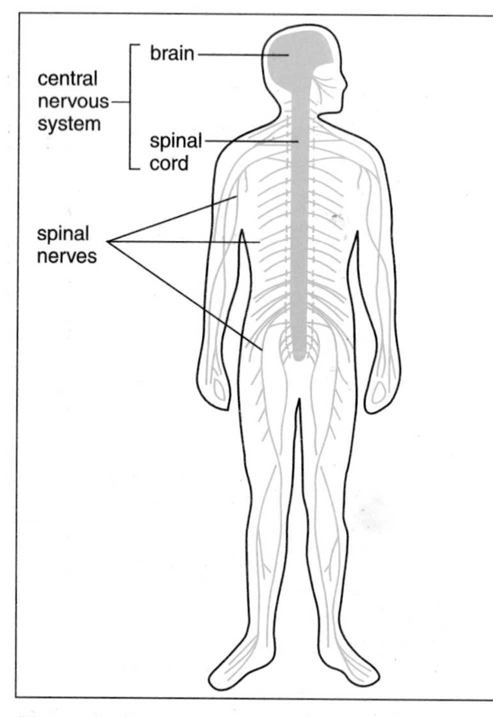

Picture 2 The main parts of the human nervous system.

Reflex action

Reading, writing and thinking thoughts are all very complicated functions of our nervous system. Let's start with something much simpler. If you accidentally touch a hot plate, you pull your hand away. This is an example of a **reflex action**.

A reflex action involves making a **response** of some kind. In the case of the hot plate, the response is pulling your hand away. But you will respond only if something makes you: this is called the **stimulus**. In the case of the hot plate, the stimulus is the high temperature of the plate.

The high temperature stimulates certain **receptors** in the skin. Messages then pass through the nervous system to the muscle in your arm. The muscle then shortens (**contracts**), and pulls your hand away from the plate (picture 1). The whole reflex takes only a fraction of a second, and this shows how quickly the messages travel through the nervous system.

When you pull your hand away from a hot plate, the structure that responds is a muscle. However, this is not the case with all reflexes. Sometimes it is a gland that responds. The general term for something which responds when it receives a message from the nervous system is **effector**.

The receptors which are stimulated when you touch a hot plate are sensitive to temperature and pain. But we have other receptors too. They are sensitive to **smell** (nose), **taste** (tongue), **touch** (skin), **light** (eyes) and **sound** (ears). Later we shall look in detail at the eye and ear.

How the nervous system is organised

The human nervous system is shown in picture 2. The main parts are the **brain** and **spinal cord.** The brain is inside the skull, and the spinal cord runs down the centre of the vertebral column (backbone). The brain and spinal cord together make up the **central nervous system**.

The central nervous system is connected to the various parts of the body by **nerves**. Some of the nerves come out of the brain, others out of the spinal cord. The nerves that come out of the brain go mainly to structures in the head, such as the eyes and jaws. Those that come out of the spinal cord go to the rest of the body. These are called **spinal nerves**.

The reflex arc

The messages which bring about a reflex action travel through the nervous system by a particular route. The route is called a **reflex arc**.

A reflex arc of the kind found in humans is shown in picture 3. Work your way round the arc, starting with the receptor and finishing with the effector. It is like a chain with three links. The links are **nerve cells** or **neurones**:

- A **sensory neurone** carries messages from the receptor to the spinal cord.
- A **connector neurone** carries messages through the spinal cord.
- An **effector neurone** carries messages from the spinal cord to the effector.

If the effector is a muscle, the effector neurone is called a **motor neurone**. The part of the neurone which carries the messages to the muscle is a long thread-like structure called an **axon**. The main part of the cell from which the axon projects is called the **cell body**. It contains the nucleus and other things that animal cells normally contain.

Only one reflex arc is shown in picture 3. In reality there would be lots of reflex arcs in a part of the spinal cord like this, and the spinal nerve would be full of neurones serving all sorts of receptors and effectors.

How are the messages carried?

The messages are tiny pulses of electricity. We call them **nerve impulses**. The axons are like electric cables, and nerve impulses pass along them very quick-

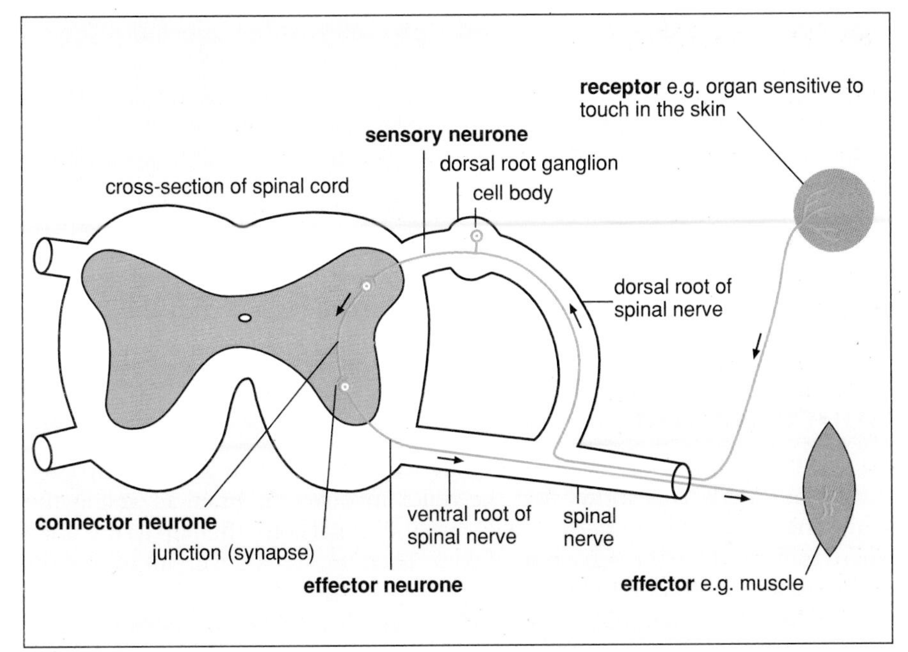

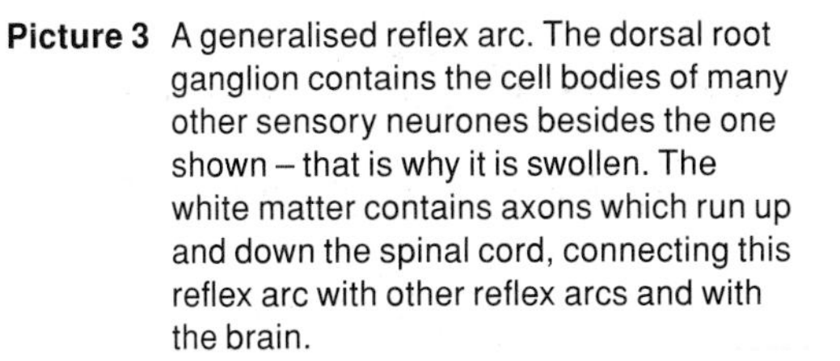
Picture 3 A generalised reflex arc. The dorsal root ganglion contains the cell bodies of many other sensory neurones besides the one shown – that is why it is swollen. The white matter contains axons which run up and down the spinal cord, connecting this reflex arc with other reflex arcs and with the brain.

ly. Nerve impulses are all separate, like cars whizzing along a road, and they travel along the axon one after the other.

In picture 3 you will see that the neurones are connected to each other by junctions in the spinal cord. These junctions are called **synapses**. At the synapse there is a tiny gap. When an impulse reaches the end of a neurone, a small amount of a chemical substance, called a **neurotransmitter**, is released into the gap. This then activates the next neurone, so that it starts carrying an impulse. The chemical can be produced only on one side of the gap, and this ensures that the impulses always travel in the right direction – *from* the receptor *to* the effector.

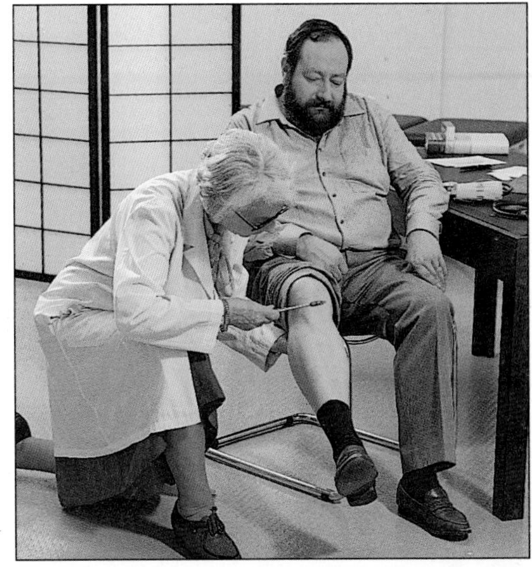
Picture 4 This doctor is testing a patient's knee jerk.

When nerve cells and synapses go wrong

All three neurones must be working properly if the impulses are to get right through the reflex arc. If one of them dies, the reflex cannot occur. Some people suffer from a condition called **motor neurone disease**. The neurones serving the muscles gradually degenerate and stop carrying impulses. As a result the muscles can't be made to contract, and the person becomes paralysed.

It is also important that our synapses should work properly. Synapses are readily affected by drugs and poisons. Some block them, others make them work too easily. One reason why drugs such as alcohol are harmful is that they interfere with synapses in the brain, slowing down our reactions. This is why it's so dangerous to drive after drinking alcohol.

Doctors can see if the various parts of our nervous system are working properly by testing our reflexes. One such reflex is the **knee jerk**: if your knee is tapped in a certain place, your leg gives a little kick (picture 4). This particular reflex involves only the spinal cord. It does not involve the brain. For this reason it is called a **spinal reflex**.

A closer look at neurones

Look at picture 6. This shows a motor neurone in detail. Notice the axon which carries impulses to the muscle. It leaves the spinal cord and, along with lots of other axons, enters one of the nerves. At the far end it breaks up into branches which go into the muscle.

The axon is enclosed in a **fatty sheath**. This insulates it and speeds up the impulses. If the fatty sheath wasn't there, the axon wouldn't be able to carry

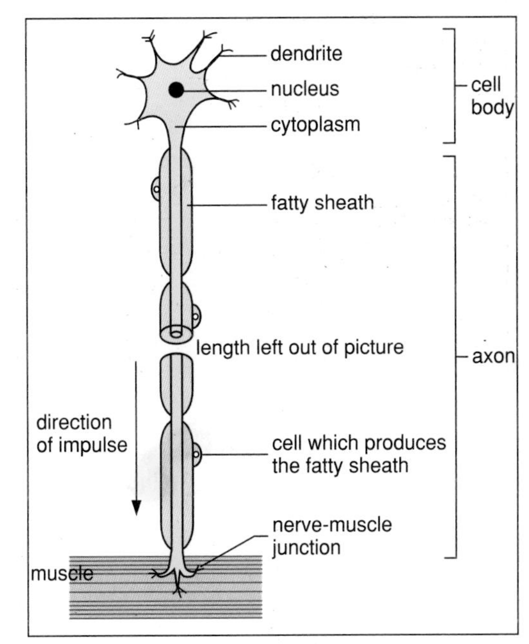

Picture 5 A motor neurone in detail. The axon runs out into a spinal nerve.

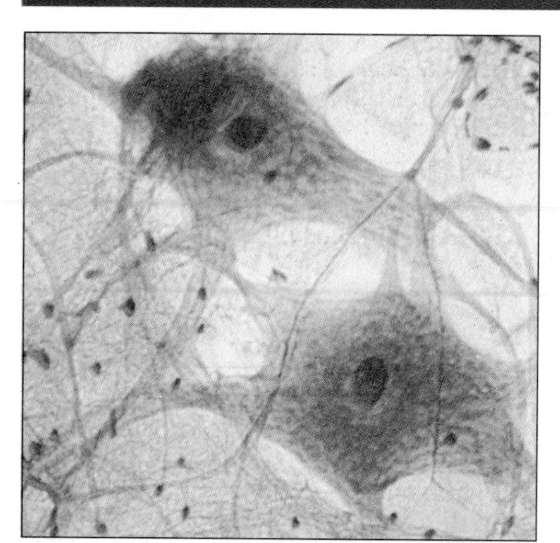

Picture 6 This is a thin section of the spinal cord seen under the microscope, greatly magnified. Two motor neurones can be seen.

impulses properly. In the disease **multiple sclerosis** the person gradually loses the use of the muscles. This is because the fatty sheaths break down, so impulses can't get to the muscles.

When the axon reaches the muscle, it splits into branches which make connection with the muscle fibres. The point where the two join is called the **nerve–muscle junction**. There is a gap here, and the message gets across by means of a chemical substance just as it does between neurones.

In pictures 5 and 6 you can see lots of 'arms' sticking out of the cell body. These are called **dendrites**, and they link up with other nerve cells to form a dense network. It is this network which enables the spinal cord and brain to carry out their important job of coordination.

Coordination

Let's go back to the hot plate. If the plate was very hot, you would jump back and let out a cry. This shows that the reflex involves the brain as well as the spinal cord. The nerve impulses travel into the spinal cord, then up to the brain where they produce the sensation of pain. Then impulses travel out to the various muscles involved in the response.

What exactly is the brain doing here? Put simply, it makes sure that the right muscles contract at the right time. This is called **coordination**. All our responses require coordination. This function is carried out by the brain and, to a lesser extent, the spinal cord..

Pulling your hand away from a hot plate is a relatively simple response. Many of our actions are more complicated than this. Think of the coordination that's necessary, for example, in walking or running. Even more coordination is needed for skilled activities like skiing and ballet dancing.

In all these cases the brain *processes* the information it receives and ensures that the right actions take place.

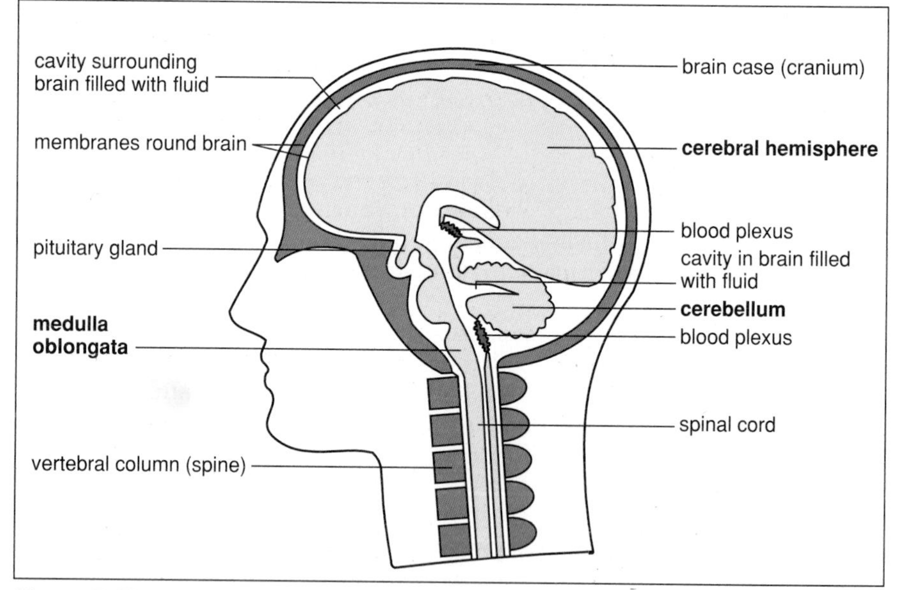

Picture 7 The human brain seen in its natural position inside the head.

The brain

There are over one thousand million neurones in the brain and each one may be connected with as many as 25 000 others. This makes the brain like an extremely complex computer and enables it to coordinate all our actions. But the brain is more than a computer for it gives us feelings and emotions – we are human beings not machines.

You can see a human brain in picture 7. Three of its most important parts are labelled in bold print:

First aid for unconsciousness

Unconscousness can occur when the brain does not work properly. Causes include a reduced oxygen supply to the brain, a severe blow to the head, or an overdose of drugs (including alcohol).

If the person stops breathing, give mouth-to-mouth resuscitation. If the person is still breathing, put him or her in the **recovery position**, like this:

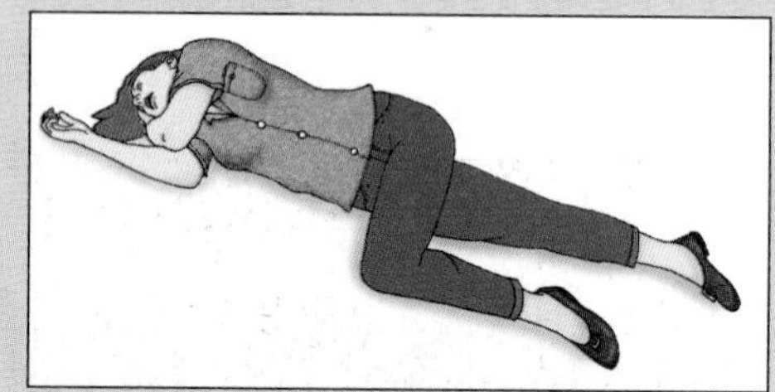

The head must be on its side so that any saliva or vomit flows out of the mouth. This is vital, otherwise the person may inhale these fluids into the lungs, which can be fatal.

The head should also be tilted slightly back. This helps to keep the airway open to the lungs.

Remove false teeth and anything else in the mouth, loosen clothes, cover the person with a blanket or coat and call a doctor.

1 In what circumstances might there be a reduced oxygen supply to the brain?
2 Suggest a reason for the arrangement of the arms and legs in the recovery position.

- The **medulla oblongata** controls various automatic processes such as breathing and the circulation.
- The **cerebellum** helps us to keep our balance and enables us to make precise and accurate movements.
- The **cerebral hemispheres** control our sensations and movements, and are also responsible for memory, thought and intelligence.

Our feelings and emotions stem mainly from the front part of the cerebral hemispheres just behind the forehead. This is the part of the brain that gives us our individuality and personality. It is the least understood part of the brain, and perhaps will never be understood.

Activities

A Some human reflexes

Work in pairs, one person acting as the subject.

1 *The knee jerk*

Sit on a table with your legs hanging loosely. With a tendon hammer, your partner should gently tap your knee just below the knee cap.

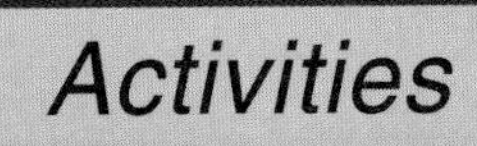

Describe the response.

2 *The ankle jerk*

Kneel on a chair and let your feet hang loosely. Your partner should tap the back of your foot just above the heel. What happens?

Repeat the above reflexes but this time make a conscious effort to prevent them taking place. Do you succeed?

What conclusions do you draw?

3 *The blink reflex*

Open your eyes and look straight ahead. Your partner should suddenly wave a hand in front of your eyes.

What happens?

4 *The swallowing reflex*

Swallow the saliva in your mouth, then immediately try swallowing again.

Is it difficult to swallow the second time?

Suggest an explanation.

Describe the route through which impulses pass in each of these reflexes.

B Measuring your reaction time

Hold a metre rule vertically. Your partner should place his hand at the bottom of the rule in readiness to catch it. Find out how far it falls before being caught by your partner. The further it falls, the longer the reaction time.

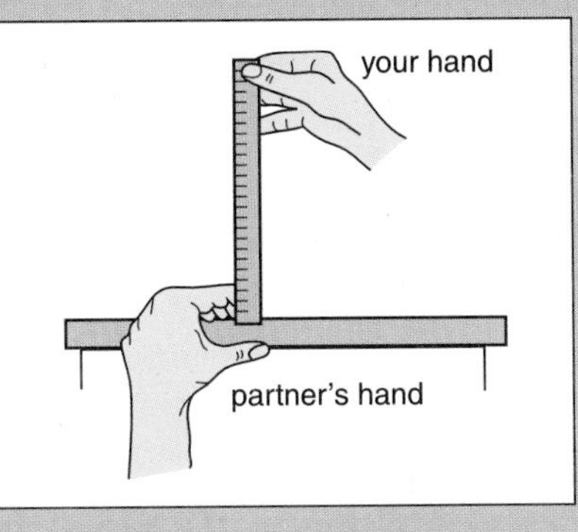

Alternatively, use an electronic device for measuring reaction time.

Describe the route through which impulses pass in bringing about the response.

Find out if the reaction time is affected by:

a closing one eye,
b catching the ruler with the other hand.

C Looking at nerve cells in the spinal cord

1 Obtain a prepared cross section of the spinal cord, which has been specially stained to show up the nerve cells.

2 Examine the slide under the low power of the microscope. Identify the grey and white matter.

Can you see nerve cells in the grey matter similar to the ones in picture 6?

3 Go over to high power, and focus on a single nerve cell. Draw the nerve cell to show its shape.

What part of the nerve cell shown in picture 5 does your drawing represent?

Questions

1 A person walks across a room in bare feet and treads on a drawing pin. He lets out a cry. Explain what happens in his nervous sytem in bringing about this response.

2 Explain the reason for each of the following in terms of reflex action.

a If you tickle a dog's tummy it 'scratches' with its hind leg.
b A person with motor neurone disease has weak responses.

3 The picture below shows the route through which messages travel in bringing about the knee jerk. When the tendon is tapped, receptors in the muscle are stretched and this causes the messages to be sent off.

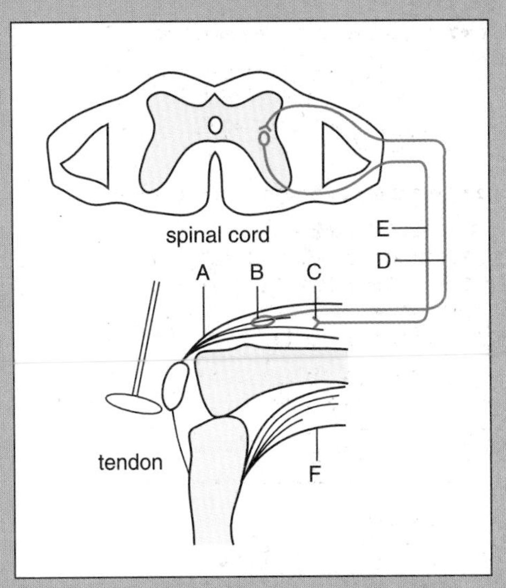

a Which structure is stimulated by the hammer?
b Which structure carries impulses away from the spinal cord?
c Which structure shortens as a result of the reflex?
d What would be the approximate length of structure E in a human?
e Assuming that the impulses travel at 100 metres per second, how long would it take for an impulse to travel through this reflex arc?
f How does the structure of this reflex arc differ from the one in picture 3 (page 195)?

4 Write a short essay on how the brain helps animals to survive. You may consider any animals you like including the human.

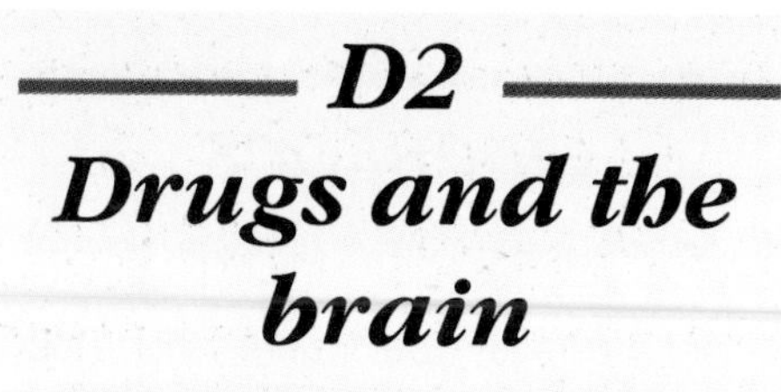

D2 Drugs and the brain

The human body is readily affected by outside influences, particularly drugs.

Picture 1 Alcohol is a sedative which can make people fall asleep in unusual places. This particular nap was brought on by an excess of wine.

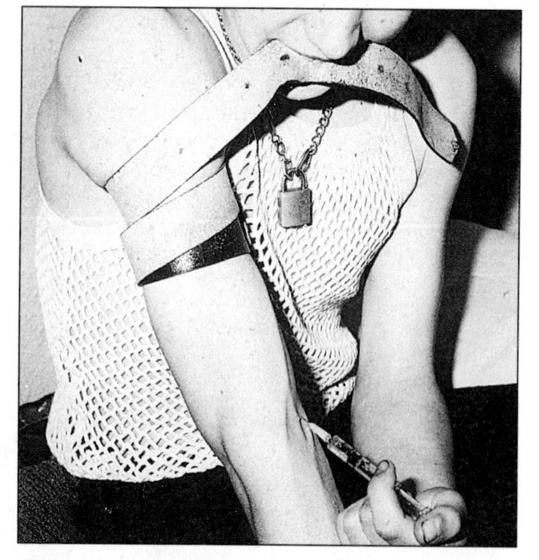

Picture 2 A drug addict injecting himself with heroin: the start of a life of misery and probably an early death.

What are drugs?

A drug is any substance which alters the way the body works. Drugs which affect the brain fall into these four main groups: **stimulants**, **sedatives**, **hallucinogens** and **painkillers**.

Stimulants

These drugs speed up the brain and make you more alert. They include **amphetamines** which were widely used at one time for relieving blocked noses. A particularly powerful stimulant is cocaine, which is obtained from the leaves of certain South American plants. Coffee and tea contain a mild stimulant called caffeine. Another mild stimulant is nicotine, the drug found in tobacco.

Sedatives

These drugs slow down the brain and make you feel sleepy. They include **tranquillisers** and sleeping pills. Tranquillisers have a calming effect and are often given to people suffering from anxiety. An example is valium. Some sedatives, for example barbiturates, are so powerful that they can send you to sleep almost immediately. They are used as anaesthetics. Alcohol is a sedative. It sedates the higher centres in the brain, making people less inhibited.

Hallucinogens

These drugs make you feel, see or hear things which do not really exist. Such experiences are called **hallucinations**. Drugs which cause hallucinations include cannabis (marijuana, nicknamed 'pot' or 'dope'). This is relatively mild. Much more powerful and dangerous is LSD (lysergic acid diathylamide). This drug can produce nightmarish hallucinations which sometimes lead to fatal accidents.

Painkillers

These drugs suppress the part of the brain which gives us our sense of pain. They are called **analgesics**. They range from mild drugs such as aspirin to very powerful ones like morphine and heroin, which are obtained from opium, a substance found in a certain type of poppy. Morphine is given to people who are in great pain after an injury or when they have a very painful illness. Because they suppress the brain, painkillers may also be sedatives and make you feel sleepy. That's why people taking painkillers are advised not to drive.

Why are drugs dangerous?

Taken under doctors' orders, certain drugs can be very helpful to sick people. However, they must be taken in the right amounts at the right times. If taken in the wrong circumstances, they may be extremely harmful. There are three main reasons for this.

■ Drugs interfere with brain function

Drugs often slow your reaction time, so you take longer to respond to a stimulus. They also impair your judgements. Such is the case with alcohol.

In an interesting experiment, some bus drivers drove their buses between two rows of posts. They then drank a small amount of whisky and drove their buses between the posts again. It was found that after drinking they knocked over many more posts than they had done before. The alcohol made them misjudge the distance between the posts. One of them, an experienced driver, tried to get his 2.44 metre wide bus through a gap of 2.03 metres.

■ Drugs damage the body

Drugs are poisons and can kill cells. For example, alcohol kills cells in the brain and liver. Prolonged drinking can cause a disease called **cirrhosis**: the liver cells are gradually replaced by useless fibrous tissue. Sniffing solvents can also be dangerous.

You can become dependent on drugs

Drugs like nicotine and cannabis are habit-forming and, once you are hooked, you crave them. In other words, you become **psychologically dependent** on them. Drugs like heroin and alcohol are even worse: they can get such a grip on the body that, if the drug is unavailable, the person develops withdrawal symptoms such as dizziness, vomiting and muscle pains. The person has become **physically dependent** on the drug, i.e. **addicted** to it.

The slippery slope towards addiction

The great danger with drugs is that what starts as a habit can quickly become an addiction. Once a person becomes addicted to a drug, it's very difficult to give it up, as picture 2 makes only too clear. And there's another problem. The body gradually gets used to the drug, i.e. *tolerant* to it. The result is that the person has to take larger and larger doses to get an effect.

It is particularly easy to become addicted to alcohol. Many heavy drinkers are on the slippery slope towards addiction though they may not realise it.

How much alcohol?

The amount of alcohol that a person drinks can be expressed in 'units'. A unit of alcohol is 10 cm^3 (10 millilitres). This is the amount present in half a pint of standard beer or lager, a glass of wine, a small glass of sherry or a single measure of spirits or vermouth.

When you drink, the alcohol goes into your bloodstream, and the more you drink the greater will be the concentration of alcohol in your blood. This is shown in the bar chart. However, other factors affect the way a person's blood alcohol rises – body mass, for example.

The bar chart also shows the legal limit for driving. If you have more than the equivalent of 5 units of alcohol in your blood, you are breaking the law. This does not mean that it is safe to drive after drinking just under 5 units. the only safe level is zero.

The alcohol we drink gets into the cells and is eventually broken down. It takes about an hour for each unit to be destroyed. An average-sized man who drinks more than 21 units a week is seriously risking his health – for women the figure is 14 units.

Because of the problems connected with alcohol, most countries have licensing laws which limit the sale of alcohol to certain hours of the day.

1 Why do you think the risk figure is lower for women than for men?

2 Some people say you should drink no alcohol at all before driving. Do you agree?

3 Do you think alcohol should be classified as a dangerous drug and banned?

4 Steve and Bob drove to a pub and drank 5 units of alcohol in exactly the same time. On the way home, both were stopped by the police, taken to a police station and given blood tests. Steve was well over the legal limit for driving, whereas Bob was just under the limit. Suggest two reasons for the difference.

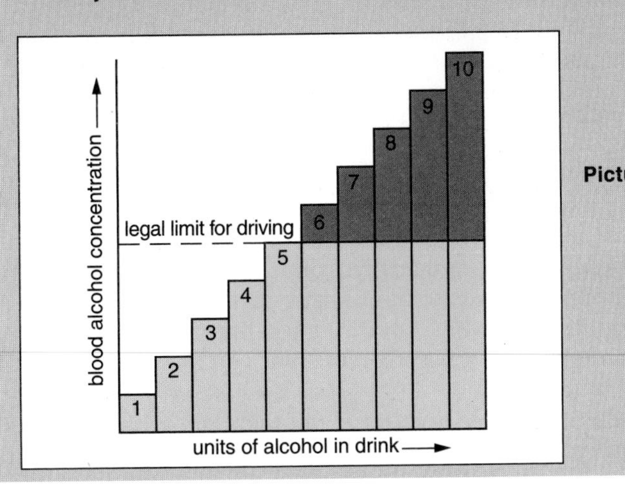

Picture 1 This bar chart shows how the concentration of alcohol in the blood rises as you drink more and more alcohol.

The danger of drinking meths

The alcohol that people drink is **ethanol** which is made by yeast in the process of alcoholic fermentation (see page 126).

A cheaper form of alcohol, **methanol**, is distilled from wood. Its effects on the human body are similar to those of ethanol, but much more severe. It particularly affects the nervous system and can damage the optic nerves, resulting in permanent blindness. This is why drinking methylated spirit ('meths') is so dangerous. Meths consists of ethanol plus approximately ten per cent methanol.

The hazards of solvent-sniffing

Solvents are liquids which dissolve things. Some solvents give off vapours which can be harmful if breathed in. They include solvents used in various glues. When sniffed, they can produce a 'high' similar to that produced by drinking alcohol.

If a mild solvent is sniffed on only a few occasions it is unlikely to cause much harm. But problems may arise if it becomes a habit, or if it leads to sniffing more toxic solvents such as aerosol fluid and lighter fuel. Inhaling the fumes from these substances has killed people. People have also died from suffocation as a result of using plastic bags and tubes for sniffing glue.

Heavy solvent-sniffing over a long period of time (e.g. 10 years) may also damage the liver, kidneys and brain.

The sensations produced by inhaling solvents come on quickly. Prolonged sniffing may cause frightening hallucinations, and in extreme cases sniffers may become stupefied and go into a coma before they realise what is happening.

1 It is more accurate to say that solvents are inhaled rather than sniffed. Why?

2 Overheard in a school: 'Glue-sniffing can't be dangerous because if it was it would be made illegal'. Discuss.

3 What is an hallucination? Why can hallucinations be dangerous?

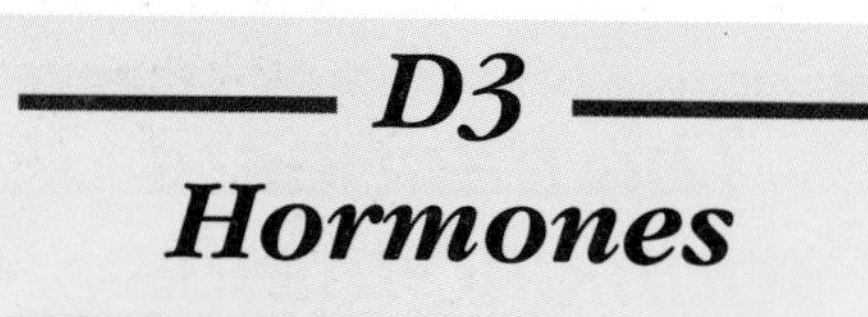

D3 Hormones

Nerve impulses are one way of sending messages from one part of the body to another. Hormones provide another way.

What are hormones?

A hormone is a chemical substance which is produced by one part of the body and has an effect on another part. Hormones are produced – or secreted – by glands.

Some glands shed their secretion into a tube, or duct, which carries it to the place where it is needed. For example, the salivary glands secrete saliva into ducts which take it to the mouth cavity. Hormone-producing glands are different. They shed their secretion, not into a duct, but into the bloodstream. For this reason they are known as **ductless glands** or, more grandly, **endocrine glands**.

Once in the bloodstream, the hormone is taken to all parts of the body. It then produces a response in certain organs.

What do hormones do?

The main endocrine glands in the human body are shown in picture 1. Each gland secretes one or more hormones, and their functions are summed up in table 1. Many of them are dealt with in other parts of this book. The index will tell you where to find them.

Most hormones produce their effects rather slowly. They bring about long-term changes in the body such as growth, sexual development and so on. However, there is one hormone which acts very quickly. This is **adrenaline**.

Adrenaline, the emergency hormone

Have you ever had that sinking feeling just before a football match, an exam or some other important event? The whole body tenses up, the heart beats faster and you feel ready for action. This effect is brought on partly by the nervous system, but also by adrenaline.

Adrenaline is secreted by the **adrenal glands** close to the kidneys. As with other ductless glands, the hormone passes straight into the bloodstream and is carried all over the body. The cells respond to it by metabolising faster and transferring more energy. At the same time the heart beats more quickly. Blood is diverted from organs that can do without it to those that really need it such as the muscles and brain (picture 2). The overall effect is to prepare the body for action.

How are hormones controlled?

It is obviously important that the right quantity of hormones should be produced at all times. A key structure in this is the **pituitary gland** at the base of the brain. This produces hormones which control other glands, telling them how much hormone to produce. It is a kind of 'master gland', and it is controlled by feedback mechanisms (page 184).

Sometimes things go wrong

Despite the control mechanisms, things sometimes go wrong and either too much or too little of a hormone is produced. For example, we have already seen what happens when too little insulin is produced by the pancreas (page 193).

The thyroid gland can also give trouble. For instance, a person may have a thyroid gland which produces too much thyroxine. As a result metabolism speeds up, and the person becomes thin, excitable and overactive. Another person may have a thyroid gland which produces too little thyroxine. In this case metabolism slows down, and the person becomes fat and sluggish (picture 3).

These are serious cases, and only a doctor can put them right. But we are all at the mercy of our hormones to a certain extent. If you get up in the morning

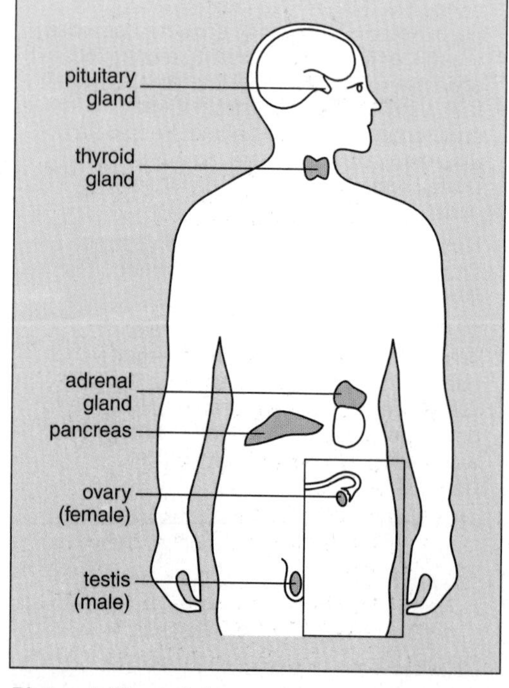

Picture 1 The main hormone-producing glands in the human body.

feeling off-colour, it may be due to a slight change in the amount of a particular hormone in your body. Normally it does not matter very much. The error will soon be corrected and your body will get back on balance.

Table 1 Summary of the human body's main hormone-producing glands and their secretions. The pituitary gland secretes other hormones besides the ones included here.

Gland	Hormone	Function
Thyroid	Thyroxine	Controls the metabolic rate
Adrenals	Adrenaline	Prepares the body for action
Pancreas	Insulin and glucagon	Regulate the amount of sugar in the blood
Ovaries	Female sex hormones	Control sexual development
Testes	Male sex hormones	Control sexual development
Pituitary	Growth hormone	Speeds up growth
	Stimulating hormones	Activate other glands

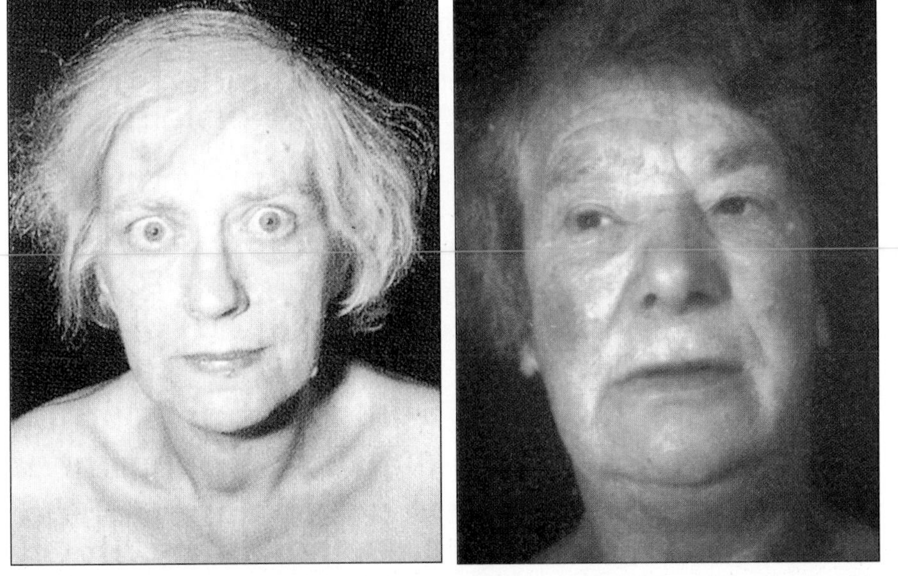

Picture 3 People with faulty thyroid glands. The person on the left has an overactive thyroid, and the person on the right has an underactive thyroid.

Picture 2 On your marks, get set, go! This picture shows how the hormone adrenaline prepares the body for action.

Questions

1 Nerves and endocrine glands both provide ways of sending messages from one part of the body to another. Write down four differences between the two systems.

2 Endocrine glands such as the thyroid contain a large number of capillaries which are located close to the cells. Why do you think this is?

3 If necessary, use the index to answer this question. The pancreas is made up of two parts: part of it secretes insulin and part of it secretes pancreatic juice.
- a Which part is functioning as an endocrine gland?
- b What is insulin and what effect does it have in the body?
- c Are insulin's effects short-term or long-term?
- d What does pancreatic juice contain?
- e Where does pancreatic juice go?

4 In table 1, which hormone
- a makes a person more active,
- b causes the male to start producing sperm,
- c prepares the body for an emergency,
- d is produced by a gland in the neck,
- e causes breasts to develop in the female?

5 What part does the diet play in helping us to have a healthy thyroid gland. (Hint: look up goitre in the index.)

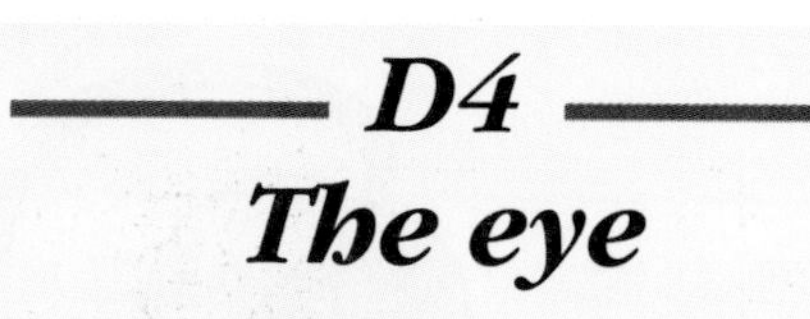

D4 The eye

Close your eyes and imagine what it must be like to live in darkness. Our eyes are amongst our most important sense organs.

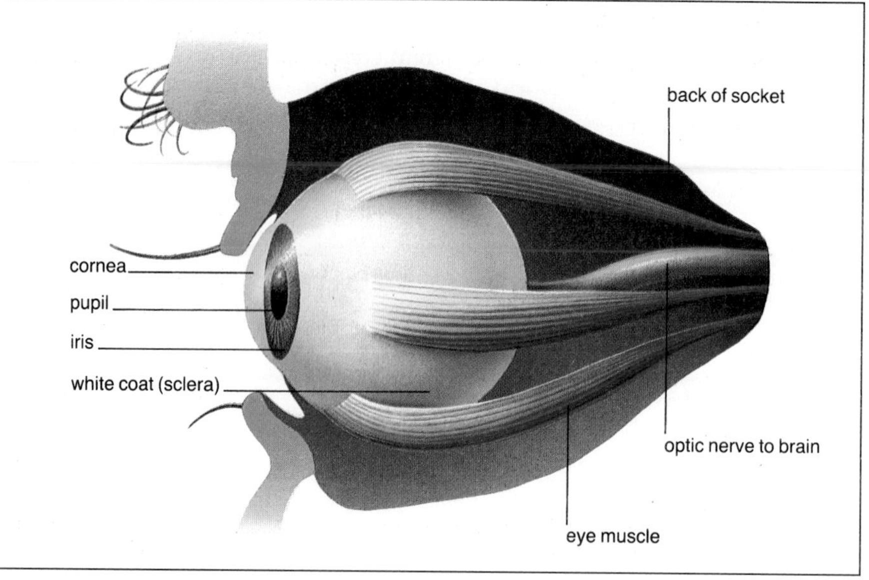

Picture 1 The eyeball in its socket. There are six eye muscles altogether, of which only three are shown here.

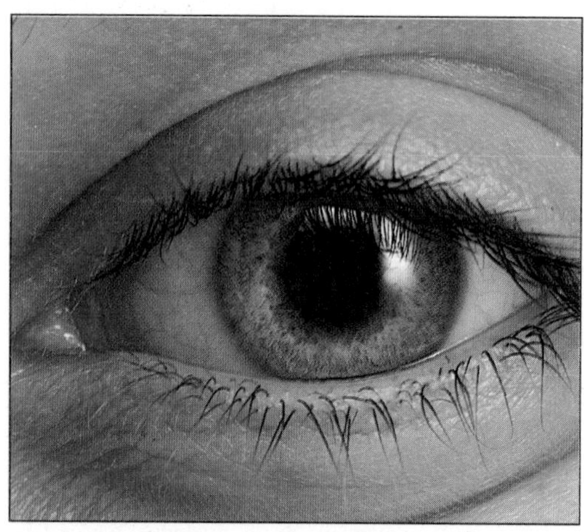

Picture 2 A close-up view of the eye. Tears are produced by a gland just under the upper eyelid on the right, and they drain away into a duct in the corner of the eye on the left.

The outside of the eye

Each eye consists of an eyeball which rests in a socket in the skull (picture 1). The sides and back of the eyeball are thick and tough – this is the 'white of the eye'. The front is transparent and is called the **cornea**. The cornea is covered by a thin and delicate membrane called the **conjunctiva**.

The conjunctiva is kept moist by **tears**, a lubricating fluid produced by a gland under the eyelid (picture 2). Tears also contain the enzyme **lysozyme** which kills bacteria.

In the centre of the eye is the **pupil**. This is surrounded by the **iris**, the coloured part of the eye.

The eyeball is held in place by muscles which can move it up and down and from side to side. A large **optic nerve** runs from the back of the eye to the brain. When you look at an apple, millions of impulses are sent off in this nerve and when they reach the correct part of the brain you see the apple.

Picture 3 The internal structure of the human eye. You must imagine that the eye has been sliced across the middle and that you are looking inside.

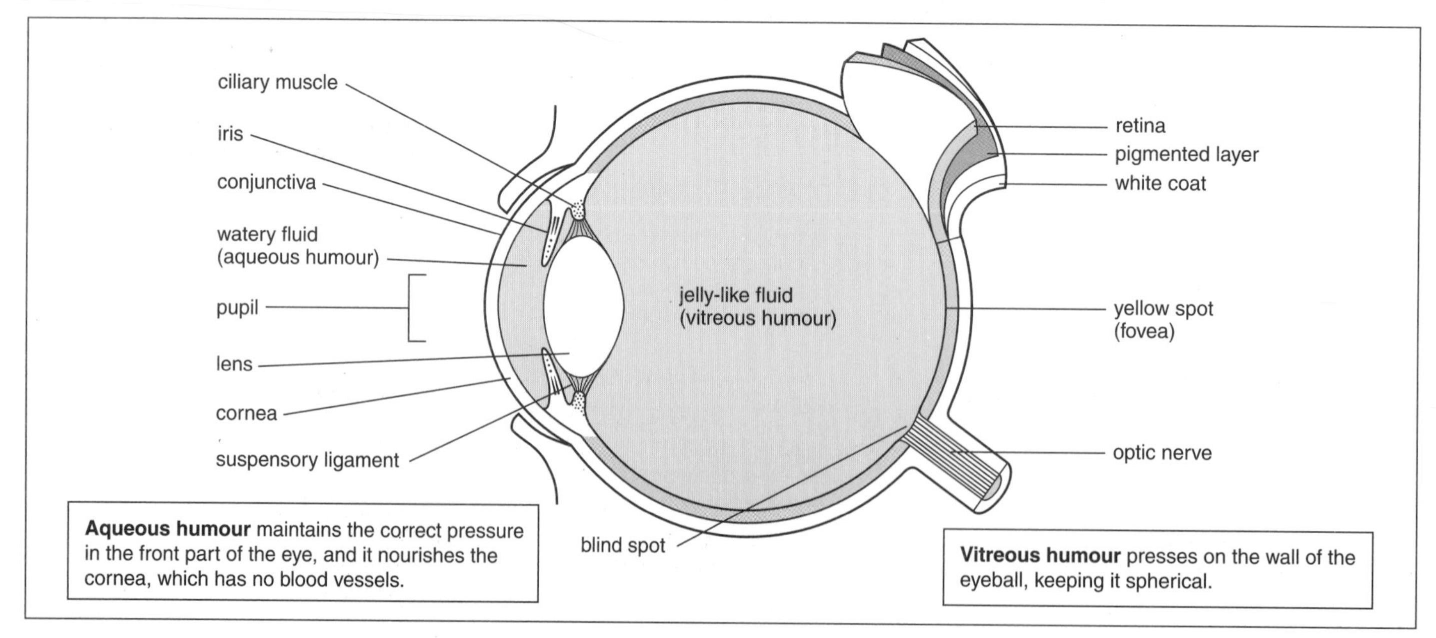

The inside of the eye

Picture 3 shows the inside of the eye. For seeing things, the two main parts are the **lens** and the **retina**.

The lens is soft and transparent – rather like a polythene bag full of water – and its shape can change. it is encircled by a ring of muscle called the **ciliary muscle**. The lens is held in position by fine threads which run from it to the surrounding ciliary muscle. The threads are called the **suspensory ligament**.

The retina lines the inside of the eyeball. It contains millions of receptor cells. These cells are sensitive to light which has entered the eye through the pupil – this is how the eye sees things. The part of the retina responsible for seeing things most clearly is right in the middle – the **yellow spot** (**fovea**).

Behind the retina is a layer of tissue containing a dark pigment. The pigment absorbs light and prevents it being reflected within the eye. Why would it be a bad thing if this happened?

The pigmented layer also contains lots of blood vessels. They supply the retina with oxygen and food substances.

The point where the optic nerve is attached to the eye is called the **blind spot.** The blind spot has no receptor cells, so it is unable to see things.

Controlling the amount of light that enters the eye

If you look at a bright light, a reflex action occurs: the pupil gets smaller and this stops too much light getting into the eye. The opposite happens in the dark: the pupil gets larger, so more light can enter the eye.

The widening and narrowing of the pupil is brought about by the iris. Picture 4 shows how it works.

How does the eye see things?

When you take pictures with a television camera, light enters the camera and is focused by a lens onto a light-sensitive film at the back. The eye works in the same kind of way.

Suppose you are looking at a dot on the wall. Light rays, reflected from the dot, enter your eye as shown in picture 5. As the light rays pass through the cornea and lens, they are bent inwards so that they meet on the retina. Here they produce an image of the dot.

For the image to be clear and sharp, i.e. in focus, the light rays must meet exactly on the retina. This is achieved by the lens which makes sure that the light rays are always bent to just the right extent. We'll come back to this in a moment.

The bending of the light rays is called **refraction**, and it plays an essential part in giving us good eyesight.

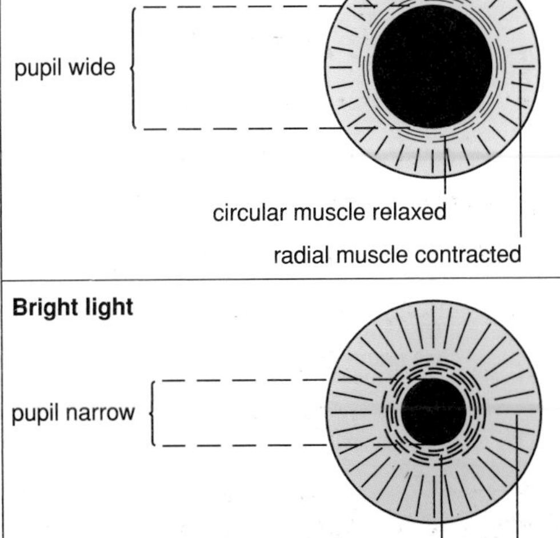

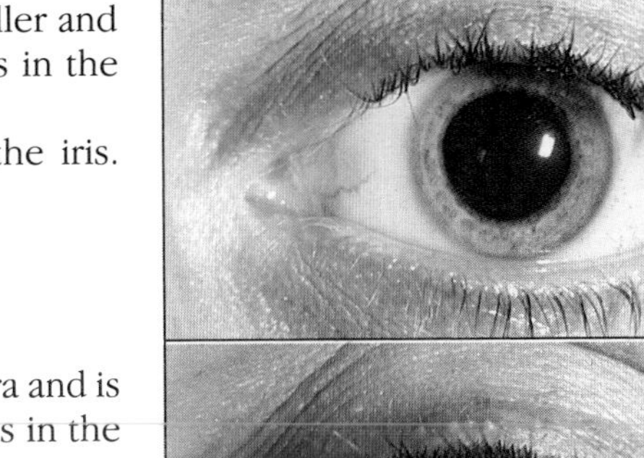

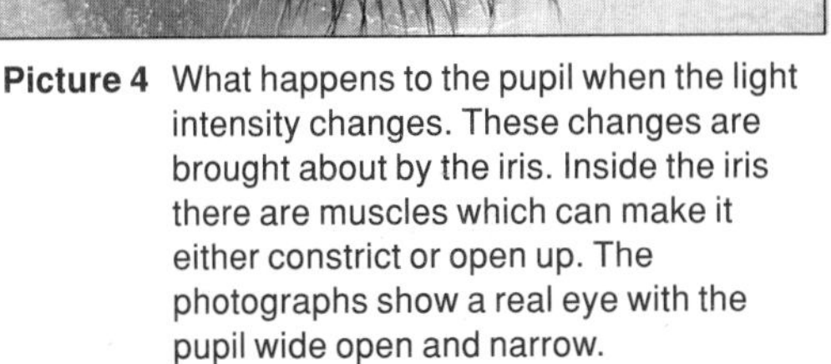

Picture 4 What happens to the pupil when the light intensity changes. These changes are brought about by the iris. Inside the iris there are muscles which can make it either constrict or open up. The photographs show a real eye with the pupil wide open and narrow.

Picture 5 How the eye focuses on a dot.

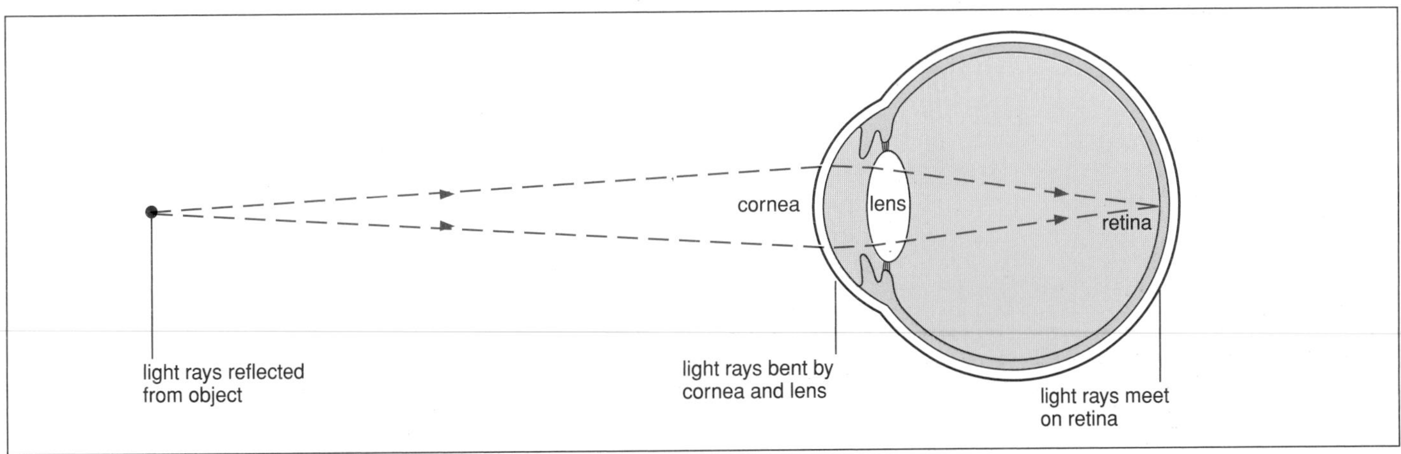

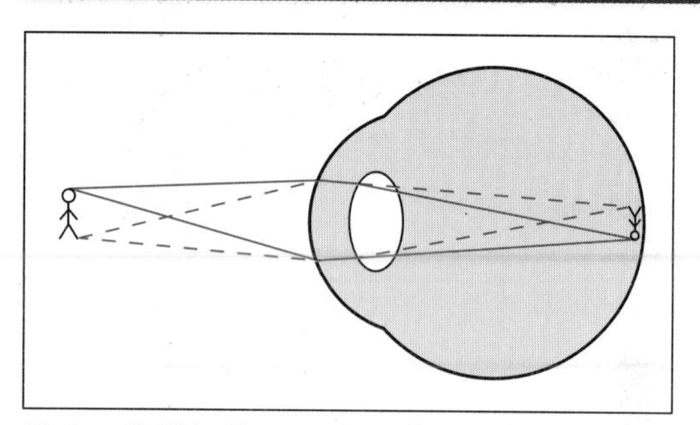

Picture 6 This diagram shows how an image is turned upside down by the lens in the eye. The same thing is done by the lens in a camera.

Seeing things the right way up

Suppose you are looking at a person. Picture 6 shows how the light rays, reflected from the person, pass into your eye. The result is that the image is upside down on the retina.

Why then don't we see everything upside down? The answer is that the brain comes to the rescue and turns the picture the right way up for us.

Some years ago an experiment was done in which a man was given special glasses that made him see everything upside down. After a while his brain made the necessary correction and he began to see things the right way up again, even though he went on wearing the glasses. What do you think happened when he took the glasses off?

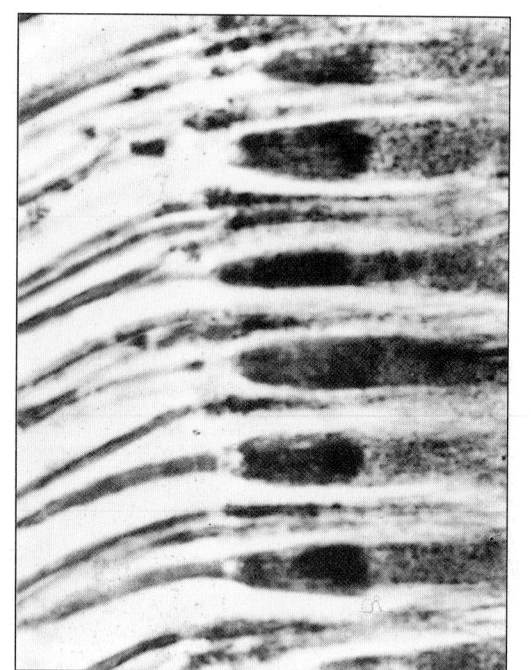

Picture 7 A small part of the retina seen under the microscope. It shows the cone cells and rod cells.

How does the retina work?

If you look straight at an object, and then look at it out of the corner of your eye, its appearance changes. From being clear and sharp, it becomes fuzzy and indistinct. Also it's hard to tell what colour it is.

How can we explain this? Well, there are two types of receptor cells in the retina. They are called **cone cells** and **rod cells** (picture 7). When you look straight at something, you are using the central part of the retina – the yellow spot. This part of the retina contains mainly cone cells, which detect things clearly and in colour.

However, when you look at something out of the corner of your eye, you are using the part of the retina further out. This contains mainly rod cells. They detect things less clearly, and in black and white.

Seeing in dim light

From what we have just said, you might have got the idea that the outer part of the retina is not much use. However, it is good at seeing things in dim light. You can prove this for yourself by looking at a faint star on a dark night. It is much easier to see it out of the corner of your eye than by looking straight at it.

The reason for this is that the rod cells are stimulated by even very small amounts of light, so they work in gloomy conditions. The cone cells, on the other hand, are less sensitive and will only work in reasonably bright conditions.

Have you noticed that when you go into a gloomy room from bright sunlight, you can't see anything at first but gradually things become visible? The reason is that the bright light causes your rod cells to lose their sensitivity. So when you go into the gloomy room, the rod cells don't work. They need time to become sensitive again. As their sensitivity returns, you begin to see things.

Picture 8 How the eye keeps a ball in focus as it gets closer.

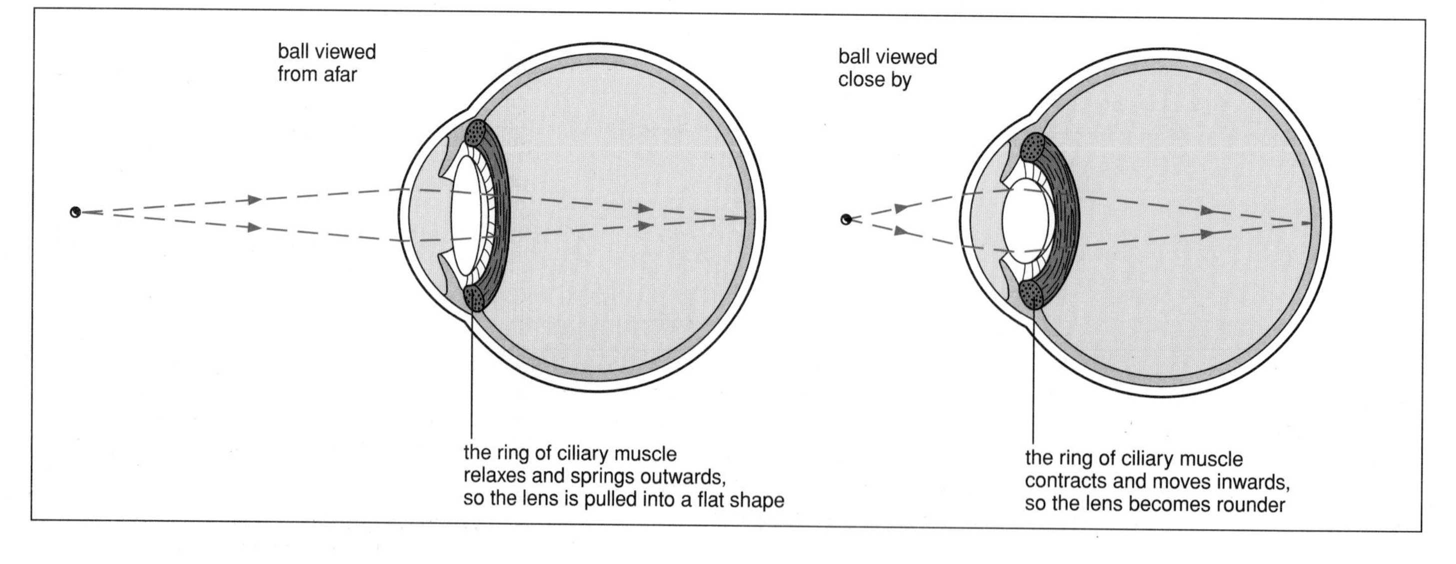

Colour vision

If you have ever been involved with stage-lighting, you will know that almost any colour can be obtained by mixing red, green and blue lights in the right proportions. These are the **primary colours**.

The same principle applies to the way we see colours. Scientists have shown that we have three different kinds of cone cells, each sensitive to one of the primary colours. The colour which we actually see depends on how many cones of each kind are stimulated.

Some people cannot see certain colours – they are **colour blind**. In rare cases colours cannot be seen at all, so everything looks black, white or grey. A more common condition is where people cannot tell the difference between red and green. Red-green colour blindness is inherited (see page 272).

Keeping things in focus

Suppose you are watching a ball hurtling towards you. If the eye did not adjust, the light rays would stop meeting on the retina as the ball got close to you – so the ball would become out of focus. However, the eye does adjust. The lens becomes rounder and bends the light rays more. So the light rays continue to meet on the retina, and the ball stays in focus.

The eye is able to keep things in focus because the lens is soft and can change its shape. The shape of the lens is changed by the ciliary muscle. This makes the lens flat or more rounded, depending on whether you are looking at something in the distance or close to (picture 8).

Despite this wonderful adjustment mechanism, many people cannot focus properly. Such people may be either **short-sighted** or **long-sighted**.

Short-sighted people

A short-sighted person can focus on things close by, but not a long way off. This is due to the lens bending the light rays too much, or to the eyeball being too long. The result is that the light rays meet in front of the retina.

Short-sightedness is corrected by wearing glasses which bend the light rays outwards before they reach the eye (picture 9).

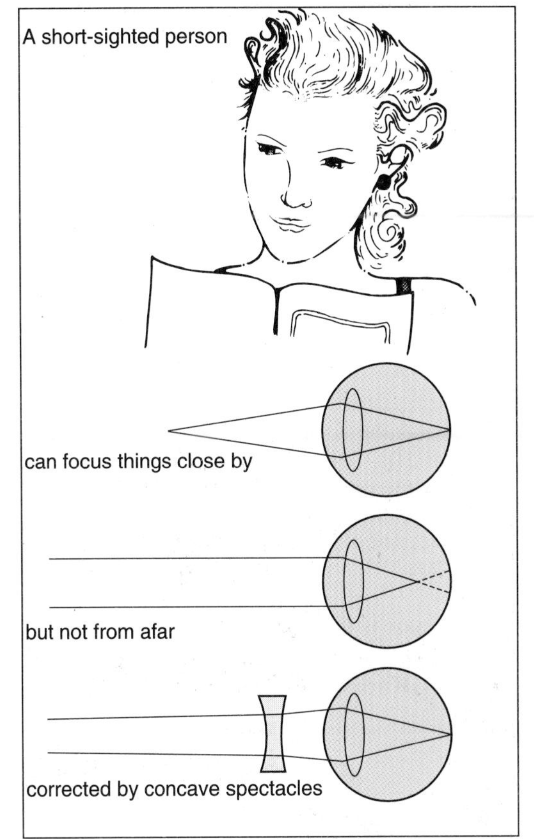

Picture 9 Short-sightedness and how it can be corrected by wearing glasses.

Long-sighted people

A long-sighted person can focus on things a long way off, but not close by. This is due to the lens not bending the light rays enough, or to the eyeball being too short. The result is that the light rays are directed to a point behind the retina.

Long-sightedness is corrected by wearing glasses which bend the light rays inwards before they reach the eye (picture 10).

Long-sightedness is also caused by the lens becoming hard, so it no longer changes its shape in the usual way. This tends to happen in old people, and is one of the main reasons why they often need glasses.

A more serious problem is that the lens may become cloudy and stop letting light through. This is called a **cataract**. The only remedy is to take the lens out in an operation and give the person very strong glasses or contact lenses. In the latest operations, the person's own lens is replaced with an artificial acrylic lens.

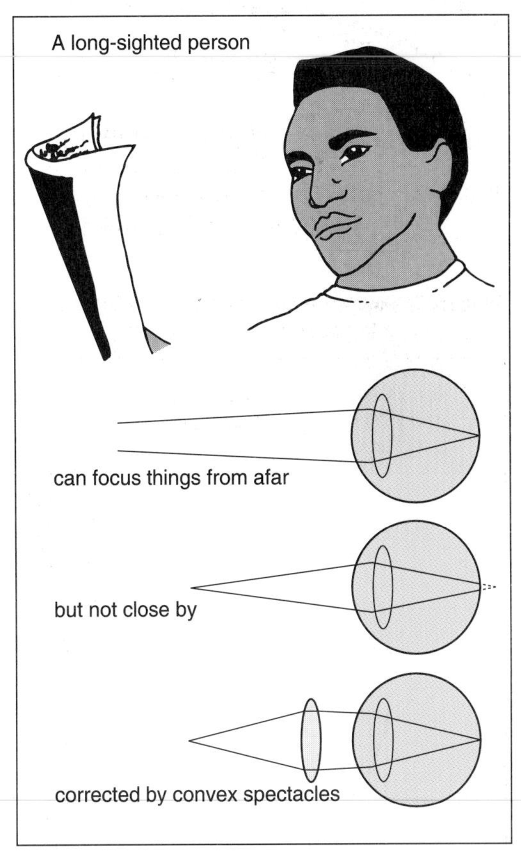

Picture 10 Long-sightedness and how it can be corrected by wearing glasses.

Three-dimensional vision

If you look at a chair, it appears to have depth. In other words you see it in three dimensions. You need two eyes for this. What happens is that each eye sees a slightly ifferent aspect of the chair. In the brain the two images are combined to give a single three-dimensional view of the chair. Seeing with two eyes is called **binocular vision**.

Binocular vision gives us a more complete view of our environment, and it helps us to judge distances. For example, if you're cycling along the road and there's a car in front of you, you know roughly how far away it is.

Activities

A The pupil reflex

Work in pairs, with one person acting as observer and the other as subject. This is what the observer should do:

1 Get the subject to close his/her eyes for ten seconds, then open them. What happens to the pupils when the eye opens?

2 Shine a torch in the subject's eye and watch the pupil. What happens to the pupil? Explain your observations.

3 Get the subject to look at an object in the distance and then nearby. What happens to the pupil when the subject does this, and why?

B Demonstrating the blind spot

1 Look at the picture below: hold it about 10 cm from your eyes.

2 Close your left eye, and look at the house with your right eye.

3 Slowly move the picture away from your eyes, keeping your right eye focused on the house all the time.

What happens to the ghost as you move the picture away from you?

How would you explain this?

4 Repeat the experiment with both eyes open.

What happens this time?

How would you explain the difference?

C How good is your eyesight?

1 On a white card draw two parallel lines one millimetre apart. Hang the card on the wall.

2 Gradually back away from the card until the two parallel lines appear as one, then stop.

How far are you from the card?

Compare your distance with that of other people in the class.

D Seeing colours

1 Obtain two cards, one red and the other green.

2 Look at the two cards out of the corner of your eye.

Can you tell which colour is which?

How would you explain your observation?

3 Obtain a set of colour blindness test cards.

Test your eyes with the cards, following the instructions carefully.

Can you see colours normally, or are you colour blind?

If you are colour blind, are you totally colour blind or are you colour blind only to red and green?

E How the brain can help

With a piece of straight-edged paper cover the top half of the following phrase:

HAPPY BIRTHDAY

Can you read it?

Now cover the bottom half of the phrase. Can you read it now?

This experiment tells us something about the part played by our brain when we look at things.

Try to explain your result.

Questions

1 The picture below shows a person wearing 'half-moon' spectacles. What sort of eye defect do you think he has, and why are these particular spectacles useful to him?

2 What are the advantages of having two eyes rather than only one?

3 Explain the reason for each of the following:

a When you go into a cinema from bright sunlight, you cannot see the seats at first, but gradually they become visible.

b If you are trying to see a faint star in the night sky, it is better to look slightly to one side of it rather than straight at it.

c When it is getting dark at night, it is impossible to make out the colours of cars on the road.

d If you look at a cinema screen out of the corner of your eye, you can see it flickering.

e If both your eyes are open and you press the side of one of your eyeballs, you see double.

4 Nocturnal animals, i.e. animals which sleep during the day and come out at night, tend to have wide pupils and lots of rods in their retinas. Suggest a reason for this.

5 Why is it important for a person to know if he or she is colour blind?

People who are red-green colour blind say that they have no difficulty telling whether the traffic lights are red or green. How would you explain this?

6 The cornea does not contain any blood vessels; it gets its nourishment from the fluid in the front of the eyeball. Use this information to explain the following fact: *a person can be given a corneal transplant with no fear of it being rejected.* (Hint: use the index to find out what causes organ transplants to be rejected.)

Living in darkness

Close your eyes for at least one minute, and imagine what it must be like to be blind. Suddenly the sense organs which make you most aware of your environment stop working.

Although blind people have serious problems, most of them manage very well. One reason for this is that they make much more use of their other senses, particularly hearing and touch. In blind people these senses become very well developed.

Take touch, for example. Blind people identify things by feeling them. Their fingertips become very sensitive, and this enables them to read Braille. This system was invented in France by Louis Braille. Each letter of the alphabet is represented by a character consisting of one to six dots embossed on thick paper (see picture 1).

Picture 1 A blind person reading Braille.

Another system was developed by an Englishman called Dr Moon. In this case the letters are represented not by dots but by shapes. They are easier to feel and to learn, but they take up more room.

1. Close your eyes. Your teacher will give you a flat shape made of wire. Feel it. Your teacher will then take it away. Open your eyes and try drawing the shape.
2. With your eyes closed, feel the characters on a Braille card with your fingertips. Do you find it difficult to tell the difference between the various characters?

   Repeat the process with a sheet of Moon. Are the characters easier to tell apart than the Braille characters? If so, why? How could you prove that they are?
3. People who go blind when they are young usually learn Braille, but elderly people who go blind usually learn Moon. Why the difference?

Lasers and the eye

A laser is an instrument which produces very intense beams of light. There are many types of laser, and they produce light at different frequencies. Laser beams can be used for all sorts of purposes. For example, they may be used to destroy or cut tissues, or to get tissues to stick together.

The eye was the first organ in the human body to be treated with lasers (see pictures 1 and 2). Let's look at an example. Sometimes a hole develops in a person's retina. Fluid in the eyeball gets through the hole and lifts the retina away from the tissue underneath. This is called a **detached retina**. If nothing is done about it, the person may go partially blind.

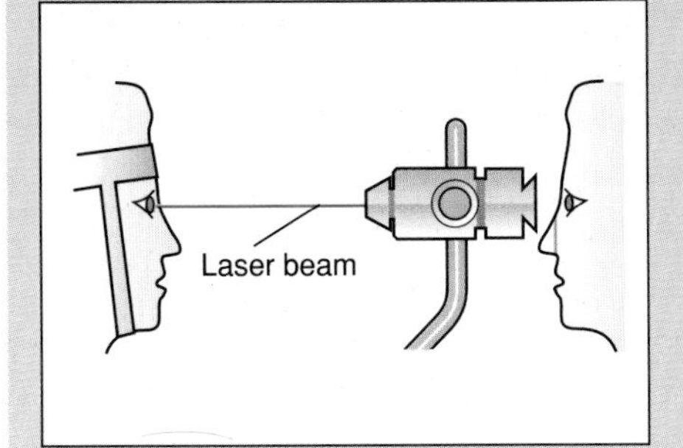

Picture 1 This diagram shows a laser beam being shone into a patient's eye. The patient wears a special kind of contact lens which allows the doctor to see into the eye and focus the laser beam on exactly the right part.

Detachment of the retina can be prevented by laser treatment. Short pulses from the laser are sent into the eye, one after the other, and focused on the retina. The pulses are not all directed at the same spot, but form a ring round the hole in the retina. It's like firing a gun at a target, but instead of aiming at the bull each time, you aim slightly to one side and make a circle round the bull. The laser beam makes the retina stick to the tissue behind it, and stops it coming off – it 'welds' it into position.

The beam is so fine that very few sensory cells in the retina are damaged – certainly not enough to affect the person's eyesight. The pulses are extremely quick – each one lasts only a tenth of a second, and you hardly feel anything at all.

But there's a snag. The treatment *must* be carried out before fluid has a chance to get behind the retina. Once the retina gets lifted off the underlying tissue, the laser can't stick it back again. The only remedy then is to perform an operation on the eye.

Lasers are a modern invention, but way back in 1949 a German professor made a sort of home-made laser to treat people with holes in the retina. He set up a tube with a system of mirrors to gather sunlight and focus it on the retina. One day he was looking down the tube and suddenly the sun caught one of the mirrors and burned the central part of his own retina. The result was that he became blind in one eye. This story illustrates the point that very intense light destroys living tissue.

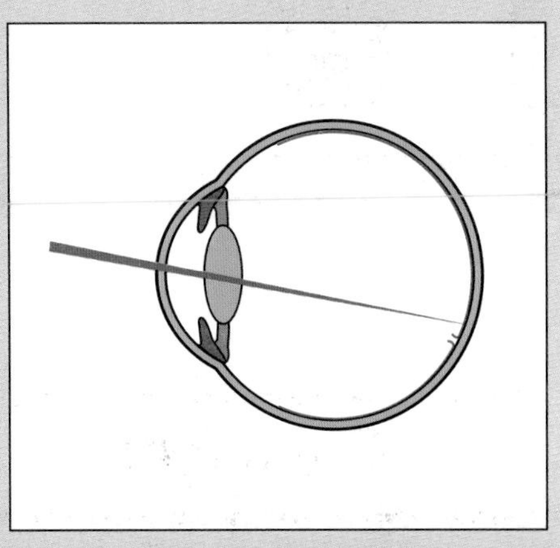

Picture 2 A laser beam being used to stick back a detached retina.

1. What advantages does a laser have over an operation for treating the eye? Think of as many advantages as you can.
2. Why do you think a laser beam makes the retina stick to the tissue underneath?
3. A person who develops a hole in the retina may not know that it has happened. How would you explain this?

D5
The ear and hearing

Block your ears with your fingers and imagine what it's like to live in silence.

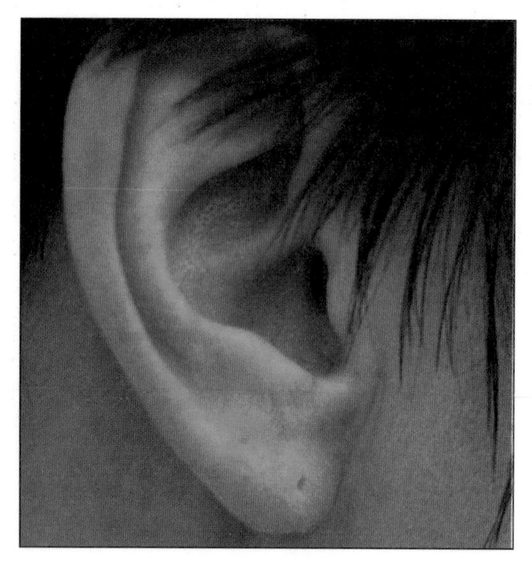

Picture 1 Our ears tell us a lot about what's going on around us.

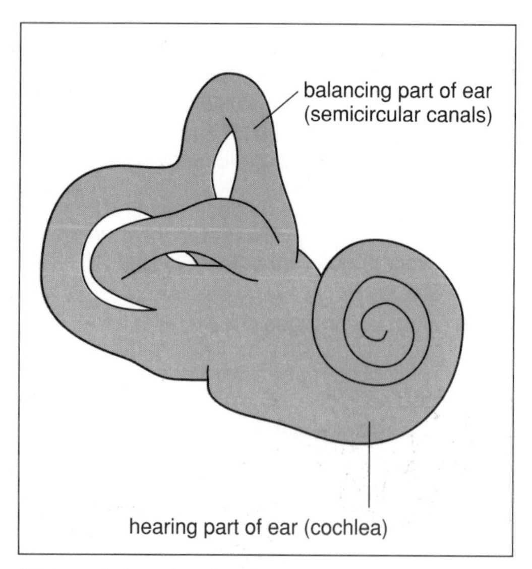

Picture 2 Imagine that the inner ear was taken out of your head. This is what it would look like.

The outside of the ear

Most people think of the ear as just a flap on the side of the head. But there is much more to it than that. The flap is simply a device for catching sounds and directing them into the **ear hole** in front. The flap is called the **pinna** and it contains gristle to keep it stiff.

Inside the ear

The 'business' part of the ear is embedded in the side of the head. Picture 2 shows it on its own outside the head. It is made up of two parts which do quite different jobs. One part helps us to keep our balance – we shall return to that later. The other part enables us to hear. The hearing part consists of a coiled tube rather like a snail's shell. It is called the **cochlea**. The ear hole is connected to the cochlea by a series of channels and chambers which are shown in picture 3.

Let's go on a guided tour of the ear, using picture 3 to help us. The hole leads into a short tube called the **outer ear channel**. The skin lining the first part of the channel secretes wax which catches germs and dust, preventing them from getting into the ear.

Stretched across the inner end of the channel is a tough membrane, the **ear drum**. On the other side of the ear drum is a chamber filled with air. It is called the **middle ear chamber**, and it contains three tiny bones called the **ear ossicles**. They are the smallest bones in the body. Because of their shape, they are called the **hammer**, **anvil** and **stirrup**. They run from the ear drum to a small hole on the other side of the middle ear chamber. This is called the **oval window**, and it leads to the cochlea which is part of the **inner ear**.

The cochlea is full of fluid and it contains receptor cells which are connected to the brain by the **auditory nerve**.

How does the ear hear?

It's Guy Fawkes night and there's a loud bang. The noise sets off **sound waves** which travel through the air. Within a fraction of a second the sound waves reach your ear, and the pinna directs them into the outer ear channel.

The sound waves pass along the channel to the ear drum. When they hit the ear drum, the drum vibrates. This moves the ear ossicles backwards and forwards, causing the foot of the stirrup in the oval window to vibrate. The vibrations of the stirrup then move the fluid in the cochlea.

The function of the ear ossicles is to transmit the vibrations of the ear drum to the cochlea. But they do more than this. They also amplify the vibrations. Why do you think this is necessary?

What happens in the cochlea?

Inside the cochlea there are two membranes stretched across from one side to the other. These membranes run the full length of the cochlea. You can see them in picture 3. The lower one has receptor cells attached to it.

Vibrations of the cochlea fluid make the cochlea membranes vibrate. When the lower membrane vibrates, it stimulates the receptor cells. The receptor cells then send off impulses in the auditory nerve. When the impulses reach the brain, we hear the sound.

That's not quite the end of the story. Look once more at picture 3. You'll see that between the middle ear chamber and the cochlea there is a hole called the **round window**. You may have wondered what it's for. Obviously the pressure which develops in the cochlea fluid has got to be taken up by something. It's taken up by the membrane covering this hole.

We can sum up by saying that the sound waves make the ear membranes vibrate, and the movements are then changed, i.e. transduced, into electrical signals which are sent to the brain.

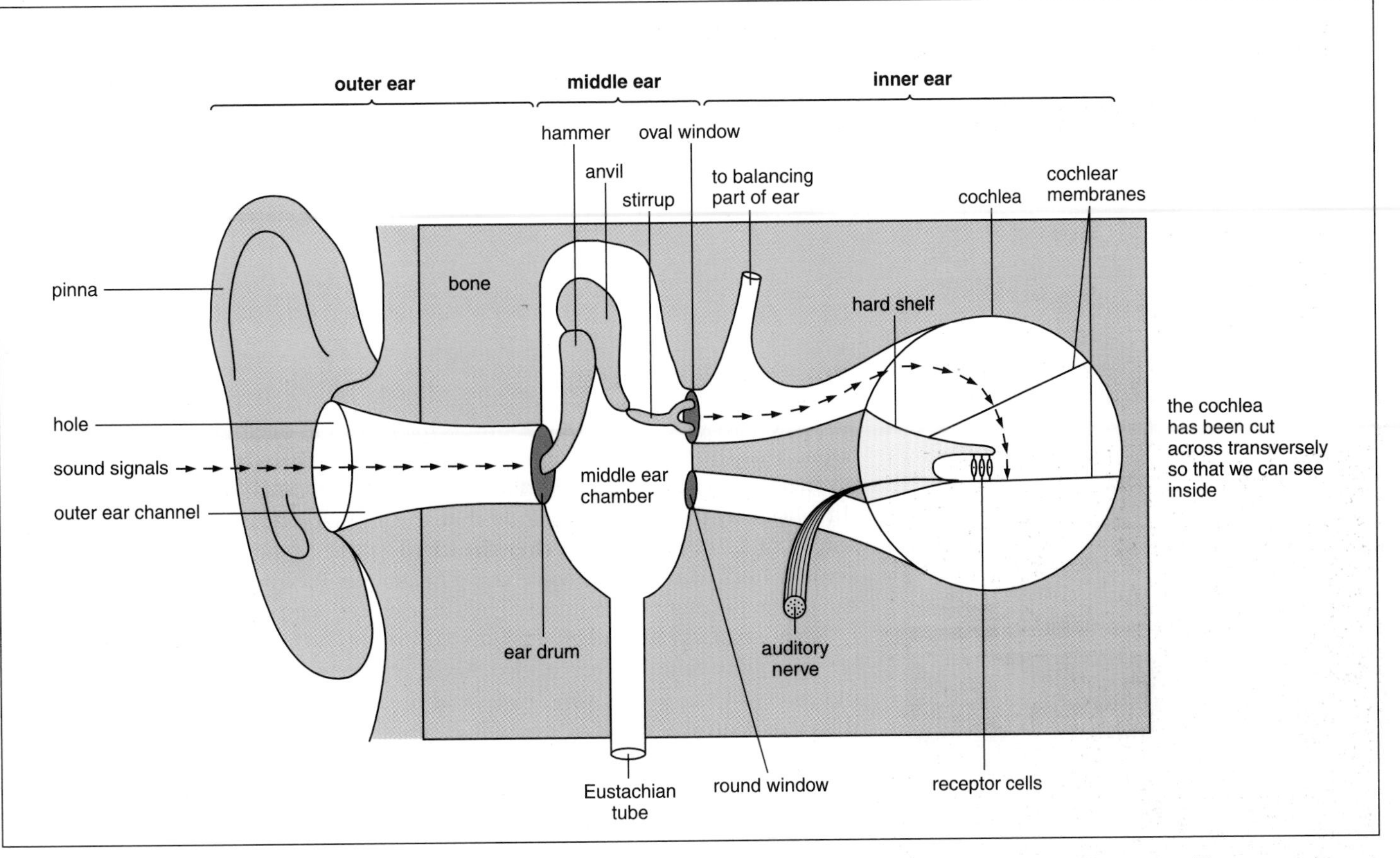

Picture 3 Inside the human ear. The arrows show how sound waves are transmitted to the receptor cells in the cochlea.

Telling the difference between loud and soft sounds

If you play a note on a guitar, its loudness depends on how hard you pluck the wire. The harder you pluck, the greater is the distance through which the wire vibrates and the louder is the sound.

The loudness of a sound is registered by the ear in the same way. Soft sounds cause small vibrations of the cochlea membranes. Louder sounds cause larger vibrations.

Telling the difference between high and low notes

With a guitar, the note depends on how rapidly the wire vibrates – in other words, the frequency. High frequency vibrations give high notes, whereas low frequency vibrations give low notes.

Although the details are different, the ear works in the same kind of way. The membrane to which the receptor cells are attached vibrates at different frequencies in different parts of the cochlea. In other words, the membrane **resonates**, and this enables different notes to be heard.

How can you tell where a sound comes from?

Normally when you hear a sound, you know where it comes from. This is because you have two ears, one on each side of the head.

Suppose you hear a sound from the right. Sound waves reach the right ear a fraction of a second before they reach the left ear (picture 4). The result is that nerve impulses are sent to the brain from the right ear slightly before they are sent from the left ear. From this the brain knows that the sound must have come from the right. Although other effects play a part, this is the basis of how we tell where sounds come from, and how we appreciate stereo music.

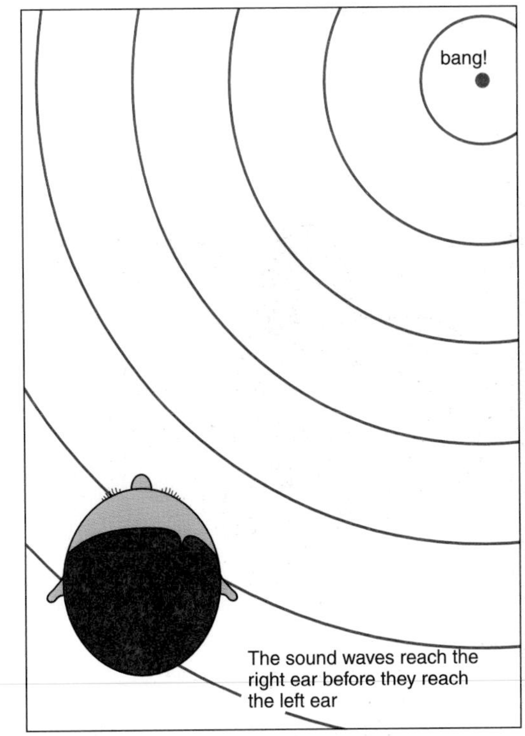

Picture 4 Where did that bang come from?

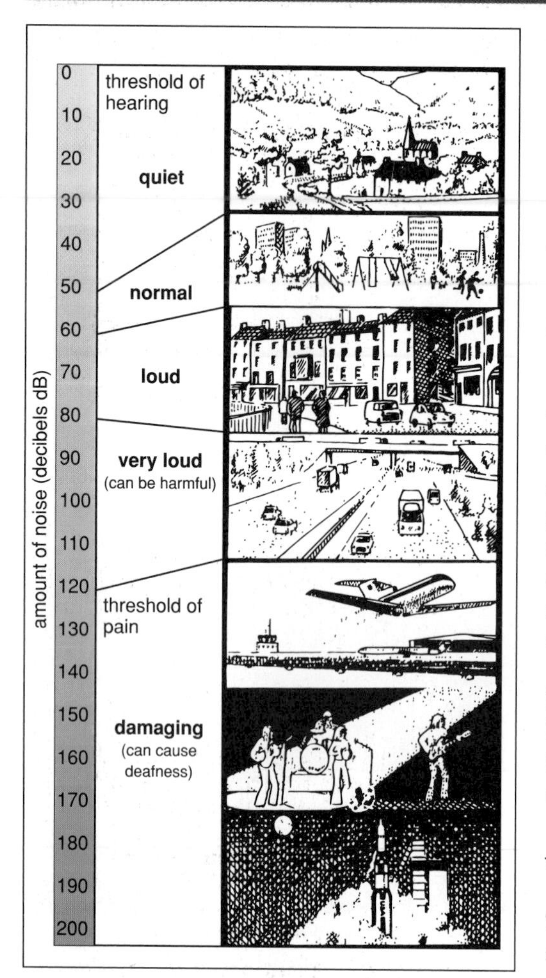

Picture 5 This illustration shows the noise scale as expressed in decibels, the standard unit of noise as measured with a sound meter.

The audible range

The frequency (pitch) of a sound is measured in cycles per second or Hertz (Hz). The human ear can detect frequencies from about 20 Hz (very low notes) to about 20 000 Hz (very high notes). This is called the **audible range**. Animals such as dogs and cats can hear higher notes than we can, and bats can hear ultrasonic sounds at a frequency of 100 000 Hz! You can hear high notes best when you are young. Elderly people lose this ability.

Within the audible range an average person can distinguish between about 2000 different notes, though a trained musician can do better than this. We are best at distinguishing between notes in the 1000 to 4000 Hz range.

How about intensity?

The intensity (loudness) of a sound is measured in decibels (dB). The quietest sound that the human ear can detect is called the **threshold of hearing** and is given a value of zero decibels. Picture 5 shows the range of intensities from the threshold to the loudest sounds that the human ear can bear. Notice that a sound of 120 dB is on the **threshold of pain**. Sounds louder than this can actually hurt your ears and give you a headache. Prolonged noise above 150 dB can cause permanent deafness. The reason is explained below.

Because of its harmful effects, it is important for noise in the environment to be conrolled. Sound levels above 90 dB are not normally allowed in factories, but many of the noises which we hear in our everyday lives are much louder than this. Can you think of examples? People who work close to noisy machinery wear ear plugs. Perhaps we should all wear ear plugs!

What causes deafness?

There are several types of deafness, depending on which part of the ear is affected.

■ Outer ear deafness

Lots of people become slightly deaf from time to time because the outer ear channel gets blocked with hard wax. This is easily removed by the doctor syringing out the ears with warm water. It helps if the wax is first softened by putting a few drops of olive oil into the ear several days beforehand.

An explosion, or a blow on the side of the head, may rupture the ear drum, causing partial or complete deafness. However, the ear drum usually heals quite quickly and then hearing returns.

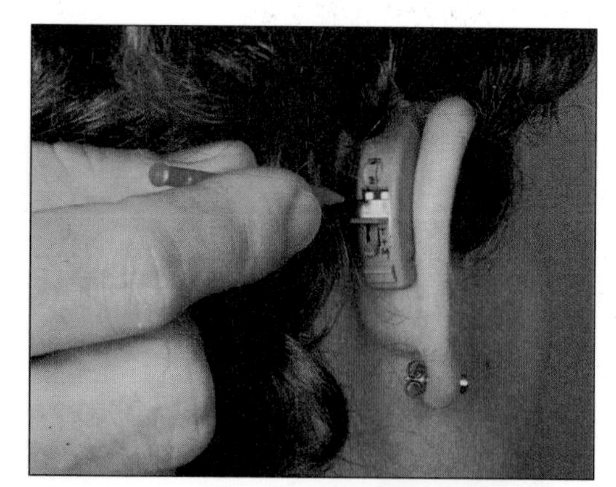

Picture 6 A hearing aid. The case contains a microphone, battery and earphone. The case fits neatly behind the ear. Sound waves are transmitted through a plastic tube to a mould which is placed over the opening of the ear.

■ Middle ear deafness

More serious deafness is caused by bone tissue growing round the stirrup in the middle ear chamber. This can prevent the stirrup moving, in much the same way as a piston may seize up with rust. If nothing is done about it, this can lead to permanent deafness. However, the person's hearing may be improved by wearing a hearing aid which amplifies the sound waves (picture 6). In severe cases the stirrup may be replaced by an artificial one made of plastic. This type of deafness runs in families, and it can begin when you are quite young.

■ Inner ear deafness

Sometimes deafness is caused by the cochlea not working properly. For example, suppose you listen to a very loud sound of a particular pitch for a long time. The cochlea membrane vibrates so much that eventually the receptor cells which detect that particular frequency get damaged. The result is that you become deaf to that particular note. Some pop singers have become deaf to certain notes because of this; so have young people who listen to very loud music through headphones. There is no cure for this kind of deafness.

People often get deaf as they grow old. This is usually caused by the auditory nerve failing to carry impulses to the brain in the usual way.

Activities

A Experiments on hearing

These experiments involve using a signal generator.

1 Your teacher will use a signal generator to compare the audible ranges of people in your class. How do people differ in their ranges, and why? (Hint: read page 284). Do you think it matters?

2 Plan an experiment to find out if a person's threshold of hearing (i.e. the quietest sound which he or she can hear) depends on the frequency of the sound?

If facilities permit, your teacher will help you to carry out the experiment.

3 It is said that females can hear higher notes than males. With the help of your teacher, test this idea on your class.

B Comparing the noise levels in different places

Using a sound meter, find the maximum amount of noise above the hearing threshold in different places such as a street corner, railway station, airport, children's playground, school dining hall, reference library, motorway, factory, park, disco.

Compare your results with picture 5, and decide whether each place is quiet, normal, loud, very loud or damaging. Do you have any difficulty in deciding? If so, why?

Questions

1 What jobs are done by each of these: the ear drum, the oval window, the receptor cells in the cochlea, the auditory nerve?

2 What is the pinna, and what job does it do? The pinna of an Alsation dog is more efficient than the pinna of a human. What makes it more efficient?

3 People who drill holes in the road or work in very noisy factories should wear ear muffs. Why?

4 Suppose someone became deaf to low notes but not to high notes.

a This is unlikely to have been caused by a ruptured ear drum. Why?

b What would be the most likely cause? Explain your answer.

5 To enjoy listening to music with stereophonic sound, we need two ears.

a How could you show that this is true?

b What is the explanation?

6 The middle ear chamber contains three ear bones. Why three rather than one? (To answer this question you'll need to find out exactly what the ear bones do. This will mean reading about them in more advanced books.)

Balance

As well as enabling us to hear, the ear helps us to keep our balance. This function is mainly carried out by three **semicircular canals** located close to the cochlea. You can see them in picture 2 on page 208.

The canals are filled with fluid, and they contain receptors which are stimulated when moved. If you move your head the fluid pulls on the receptors. As a result messages are sent to the brain. The brain then causes certain reflexes to take place so that you keep your balance and don't fall over.

The picture on the right shows the positions of the three canals. Notice that they are at right angles to each other. This means that movement of the head in any plane can be detected by the receptors.

1 What sort of reflexes take place when the receptors are pulled, and how do they help you to keep your balance?

2 Why do you think you feel dizzy after going on a roundabout?

3 What part do our eyes play in helping us to keep our balance?

4 We have receptors in the soles of our feet which are sensitive to pressure. How might these pressure receptors help us to keep our balance?

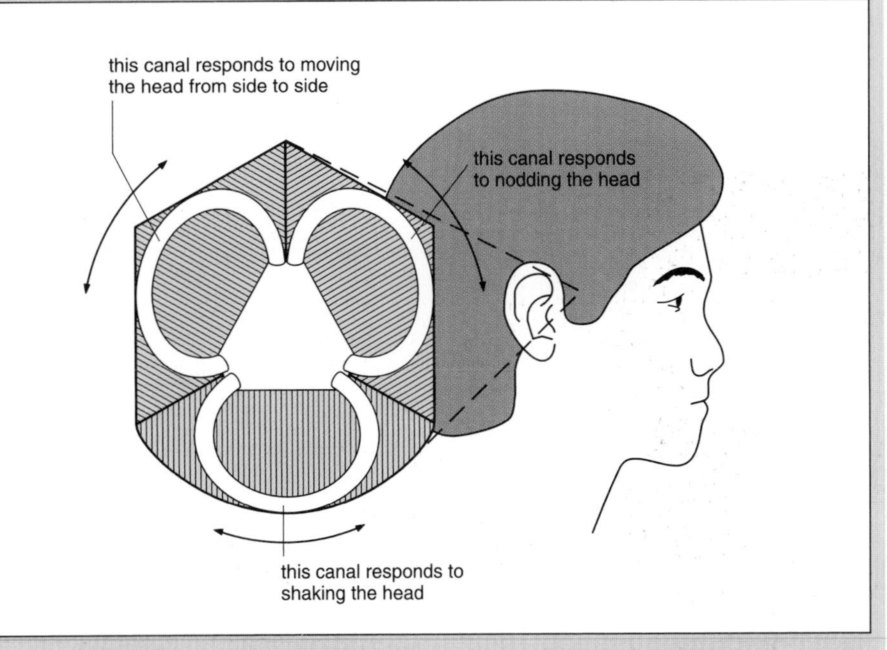

Picture 1 The three semicircular canals are at right angles to each other, so movement of the head in any plane can be detected by the sense organs.

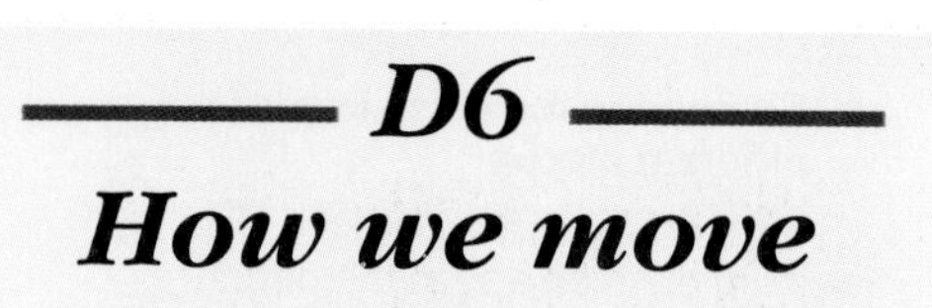

Can you imagine what life would be like if you couldn't move?

What do we need in order to move?

To move we need three basic things: **nerves**, **muscles** and a **skeleton**. The nerves carry messages (nerve impulses) to the muscles, which respond by shortening. The shortening of a muscle is called **contraction**.

Our muscles are attached to the skeleton. The skeleton is made mainly of **bone**. This is a living tissue, and it is very hard. It is hard because it contains minerals. The main mineral is calcium phosphate.

The individual bones that make up the skeleton are held together by **ligaments**. Ligaments are tough but elastic, so they will stretch when pulled.

A muscle is made up of hundreds of thread-like **muscle fibres** each of which can contract individually. The muscle is attached to the skeleton at both ends by a **tendon**. The tendons are tough and not very elastic, so they don't stretch much when they are pulled.

When a muscle contracts, it may pull one part of the skeleton towards another. To see how this produces movement, let's look at the arm.

Picture 1 A ballet dancer caught in mid-action.

The arm, an example of how we move

Two main muscles move the arm: the **biceps** and the **triceps**. The biceps bends the arm at the elbow: it is a **flexor muscle**. The triceps straightens the arm: it is an **extensor muscle** (picture 2). These two muscles produce opposite effects, so obviously they must not contract at the same time. If they did, the arm wouldn't move at all! The nervous system makes sure that this doesn't happen. Each muscle has its own nerve supply. when messages are sent to the biceps telling it to contract, they stop being sent to the triceps – and vice versa. This is achieved by the brain which coordinates the actions of the muscles (see page 196). Muscles which produce opposite effects are called **antagonistic muscles**.

The biceps and triceps muscles simply bend and straighten the arm. Other muscles move the arm in other directions. The same principle applies to the legs. Together, these muscles enable us to walk, run, play games and do all sorts of other things.

For our muscles to do their job properly, the bones must move easily against each other, i.e. **articulate** smoothly. This happens at the **joints**.

Picture 2 The biceps and triceps muscles move the arm at the elbow joint.

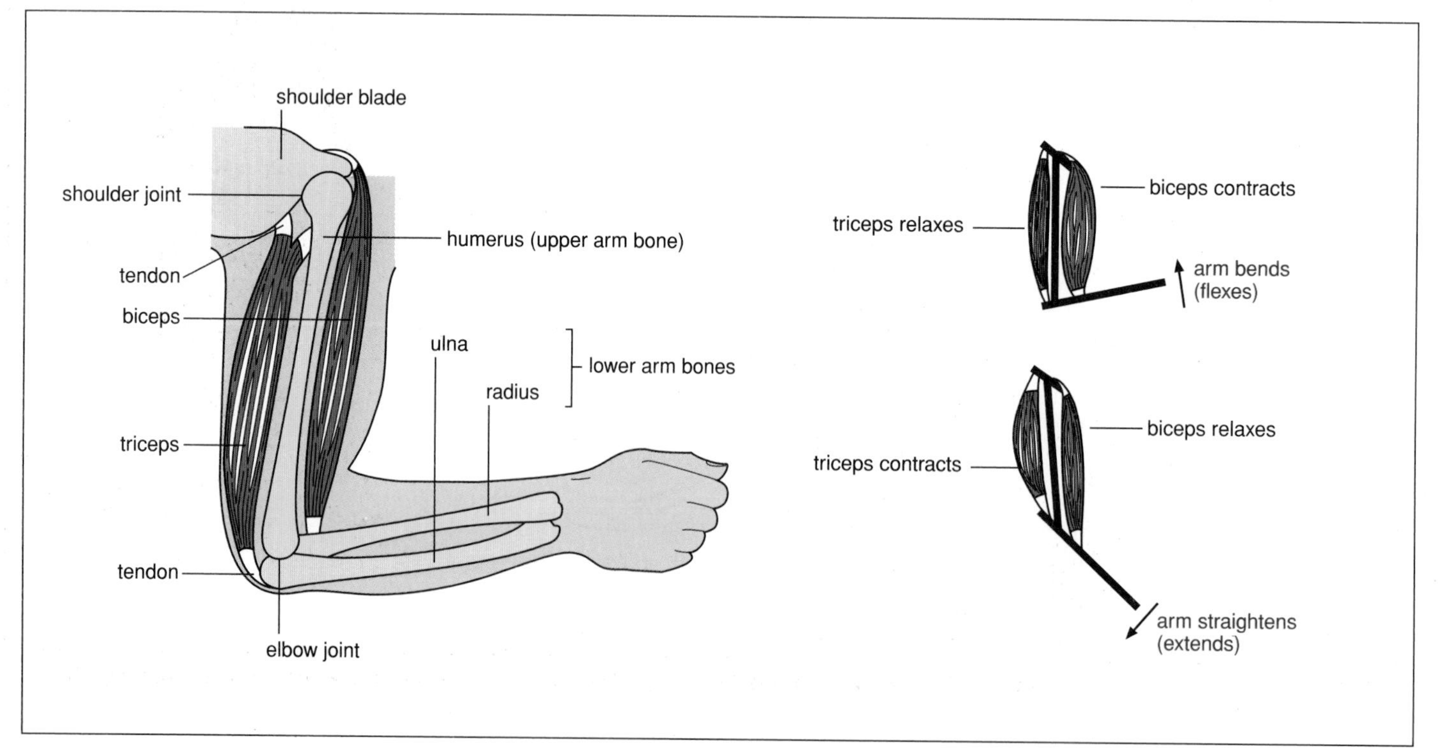

Joints, movement with minimum friction

A joint is shown in picture 3. Notice how the two bones fit together. The two bones are separated by a special fluid called **synovial fluid**. This serves as a lubricant, enabling the knob to move smoothly inside the socket with very little friction. It's like the oil between the moving parts of a machine. Synovial fluid is the best lubricant in the world: manufacturers of artificial lubricants have never been able to better it.

The ends of the two bones are made of **cartilage**. This is softer than bone and slightly springy. It helps to prevent jarring when the two bones move against each other.

Joints are weak points in the skeleton, and they must be protected. The joint is surrounded by a tough **fibrous capsule**. Knees are particularly vulnerable because of their exposed position. The knee is protected in front by a small bone called the **knee cap**. Behind and in front of the knee cap there are cavities filled with synovial fluid which cushion the knee.

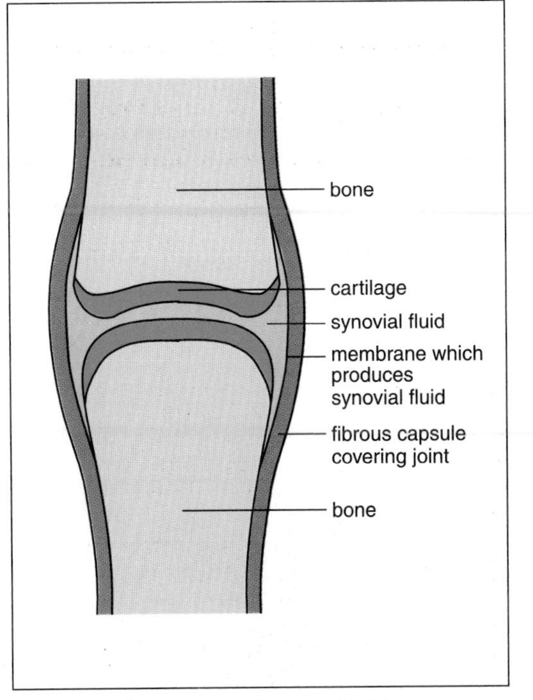

Picture 3 The structure of a typical joint.

Different kinds of joints

The hip joint is called a **ball and socket joint** (picture 4A). This kind of joint allows movement in any direction. Test this for yourself by standing up and moving your leg around at the hip. You can move it forwards and backwards and from side to side, and you can rotate it.

Now try bending your leg at the knee. You will find that you can bend it in only one direction: backwards and forwards. This is because of the way the knee joint is constructed. It consists of two knobs which fit into two cups (picture 4B). This is called a **hinge joint** because it operates rather like the hinge on a door.

What sort of joint is the shoulder joint? And the elbow? How can you tell what sorts of joints these are?

Holding the bones together

If the bones were not held together in some way, they would fall apart at the joints. They are held together by the fibrous capsules which cover the joints, and by the muscles and ligaments which run from one bone to another. Some of the ligaments are inside the joints, others outside. You can see ligaments inside the joints in picture 4.

The ligaments limit the amount of movement which is possible at a joint. Collectively, though, the joints enable the human skeleton to be amazingly flexible. Flexibility of the body is an aspect of fitness (see page 162).

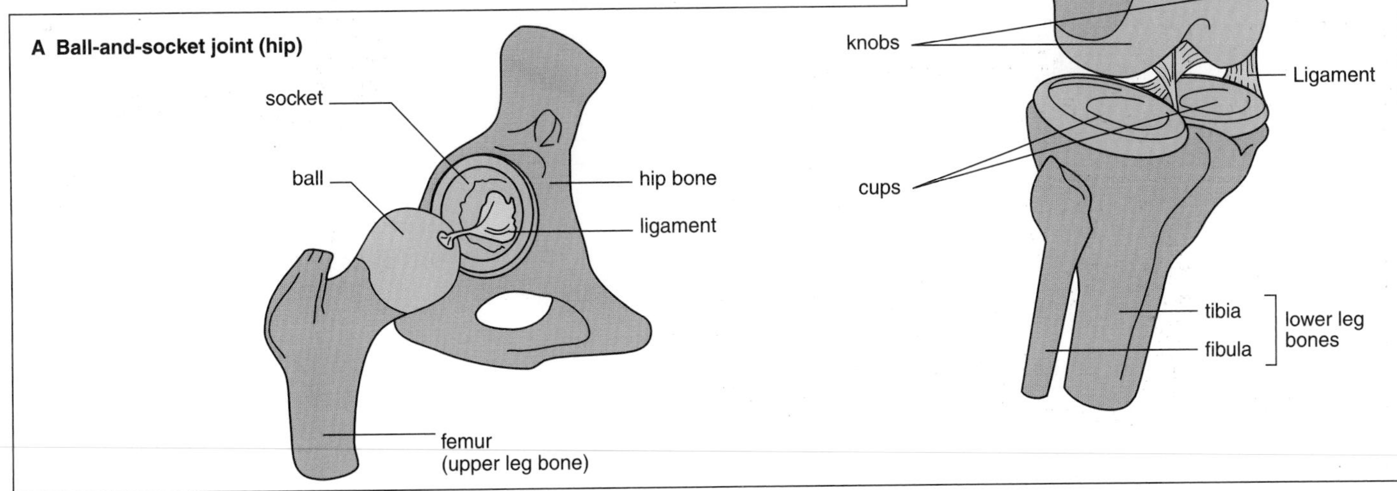

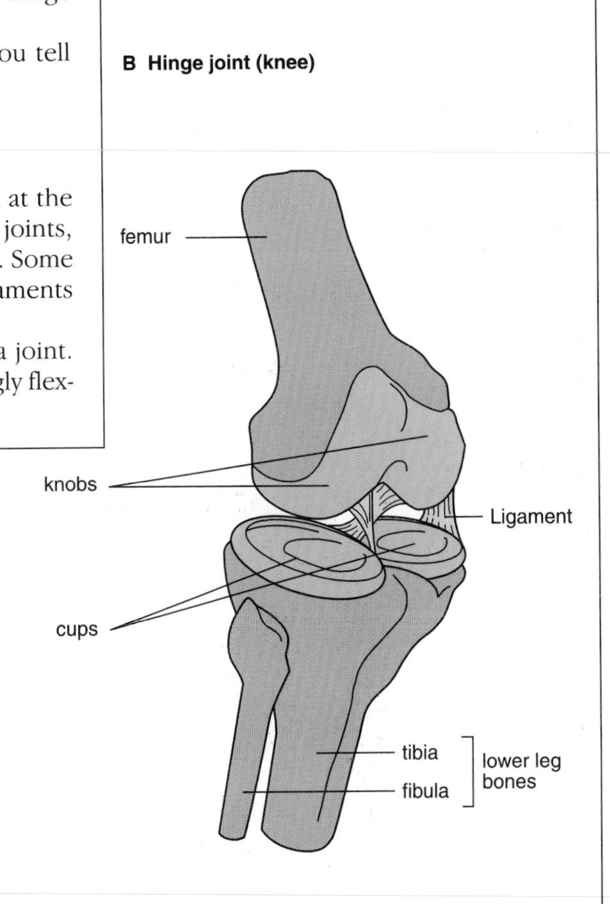

Picture 4 A ball and socket joint and a hinge joint. In each case the fibrous capsule has been removed and the two bones pulled apart slightly.

Questions

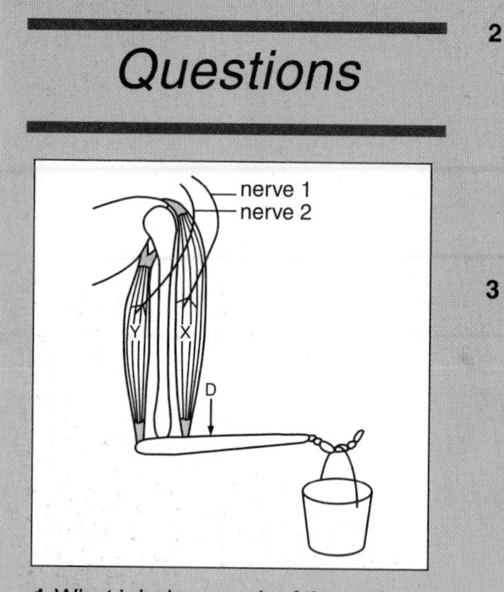

1 What job does each of these do: muscle, tendon, ligament, joint.

2 The picture shows some of the muscles, bones and nerves in the human arm. When muscle X contracts, muscle Y relaxes. Explain briefly how this is achieved. Include the following words in your account: nerve 1, nerve 2, brain, spinal cord, coordination, coordinator.

3 In lifting the bucket the arm in the picture is working as a lever with a load, effort and pivot.

a With the help of diagrams, explain how the arm works as a lever.

b If muscle X was attached to the lower arm bone at point D, would it require more or less effort to raise the bucket?

c Why do you think muscle X is attached so close to the elbow joint?

Activity

The human skeleton and how it works

Your teacher will show you a human skeleton or a model of one. Use the diagrams below to help you find these parts: the skull, lower jaw bone, vertebral column (backbone) made up of vertebrae, ribs, breast bone, shoulder girdle, collar bone, hip girdle (pelvis), arm bones (humerus, radius and ulna), and leg bones (femur, tibia and fibula).

Now move the arms and legs at the joints in the same way as you can move your own arms and legs. Work out where the muscles would have to be attached to the skeleton to produce these movements.

Make a diagram to show where you think the muscles should be.

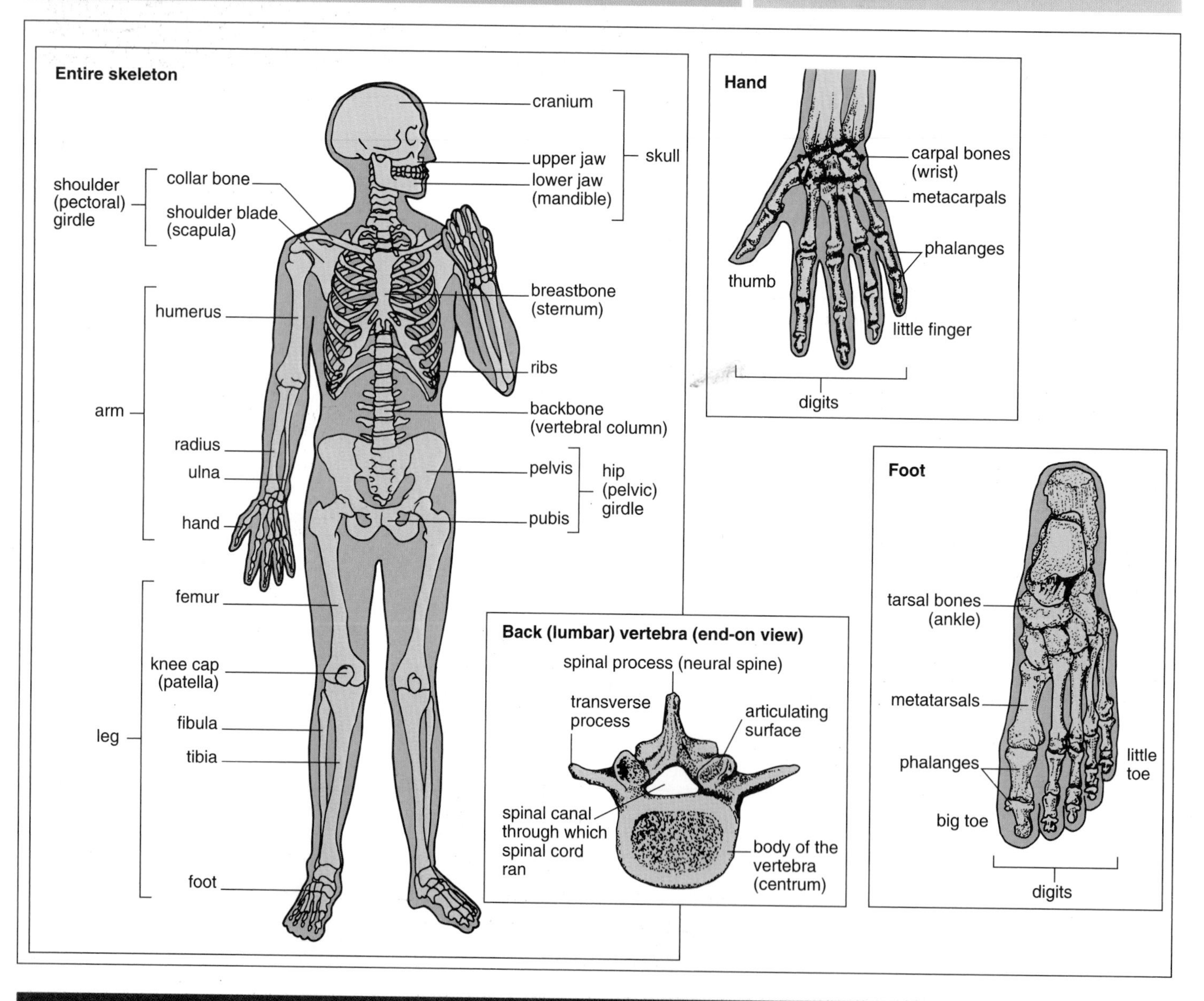

The structure of a bone

Picture 1 shows the top part of the femur. It consists of two main parts, the **shaft** and the **head**. The head articulates with the hip bone.

The outer part of the femur consists of dense bone tissue which is particularly thick in the shaft. The head contains a network of bony fibres which do the same job as the metal lattice in a crane. They enable the head to bear the full mass of the body.

In the centre of the bone there is a cavity filled with **bone marrow**. The marrow in the shaft consists mainly of fat. The marrow at the two ends of the bone is **red bone marrow** where blood cells are made (see page 157).

1 Draw a cross section of a femur through the shaft, and another one through the head.

2 What makes the femur so strong?

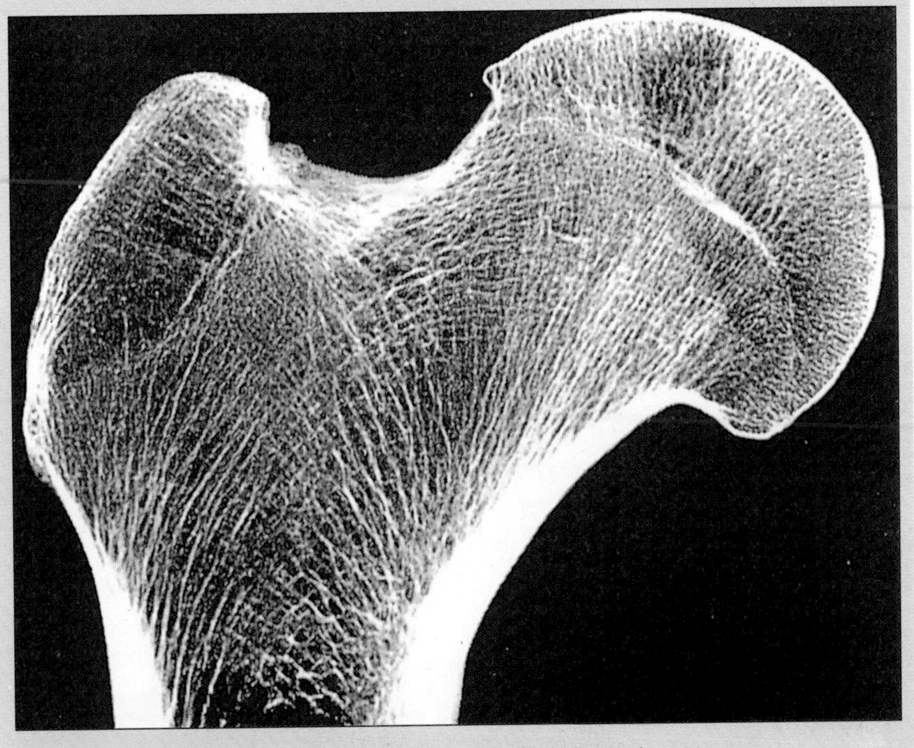

Picture 1 Radiograph of the top part of a femur showing its internal structure.

How a muscle contracts

The picture shows the inside of a muscle. The whole muscle is made up of **muscle fibres**. These in turn are composed of very fine strands called **muscle fibrils.**

By using the electron microscope scientists have shown that each muscle fibril is made up of bundles of extremely fine **filaments**. Some of these are made of a protein called **actin**, others of a protein called **myosin**. When a muscle contracts the actin and myosin filaments slide between each other, as shown at the bottom of the picture. The result is that the whole muscle shortens or develops tension.

1 Say what you can about the possible mechanism by which the myosin and actin filaments in a muscle are made to slide as shown in the picture.

2 Give examples of when and where in your body muscles develop tension but do not shorten.

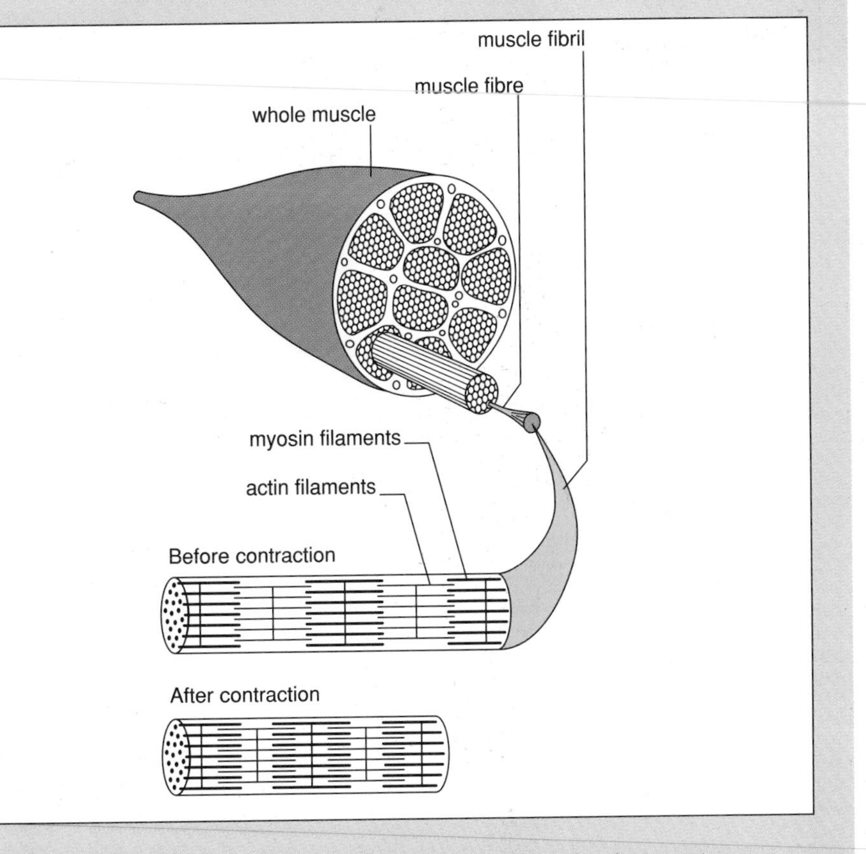

Picture 1 The internal structure of a muscle and what happens when it contracts.

Aches, pains and broken bones

A break in a bone is called a **fracture**. There are many kinds of fracture. Some are shown in picture 1.

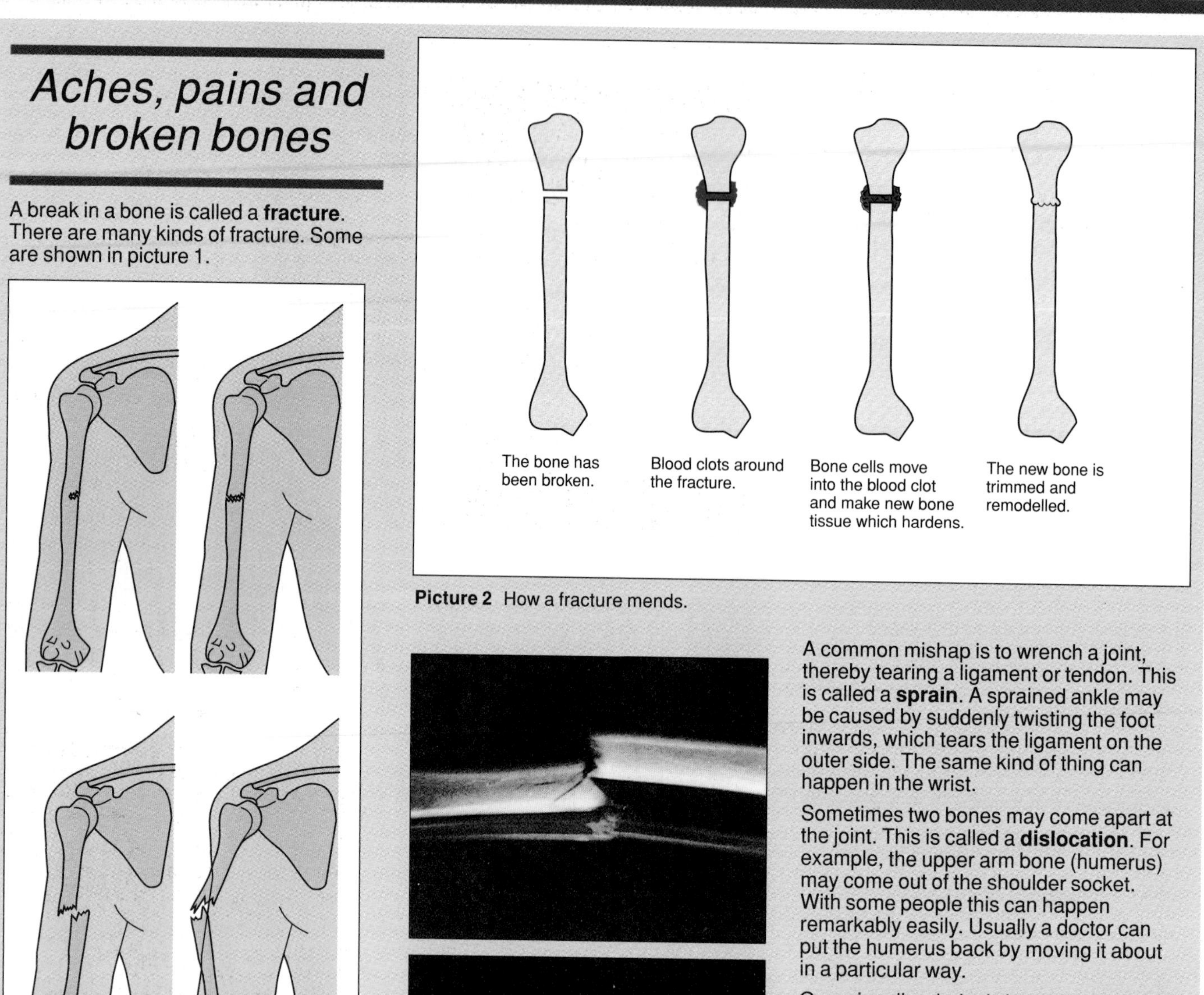

Picture 1 Four ways in which the upper arm bone (humerus) can be fractured.

Picture 2 How a fracture mends.

Picture 3 The top X-ray shows a fractured tibia and fibula (the lower leg bones) just after a car accident. The bottom X-ray shows the same bones six months later, after the leg had been in plaster.

Let's take the type of fracture where a limb bone breaks right across. Picture 2 shows how such a fracture mends. For the bone to heal neatly the two ends must be in contact but not allowed to move against each other. This is achieved by putting the limb in plaster or holding it in position with a splint.

Fractures heal much more slowly than cuts in the skin. This is because it takes a long time for bone tissue to grow and harden. However, if all goes well the final mend is almost undetectable in an X-ray (picture 3).

A common mishap is to wrench a joint, thereby tearing a ligament or tendon. This is called a **sprain**. A sprained ankle may be caused by suddenly twisting the foot inwards, which tears the ligament on the outer side. The same kind of thing can happen in the wrist.

Sometimes two bones may come apart at the joint. This is called a **dislocation**. For example, the upper arm bone (humerus) may come out of the shoulder socket. With some people this can happen remarkably easily. Usually a doctor can put the humerus back by moving it about in a particular way.

Occasionally a baby is born with the head of the femur outside the hip socket (picture 4). In this case the doctor moves the legs about in such a way that the head of the femur is forced into its socket. The child is then put in plaster with the legs pushed wide apart. After many months the plaster is removed and the head of the femur stays in its socket.

Dislocated hips run in families. Nowadays a simple test is carried out on babies immediately after birth to see if their hip joints are working properly.

Many people suffer from **arthritis**. In one type of arthritis the cartilage wears away and the joint surfaces become roughened. In another type of arthritis fibrous tissue grows over the joint surfaces and destroys them. Either way, movement becomes painful and difficult.

Most of us get **cramp** from time to time. This is caused by a muscle suddenly contracting so powerfully that it hurts. Cramp is brought on by cold, or by using a muscle a great deal. **Stitch** is a type of

(*Continued on next page*)

cramp which occurs in the abdominal muscles, usually after a hard bout of exercise.

1 Look at picture 4. Make a drawing of a normal hip joint.

2 What sort of injuries are likely to arise from different sports, and why?

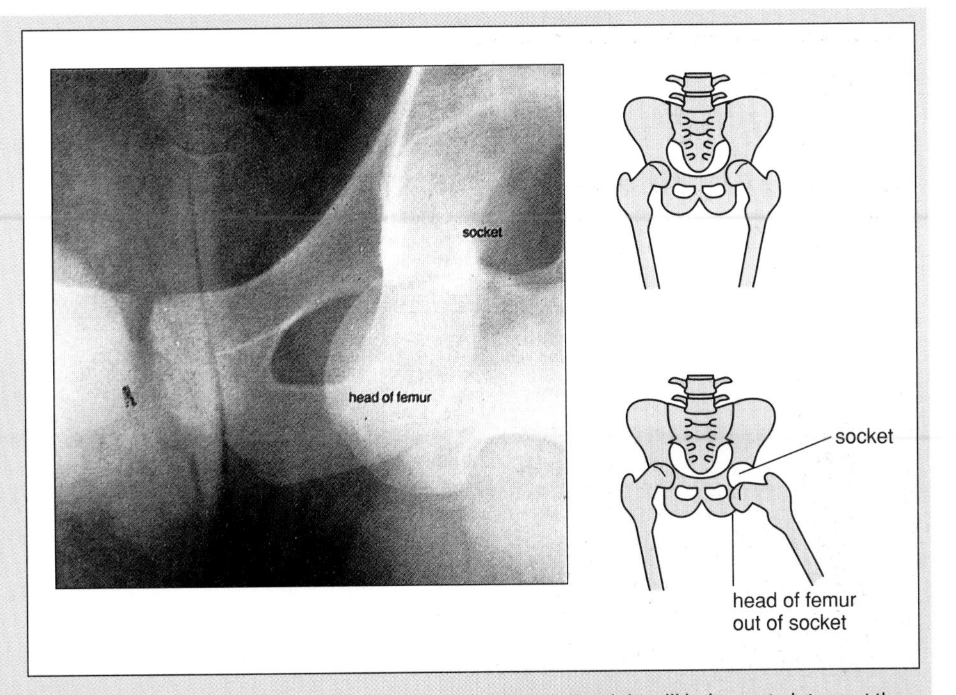

Picture 4 X-ray of a dislocated hip joint. The diagram on the right will help you to interpret the X-ray.

Replacement surgery

You probably know someone who has had a hip or knee replacement operation because they were suffering from severe arthritis.

Hip replacement is one of the commonest types of **replacement surgery**. In the operation the surgeon cuts off the head of the femur and replaces it with a stainless steel ball. The ball is held in place by a metal rod which is pushed into the marrow cavity of the bone and fixed securely with an acrylic cement.

The socket in the hip bone is usually lined with a plastic cup, and this too is fixed securely with an acrylic cement. The stainless steel ball is then inserted into the plastic cup in which it should fit snugly.

A person's life can be transformed by a hip replacement. People who could barely get around with two sticks find they can walk, and even run, again – and without pain.

1 What sort of things might go wrong during, or after, a hip replacement operation? Think of as many possibilities as you can.

2 A knee replacement is more difficult to do than a hip replacement. Why do you think this is?

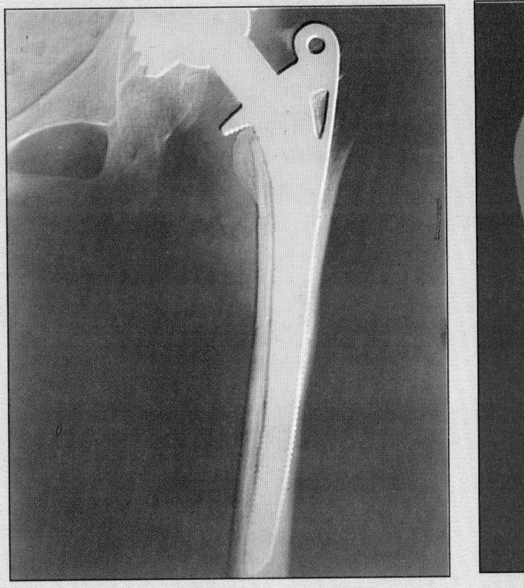

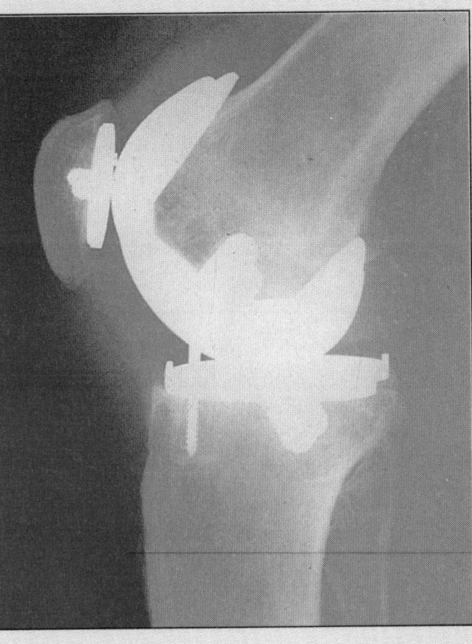

Picture 1 X-rays of an artificial hip joint (far left) and an artificial knee.

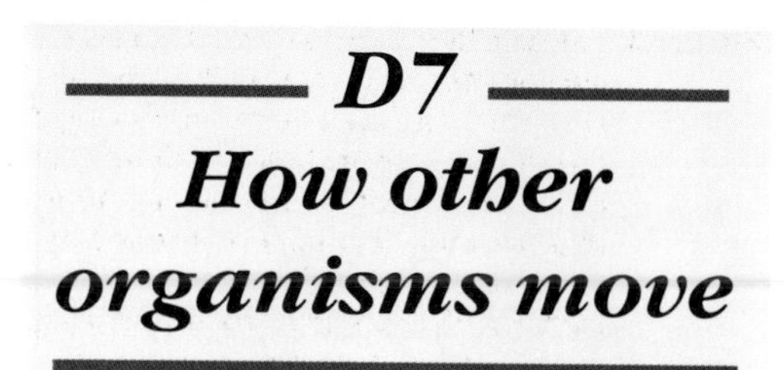

D7 How other organisms move

Organisms move in different ways, and some don't move at all.

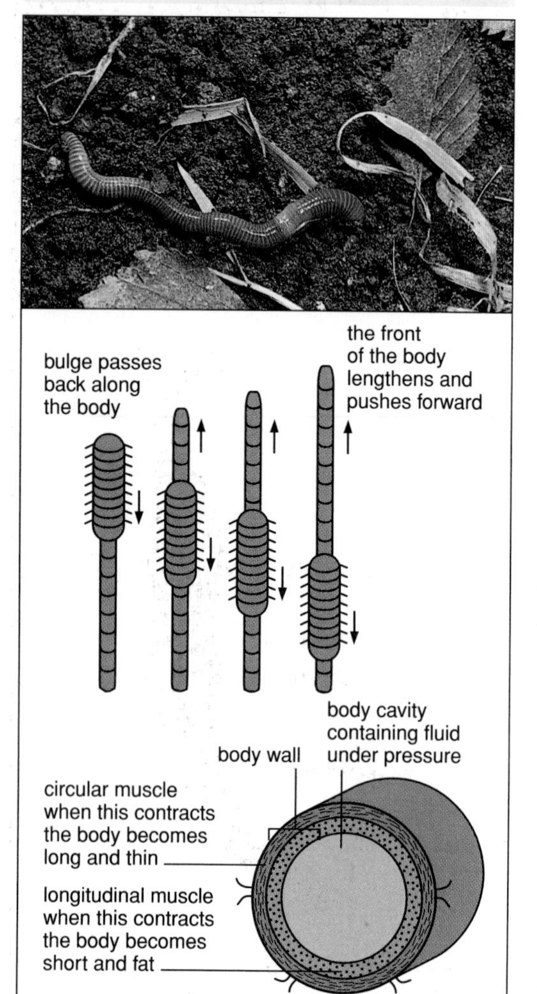

Picture 1 How an earthworm moves.

Why is movement necessary?

In general organisms need to move to search for food, escape from predators and find a mate.

Some organisms are fixed to the ground and do not move from place to place. They are described as **sessile**, a Latin word meaning 'sitting'. Plants are sessile. They do not need to search for food since they feed by photosynthesis (see page 166). Fungi too are sessile – think of a mushroom, for example.

Certain animals are sessile. Examples are sea anemones and mussels. However, most animals can move from place to place. They are described as **motile**.

Movement without legs

We associate movement with legs. But not all animals have legs. Take the earthworm, for example. This animal burrows through the soil. Picture 1 shows how a worm moves. The process depends on the fact that the animal can change its shape, becoming long and thin or short and fat. These changes are brought about by muscle tissue in the soft body wall. The muscles contract against the fluid in the body cavity. The fluid is under pressure. It gives the muscles something to work against and is therefore functioning as a skeleton. A fluid skeleton of this kind is called a **hydrostatic skeleton**.

When the worm becomes short and fat, little bristles stick out from the body wall. They grip the soil as the worm burrows through it.

Movement with legs

Arthropods

Arthropods include crabs, woodlice and insects. These animals all have a **hard cuticle**. This protects the body, but it also serves as a skeleton. In this case the skeleton is outside the muscles which work it, so it is called an **exoskeleton**.

Picture 2 shows how an arthropod's leg works. The leg is jointed, and at the joints there is a ball and socket pivot which works like a see-saw. Inside the leg there are two main muscles: an extensor muscle straightens the leg, and a flexor muscle bends it.

When the leg is straightened by contraction of the extensor muscle, the tip of the leg pushes against the ground. This has the effect of propelling the body forward. The leg is then bent and the tip is placed on the ground further forward. The cycle is then repeated.

If all the legs went through this cycle at the same time, the animal would move forward in a series of lurches. In fact the legs are out of phase with each other, so the animal moves smoothly at a steady speed.

Tetrapods

Tetrapods are vertebrates with four limbs; they include amphibians, reptiles, birds and mammals. A tetrapod moves by means of muscles attached to the limb bones. In a typical tetrapod leg, when the correct muscles contract, the foot pushes against the ground and the body is thrust forward.

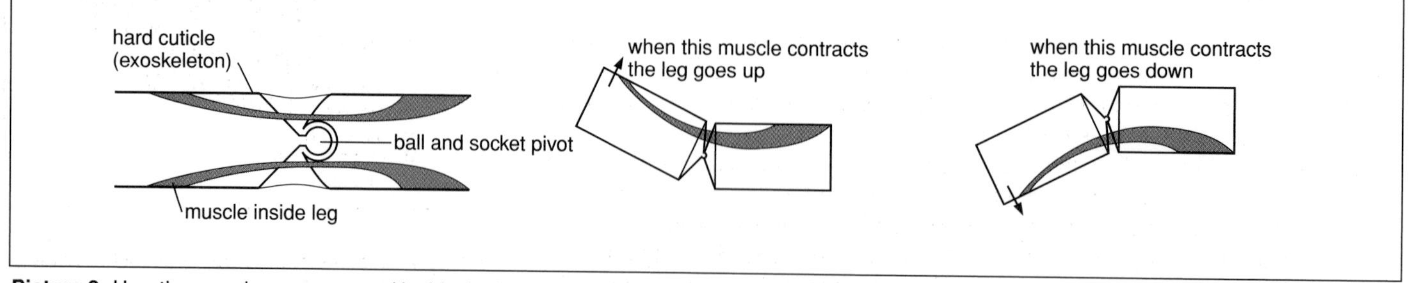

Picture 2 How the muscles are arranged inside the leg of an arthropod, and how they make the leg move.

Picture 3 The hind leg of a horse showing, very diagrammatically, some of the muscles which move it. The muscles are coloured brown. They are really much thicker than shown here. Notice the foot which is long and has fewer bones than ours does. In effect the animal stands on a greatly lengthened middle toe. This gives the leg extra length, making it more efficient for running.

Picture 3 shows the hind leg of a horse. Notice that the skeleton is internal to the muscles, the reverse of what it is in arthropods. It is called an **endoskeleton**. This kind of skeleton is found in all vertebrates, including ourselves.

A tetrapod such as a horse can move with great skill and speed. As with arthropods, the legs are out of phase with each other so forward motion is smooth and steady. A galloping horse has all four hooves off the ground at times, and yet it keeps perfect balance.

Swimming

Fish are expert swimmers. A bony fish such as a mackerel or carp propels itself forward in the water by lashing its tail from side to side (picture 4). The **tail fin**, with its large surface area, increases the forward thrust. The streamlined shape of the body minimises resistance.

Like all vertebrates, fish have an endoskeleton. The tail movements are brought about by contraction of muscles on either side of the vertebral column. The muscles contract on one side first and then on the other.

Fish constantly change direction, or slow down and stop. Steering and braking are achieved mainly by movements of the **pectoral fins**. These, and other

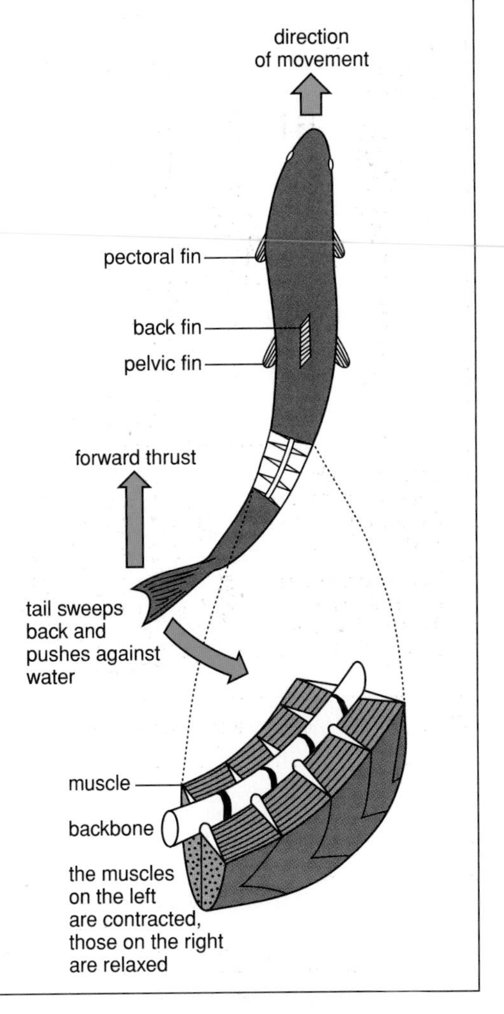

Picture 4 A fish swims by lashing its tail from side to side, as shown above.

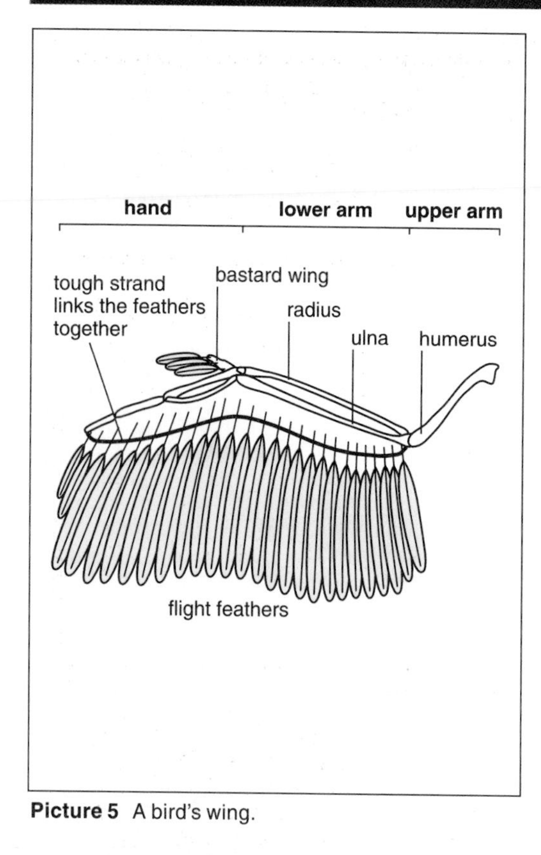

Picture 5 A bird's wing.

fins that stick out from the body, prevent the fish rocking about when it is swimming. They work like the feathers at the back of an arrow or dart.

Most fish have a **swim bladder** which keeps them up in the water. This is a sausage-shaped bag full of air situated towards the upper side of the body.

Flight

Only three groups of animals are able to fly actively: insects, birds and bats. Let's consider birds.

To fly, birds depend on their **wings** which are covered with **feathers** (picture 5). A bird flies either by flapping its wings, or by gliding.

When gliding the wing acts as an **aerofoil**: air flows over it in such a way that the pressure below the wing is raised while the pressure above it is lowered (picture 6). The result is that the bird is given **lift**. The bastard wing smooths the flow of air over the top of the wing, preventing turbulence which could make the bird stall. When gliding, birds make use of rising air currents to hold them up.

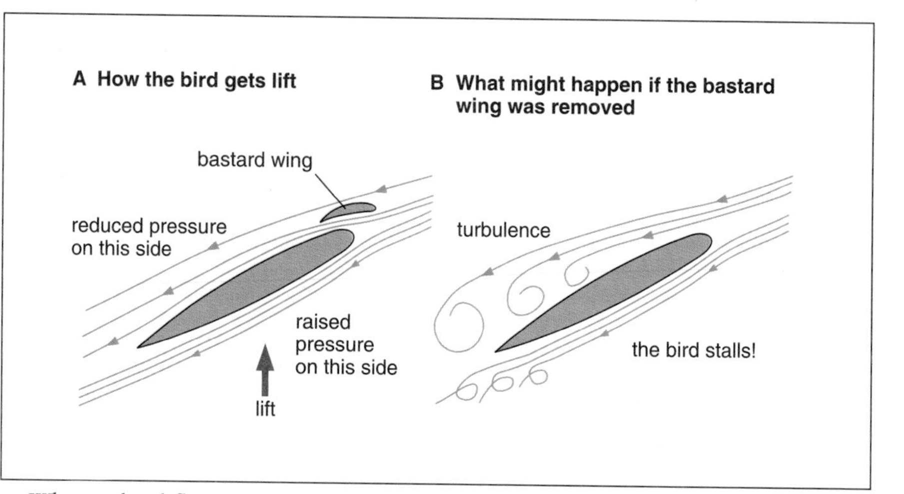

Picture 6 End-on view of the wing of a gliding bird. The leading edge of the wing is to the right. The arrows show the airflow.

When a bird flaps its wings, the feathers behave like the slats of a Venetian blind: they close during the downbeat and open during the upbeat. In this way the body is given lift as the wings go down, but is not dragged downwards when the wings go up.

Each wing is pulled down by a powerful muscle which runs from the lower side of the humerus to the breastbone (sternum). The breastbone has a deep **keel** which increases the area for the attachment of the large flight muscles.

Birds have other features which help them to fly. For example, they are streamlined and light. They are made light by **air sacs** in the body, and their bones are hollow. Their tail feathers help them to keep their balance.

Activities

A Watching animals moving

Watch these animals moving: earthworm, woodlouse, carp. Set them up in the best way for observing how they move, but do not harm them in any way. Describe, *from your own observations*, how each animal propels itself forward.

B Gliding on land

A type of movement not mentioned in this Topic is gliding on land. Find out as much as you can about this method of movement. What sort of animals do it, and how? Report fully on your findings.

C Bird flight

Find out as much as you can about the structures involved in bird flight. Examine a wing and work out the extent to which the feathers increase its surface area. Examine part of a feather under the microscope and see how it is adapted for flight.

Questions

1 Sea anemones and mussels are sessile. Find out how they manage to feed without having to move in search of food.

2 What is meant by *antagonistic muscles* and why are they important? Illustrate your answer by referring to pictures 1 to 4 on pages 218 to 219.

3 These questions relate to picture 3 on page 219. In your answers refer to the muscles by their numbers.
   a Describe the effect on the bones when each muscle contracts.
   b Which muscles thrust the body forward when they contract?
   c Describe *one* muscle which must exist but is not shown in the diagram. Which bones might it be attached to, and what effect would it have when it contracted?
   d Why is a *long* leg more efficient at running than a *short* leg?

How limbs are adapted for different jobs

Picture 1 shows the forelimbs of three tetrapods. Notice how they differ. Each is adapted to do a particular job. For example, the bird's forelimb is adapted for flight, whereas the horse's is adapted for high-speed movement on land.

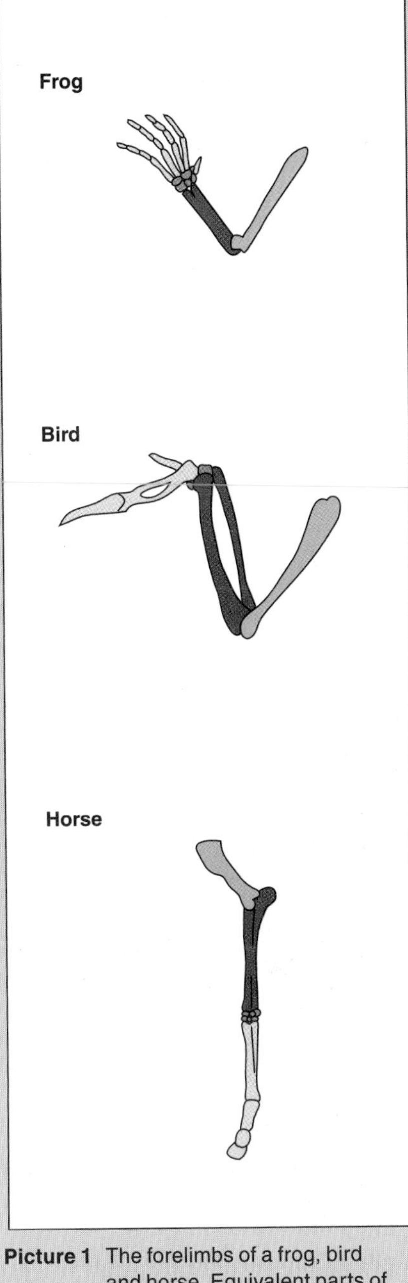

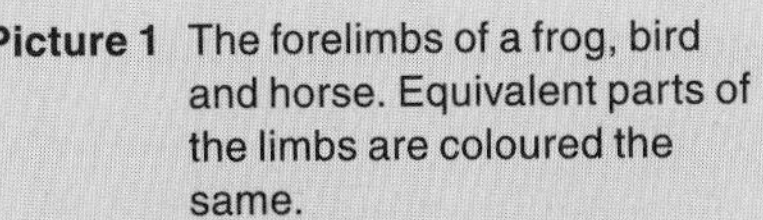

Picture 1 The forelimbs of a frog, bird and horse. Equivalent parts of the limbs are coloured the same.

Although these three limbs differ, they are all built on the same basic plan. This is shown in picture 2. It is called the **pentadactyl limb** – *pentadactyl* comes from Greek and literally means 'having five digits' (fingers or toes).

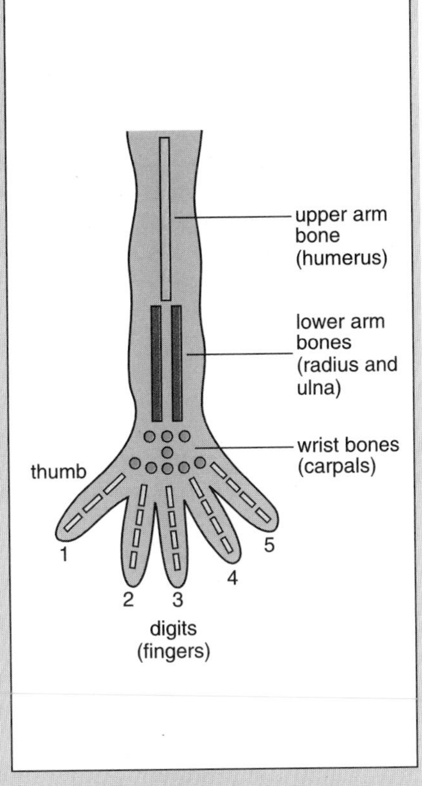

Picture 2 The basic plan of the pentadactyl limb. The parts are coloured the same as in picture 1.

The pentadactyl limb is found in amphibians, reptiles, birds and mammals. It is often used as evidence that these animals have evolved from a common ancestor.

1. Describe in your own words how each of the three limbs in picture 1 differs from the basic plan in picture 2.
2. Which limb is most like the basic plan? Suggest a reason for the similarity.
3. What would you expect a frog's *hind* limb to look like? Explain your reasoning.
4. In what way does the pentadactyl limb provide evidence for evolution from a common ancestor? Suggest *one* other explanation of the fact that amphibians, reptiles, birds and mammals all share the same kind of limb.

Movement of a single celled organism

We shall consider the single celled organism *Amoeba*. It is very small – a fully grown amoeba is about the size of a pinhead. It lives in fresh water ponds.

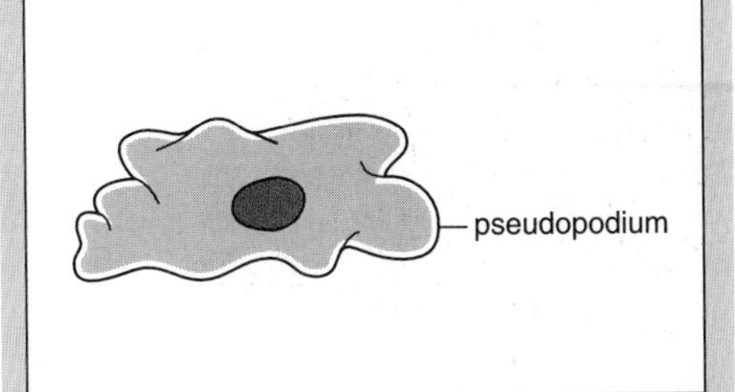

Picture 1 An amoeba.

Amoeba moves very slowly over stones and weeds. It constantly changes shape and appears to 'ooze' along. Its cell surface membrane is thin and flexible, and the cytoplasm can flow from one part of the cell to another. This enables the organism to send out a bulge called a **pseudopodium** ('false foot'). Picture 2 shows how this enables the organism to move in a particular direction.

The cytoplasm is fluid in the middle of the cell (**endoplasm**) and jelly-like towards the edge (**ectoplasm**). When the organism moves, the endoplasm in the middle of the cell flows into a pseudopodium. At the tip of the pseudopodium the endoplasm changes into ectoplasm. At the rear the ectoplasm changes back into endoplasm. So the organism appears to flow along in a tube of its own making.

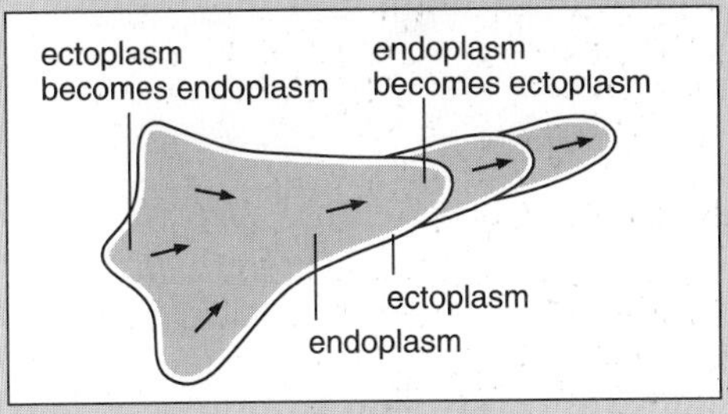

Picture 2 How *Amoeba* moves.

Certain of our white blood cells (the phagocytes) move like *Amoeba* (see page 157). It is called *amoeboid movement*.

1. The way our phagocytes move suits them for their job. Explain.
2. We have white blood cells like little amoebas. What conclusion might be drawn from this observation?

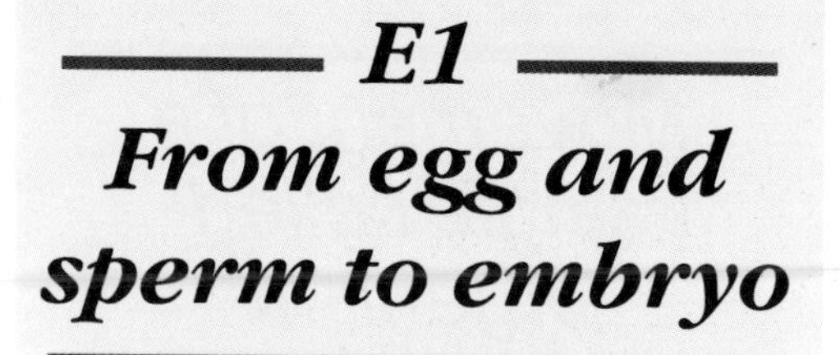

E1
From egg and sperm to embryo

Here we shall look at the way an embryo is formed in the human.

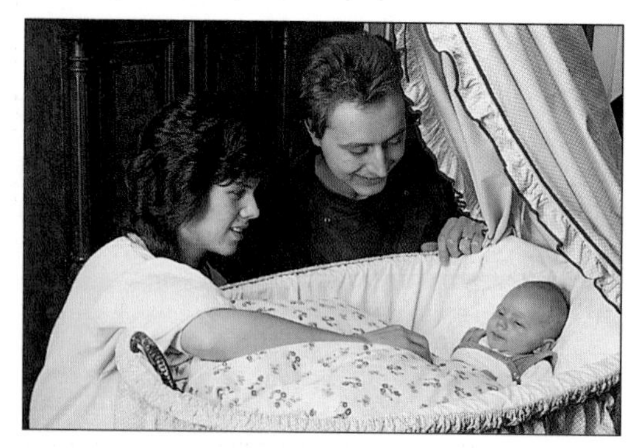

Picture 1 Human reproduction brings delight to parents and it keeps the species going.

The human reproductive system

The organs which enable us to reproduce make up the **reproductive system**. The reproductive system does three things:

- It manufactures eggs and sperm. These are our 'sex cells' or **gametes**. The eggs are produced by a pair of **ovaries** in the woman, and the sperm are produced by a pair of **testes** in the man.
- It brings the eggs and sperm together. The combining of a sperm with an egg is called **fertilisation**. It is made possible by the **penis** which transfers sperms from the man to the woman during **mating** (**sexual intercourse**).
- It protects and nourishes the embryo. The organ which does this is the **uterus** or 'womb'. Here the **embryo** develops into the baby.

Keep these functions in mind as we look in detail at the reproductive system (pictures 2 and 3).

The female system

The two ovaries are on either side of the abdomen. Once every 28 days or so, one of the ovaries sheds an egg into the **oviduct**. The oviduct is also known as the **Fallopian tube** after the Italian anatomist, Gabriello Fallopio, who discovered it. The egg passes slowly down the oviduct towards the uterus. If it is not fertilised within a day or so, it will die.

The uterus is like a bag with a thick muscular wall. Its inner lining is very special, because it is here that the embryo develops. We shall have more to say about it later. The lower end of the uterus is called the neck or **cervix**, and it leads to the **vagina**.

In picture 2 you will see that the vagina opens to the outside close to the urinary opening. Just above the urinary opening is a small bump called the **clitoris**.

The external parts of the female genital organs together comprise the **vulva**. At first the vaginal opening is partly covered by a fold of tissue called the **hymen**. This becomes stretched by wearing a tampon or when sexual intercourse takes place.

Picture 2 The reproductive system of the human female. The bladder is not shown.

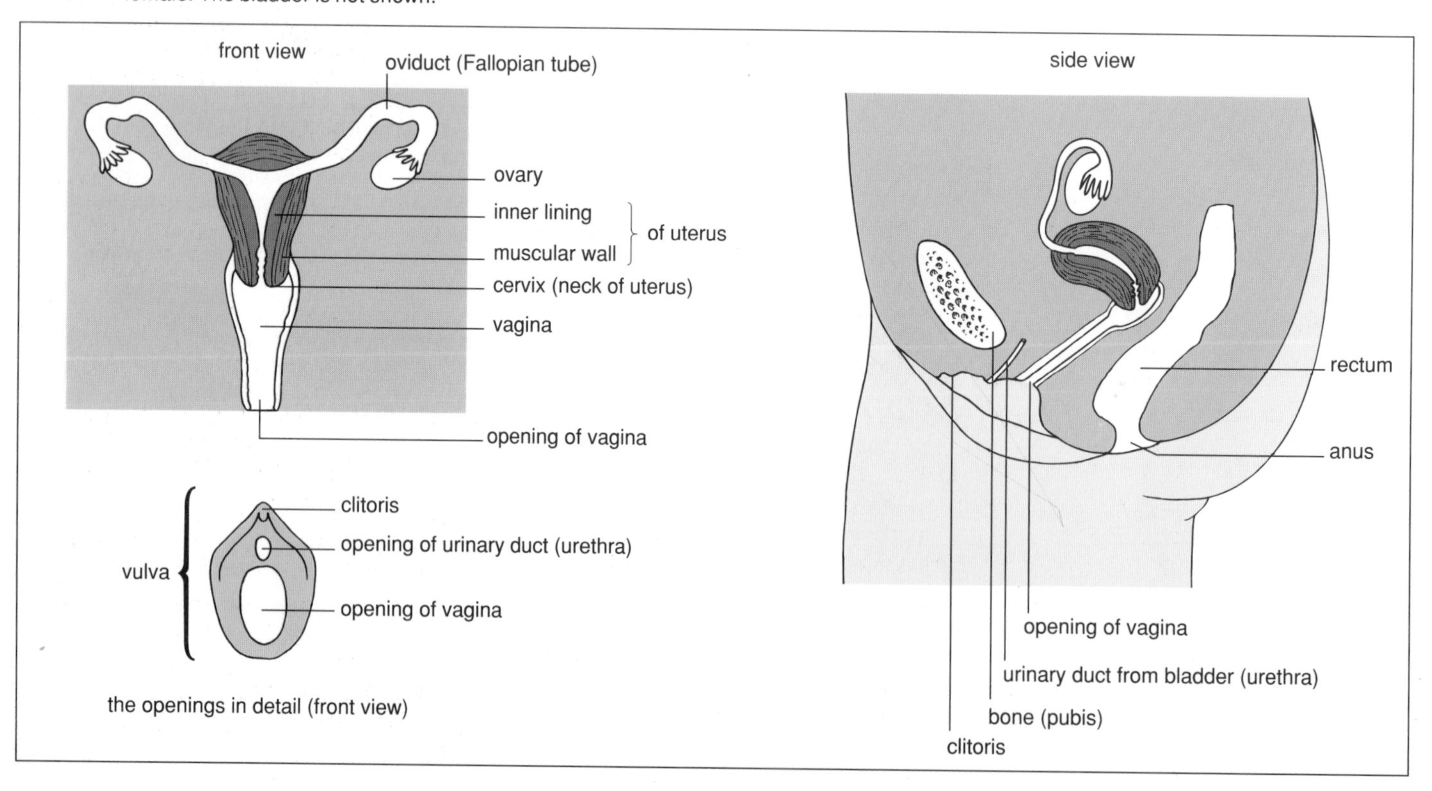

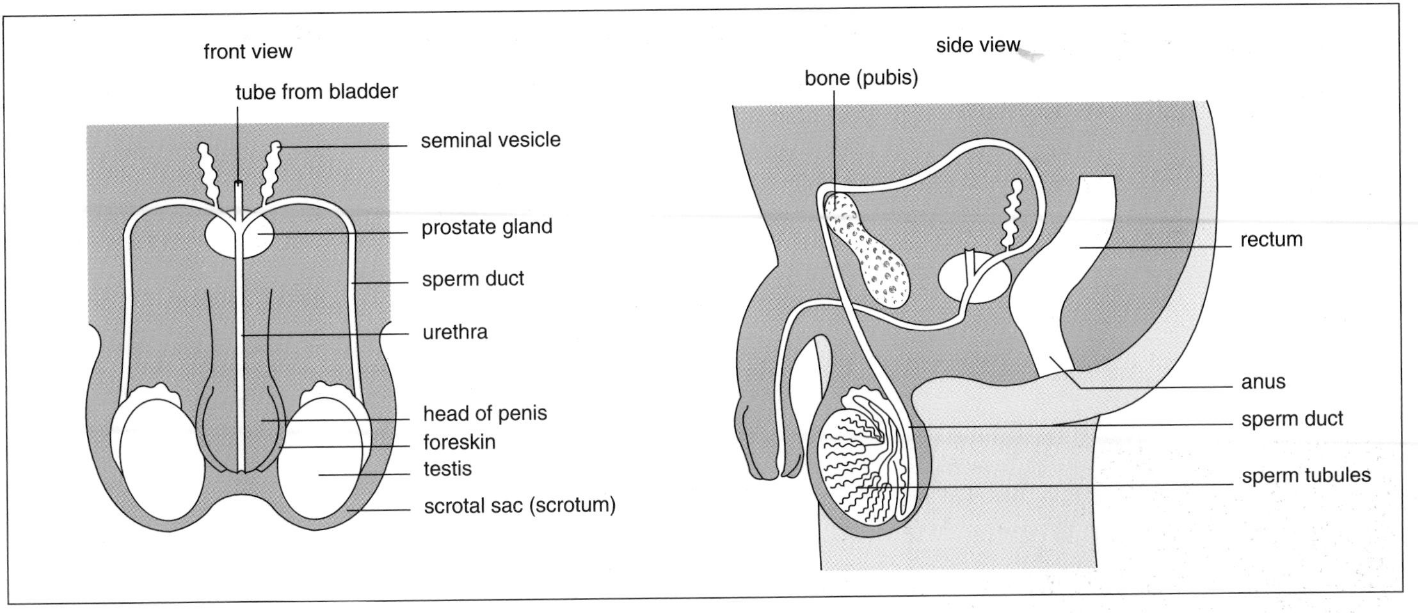

Picture 3 The reproductive system of the human male. The bladder is not shown.

The male system

The testes are inside the **scrotal sac** just behind the penis. They are slightly cooler than the rest of the body, because of their position outside the abdomen. This is necessary because the testes make larger numbers of sperm in cool conditions. When a baby boy is developing, the testes start off in the abdomen, but later they move down into the scrotal sac. Normally this happens before, or soon after, birth.

Each testis contains lots of narrow tubules where the sperms are made. Placed end to end, these tubules would stretch for over 500 metres – that's long enough to go right round a football pitch! This gives the testis a high production rate – it is truly a sperm factory.

Sperm are produced all the time, not just once a month like eggs are. As the sperm are produced, they move into a coiled tube alongside the testis where they are stored. This tube leads to the **sperm duct** which in turn leads to the **urethra**. The urethra runs down the centre of the penis. Notice the **seminal vesicles** and **prostate gland** at the top end of the urethra – we shall see what they do in a moment.

The head of the penis is very sensitive and is protected by the **foreskin**. The foreskin is sometimes removed in the simple operation **circumcision**. This may be done for religious reasons (for example, Jews and Muslims usually have it done), or because the foreskin is too tight. The operation is carried out at an early age, and is in no way harmful.

For sperm to leave the body, the penis has to become stiff. This is called an **erection**, and it is brought about by blood being pumped into a special spongy tissue in the penis. Rubbing of the erect penis results in **ejaculation**. Ejaculation is a reflex in which two things happen. Firstly, the seminal vesicles and prostate gland pour fluid into the sperm ducts and urethra. This fluid, together with the sperm, makes up **semen**. At the same time, repeated contractions of the muscles surrounding the sperm ducts and urethra sweep the semen out of the penis.

Ejaculation is accompanied by a pleasurable feeling called an **orgasm**. Women have orgasms too. They are brought about mainly by repeated rubbing of the clitoris which can become erect, rather like the penis.

If you look at picture 3 you will see that the bladder and sperm ducts both open into the urethra. However, urine and semen never pass down the urethra together. This is because it is impossible for urination and ejaculation to occur at the same time.

Other animals

It is interesting to compare our reproductive system with that of other animals. Take snails and earthworms, for example. These animals are **hermaphrodites**: each individual has both male and female organs, and when they mate sperm pass from each individual into the other. The trouble is that an individual's eggs may get fertilised by its own sperm. This is called **self-fertilisation**. Self-fertilisation has disadvantages and most hermaphrodites have ways of preventing it. It is better for the sperm to come from another individual, as it does in humans and most other animals. This is called **cross-fertilisation.**

Some water-dwelling animals have separate sexes but the male has no penis and cannot put his sperm into the female. So the eggs and sperm are released into the surrounding water where, hopefully, fertilisation will take place. This is called **external fertilisation**, and it is what happens in frogs, toads and many types of fish. In most animals, including us, fertilisation takes place inside the female. This is called **internal fertilisation**.

1 What are the disadvantages of (a) self-fertilisation, and (b) external fertilisation?

2 A cod may produce eight million eggs in a single spawning. Why so many?

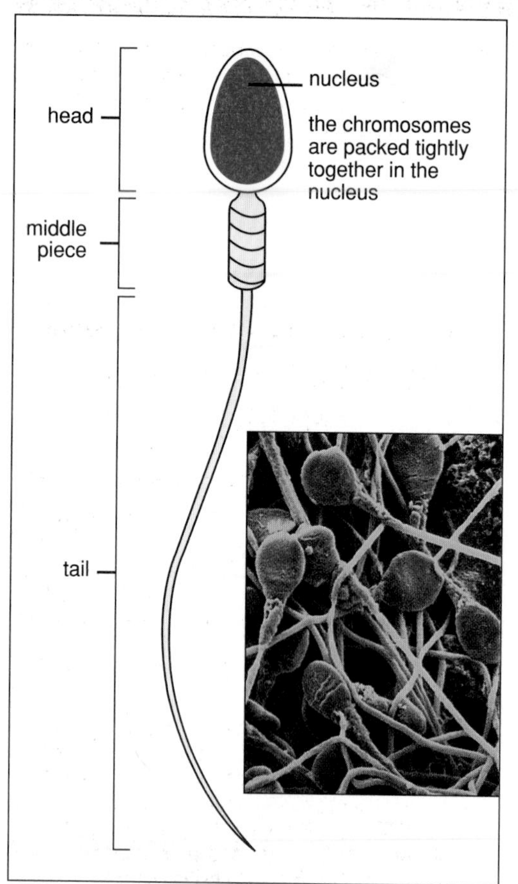

Picture 4 Human sperm. The head is about 3 μm wide. The tail lashes from side to side and the energy for this comes from the middle piece. On the right are sperms as seen in an electron microscope.

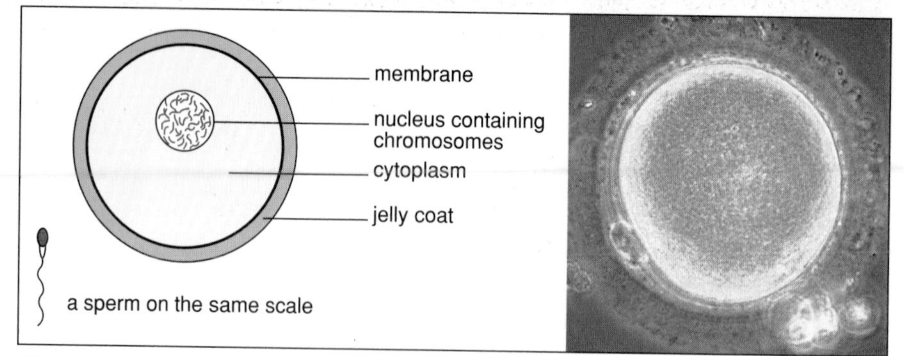

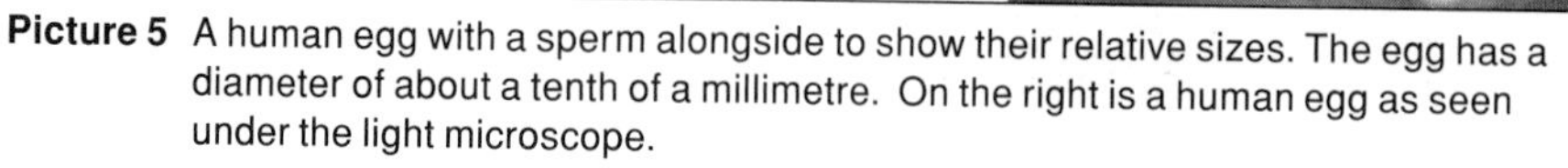
Picture 5 A human egg with a sperm alongside to show their relative sizes. The egg has a diameter of about a tenth of a millimetre. On the right is a human egg as seen under the light microscope.

Eggs and sperm

Normally the male produces about 4 cm^3 of semen when he ejaculates – that's a small spoonful. This may not seem much, but it may contain as many as 500 million sperm. They are kept alive and active by substances in the semen.

The sperm are very small, and you need a microscope to see them. One is shown on a large scale in picture 4. It is shaped like a tadpole, with a head and tail. It is a single cell, with the nucleus packed tightly into the head. The sperm swims by waving its tail from side to side.

An egg is shown in picture 5. It too is a single cell, but it is much larger than the sperm. It is round like a ball, and is surrounded by a thin membrane and a layer of jelly. The nucleus is towards the centre.

The nuclei of the sperm and egg contain chromosomes, which carry the genes (see page 274). In the sperm nucleus, the chromosomes are so tightly packed that you can't see them individually. In the egg nucleus they are more spread out.

Sexual intercourse

The erect penis is placed in the vagina, and moved in and out repeatedly. Drops of fluid, secreted by glands lining the urethra, come out of the penis and serve as a lubricant. Mucus in the vagina secreted by the cervix performs the same function.

After ejaculation, the sperm swim through the mucus into the uterus. They then swim the full length of the uterus, and up the oviducts (picture 6).

Picture 6 During sexual intercourse sperm pass from the male to the female, as shown by the arrows in the diagram below. The pictures on the right show fertilisation taking place.

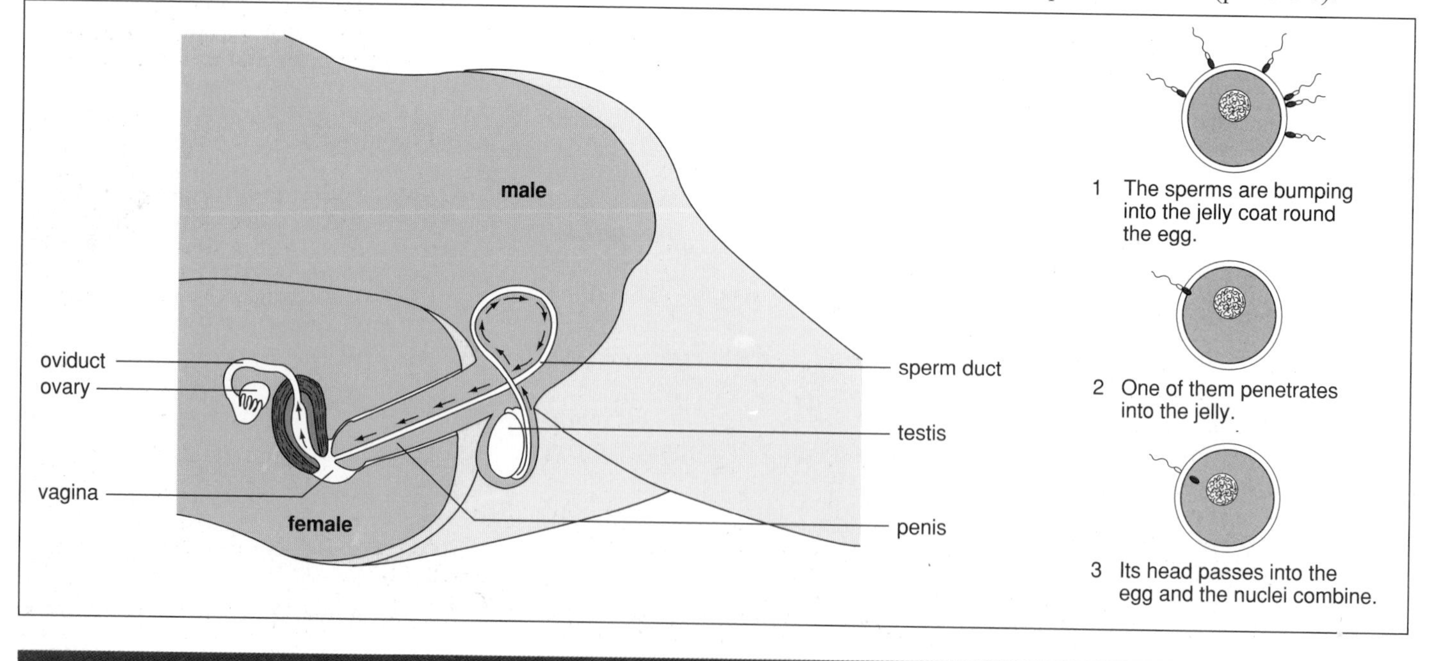

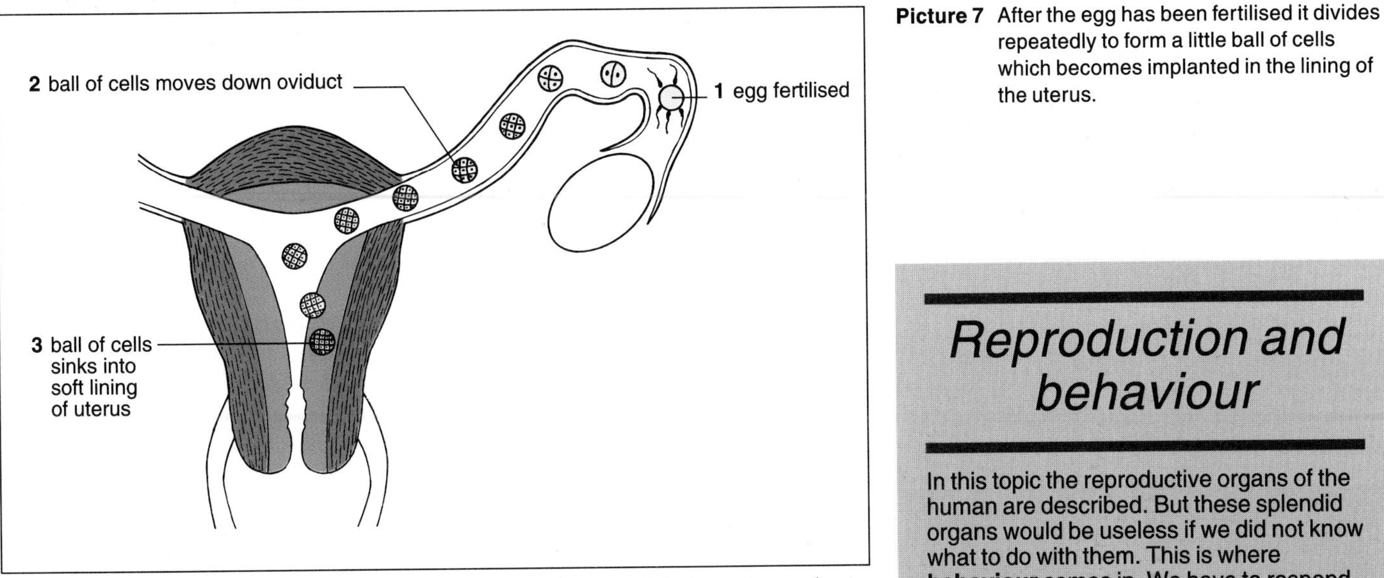

Picture 7 After the egg has been fertilised it divides repeatedly to form a little ball of cells which becomes implanted in the lining of the uterus.

The sperm are very small and the journey is not easy. Most never make it. That is why so many are produced – it raises the chance of one of them getting through. A few days before an egg is shed from the ovary, the cervical mucus becomes more runny and plentiful and this helps the sperm to swim through it.

Fertilisation

If there is an egg in the oviduct, one of the sperm may bump into it and fertilise it. The head of the sperm penetrates the egg, and its nucleus combines with the egg's nucleus. The fertilised egg is called a **zygote**.

What happens if there is no egg in the oviduct? The sperm remain capable of fertilising an egg for two or three days. If an egg is released by the ovary within this time, it may get fertilised.

What happens if an egg is produced and there are no sperm to fertilise it? The egg is capable of being fertilised for about a day after it has been released into the oviduct. If intercourse takes place within this time, fertilisation may occur.

What happens to the fertilised egg?

After the egg has been fertilised, the zygote divides into a little ball of cells. This moves down the oviduct to the uterus (picture 7). It then sinks into the soft lining of the uterus. This is called **implantation**.

Once fertilisation has happened, **conception** has been achieved and **pregnancy** begins. The ball of cells is the **embryo**. We shall see what happens to the embryo in the next topic.

Questions

1 Why is it important that a very large number of sperm should be present in the semen?

2 Why is it an advantage for the testes to be situated in the scrotal sac outside the main body cavity? Can you think of any disadvantages?

3 Why do you think human eggs and sperm are so different in size?

4 Which structures in the female are equivalent to these structures in the male:

a penis,
b testes,
c sperm ducts,
d urethra?

In each case say in what respect the structures are equivalent.

Reproduction and behaviour

In this topic the reproductive organs of the human are described. But these splendid organs would be useless if we did not know what to do with them. This is where **behaviour** comes in. We have to respond and behave so that mating occurs in the right way and the offspring are produced successfully.

To begin with it is important that males and females should be attracted to each other and that they should want to mate. All sorts of stimuli bring the sexes together. In humans the stimuli are very varied, but in other animals they are easier to identify. The sight of the opposite sex is often important – for example, in many species of birds the female is attracted to the male by his bright colours. Smell also plays a part, as anyone who has kept a dog will know. In such cases the female produces a chemical substance which attracts the male. In animals as a whole, chemical stimuli are very important in starting up particular types of behaviour.

In some animals, especially birds, the male and female display to each other in various ways before they mate. This is called **courtship**. This kind of behaviour ensures that both partners are ready to mate at the same moment, when their reproductive organs are working fully.

Behaviour may continue to be important after the young are born. For example, many animals look after their offspring. The parents feed and protect them, and may teach them to fend for themselves. This **care of the young** is particularly well developed in mammals and birds and reaches a peak in the human, as we shall see in the next topic.

1 In what ways is behaviour important in *human* reproduction? Make a list of as many ways as you can think of.

2 Some male birds, robins for example, find a piece of territory before they mate, and defend it against intruders. Why do you think it is a good idea for a pair of breeding birds to have their own territory?

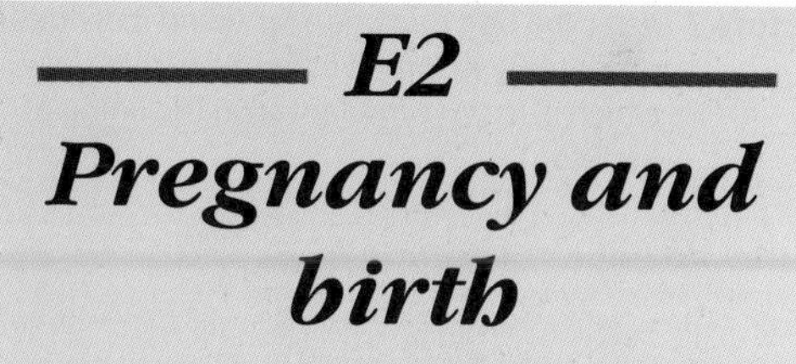

E2 Pregnancy and birth

In the last topic we finished up with an embryo implanted in the lining of the uterus.

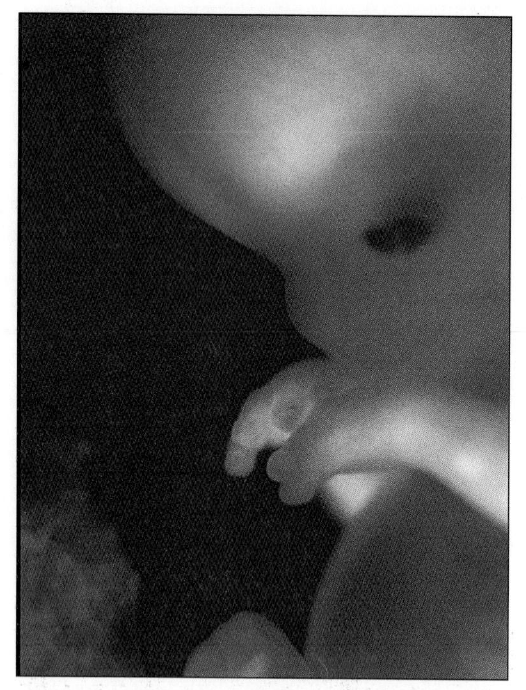

Picture 1 Photograph of a ten week old fetus. All the organs have been formed by this stage.

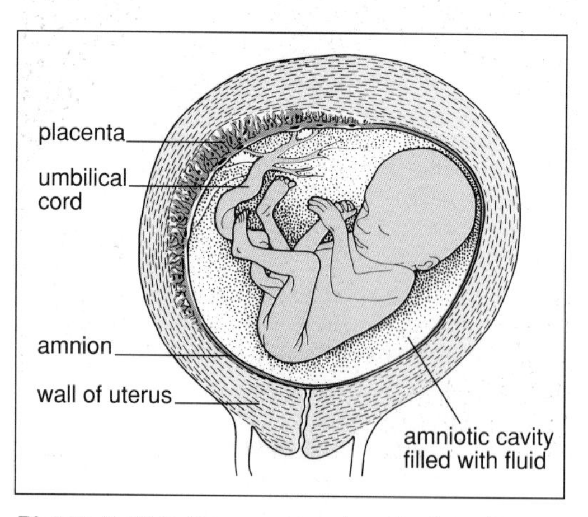

Picture 2 This diagram shows how the fetus is attached to the wall of the uterus by the umbilical cord and placenta.

What happens to the embryo?

In the next few weeks, the cells of the embryo multiply and start developing into different kinds of tissue. Gradually the embryo grows into something that looks vaguely like a miniature human being. It is called a **fetus** or foetus.

The private pond of the fetus

As the embryo develops in the mother's uterus, it becomes surrounded by a thin membrane called the **amnion**. The amnion forms a kind of bag round the embryo, and it is filled with watery fluid called **amniotic fluid**.

As development goes on, the amniotic cavity expands like a balloon until it fills the entire uterus. The fetus floats in the middle of it, in a kind of 'private pond' (picture 1). The amniotic fluid cushions the fetus, protecting it from being bumped and damaged, for example when the mother runs for a bus.

How is the fetus kept alive?

Attached to the belly of the fetus is a flexible strand called the **umbilical cord**. This is its lifeline, like the tube which connects a diver with the surface. The umbilical cord brings the fetus everything it needs, such as oxygen and food substances. It also takes away unwanted substances such as carbon dioxide and excretory waste.

The umbilical cord runs from the fetus to a structure called the **placenta** which is attached to the lining of the uterus (picture 2). The placenta is shaped like a plate, and it has lots of finger-like **villi** which stick into blood spaces in the wall of the uterus. The villi contain blood capillaries which are connected to the fetus by an artery and vein in the umbilical cord: the **umbilical artery** and **vein**. A detailed picture of the placenta is shown in picture 3.

The barrier separating the fetus's blood from the mother's blood is very thin, and substances can diffuse across it easily. As the fetus's blood flows through the placental villi, it picks up oxygen and soluble food substances from the mother's blood. At the same time, waste substances like carbon dioxide and urea pass into the mother's blood.

It is not just food and oxygen that pass from the mother to the fetus. Antibodies do so too. These help to protect the newborn baby from diseases until it has had a chance to make its own antibodies.

Although the bloodstreams of the fetus and mother come very close to each other, they never mix. If the two bloodstreams were joined, the mother's blood pressure might burst the fetus's blood vessels. What's more, if the two lots of blood belonged to different blood groups, clumping of the red blood cells could occur with disastrous results (page 161).

Growth of the fetus

By the end of the third month, the fetus is fully formed right down to the fingers and toes. It now grows until it fills the uterus. Meanwhile, the uterus expands greatly as the fetus grows, and its wall becomes thicker and more muscular in readiness for birth. As the baby grows, it gets more active. It moves its arms and legs, and by the end of the fourth month the mother may feel it moving inside her.

Care during pregnancy

A pregnant woman should eat the right kinds of food, and do nothing that might injure her baby. From time to time she is examined by a doctor or midwife to make sure that everything is all right. At the same time she is given advice on how to keep fit and prepare for the birth of her baby.

Being pregnant has its problems. As well as the extra load to be carried, changes occur in the hormones and occasionally the mother may feel unwell.

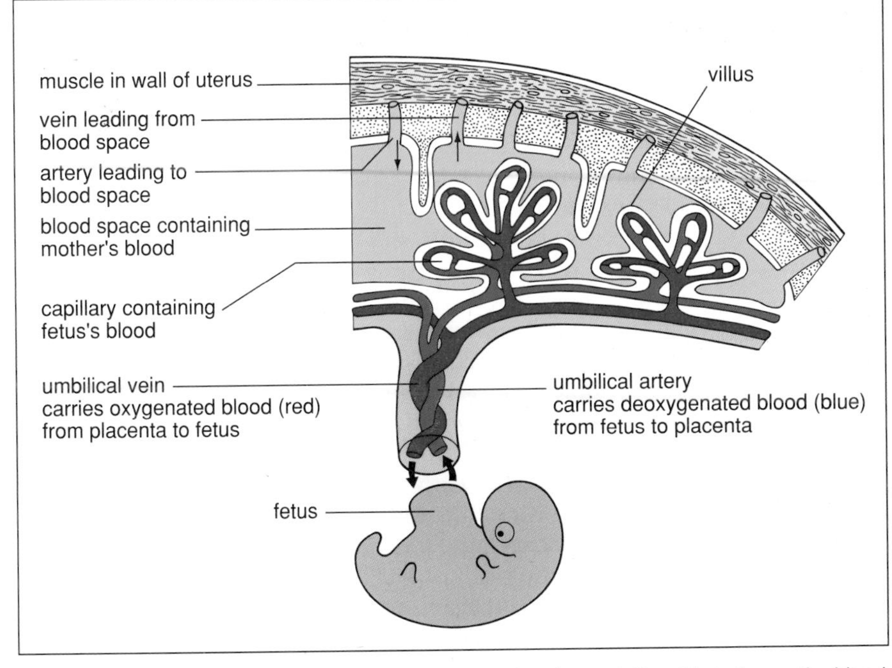

Picture 3 This diagram shows the placenta. Notice the close relationship between the blood of the fetus and the blood of the mother.

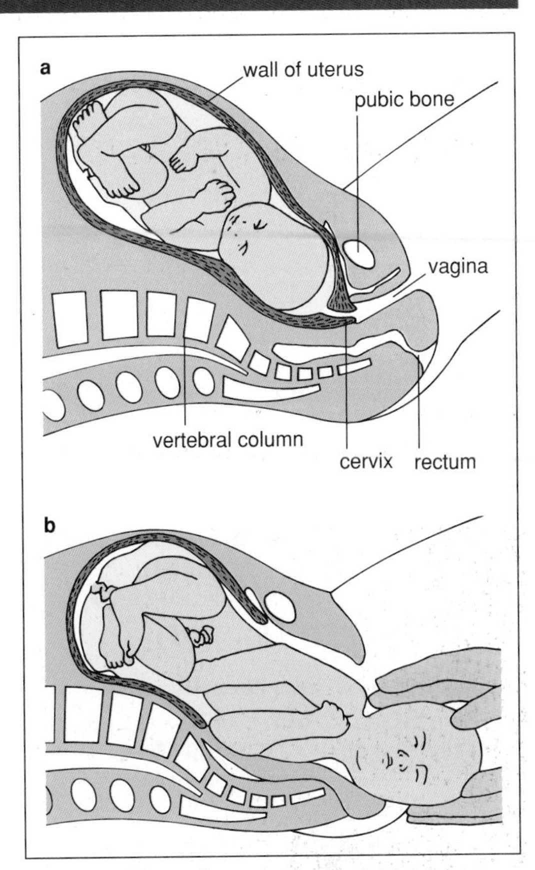

Picture 4 These diagrams show how birth takes place.

If necessary the doctor may give her some medicine to help. However, doctors do not prescribe drugs unless they have been very thoroughly tested beforehand. In the 1960s a number of pregnant women in Britain were given a tranquilliser called thalidomide. No-one realised the harm the drug was doing. Some of the mothers gave birth to babies with severe deformities such as no arms or legs.

Smoking and alcohol can also affect the health of the fetus (page 134). The placenta is very good at supplying the fetus with the things it needs, but unfortunately it does not always hold back things which the fetus does *not* need.

Certain germs may get across the placenta and harm the baby. Such is the case with the virus that causes German measles (rubella). This is a mild disease in adults, but if a pregnant woman is infected it can harm the baby. For this reason, girls who haven't had this disease are always immunised against it. Germs that cause sexually transmitted diseases can also get across the placenta (page 238).

Birth

Approximately nine months after conception the baby is ready to be born. The first sign of birth is that the uterus gives occasional contractions which become more and more frequent and strong. This is called '**going into labour**'. At about this time, the amnion bursts and the fluid escapes through the vagina, the so-called '**breaking of the water**'. Soon afterwards, the uterus contracts very powerfully and the cervix opens up (dilates). As a result, the baby is pushed through the vagina, usually head first (picture 4).

Once the baby has come out, it starts to breathe. If it does not start breathing of its own accord, the doctor or nurse may give it a tap on the bottom which makes it take a breath in. The umbilical cord is no longer needed, so it is tied and cut. The scar becomes the **navel** ('belly button'). Meanwhile the placenta comes away from the wall of the uterus, and passes out through the vagina. It is called the **afterbirth**. On average a newborn baby weighs just over 3 kg. The birth, or delivery as it's called by doctors and nurses, is now over.

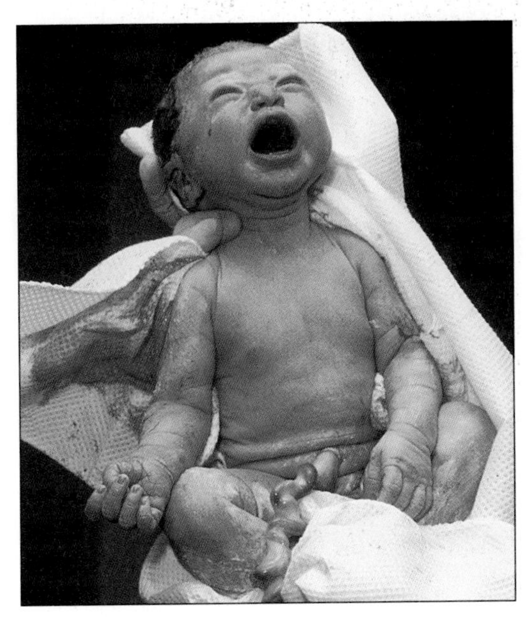

Picture 5 This baby has just been born. The umbilical cord is just about to be tied and cut.

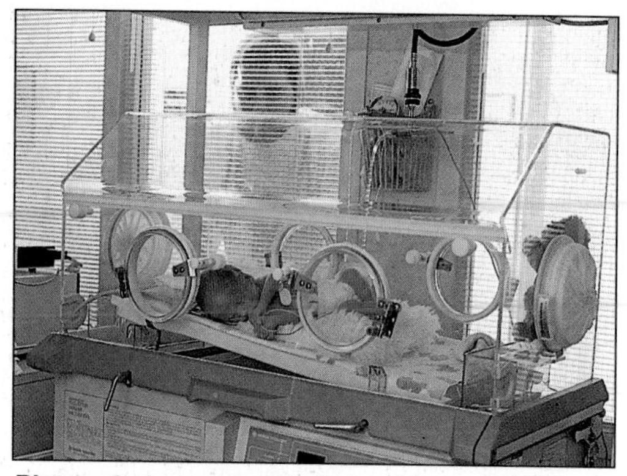

Picture 6 A premature baby in an incubator. The temperature, oxygen supply and humidity are carefully controlled.

Factors affecting birth weight

Some babies are born under-weight, even though they are born at the right time. Why? Here are the main reasons:

The mother

- had a poor diet during pregnancy,
- was ill during pregnancy, for example she may have had anaemia,
- has parasites in her body,
- smoked during pregnancy,
- drank alcohol during pregnancy,
- took drugs during pregnancy (even medical drugs have to be carefully controlled),
- was under 16, or over 35, years old (the babies of such mothers tend to be small).

The placenta

- was diseased and had a reduced blood supply,
- was unusually small.

The baby

- was a twin (multiple birth babies are usually smaller than single-birth babies,
- possesses genes that cause it to be naturally small,
- has a chromosome abnormality, e.g. Down's syndrome.

1 Explain, as far as you can, why each of the above results in an under-weight baby.

2 Which of the factors in the list have a genetic cause, and which ones are caused by the environment?

3 In what ways may a mother's diet be poor? What are the possible causes of this?

Sometimes there are problems

In Britain women usually have their babies in hospital. Most babies are born quite easily, but sometimes the baby needs to be helped out with forceps. They are like large tongs, and are used to grasp the baby's head as it emerges. In particularly difficult cases the medical staff may give the mother an anaesthetic and take the baby out by cutting into the abdomen and opening up the uterus. This operation is called a **Caesarian section**.

Sometimes birth occurs before the ninth month, and the baby is **premature**. If the baby is very small and weak, it may have to be kept in an **incubator** until it can support itself (picture 6). In a modern incubator the temperature, oxygen supply and humidity are carefully regulated. The temperature is controlled by a feedback mechanism: the baby's body temperature is monitored by a thermocouple attached to the skin, and this tells the heater in the incubator whether to come on or go off. Thanks to incubators, babies born seven months or less after conception can survive.

Occasionally things go wrong at an early stage of pregnancy, and the fetus is expelled from the uterus. If it is not already dead, it dies almost immediately afterwards. This is called a **miscarriage**. It is very distressing for a woman who has been looking forward to the birth of her baby. However, there is usually nothing to stop her getting pregnant again, and next time she will be given special help to assist her through her pregnancy.

Caring for the newborn baby

A woman's breasts contain **mammary glands** which secrete milk. During pregnancy the breasts enlarge and the mammary glands get ready to produce milk. Soon after birth the baby sucks its mother's nipples. This stimulates the breasts to release the milk, and make more.

Milk is the baby's only food for the first few months of life. The baby cannot cope with solid food at this stage. It has no teeth to chew it with, and its gut would be unable to digest it. The baby doesn't *have* to get milk from its mother – it can be fed on cow's milk from a bottle with a teat. However, there's a lot to be said for breast-feeding. The mother's milk is a perfect food, containing all the substances the baby needs at just the right temperature. It also contains antibodies which help to protect the baby from certain diseases and allergies. The milk produced for the first few days after birth (called **colostrum**) is particularly rich in antibodies.

Breast-feeding also allows close contact between the mother and her baby, which is good for both of them, physically and emotionally. And it's cheaper than bottle-feeding!

After several months the baby can be given semi-solid food, and gradually it moves on to a normal diet. When the baby's milk diet has been replaced by solid food, the baby is said to be **weaned**.

Parents should make sure that their babies are kept warm, particularly in cold weather. The reason is that human babies can't control their own body temperature in the way that adults can. An adult adjusts to the cold by shivering and so on, but a baby cannot yet do this efficiently.

These are just a few aspects of looking after a baby. For ten days or so after the birth, the midwife visits the mother at home every day. After that a health visitor gives regular advice and support to the family. About six weeks later the mother and baby are examined by the doctor. After that the health visitor sees them in the baby clinic, and visits the family at home as well. The health visitor remains in contact until the child goes to school.

Twins and multiple births

Normally in the human only one embryo develops in the uterus at a time. However, two may sometimes be present together, each with its own placenta and umbilical cord. These are **twins**.

There are two kinds of twins: **identical** and **non-identical.** Identical twins are exactly alike, whereas non-identical twins are not.

This is how identical twins arise. An egg is released from one of the ovaries and fertilised in the usual way. It then splits into two cells, each of which develops into a baby. The two cells have exactly the same genes, so the two babies will be exactly alike and of the same sex.

What about non-identical twins? In this case two eggs are released from the ovaries at the same time, and both are fertilised. Although the two babies will be born together, they don't have exactly the same genes and will be no more alike than brothers or sisters. Twins of this sort are known as **fraternal twins**, and they may be of different sexes.

Occasionally three or more eggs are produced by the ovaries at the same time, resulting in triplets (three babies), quadruplets (four babies) or quintuplets (five babies). There have even been cases of sextuplets (six babies). Such **multiple births** tend to happen in women who have been given a fertility drug to help them get pregnant (page 235). As the fetuses grow, it gets more and more difficult for the uterus to contain them. As a result, the mother gives birth early, often around the seventh month. The babies are put in incubators, and usually at least some of them survive.

Multiple births are uncommon in humans, but they are quite usual in other mammals such as dogs, cats and mice. The offspring from one pregnancy make up a **litter**.

Picture 7 Twins! The top pair are identical, and the bottom pair are non-identical.

Activities

A At what age do children do different things?

Find out from books, or by asking parents, when approximately a child starts:

- eating semi-solid food,
- eating completely solid food,
- feeding itself,
- focusing the eyes,
- recognising faces,
- crawling,
- standing up,
- walking,
- smiling,
- laughing,
- understanding the word 'no',
- saying 'mum',
- saying 'dad',
- drawing,
- writing.

You might like to add some further questions of your own. For example it is interesting to find out when children start gaining skills such as piling bricks up and making things.

B What are your first memories?

Think back to your childhood. Make a list of your earliest memories. Try to find out how old you were at the time.

If everyone in your class does this, you can compare results. On average how old does a child have to be before experiences go into the long-term memory system? Why do you think earlier events are not remembered?

Questions

1 What functions are performed by:

a the muscle in the wall of the uterus,
b the amniotic fluid,
c the umbilical cord,
d the mammary glands?

2 Name five jobs which are carried out by the placenta. what is it about the structure of the placenta which makes it ideally suited to do these jobs?

3 Babies are usually born head first. What advantages are there in being born this way? What changes take place in the baby's body soon after it is born?

4 Some of the daily food requirements of an adult woman with a body mass of 55kg, and of the same woman in an advanced state of pregnancy are given in the table:

	non-pregnant	pregnant
Energy	9200 kJ	10700 kJ
Protein	29 g	38 g
Vitamin A	750 μg	750 μg
Vitamin D	2.5 μg	10 μg
Vitamin C	30 mg	30 mg
Calcium	0.5 g	1.2 g
Iron	20 mg	28 mg

a Suggest one reason why the woman requires more energy when she is pregnant.
b Name two kinds of food from which she is likely to obtain most of this energy.
c Suggest one reason why she requires extra protein when she is pregnant.
d Why do you think she needs extra vitamin D and calcium when she is pregnant?
e What are the percentage increases in the amounts of energy, protein, vitamin D and calcium which she needs when she is pregnant?
f Which substances in the table do not need to be increased during pregnancy?
g Why does she need extra iron?

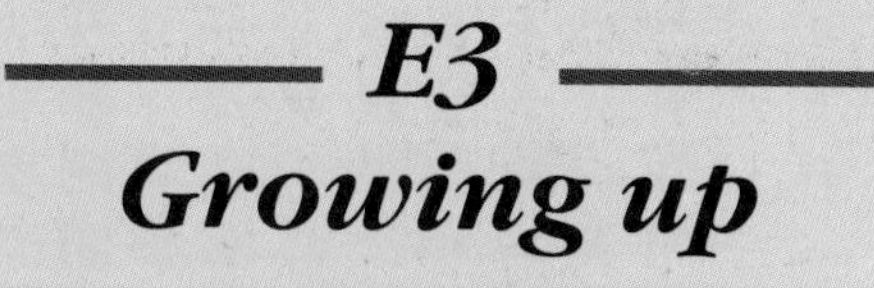

E3 Growing up

Think of the changes that have taken place in your body since you were a baby.

Picture 1 Yvonne aged 1, 8, 15 and 21.

Childhood

A helpless baby becomes a child. The main difference between a baby and a child is that the child is able to do things for itself. An important process is **conditioning**. This is explained on the opposite page.

During childhood we learn all sorts of skills. Some of these skills are physical, others are mental. One of the greatest tests of physical skill is riding a bicycle, which most children learn quite early in life. In gaining such skills, **practice** is essential.

Learning is important not only in childhood but throughout our lives. To begin with we learn from our parents and grandparents. Later on, teachers and older friends play a part. We also learn from people of our own age – our **peers**.

A healthy **balanced diet** is essential for a growing child (see page 110). Throughout childhood, growth takes place slowly and steadily. Round about the age of eight to ten years, growth slows down slightly. Then a change takes place which has a great effect on the body.

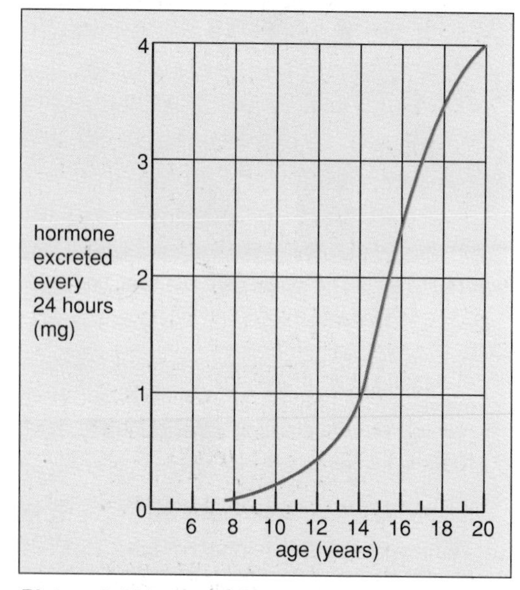

Picture 2 This graph shows how the amount of sex hormone produced by boys increases. The data was obtained by measuring the total amount of hormone excreted in the urine every 24 hours.

Puberty, a time of change

A baby has a complete set of sex organs. However, the testes cannot make sperm, and the ovaries cannot produce eggs – though thousands of immature eggs are already present, ready and waiting.

Between the ages of about twelve and fourteen, the sex organs become active. The testes start making sperm, and the ovaries start releasing eggs. This period in a person's development is called **puberty**. The time of puberty varies from person to person. It usually occurs slightly earlier in girls than in boys, and interestingly it occurs earlier now than it did fifty years ago.

What causes the ovaries and testes to start working? The answer is hormones. These particular hormones are secreted by the pituitary gland at the base of the brain (see page 200). Once the ovary and testes become active, they secrete **sex hormones**. the main sex hormones are **oestrogen** in the female, and **testosterone** in the male (picture 2).

The sex hormones bring on other changes at puberty. In both sexes pubic hair develops in the genital region. In girls the breasts start developing, and fat is laid down in the thighs. In boys the penis gets larger, the voice breaks and hair starts growing on the legs, chest and face. If a male is castrated (has his testes removed) before reaching puberty, these changes do not happen. At one time choirboys were castrated to prevent their voices breaking.

The features which have just been described make up a person's **secondary sexual characteristics**. At this stage growth speeds up, and boys and girls become interested in the opposite sex.

You will know from your own experience that these changes do not occur all at once. They take place gradually over several years. What does happen suddenly is the production of eggs and sperm. With boys this may start as a '**wet**

Activity

Care of the young

The human species is remarkable for the care it takes of the young. Other animals show this too, particularly mammals and birds, but the human spends longer than any other animal caring for the young.

Work in a group, pooling your ideas. First, decide what 'caring for the young' really means, and who is responsible for it. Then make a list of all the ways you have been 'cared for' during your life. Include your teenage years, as well as your childhood. Finally, try to decide when a human becomes self-supporting.

Find out, from books or talking to people or from your own observations, how other mammals such as dogs, cats or hamsters care for their young.

dream'. Sperm start being made and semen builds up; it has to go somewhere, so it is discharged spontaneously during sleep. Not everyone has a wet dream – boys who masturbate find that semen is produced when they have an orgasm.

Girls start having '**periods**'. The first sign is a small amount of blood passing out of the vagina. This is called **menstruation**, and the reason why it happens will be explained in the next topic. Periods tend to be irregular at first, but eventually they settle down and take place regularly once a month. Generally the bleeding goes on for several days. During this time pads may be used or a cotton wool tampon may be placed in the vagina. This absorbs the blood and enables you to go on doing all the usual things, even swimming and playing games.

In the next topic we shall see that menstruation is controlled by the sex hormones. Changes in the amounts of these hormones make some women feel tense and off colour just before their periods, and there may be slight pain during the period itself.

Adolescence

When you reach puberty you become an **adolescent**. Adolescence continues until the age of about eighteen, when growth ceases.

During adolescence the sex urge increases. Having intercourse too early in adolescence may be a risk to health, and can cause emotional problems. In fact it is against the law to have intercourse before the age of sixteen.

The changes which occur during adolescence may cause emotional swings that are sometimes difficult to handle. With these changes comes a desire for greater freedom and self-expression. A conflict may develop between the desire for independence and the need for security. Teenagers have to cope with these two opposing forces. Most manage very well, emerging from adolescence as responsible young adults. That's when you get the vote and start paying taxes!

Sexual decline

A woman will go on producing an egg every month until she reaches the age of 45 to 50. Her ovaries then stop producing eggs, and her menstrual periods cease. She can now no longer become pregnant. These changes occur because her sex hormones stop being produced in sufficient amounts. She has reached the **menopause** or 'change of life', and it may cause her to feel tired and run-down for a while.

Men don't go through a change of this kind. Normally a man goes on producing sperm for the whole of his life. However, the amount of sex hormone which he produces gradually falls, and he may find that his sex drive decreases.

What is conditioning?

When a dog sees its owner going to the cupboard for a tin of dog food, its mouth waters. Now you would expect a dog's mouth to water when it is given food, because salivation is a reflex (see page 194). But in this case salivation occurs before the food is given. This is because the dog has learned to associate the cupboard with food. The response is called a **conditioned reflex** and the process by which it becomes established is called **conditioning**.

Conditioning is a type of **learning**. It is made possible because the brain can store information or, to put it simply, can remember things. Many examples of it are shown by humans and other animals. It is how dogs and cats learn not to make messes in the house, and how children learn to do certain things and not others. For example, a child who trips over the bottom step every time he or she goes upstairs soon learns to be more careful.

1 Much of our behaviour is the result of conditioning. Suggest some examples. Do you think conditioning has influenced your opinions and outlook on life?

2 A quite different kind of behaviour is **instinct**. You are born with this kind of behaviour and it does not need to be learned. Can you think of any examples of instinct in humans?

3 Give examples of conditioning and instinct in wild animals. how do these types of behaviour help young animals survive?

Questions

1 In a certain town in Canada there is a horizontal line by the door of buses. If the top of the passenger's head comes below this line, only a half fare is charged. The bus company has had to raise the level of this line twice during the last 30 years.

a Why do you think they have had to do this?
b Do you think this is a fair system?

2 The Smith family consists of father, aged 50, mother 45, John 17 and Wendy 16. During the past three years the number of family rows has increased by 30 per cent. Suggest reasons.

3 Professor JM Tanner has estimated the relative rates of growth of the brain, the body in general, and the reproductive organs in humans. His findings are shown in the graph (right). Explain what the graph shows, and then suggest reasons why the three curves are different.

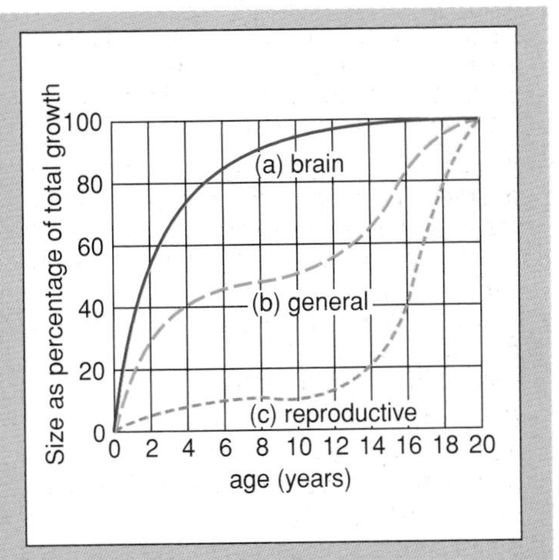

E4
The menstrual cycle

Menstruation is part of a cycle of changes which occurs in a woman's body.

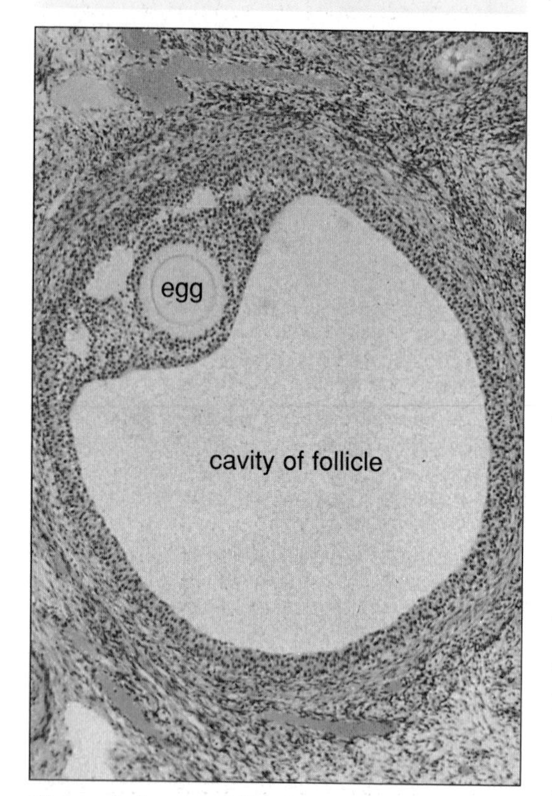

Picture 1 A mature follicle in the ovary, as seen in section under the microscope. The follicle contains a large cavity full of fluid. It is about to burst and release its egg from the ovary.

What happens during the cycle?

Let's begin with menstruation. During menstruation the lining of the uterus breaks down, and a small amount of blood passes out through the vagina. This may go on for several days.

Immediately after menstruation, an immature egg in one of the ovaries starts maturing. As it matures, it becomes enclosed in a protective case. This is called an **ovarian follicle**, and it gets steadily larger (picture 1). About two weeks after the beginning of menstruation, the follicle moves to the edge of the ovary and the egg pops out of it into the oviduct. This process is called **ovulation**.

While the follicle has been developing in the ovary, the uterus has been healing and building itself up again following menstruation. By the time ovulation occurs, the lining of the uterus is ready to receive a fertilised egg.

When the egg is shed from the ovary, the follicle stays behind and develops into a solid object called a **yellow body**. We shall see what it does in a moment. Meanwhile the lining of the uterus continues to develop – it thickens, and lots of blood vessels grow into it. What happens next depends on whether the egg is fertilised or not. If it isn't, the yellow body withers away. At the same time the lining of the uterus breaks down, and menstruation occurs. The cycle then begins all over again.

The whole point of the cycle is to make sure that the uterus is in the right state when the ovary produces an egg. So what happens in the uterus must fit in with what happens in the ovary. This means that the cycle must be *controlled* in some way.

How is the menstrual cycle controlled?

The cycle is controlled by hormones. The control centre is at the base of the brain where the **pituitary gland** is situated (see page 200).

After menstruation, the pituitary gland secretes a hormone which causes a follicle to develop in the ovary. The follicle in turn secretes the hormone **oestrogen**. This causes the lining of the uterus to repair itself following menstruation.

Shortly before ovulation, the pituitary gland secretes another hormone. This brings about ovulation and causes the follicle to change into a yellow body. The yellow body then secretes the hormone **progesterone**. This causes the lining of the uterus to thicken, and more blood vessels to grow into it.

The two hormones from the ovary (oestrogen and progesterone) *prepare* the lining of the uterus for receiving an embryo. If the egg does not get fertilised, the two hormones stop being secreted. As a result, the lining of the uterus breaks down and menstruation occurs.

What happens during pregnancy?

If a woman becomes pregnant, menstruation stops until after the baby has been born. In fact the sign that a woman is pregnant is that she misses her usual period.

What causes menstruation to stop? The presence of the embryo in the uterus alters what happens to the yellow body in the ovary. Instead of withering away, it stays in the ovary and goes on secreting progesterone. Oestrogen goes on being secreted too. As a result, the lining of the uterus remains intact. Indeed it continues to thicken, and gets even richer in blood vessels. So the hormones from the ovary prevent menstruation. They also stop further eggs being produced.

Once the placenta has been formed, it takes over the job of secreting these hormones. They continue to be secreted until just before birth. Birth is brought on by a sudden drop in these hormones. Another hormone from the pituitary gland helps the uterus to contract. And yet another pituitary hormone causes the breasts to make milk.

The contraceptive pill

Oestrogen and progesterone prevent eggs being produced by the ovaries while a woman is pregnant. This fact was used by scientists to develop the contraceptive pill. The pill consists of a mixture of synthetic equivalents of progesterone and oestrogen (or progesterone only). If taken by a woman on a regular basis, it will prevent eggs being produced by her ovaries. There is more about this and other contraceptive methods in the next Topic.

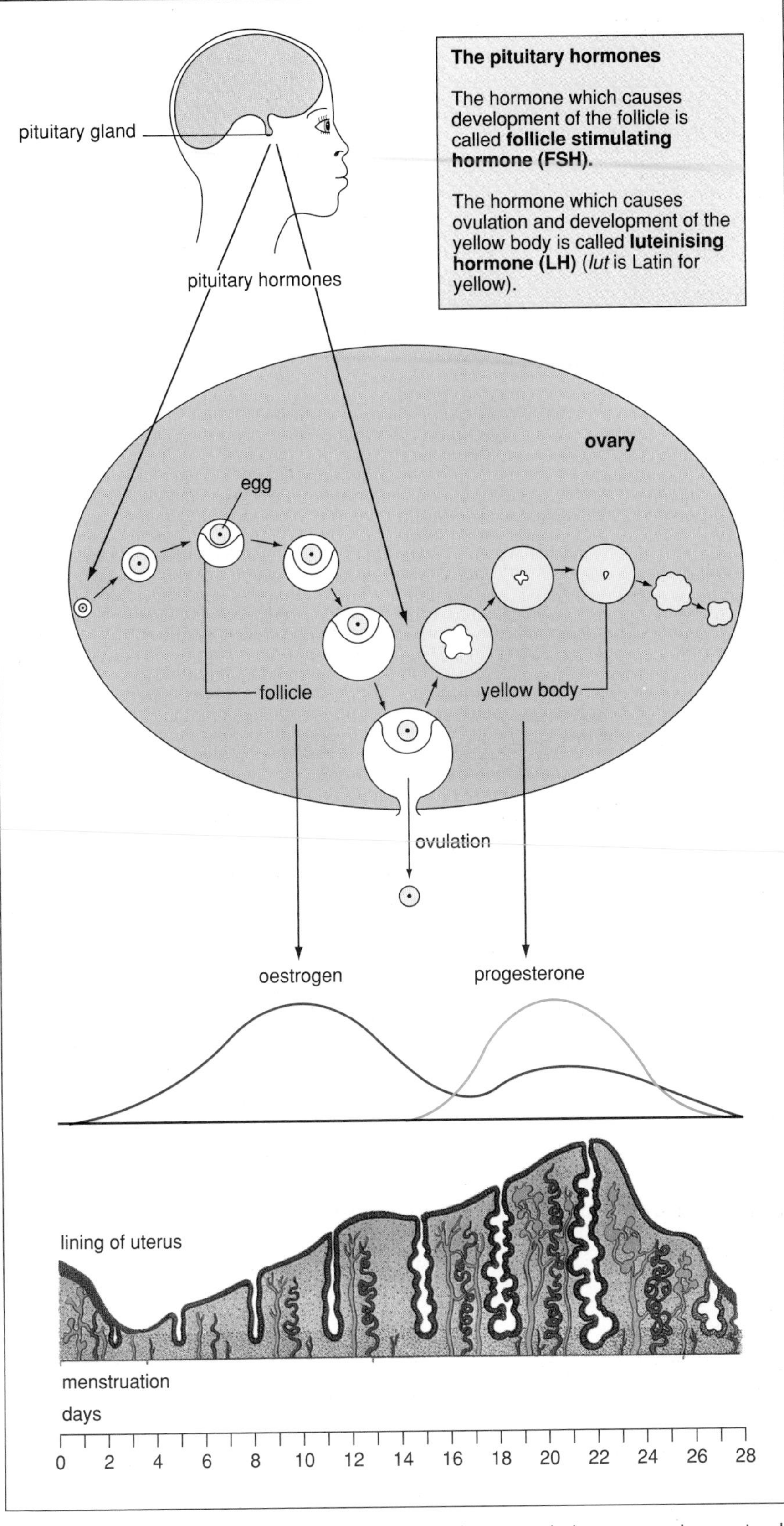

Picture 2 This picture summarises the main events that occur during a woman's menstrual cycle. The graph shows the relative concentrations of oestrogen and progesterone in the bloodstream. The amounts of these two hormones is regulated by negative feedback. When their concentrations reach a certain level, they inhibit the pituitary gland so that it stops secreting its hormones.

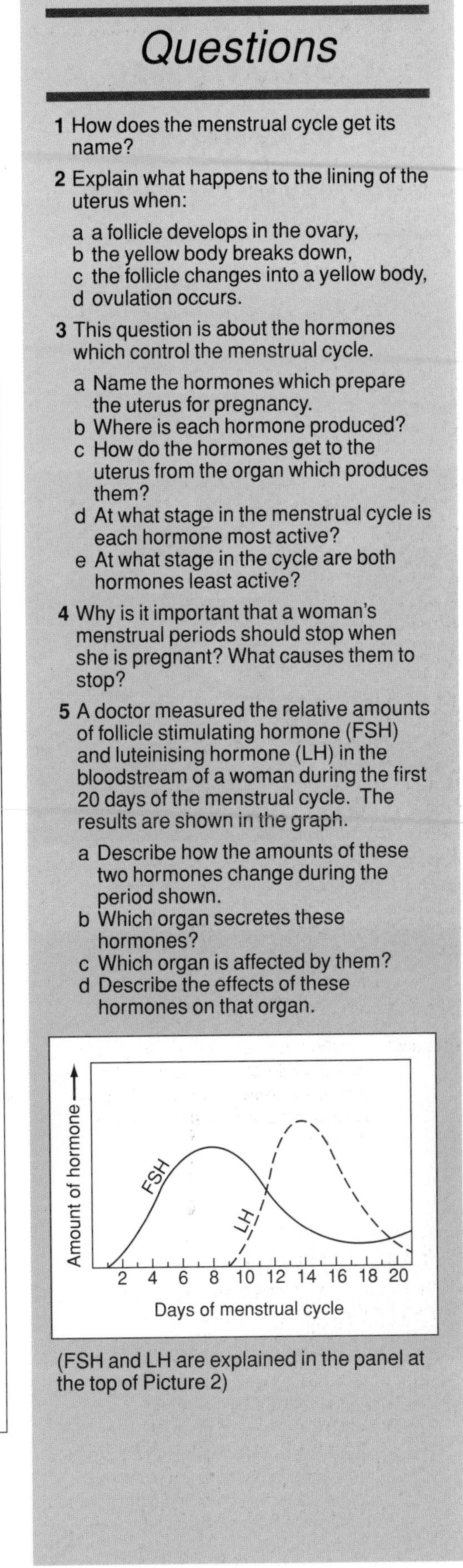

Questions

1 How does the menstrual cycle get its name?

2 Explain what happens to the lining of the uterus when:

a a follicle develops in the ovary,
b the yellow body breaks down,
c the follicle changes into a yellow body,
d ovulation occurs.

3 This question is about the hormones which control the menstrual cycle.

a Name the hormones which prepare the uterus for pregnancy.
b Where is each hormone produced?
c How do the hormones get to the uterus from the organ which produces them?
d At what stage in the menstrual cycle is each hormone most active?
e At what stage in the cycle are both hormones least active?

4 Why is it important that a woman's menstrual periods should stop when she is pregnant? What causes them to stop?

5 A doctor measured the relative amounts of follicle stimulating hormone (FSH) and luteinising hormone (LH) in the bloodstream of a woman during the first 20 days of the menstrual cycle. The results are shown in the graph.

a Describe how the amounts of these two hormones change during the period shown.
b Which organ secretes these hormones?
c Which organ is affected by them?
d Describe the effects of these hormones on that organ.

(FSH and LH are explained in the panel at the top of Picture 2)

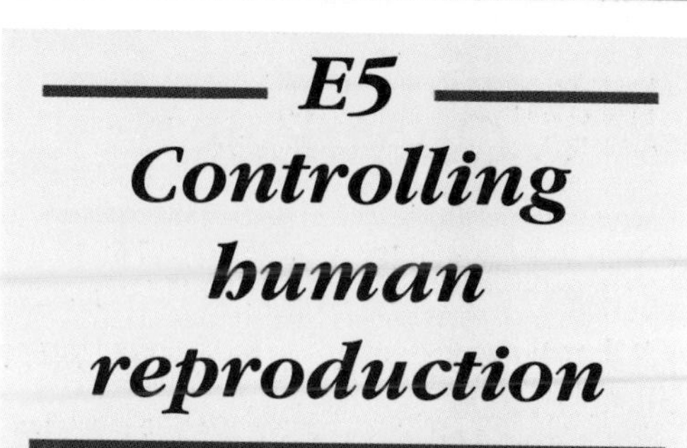

E5 Controlling human reproduction

Reproduction creates new life and carries responsibilities which everyone should consider.

Picture 1 A family planning clinic where people can obtain advice about birth control.

Picture 2 A technician examines a petri dish containing human eggs and sperm to see if fertilisation has taken place. The sperm were introduced with the pipette. The whole procedure is carried out under carefully controlled conditions in a dust-proof cabinet to prevent contamination.

Coping with our sex drive

Sex is one of the most powerful forces in animals, and humans are no exception. Most animals mate, or may try to mate, whenever a partner is available and they feel the urge. With humans it's different – or should be. In human society the **sex drive** has to be controlled.

The sex drive can be a particular problem during adolescence when it is developing. Adults are usually not short of advice on the subject, but they often disagree. However, most people would agree about one thing: sex should be part of a loving and caring relationship. This means not doing anything which may harm the other person, either physically or emotionally. Such a relationship stands the greatest chance of being stable and long-lasting.

How many children?

One of the most important decisions that faces every married couple is how many children to have. A sensible couple will want to give each child the support and love it needs, and this means not having too many. There may therefore be times when the couple wish to prevent pregnancy. This is called **family planning** or **birth control**.

In Britain and many other countries there are **family planning clinics** where people can get advice about birth control methods. These methods aim to prevent conception, so they are called **contraception**. One of the methods is natural and does not depend on the use of artificial devices. Other methods involve using a device called a **contraceptive**.

The **natural method** is simply not to have intercourse at times when the woman is likely to become pregnant. There is more about this on page 236.

One of the most widely used contraceptive devices is the **condom**, a rubber sheath which is put over the penis during intercourse (picture 2). The condom prevents sperms getting into the vagina, and is a **barrier method** of contraception. A quite different method – a chemical one – is for the woman to take the **contraceptive pill** which prevents ovulation.

Some people who are quite sure they do not want any children (or any *more* children) may choose to become **sterilised**. In the man the sperm ducts are cut, so there are no longer any sperm in the semen. If a woman wants to be sterilised, the oviducts are cut.

Family planning is extremely important, particularly in over-populated countries which are short of food. Every couple should decide how many children they can raise. They should then consider which birth control methods, if any, are right for them and in keeping with their conscience. There is a summary of the different methods on page 237.

Abortion

Despite the various birth control methods available, women may become pregnant when they do not want to. The only way to avoid having a baby then, is to destroy the embryo or fetus and terminate the pregnancy. This is known in everyday language as **abortion**.

There are various ways of carrying out an abortion, and it should always be done in hospital by a doctor. It is extremely dangerous, and indeed illegal, for it to be done in any other way.

In Britain, abortion is normally only allowed if two doctors consider that by continuing the pregnancy the woman's physical or mental health is at risk. Some people feel that abortion should be made more easily available. Others feel that it should not be allowed at all. The present law is that normally an abortion may only be carried out up to the 24th week of pregnancy. Some people have been pressing hard to get the time limit reduced to 18 weeks.

Abortion is a controversial matter which should be thought about very carefully. Many moral issues are involved.

Helping childless couples

Some couples are unable to have children, however hard they try. A common cause is that either the man cannot produce sperm, or the woman cannot produce eggs. The person is **sterile**. Sometimes the man does have sperm in his semen but there are not enough of them to ensure fertilisation, or he may not be able to get an erection and ejaculate. In the case of the woman, the oviducts may be blocked, making it impossible for the sperm to reach the egg.

A lot can be done to help childless couples these days. For example, blocked oviducts can be opened, and a man's 'sperm count' can be increased. If this fails, the woman can have semen placed in her vagina with a syringe. This is called **artificial insemination**. The semen may be taken from the husband, or it may come from a **sperm bank**, having been supplied by an anonymous donor. In a sperm bank the semen is stored in a frozen state until it is needed.

A woman who is not producing eggs can be given a **fertility drug**. This is a hormone preparation which stimulates her ovaries to start working. It is similar to the pituitary hormone FSH (see page 233). On occasions this treatment has proved more successful than either the doctor or the patient bargained for, and the woman has produced quadruplets, quintuplets or sextuplets.

Nowadays scientists can fertilise human eggs outside the body. An egg is taken from the ovary in a small operation, and fertilised by sperm in a dish (picture 2). This is called ***in vitro* fertilisation**. The fertilised egg divides to form a tiny embryo, and this is put into the uterus where it develops in the usual way (picture 3). Babies conceived this way are sometimes called **test tube babies**.

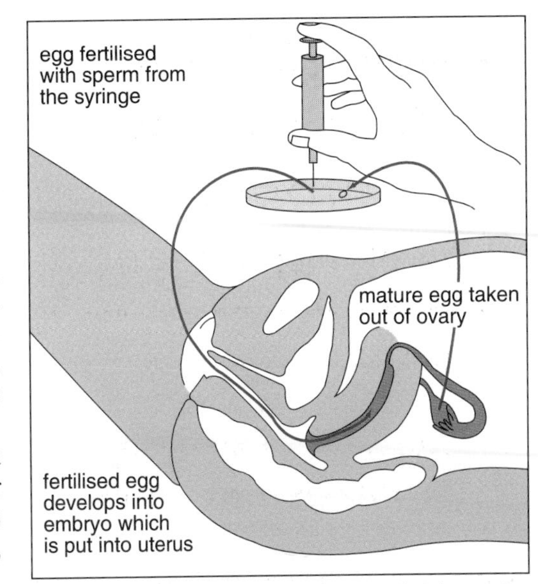

Picture 3 In *in vitro* fertilisation an egg is taken out of the ovary, fertilised in a dish and then placed in the uterus. The embryo can be placed in the uterus of the same woman or, if necessary, a different woman. In the latter case the woman is known as a surrogate mother. Under what circumstances do you think a surrogate mother would need to be used?

Research on human embryos – right or wrong?

Many people have strong views about this. Some approve, and others disapprove. You probably have opinions on this yourself, especially now that you are studying biology. But it's no use having opinions if you don't know the facts. So here, very simply, are the facts.

It all starts with a woman who can't get pregnant and chooses to have a 'test tube baby'. Now fertilising human eggs outside the body isn't easy, and sometimes it doesn't work. So the doctors don't just take one egg from the ovary, they take several. The woman is given hormone treatment beforehand to make her ovaries produce several eggs at the same time.

All the eggs are mixed with sperms, and if fertilisation is successful they may all develop into embryos. But only one embryo is needed. So what happens to the rest? The answer is that they are frozen and stored for future use. The woman herself might need them, if the first pregnancy isn't successful. Alternatively they may be used for research.

Research on what? Well, three things in particular. Firstly, doctors want to learn how to help the sperm to fertilise the egg. Secondly, doctors want to improve ways of rearing the embryos and transferring them to the mother's uterus. Thirdly, doctors want to devise methods of telling if an embryo has an inherited disease such as cystic fibrosis. A healthy embryo can then be selected for transferring to the mother.

But a basic question arises. Suppose you have an embryo, and you wish to do research on it. When should the research stop, and the embryo be destroyed? In the early 1980s a committee, headed by Baroness Warnock, was set up to look into this question and make recommendations to the government. After a great deal of deliberation, the Warnock committee recommended that the limit should be fourteen days after fertilisation.

Why fourteen days? At this stage the embryo is still very simple. The cells are all similar, and have not yet developed into particular kinds of cells. So at this stage the embryo is a bunch of undifferentiated cells. Each cell, if separated from the rest, is potentially able to develop into a new human being. So, argued the committee, the embryo at this stage can't be regarded as a single individual. Another important thing is that the nervous system has not yet started to form, and indeed won't do so for at least another week. This means that there is no possibility of the embryo feeling pain.

The fourteen day embryo is clear and easy to recognise, and a scientist would never have any difficulty knowing how old it is. At this stage a groove starts forming along the top. It's called the **primitive streak**. When the Warnock Report was first published in 1983, the expression 'primitive streak' hit the headlines. The anatomist who originally named this part of the embryo could not have guessed the publicity it would later receive!

In April 1990 the Warnock Report was debated in the House of Commons. It was decided by a majority of 362 votes to 189 that embryo research should be permitted up to the fourteenth day after fertilisation.

1. Some people (including some scientists) feel that research on human embryos should not be allowed at all. What do *you* think?
2. Some scientists would like to do research on embryos up to the stage at which the nervous system is formed. What do *you* feel about this?
3. Views about embryo research (and abortion) depend on when you feel a human life begins. Does it begin at conception, or at fourteen days after conception – or perhaps at a later stage? What do *you* think?

Methods of contraception

Life is full of choices. One choice which many people have to make is what sort of contraception to use. Doctors and family planning clinics help people to make the right choice. In making a decision, these questions have to be asked:

Is the method

- reliable at preventing pregnancy?
- likely to make sex less enjoyable or fulfilling for either partner?
- uncomfortable or difficult to use, and are there any health hazards?
- in keeping with one's religious and ethical beliefs?

The main methods of contraception are summed up in the table on the opposite page.

Other contraceptive methods

Research is always going on into new methods of contraception. For example, since 1992 a **female condom** that can be worn by women has become available. This is a soft polyurethane sheath which lines the vagina and the area just outside. Like the male condom, it also helps to prevent the transmission of AIDS and other sexually transmitted diseases.

New kinds of pill are available too. The pill described in the table is called the **combined pill** because it contains oestrogen as well as progesterone. Occasionally the combined pill may cause thrombosis in older women, particularly if they smoke. For such women a **progesterone-only pill** is available. If taken regularly according to instructions, this stops sperm entering the uterus, or prevents the fertilised egg implanting in the uterus. In some women it prevents ovulation. Its success rate is almost as good as that of the combined pill.

The trouble with pills that have to be taken regularly is that it is easy to forget to take them. This can be overcome by using an **injectable contraceptive**: a single injection given once every few months releases progesterone very slowly into the bloodstream. Or you can have an **implant**: small soft tubes placed under the skin on the inner side of the upper arm slowly release progesterone into the bloodstream for up to five years. The success rate for injectables and implants is slightly better than for the combined pill.

Research is going on into a **male contraceptive pill.** This works by preventing the testes producing sperm though semen is still produced as normal.

The natural method

Despite all the types of contraception available, some people prefer the natural method. Its advantages are that it does not involve using any artificial devices or substances, and there are no objections to it on religious grounds.

For the natural method to work, the woman needs to know when she is fertile, and this means knowing when she is going to ovulate. There are two main ways of telling when this is:

- **The calendar method**. Ovulation normally occurs about half way between one menstrual period and the next – usually between days 12 and 16 in the menstrual cycle (see page 233). If you know the dates of your menstrual periods, you should be able to work out when ovulation is likely to occur.
- **The temperature method**. Just after ovulation the body temperature rises slightly (see picture 1). By recording the body temperature every day over several months, you can work out when ovulation has occurred in the past and when it is likely to occur in the future.

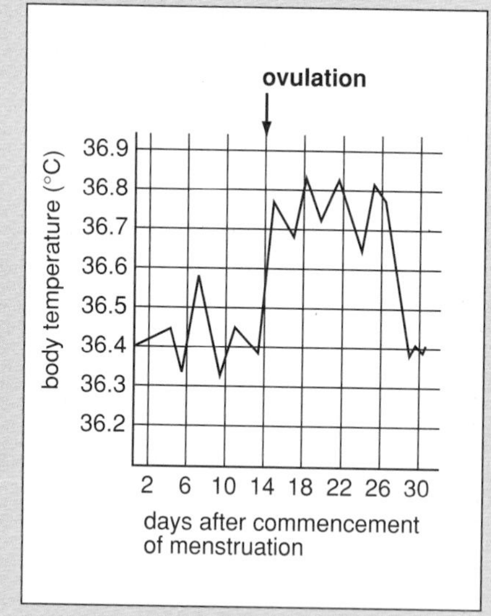

Picture 1 The basal body temperature of a woman during one menstrual cycle. The basal body temperature is the temperature at rest. It is taken at the same time each day, just after waking in the morning. Notice the slight rise in temperature following ovulation.

The time of ovulation and the time between menstrual periods may vary, particularly in teenagers, so this makes the calendar method difficult to use. The temperature method is more accurate, particularly if it is combined with other signs of ovulation. For example, the changes in the cervical mucus mentioned on page 225 are a sign that ovulation will soon take place.

In practice a combination of methods is used. Armed with all the necessary data, it is then possible to work out when it is safe to have intercourse without the risk of becoming pregnant. This is called the **safe period**.

To be successful, natural contraception requires expert guidance from a specially trained teacher. It is also important that *both* partners, i.e. the woman *and* the man, should want it to succeed.

1 What is the *scientific* basis of the natural method of contraception? Relate your answer to the menstrual cycle.

2 List as many things as you can think of which might affect the time and duration of the safe period.

3 There are cases of women becoming pregnant having had intercourse just before, or just after, menstruation. Suggest reasons.

4 In the list of birth control methods in the table on the opposite page, which ones

a stop sperm reaching the egg,
b stop eggs being produced,
c stop the fertilised egg implanting in the uterus,
d are barrier methods,
e are chemical methods?

5 What does the combined pill consist of and how does it work?

6 Look at the right hand column of the table on the opposite page. It tells you how many women in one hundred are likely to become pregnant if they use each method of contraception.

a Convert the information into *percentage reliability* for careful and less careful use of each method.
b Suppose you were put in charge of a research project to assess the reliability of a particular method of contraception. How would you go about it?
c Suggest possible sources of inaccuracy in assessing the reliability of contraceptive methods.

Table 1 The main methods of contraception. The reliability figures in the right hand column are taken from *Your Guide to Contraception*, published by the Family Planning Association. The figures are based on an evaluation by a panel of experts of the results of different investigations carried out in recent years. Individual investigations into the reliability of contraceptive methods often give widely different results. For example, some suggest that the condom is more reliable, and the natural method less reliable, than the figures in this table would suggest. Why do you think different investigations give different results?

Method	How it works	Reliability
Natural method	The woman avoids having intercourse when she is fertile, i.e. during the part of the menstrual cycle when ovulation is likely to occur.	With careful use under expert guidance, 2 women in 100 get pregnant in a year. With less careful use up to 20 women in 100 get pregnant in a year.
Condom	Thin rubber sheath placed over the erect penis. Stops sperm entering the vagina. Also helps to prevent transmission of sexually transmitted diseases, including AIDS.	With careful use 2 women in 100 get pregnant in a year. With less careful use up to 15 women in 100 get pregnant in a year.
Diaphragm (cap) with spermicide	A flexible rubber cap is inserted into the vagina to cover the cervix, then a spermicide is squirted into the vagina. Stops sperm entering the uterus.	With careful use, 4 to 8 women in 100 get pregnant in a year. With less careful use up to 18 women in 100 get pregnant in a year.
Intra-uterine device (IUD)	A small plastic and copper object is placed in the uterus by a doctor. Stops sperm reaching the egg and/or the fertilised egg implanting in the uterus.	Fewer than 2 women in 100 gets pregnant in a year.
The combined pill	Hormone preparation, consisting of synthetic equivalents of oestrogen and progesterone, taken regularly according to doctor's instructions. Prevents ovulation (see page 232).	With careful use fewer than 1 woman in 100 gets pregnant in a year. With less careful use 3 or more women in 100 get pregnant in a year.
Female sterilisation	The oviducts are tied and cut so that sperm can never reach an egg.	The failure rate is 1 to 3 in 1000.
Male sterilisation (vasectomy)	The sperm ducts are tied and cut so that sperm are not present in the semen when ejaculation occurs.	The failure rate is about 1 in 1000

Other methods of contraception are discussed on the opposite page.

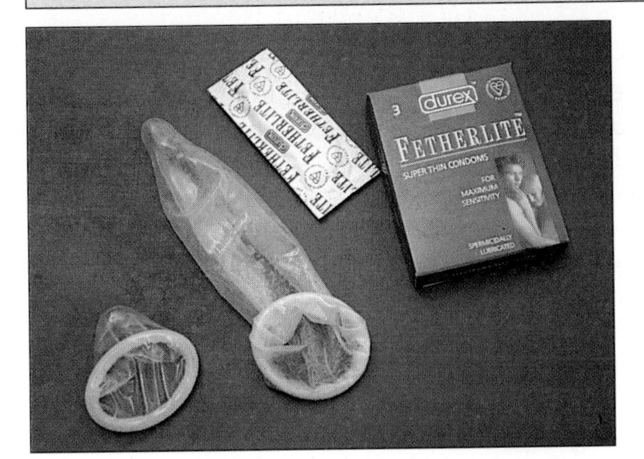

Picture 2 The condom, a widely used contraceptive. It is unrolled over the erect penis, and the semen is caught in the teat at the end. The condom may be coated with a lubricant which also acts as a spermicide (kills sperm). Oil-based lubricants should not be used as they can damage the rubber.

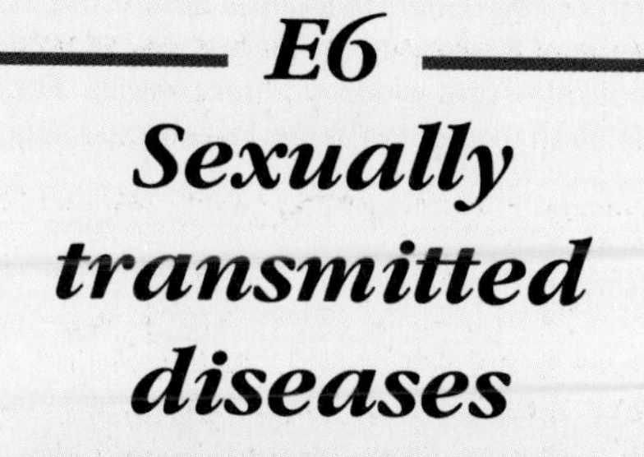

E6
Sexually transmitted diseases

Sex involves intimate contact between people. It is therefore an easy way for germs to get from one person to another.

What are sexually transmitted diseases?

Any infectious disease can be passed from one person to another during sexual contact. However, certain diseases are particularly likely to be passed this way. These are called **sexually transmitted diseases (STDs)**.

Some STDs are relatively minor infections of the genital organs. Examples include **thrush** and **genital warts**. Others are more serious and their effects more widespread, Examples are **herpes**, **non-specific urethritis** (infection of the urethra) and **hepatitis B** (viral infection of the liver).

Not all STDs are caught only by sexual contact. Some, such as thrush and hepatitis B, are caught in other ways as well.

Here we shall concentrate on three very harmful STDs: **syphilis**, **gonorrhoea** and **AIDS**.

Syphilis

Syphilis, or 'the pox' as it used to be called in Shakespeare's day, may start with a sore on the genital organs and spots on the skin (picture 1). But these soon clear up. However, the germs remain in the bloodstream. Eventually, perhaps after many years, they may attack the brain, causing the person to go blind and insane.

Syphilis is caused by a bacterium which can be caught only by close sexual contact. The bacteria can pass across the placenta, so a mother can infect her baby. The baby may be born dead, or be crippled with the disease later.

Syphilis used to be a major killer. Today people can be cured by treatment with antibiotics such as penicillin, provided that treatment is started at an early enough stage. This disease is now rare in Britain.

Gonorrhoea

This is also caused by bacteria. As with syphilis, the early signs are hardly noticeable: a slight discharge from the reproductive opening and a burning sensation when you urinate. But the germs may spread to other parts of the body, causing painful joints and illness. Gonorrhoea can cause sterility.

A baby can get infected from its mother while it is being born. The baby may develop sore eyes which, if untreated, can lead to blindness.

As with syphilis, gonorrhoea can be cured by antibiotics if the treatment is started soon enough. The disease is still quite common in Britain.

AIDS

Syphilis and gonorrhoea have been around for thousands of years. AIDS is a new disease. It was discovered around 1980 mainly amongst male homosexuals in America. It also occurs in heterosexuals, both male and female, and is spreading through the world.

AIDS stands for **acquired immune deficiency syndrome**. It is caused by a virus called **HIV (human immunodeficiency virus)**. The virus attacks the immune system (see page 158). The result is that the person succumbs very easily to diseases like pneumonia and a type of skin cancer called Kaposi's sarcoma. The virus also attacks the brain. Although people may keep recovering for a while, the disease usually wins in the end. Most patients waste away and eventually die of an infectious disease such as pneumonia, which they can no longer fight.

You can only catch AIDS if an infected person's blood or other body fluid gets into your blood. This can happen during intimate sexual contact, both heterosexual and homosexual. It can also happen if drug users share a needle, or if someone is given contaminated blood in a transfusion. You cannot catch AIDS by shaking hands, hugging, kissing, or drinking from the same cup as an infected person.

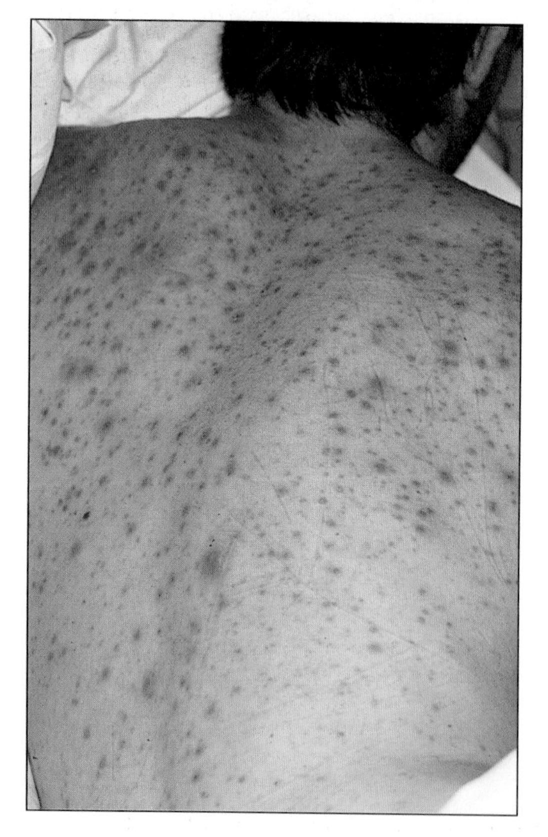

Picture 1 The spots on this person's back are a symptom of syphilis.

Picture 2 An AIDS hospice where patients receive treatment and care in pleasant, friendly surroundings.

A person can carry the virus for years without having any symptoms of the disease. However, carriers can pass the virus to other people, and a woman can pass it to her baby across the placenta or in her milk.

If the virus gets into your blood, your white blood cells produce antibodies against it. However, the antibodies don't manage to destroy the virus, which remains hidden in your cells. Doctors test people for the virus by finding out if their blood contains antibodies against it. A person with the antibodies is described as **HIV positive**.

At present there is no cure for AIDS, though various drugs can help to keep the virus under control. Scientists are trying to develop a vaccine which will protect people against it.

How to avoid sexually transmitted diseases

The only sure way of avoiding sexually transmitted diseases is to be certain that you do not have an infected partner. Obviously this isn't always easy, but having casual sex with all sorts of different people is asking for trouble.

Using a condom helps to prevent the spread of most sexually transmitted diseases, including AIDS. But it doesn't give total protection.

Most large hospitals have a Special Clinic where people can be examined to find out if they have got any sexually transmitted diseases. The examination includes a urine and blood test, and the results are confidential. When the tests have been done, treatment can, if necessary, be given. If possible, the person's sexual contacts should also be traced and treated. This helps to prevent sexually transmitted diseases spreading through the community.

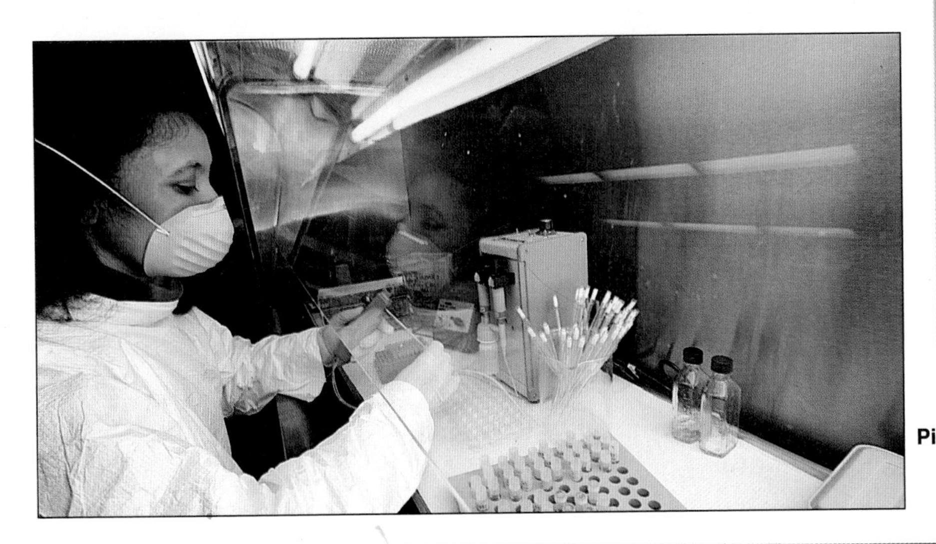

Picture 3 A technician preparing samples of human blood serum to be tested for the presence of AIDS antibodies.

Activity

AIDS: fact or fiction?

How accurate is the information which we receive about AIDS? The trouble is that what we read in newspapers and magazines is not always completely true.

Your teacher will give you a selection of newspaper articles about AIDS. Read them carefully, and try to answer the following questions.

1. What appears to be the purpose (if any) of each article? For example, does it seem to want to make people more scared, or less scared? Is it simply passing on information, or opinions as well?
2. On what points, if any, do the articles disagree in their information? (If two articles disagree, one of them may be wrong.)
3. Do any of the articles make factual statements without backing them up with evidence? (It is not always possible to give evidence for every statement, but some articles expect us to take a lot on trust.)
4. In your opinion which are the 'best' and 'worst' articles. Give reasons for your choices.

The idea of this activity is to help you to read newspaper and magazine articles in a critical and scientific way. This applies not only to articles on AIDS, but to articles on any scientific matter.

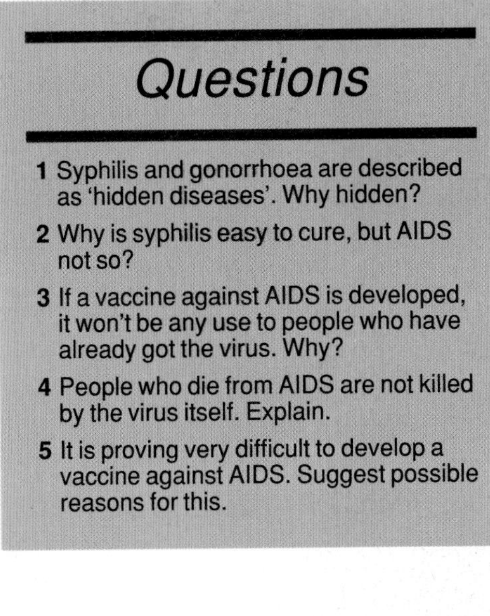

Questions

1. Syphilis and gonorrhoea are described as 'hidden diseases'. Why hidden?
2. Why is syphilis easy to cure, but AIDS not so?
3. If a vaccine against AIDS is developed, it won't be any use to people who have already got the virus. Why?
4. People who die from AIDS are not killed by the virus itself. Explain.
5. It is proving very difficult to develop a vaccine against AIDS. Suggest possible reasons for this.

E7 From flowers to fruits

In flowering plants the part of the plant responsible for sexual reproduction is the flower.

The ins and outs of a flower

The main parts of a flower can be seen very easily in plants like cherry, plum and hawthorn. This kind of flower is shown in the pictures below.

The flower is made up of a series of rings of structures. The outermost ring consists of several small leaf-like **sepals**. Then come the **petals** which are often brightly coloured and strongly scented. Next come the **stamens** which look rather like pins. And finally in the centre there is a club-shaped **carpel**.

At the base of each petal you can see an area which is slightly thicker than the rest. This is called the **nectary**, and it produces a sugary liquid called **nectar**.

The stamens are the male part of the flower. Each one consists of a slender **filament** with a knob at the top called the **anther**. The anther contains four **pollen sacs** in which **pollen grains** are formed. The pollen grains are equivalent to an animal's sperm.

The carpel is the female part of the flower. It consists of three parts: a slightly swollen **stigma** at the top, then a slender stalk called the **style**, and a swollen **ovary** at the bottom. Inside the ovary there is a small body called an **ovule** which is attached to the wall of the ovary by a short stalk. There is a small hole in the wall of the ovule, leading to the inside. In the middle of the ovule, surrounded by a little bag, is an **egg cell**. This is equivalent to an animal's egg.

Variations on the theme

Not all flowers are like the one just described, though they all follow the same basic plan. In particular their shapes and symmetry differ (picture 3). So do the numbers of the parts. For example, the buttercup has many carpels, each containing one ovule, whereas the sweet pea has one carpel containing a row of ovules.

Flowers differ in other ways too. For example, in a daffodil the petals may be joined to form a 'trumpet', and in tulips the sepals are missing.

Picture 1 This flower has been sectioned down the middle to show the inside.

Picture 2 The structure of a typical flower.

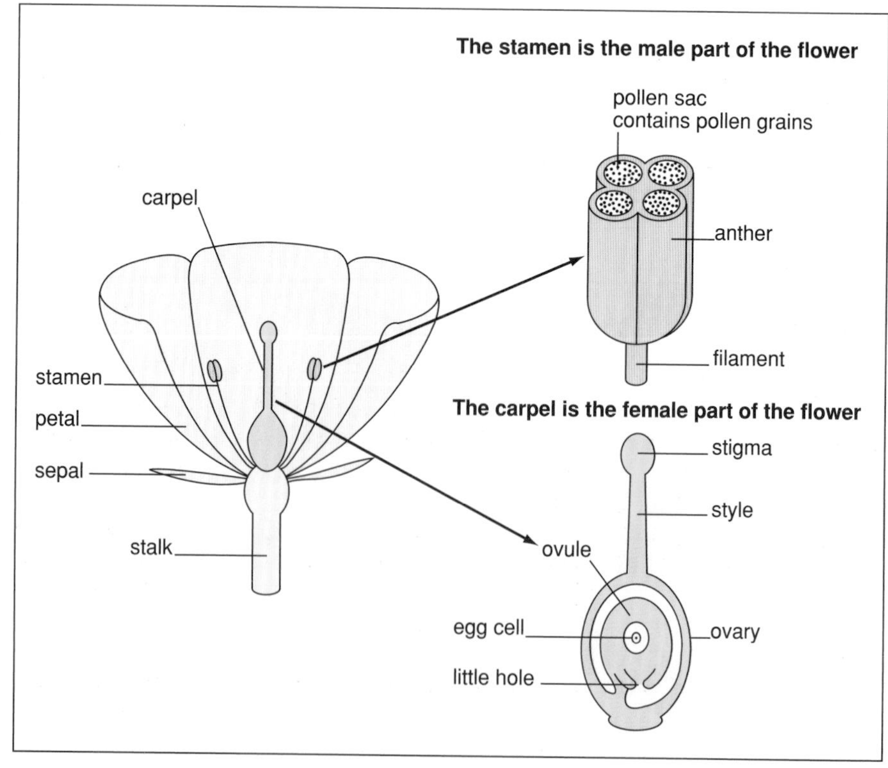

Picture 3 There is an infinite variety of flowers. This orchid attracts insects by looking like a potential mate.

Pollen and pollination

Pollen grains are very small, like little specks of dust. Their job is to see that the egg cells become fertilised.

The pollen grains develop inside the anthers. When the anther is mature, it splits open down each side and the pollen grains are set free. They are then carried to another flower, and if one of them gets onto a stigma it sticks to it. The process by which the pollen grains reach the stigma is called **pollination** (picture 4).

Why is the pollen carried to another flower? Why not let it fall onto the stigma of the same flower? Actually this does sometimes happen – it is called **self-pollination**. But it isn't good for the species (see page 252). It is much better if the pollen is transferred to another flower on a different plant of the same species. This is called **cross-pollination**.

The pollen grains are usually transferred by wind or insects. In the tropics some flowers are pollinated by humming birds and bats. Of course, pollination can also be carried out artificially by humans.

An example of a **wind-pollinated plant** is hazel. The familiar hazel catkin is a clump of very small male flowers. The catkin hangs down and is shaken by even the slightest gust of wind. This ensures that the pollen is scattered over a wide area. The pollen grains themselves are small and light.

An example of an **insect-pollinated plant** is the rose. Insects such as bees visit the flowers to feed on the nectar. As the insect pokes its head into the flower, its hairy body gets covered with pollen (picture 5). When the insect visits another flower, some of the pollen gets onto the stigmas, pollinating them. The pollen grains often have sculptured or spiky walls which help them to stick to the insect (picture 6).

Experiments have shown that insects such as bees are attracted to flowers by their colour, shape and smell. Some flowers are wonderfully adapted to attract insects and to make it easier for them to collect the pollen. For example, many flowers have lines or spots on the inner side of their petals. These markings show bees where to enter the flower and reach the nectar.

Fertilisation

Once a pollen grain has landed on a stigma, it sends out a snake-like outgrowth called a **pollen tube**. This grows into the stigma and down the style. The course that it normally takes is shown by the arrow in picture 3. Towards the tip of the pollen tube there is a nucleus. This is equivalent to the nucleus in the head of an animal's sperm.

When the pollen tube reaches the ovary it pushes its way into the ovule, usually through the little hole in the wall. It now grows towards the egg cell in the centre. Then the pollen nucleus fuses with the egg cell nucleus. This is **fertilisation**, and it is equivalent to the fertilisation of an egg by a sperm.

What happens next?

The fertilised egg now divides into a little ball of cells which becomes an **embryo**. This remains in the centre of the ovule where it becomes surrounded by a special tissue called the **endosperm**. The endosperm supplies the embryo with food. Humans have a particular interest in the endosperm – you probably ate it for breakfast (see page 246).

Meanwhile the ovule itself becomes the **seed**, and the tissue round it forms a protective **seed coat** (**testa**). As this happens, water is drawn out of the seed, with the result that it becomes very dry. In this state the embryo becomes dormant, and can survive bad conditions such as drought or cold.

While the seed is forming, the ovary expands into a **fruit**. So the seed becomes surrounded by a fruit. If you cut fruits open, you can expect to find seeds inside.

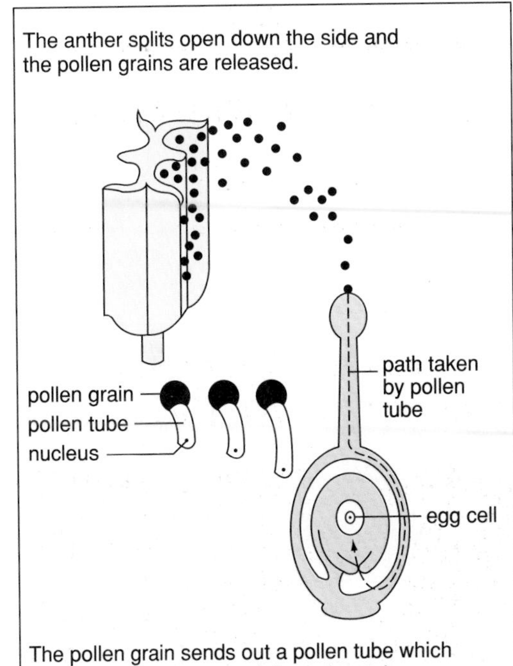

Picture 4 In pollination pollen grains are transferred from an anther to a stigma. A pollen tube then grows down the style to the ovary, as indicated by the arrow.

Picture 5 A bee collecting pollen from a flower.

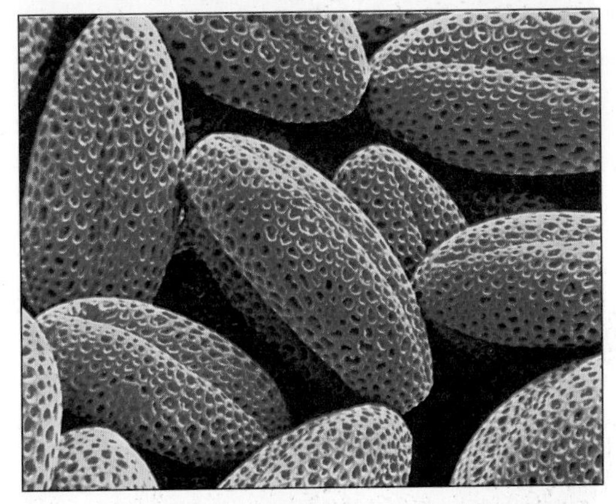

Picture 6 Pollen grains of the rape plant *Brassica napus* as seen greatly magnified by the scanning electron microscope. Notice the sculptured walls. They help the pollen to stick to insects.

Picture 8 The fruits of the sycamore tree have 'wings' which enable them to whirl through the air.

Picture 7 Some edible fruits

Picture 9 Dandelion fruits are like little parachutes and float through the air.

Fruits and dispersal

The job of the fruit is to help spread the seeds. This is called **dispersal**. Efficient dispersal is important to all organisms, not just plants. It ensures that the species is distributed over a wide area. This helps to avoid overcrowding, and increases the chances of survival.

Fruits disperse their seeds in many different ways. Some are eaten by animals, particularly birds. The soft part of the fruit is digested but the seeds are protected from the action of the animal's digestive juices. Eventually they pass out with the faeces, probably a long way from where the fruit was eaten. These sorts of fruit are often highly coloured, and this attracts animals. Think of bright red cherries, for example. They also taste good – which is why humans are interested in them (picture 7).

Other fruits disperse their seeds by floating through the air, clinging to the fur of animals or floating in water. These fruits are usually dry and hard, and they release their seeds by splitting open. In some cases the fruit splits open with such force that the seeds are thrown out quite a long way. In other cases the seeds may be shaken out of the fruit, like pepper from a pepper pot. Examples of fruits, and the way they disperse their seeds, are given in pictures 8 to 12.

Picture 10 The fruits of burdock, called burs, have hooks which cling to the fur of animals and to people's clothes.

Picture 11 The fruit of the coco-de-mer palm from the Seychelles contains air spaces which enable it to float in the sea.

Picture 12 The fruit of a poppy, called a capsule, scatters its seeds when it is shaken by the wind.

Activities

A Looking at the structure of a typical flower

1. Your teacher will give you a flower from, e.g. a cherry or a hawthorn tree.
2. Identify its parts, using picture 2 to help you.
3. Make an accurate drawing of the flower and label its parts.
4. Pull off a sepal, petal, stamen and carpel, and lay them on a piece of paper.
5. Examine each one under a hand lens. Draw them in outline.
6. Cut open the carpel.

   Can you see ovules inside it?
7. Cut open an anther.

   Can you see pollen grains inside?

B Looking at other flowers

1. Your teacher will give you up to six different kinds of flowers.
2. Examine each one carefully and write down how it differs from the flower which you looked at in activity A.

   Suggest reasons why each kind of flower has its own characteristic shape and form.

C Exploring the differences between wind- and insect-pollinated flowers

1. Your teacher will give you one wind-pollinated flower, and one insect-pollinated flower.

   Which do you think is which? How do you know?
2. Examine the insect-pollinated flower in detail.

   What sort of insect do you suppose pollinates it?

   Give reasons for your suggestion.
3. If possible look at flowers being visited by insects. In each case observe what the insect does, and note any special adaptations which the flower has for being pollinated.

D What makes pollen grains send out a tube?

1. With a pipette put a drop of sugar solution (15% sucrose) on a slide.
2. Obtain a flower, e.g. of nasturtium, that has ripe pollen.
3. With a paintbrush pick up a few pollen grains from the anther and place them in the sugar solution. Put on a coverslip,.
4. Set up a second slide, but put the pollen grains into a drop of water instead of sugar solution.
5. Place your two slides in a warm, dark place for about 30 minutes.
6. After about 30 minutes look at the slides under the microscope (low power).

   How does the appearance of the pollen grain differ in the two slides?

   What effect has the sugar had?

   What does this suggest about the stigmas in a flower?

   How could your suggestion be tested?

E Looking at fruits

Your teacher will give you several different fruits. Some of them may be unknown to you. Don't worry! Examine the fruits in as much detail as you can. Cut them open to see what's inside. If you like, do simple experiments on them but be sure to check any plans with your teacher first.

Try to answer these questions:

1. How many seeds are there inside the fruit?
2. How are the seeds arranged?
3. How do the seeds get dispersed?

Which of the fruits would disperse its seeds furthest, do you think? Give reasons for your answer.

In fruits such as cherries and plums the 'stone' consists of the inner part of the fruit with the seed inside. Why is the inner part of the fruit so hard? What are the functions of the middle and outer parts of the fruit?

Questions

1 Each of the words in the left-hand column is related to one of the words in the right-hand column. Write them down in the correct pairs.

sepal	colour
petal	egg cell
pollen	sugar
nectary	sperm
ovule	leaflet

2 Explain the difference between pollination and fertilisation.

Why do plants generally produce very large numbers of pollen grains?

3 'One year's seeding, seven year's weeding'.

What do you think this saying means, and what should gardeners do about it?

4 The flower shown diagrammatically below is pollinated by wind.

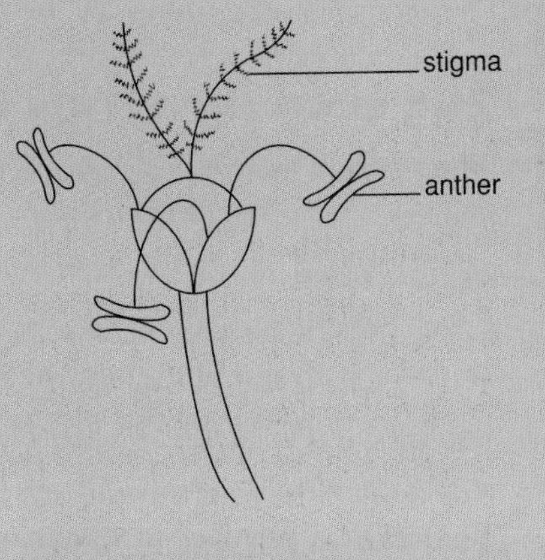

How might this arrangement of flower parts favour cross-pollination rather than self-pollination.

5 The fruit of the coco-de-mer palm (picture 11) can weigh over 20 kg but a dandelion fruit only weighs about a milligram. Why does the dandelion fruit have to be so light, when the coco-de-mer fruit can be so heavy?

6 The flowers of primroses are of two kinds. Some flowers have their anthers high up and their stigmas low down, whilst others have their stigmas high up and their anthers low down, as shown in the illustration below. They are pollinated by bees.

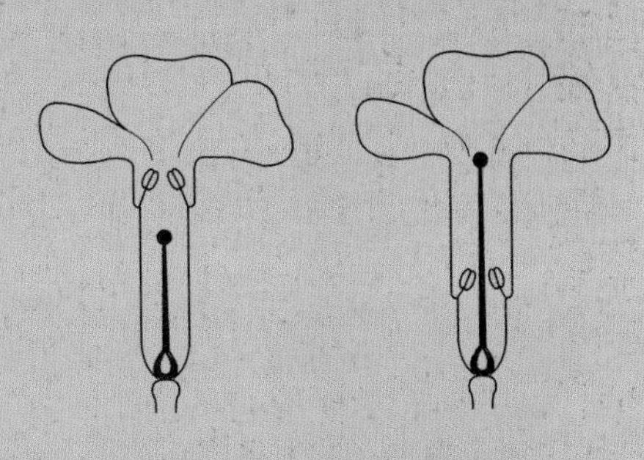

How might this arrangement of flower parts favour cross-pollination?

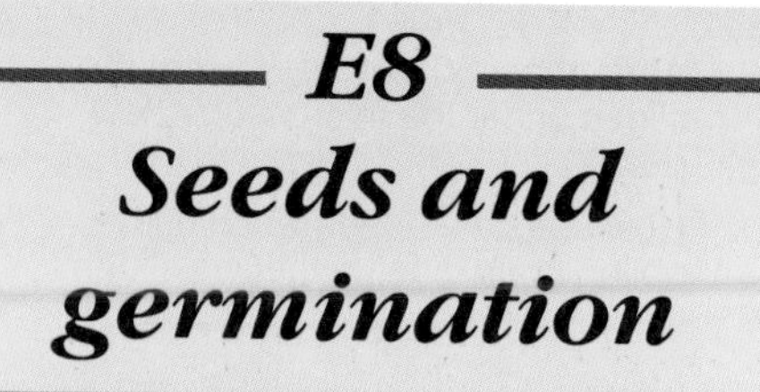

E8 Seeds and germination

Here we look at seeds and what happens to them after they have been dispersed.

Picture 1 A selection of seeds, all used in cooking.

Inside a seed

If you have broad beans with your lunch, they will be soft (or should be!). But in their natural state, they are hard and dry. This is true of seeds in general. In this state they survive the winter – or the dry season in the tropics.

Picture 2 shows a broad bean seed. It is surrounded by the **seed coat** (**testa**). If you want to see the inside of the seed, you need to soften it first. This can be done by soaking it in water for a day or so. You can then slice it down the middle. Inside is the **embryo**.

The embryo is made up of three parts:

- a baby shoot (called the **plumule**),
- a baby root (called the **radicle**),
- a pair of thick wing-like **cotyledons** ('seed leaves'),

Not all seeds are like the broad bean seed. The cotyledons are particularly variable. Sometimes they are very small, and there may be only one. But the broad bean is a good seed to study because it is so large. Let's see what happens to it.

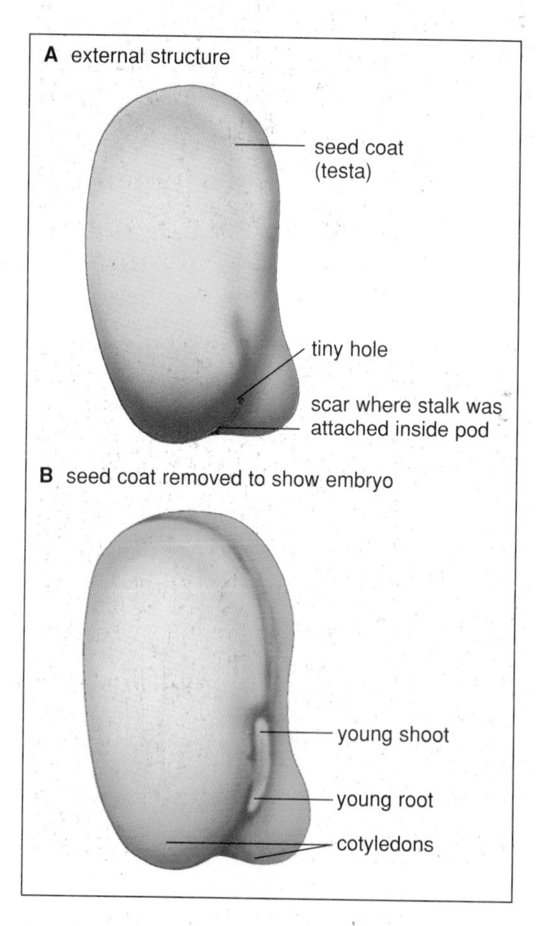

Picture 2 The structure of a broad bean seed.

The seed develops into a new plant

When spring arrives, the seed bursts open and a new plant grows out. This process is called **germination**.

Picture 3 shows stages in the germination of the broad bean. First the seed takes up water, mainly through the little hole in its wall (see picture 2). This makes it swell. As a result, the seed coat splits, and the new plant starts growing out – the root first, then the shoot. The root grows downwards, and the shoot upwards. The shoot is bent back on itself, like a hook. This protects its delicate tip as it pushes up through the soil.

The tip of the root is protected by a slimy mass of loosely packed cells called the **root cap**. This prevents it being damaged as it grows down into the soil. As it grows, the root gives off side branches which help to anchor the young plant and absorb water and mineral salts from the soil (picture 4). Slender root hairs increase the surface area for absorption.

The shoot eventually breaks through the surface of the soil. It then straightens, and the first leaves open out and turn green. Germination is complete, and we now have a young plant or **seedling**.

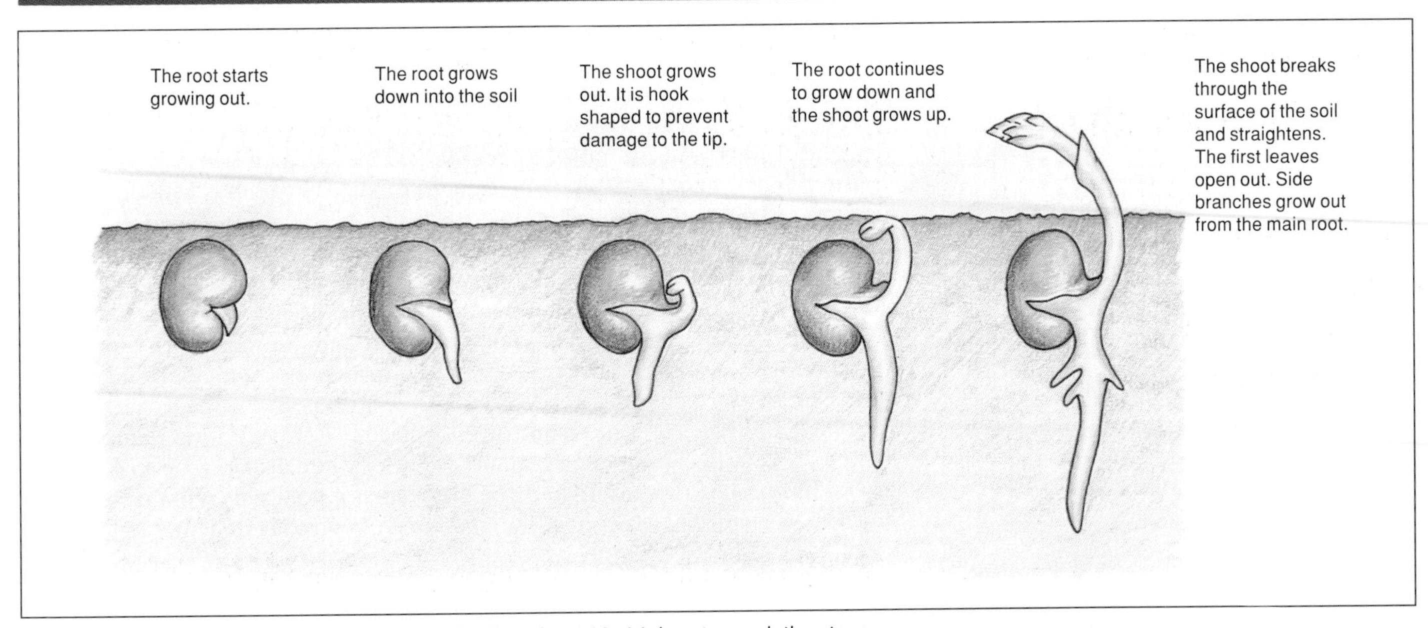

Picture 3 Germination of a broad bean seed. It takes about 12–14 days to reach the stage shown on the far right.

Not all plants germinate like the broad bean, but it is typical of many. Wheat germinates like the broad bean but the shoot, instead of being hooked, points straight up (picture 5). Its delicate tip is protected by a sheath called the **coleoptile**. When the leaves open out, they break through the coleoptile. Wheat is a type of grass, and all grasses have a coleoptile.

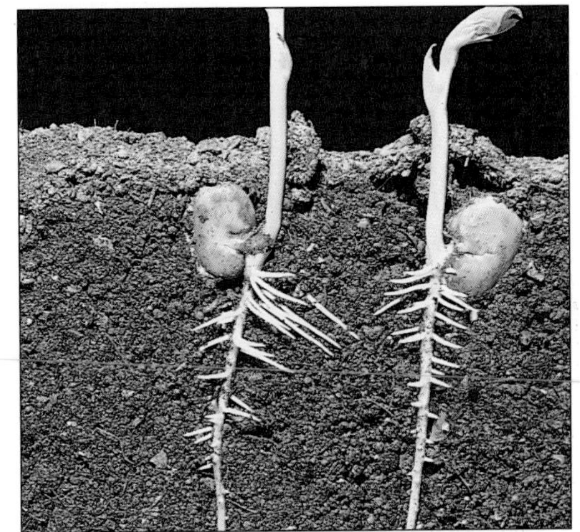

Picture 4 In this picture of young broad bean seedlings notice the side roots.

Where does the germinating seed get its food from?

For the embryo to grow, food is needed. In the broad bean, the food comes from the cotyledons. They remain inside the seed coat beneath the soil. The insoluble starch in the cotyledons is turned into soluble sugar. This is then transported to the tips of the shoot and root where growth takes place.

Once the seedling has formed its first green leaves, it can make its own food by photosynthesis. It is then self-supporting. The cotyledons are no longer needed; their food store has been used up, and they wither away.

What conditions are needed for germination?

It is very annoying when you plant seeds in the garden and they don't grow. This is usually because the seeds lack the conditions they need in order to germinate.

What are these conditions? We can find out by trying to germinate seeds in different conditions – you can do this for yourself in activity D.

From experiments of this sort we can say that these three conditions are needed for germination:

- **Water** is needed for the seed to swell and burst open. It is also needed for the stored food to be made soluble and moved to the growing embryo.
- **Oxygen** is needed for the embryo to respire. Respiration supplies energy for the embryo to grow and develop.
- **Warmth** is needed by most seeds. This is why seeds do not normally germinate till the spring or summer. The degree of warmth varies from one type of plant to another.

What about **light**? Most seeds will germinate in the light or dark. However, some germinate only in the dark. Others require light. The amount of light needed may be very small – one quick flash is enough in some cases.

All plants need light once the young shoot breaks through the surface of the soil. This is because light is needed for the leaves to open out and make chlorophyll. Only then will they start photosynthesising and feeding the plant.

Picture 5 The shoot of this wheat seedling is covered by a sheath-like coleoptile from which the first leaf can be seen emerging.

Picture 6 Wheat grain store, Iowa, USA.

Food from seeds

Seeds contain a store of food for feeding the new plants until they can support themselves. This makes them a good source of food for humans. In peas and beans the food is in the **cotyledons**. In cereals like wheat the food is in the **endosperm** (see page 241). Some seeds contain oil which is used for various purposes. Such is the case with rape, the bright yellow crop that adds splashes of colour to the British countryside in the spring. We also get oil from the seed (and fruit) of the oil palm. Its story is told on page 252.

Wheat and rice seeds are particularly useful because they form the main part of the diet of so many people. Starch, protein and a number of other useful nutrients are packed into the ripe seeds (the **grain**). From the grain all sorts of foods are made, including bread.

Wheat and rice seeds are a more concentrated source of food than, say, potatoes. This is because seeds contain so little water. In fact one kilogram of wheat has more food in it than three kilograms of potatoes. A grain store, like the one in picture 6, is a concentrated source of naturally dehydrated food.

Wheat seeds and bread-making

The picture shows the inside of a wheat seed or **grain**. It is surrounded by a coat called the **bran**. inside is the embryo and a mass of tissue called the **endosperm**.

Wheat grain is used for making flour from which, of course, bread is made. For making white bread the bran and the embryo are removed first, leaving only the endosperm. This consists mainly of starch. It is ground up (milled) into **white flour**.

For making wholemeal bread the entire grain, including the bran, is ground up. This gives **brown flour**. The brown colour is caused by a pigment in the bran. You have probably been told that wholemeal bread is good for you. It is. The reason is that the whole wheat grain contains not only starch, but other useful substances such as cellulose and vitamins. The cellulose makes wholemeal bread coarser than white bread, and provides fibre in the diet.

Nowadays white flour has some of the missing nutrients added to it after milling. So it's good for you too. But it doesn't contain as much fibre.

Wheat grain contains a protein called **gluten**. This is present in flour, and it makes dough sticky. When you make a loaf and the dough rises, the gluten holds it together and stops the gas escaping (see page 126). Gluten is therefore important in baking bread. Wheat is the only cereal that contains gluten, which is why wheat is so good for making bread. However, some people are allergic to gluten, and it can make them very ill. They have to eat gluten-free bread, e.g. bread made from rye.

Wheat grown in Europe doesn't contain as much gluten as wheat grown in North America. English bakers use a mixture of home-grown and imported wheat.

However, in France they use only homegrown wheat, so their bread doesn't hold together so firmly. French bakers get round this by making their loaves long and thin.

1 Why is it a good idea to include fibre in our diet?

2 Name the gas that is given off in dough? Where does it come from?

3 Do you think bread could be made from broad beans? Explain your answer.

4 You can buy 'French loaves' in England, but people say that it is not as good as the bread you buy in France. Why the difference?

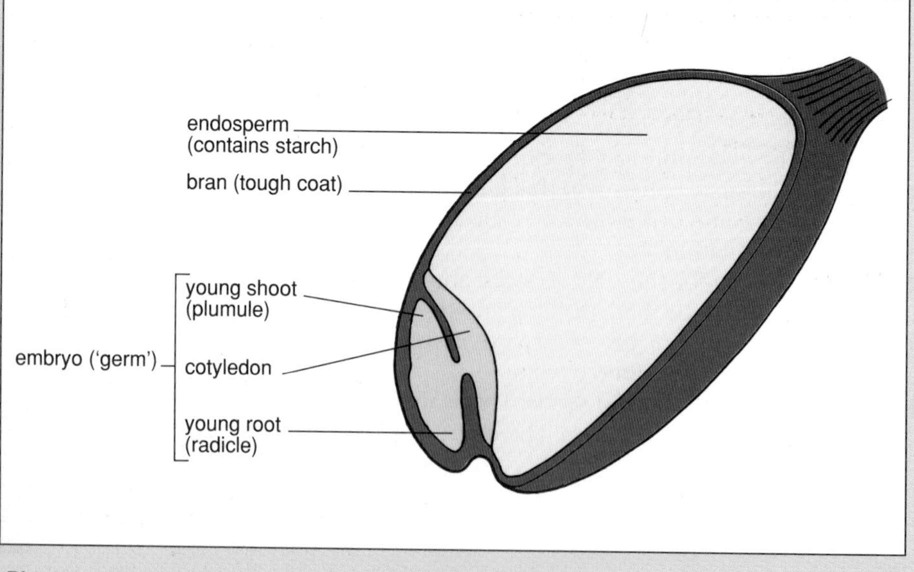

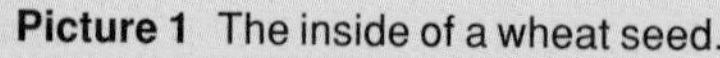

Picture 1 The inside of a wheat seed.

Activities

A Looking at seeds

1 Split open a bean pod and notice the row of beans inside. The beans are the seeds, and the pod is the fruit.

2 Look at the outside of a broad bean seed. Notice the structures shown in picture 2(A).

3 Take a broad bean seed which has been soaked in water and carefully remove the coat. Can you see the structures shown in picture 2(B)?

4 Pipette a drop of dilute iodine solution onto one of the cotyledons inside the seed. You may need to scratch the surface so that the iodine solution can get inside. What colour does the cotyledon turn? What does this tell you?

5 Examine the seeds of other plants. How do they differ from the broad bean?

B How do seeds germinate?

Do all seeds germinate the same way, or are there differences between species?

Investigate this by watching seeds germinate. Try broad bean, cress, mustard, wheat, pea, etc. The seeds must be given the right conditions for germination, particularly moisture.

What general conclusions would you draw about the way seeds germinate?

C The effect of water on a seed

The purpose of this activity is to compare a dry seed with a seed that has taken up water.

First decide what type of seed you'd like to investigate. Then decide how you should compare the seeds before and after taking up water. For example, should you simply look at them, or should you compare a particular feature such as their mass? Then decide what measurements, if any, you are going to make.

After discussing your plan with your teacher, carry out the investigation. What is the significance of your observation in the life of the seed?

D What conditions are needed for germination?

1 Push some cotton wool into the bottom of five large test tubes.

2 Pour a little water into four of the test tubes, so as to moisten the cotton wool. Leave the other one dry.

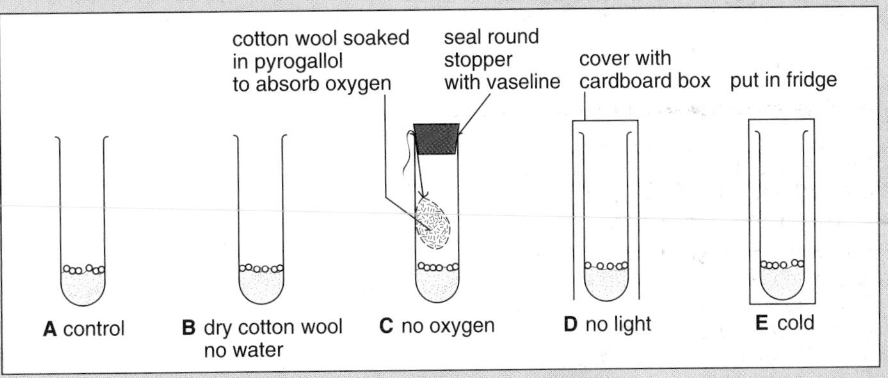

3 Sprinkle some cress seeds onto the cotton wool in each test tube.

4 Set up the test tubes as shown in the picture. Be careful with pyrogallol – see note below.

5 Observe the test tubes at intervals during the next few days.

In which tubes does germination take place, and not take place?

What conclusions do you draw about the conditions needed for germination?

What is the purpose of the control tube?

CORROSIVE

CARE! Pyrogallol is corrosive and can hurt your skin. The cotton wool bundle has been prepared for you beforehand. Handle it with forceps only. Any spills should be wiped up immediately.

Questions

1 Peas were placed on moist cotton wool in a special flask and set up as shown in the picture below. The flask was then left for two days:

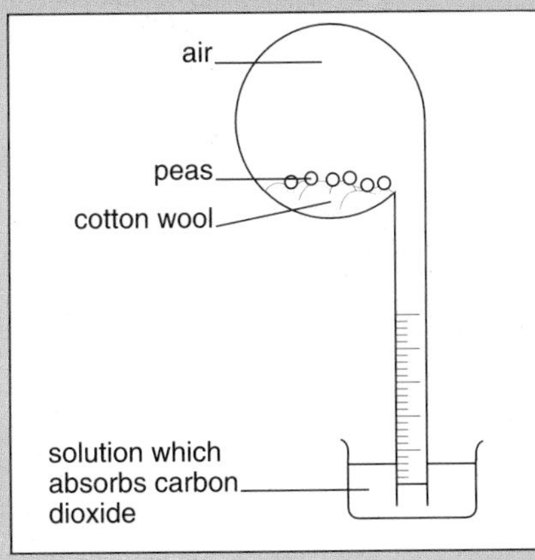

a What would you expect to have happened to the level of the solution after two days?
b What would have happened to the composition of the air in the flask?
c What was the person who did this experiment trying to find out?
d By means of a diagram show the control which is needed in this experiment.

2 Seeds which are planted too deep in the soil will not germinate. Suggest *two* possible reasons for this. Describe an experiment which you would carry out to test one of your suggestions.

3 The graph shows how the dry mass of a germinating seed (and seedling) changes from the moment germination starts. (The dry mass is estimated by getting rid of all traces of water from the plant and then weighing it.)

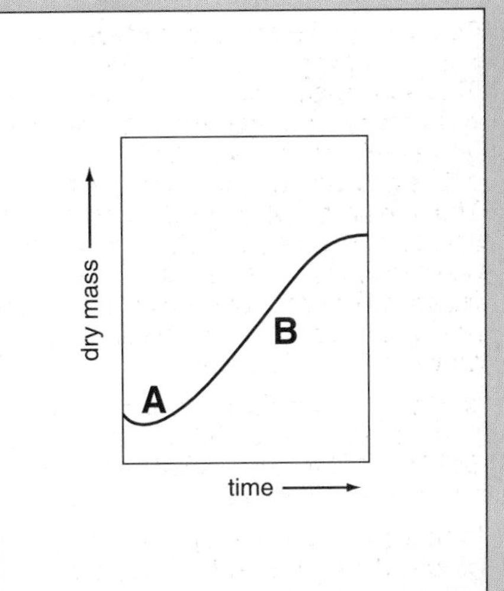

a How do you think this experiment was actually carried out?
b Explain what is happening at points A and B on the graph.

E9 Reproduction without sex

Some organisms can reproduce on their own without the help of another individual.

Asexual reproduction

Can you imagine your arm dropping off and develping into a new you! This may seem a weird idea, but it is the kind of thing that happens with some organisms. It is called **asexual reproduction** – reproduction without sex.

At its simplest, asexual reproduction takes place by the organism splitting in two (**binary fission**). Many single-celled organisms do this, for example *Amoeba* (picture 1). Bacteria do it too. In suitable conditions bacteria can split once every twenty minutes or so. This may not seem very fast, but try working out how many bacteria would be formed from one original cell after 24 hours. No wonder bacterial diseases can spread so quickly.

Another quick way of reproducing is by **budding**. This is what yeast does. Yeast is a single-celled fungus. When it buds, it sends out a small outgrowth which gets steadily larger and then breaks away as a new cell (picture 2).

Many organisms produce **spores**. Mosses, ferns and fungi all do this (see page 147 for example). It is a very fast way of reproducing. For example, a single mushroom can produce millions of spores in a very short time. Fungus spores are very light, and can be carried over long distances by the wind. This makes fungal diseases like potato blight difficult to get rid of.

Plants reproduce sexually – that's what flowers are for. However, many of them reproduce asexually as well. This is called **vegetative propagation**.

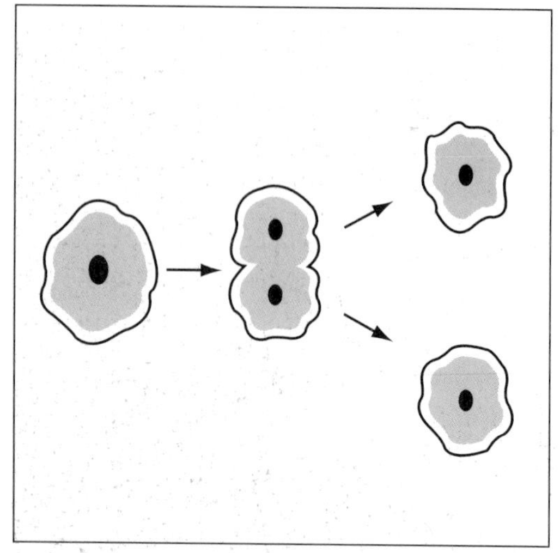

Picture 1 An amoeba reproducing by binary fission. The organism rounds off like a ball. Then the nucleus (black blob) splits in two, followed by the cytoplasm. The two small amoebas then feed and grow to full size.

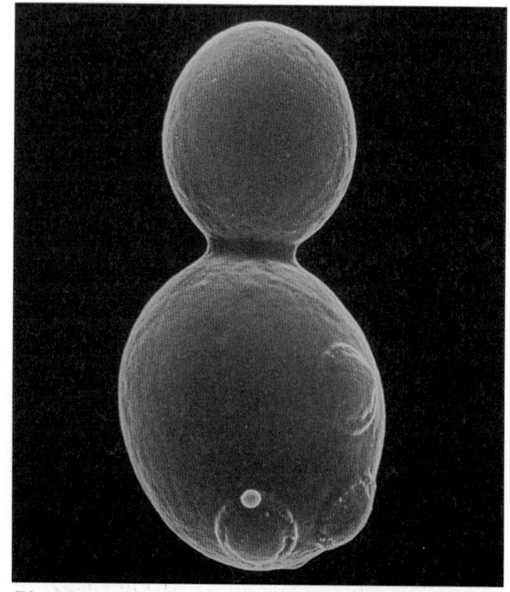

Picture 2 A yeast cell budding. Sometimes a cell produces several buds at once or a bud starts budding before it has broken off.

Vegetative propagation

You may have noticed that many garden plants die down in the autumn, but the following spring they grow up again in the same place. During the summer they form underground **storage organs** which fill up with food substances such as starch. The organ remains dormant in the soil during the winter, and next year a new plant grows out of it (picture 3). Some well-known storage organs are shown in picture 4.

These organs enable plants to carry on from one year to the next. For this reason, they are called **perennating organs**. But they are also a way of reproducing. This is because a plant may produce not just one but several storage organs. Each then gives rise to a new plant. A potato plant, for example, may produce a whole bunch of potato tubers (picture 5).

These storage organs are very important for humans, because we use them as food. Go into any large supermarket and you will see hundreds of them.

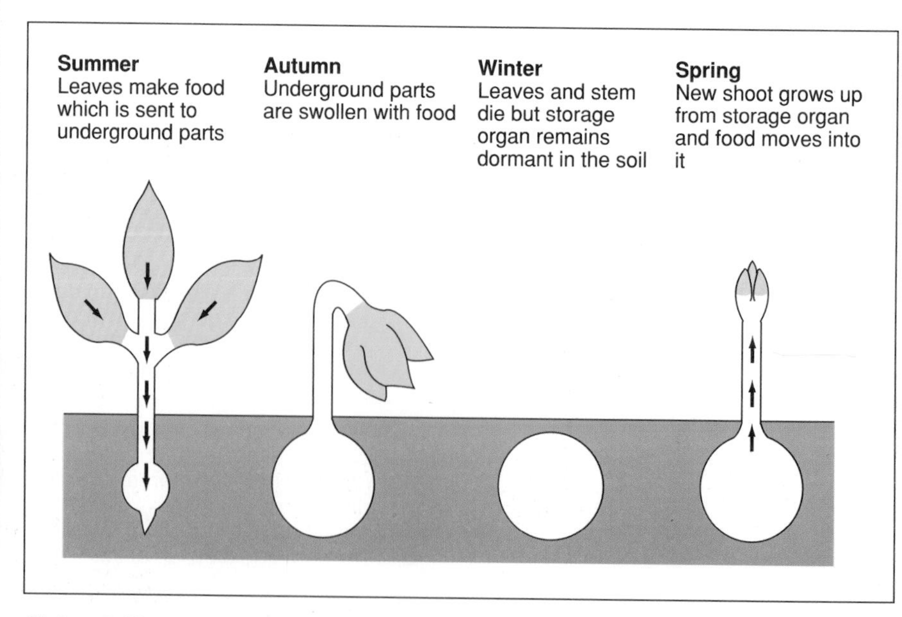

Picture 3 How a storage organ enables a plant to survive the winter and come up again in the spring. Many herbaceous perennials behave like this.

Picture 4 A selection of plant storage organs. These particular ones are all used by humans for food.

Picture 5 A single potato plant with tubers. The tubers are 'new' potatoes.

Other methods of vegetative propagation

Vegetative propagation does not always involve forming a storage organ. Many plants do it in other ways. For example, part of a plant – a leaf perhaps – may drop off and take root in the soil.

Some plants send out, from the base of the stem, a side branch which grows along the surface of the soil. This is called a **runner**. Roots grow down from it at the nodes and new plants develop at these points, as shown in picture 6. Strawberry plants spread this way. So, unfortunately, do certain weeds – which makes them very difficult to get rid of.

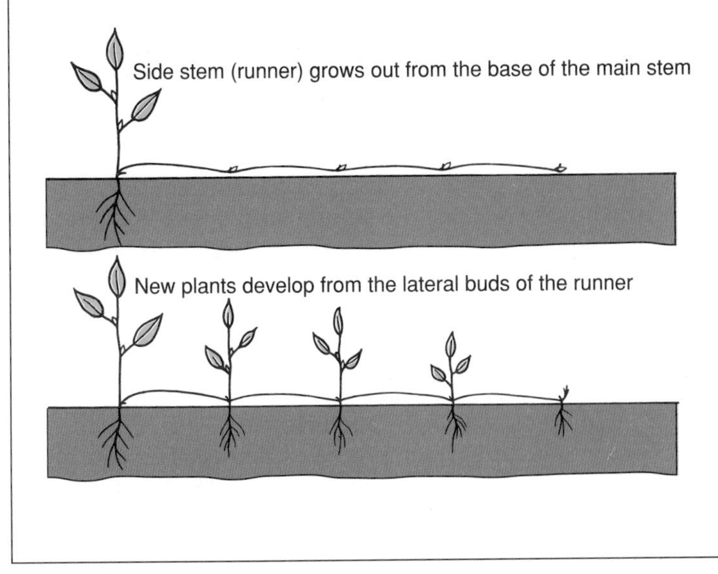

Picture 6 How a plant produces a runner. The photograph shows a young strawberry plant growing from a runner.

Picture 7 A spider plant. Notice the new plants developing from the ends of the hanging side branches.

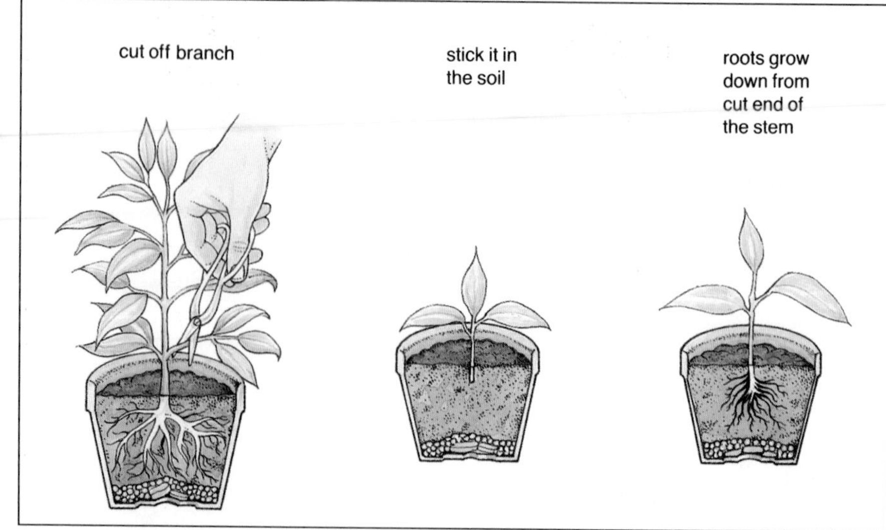

Picture 8 Taking a cutting.

You have probably seen the plant in picture 7 in people's houses. Perhaps you have got one in your house. It is called a spider plant. This sends out shoots which hang down and produce new plants at the ends. Gardeners get other plants to do the same kind of thing. A branch of a shrub is bent down and pressed into the soil at one of the nodes. Roots grow out and eventually a new plant is produced. This is called **layering**.

People often reproduce their favourite plants by taking **cuttings**. You cut off a healthy young branch, preferably just below a node, and remove some of its leaves. You then stick the cut end into some good soil. Roots grow out, and the cutting becomes established as a new plant (picture 8).

Grafting

Suppose you have an apple tree in your garden, and you want to have another one exactly like it. how could you produce a second apple tree just like the first one? One way would be to cut a young twig off the apple tree and join it to the trunk or branch of another similar tree. This is called **grafting**.

If the two plants are to join properly, the cut surfaces must be in close contact with each other. Picture 9 shows one way of doing this. The two plants are bound together with tape or raffia, and the joint is covered with wax. This prevents evaporation, and stops microbes getting in (picture 10).

The aim in grafting is to get the living tissues of the two plants to grow together. Once this happens, water and nutrients will pass freely from one to the other.

Grafting is carried out a lot by gardeners and growers of trees and shrubs. The plant you want to progagate is called the **scion**, and the one you join it to is called the **stock**. The stock is chosen for its good roots and resistance to disease. The scion is chosen for its good flowers or fruits. The new plant combines the best qualities of both.

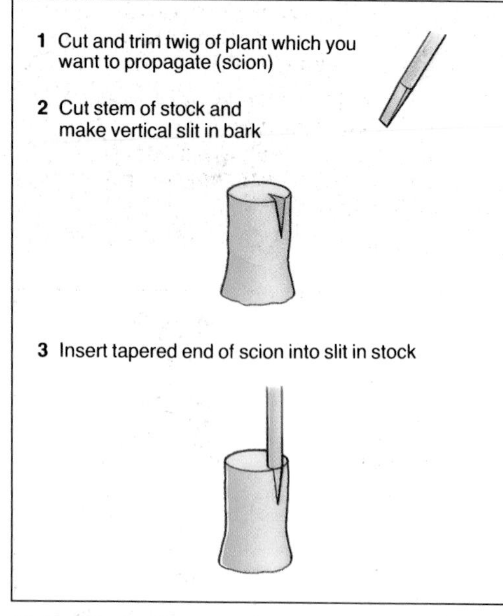

Picture 9 Grafting one plant onto another. This is just one of a number of methods that can be used.

Picture 10 Grafting young trees.

Cloning

Producing genetically identical organisms is called **cloning**. This name can be given to any method of reproduction which produces genetically identical offspring. So taking cuttings, layering and grafting are all examples of cloning. But recently the word has taken on a rather special meaning as a result of two very interesting experiments.

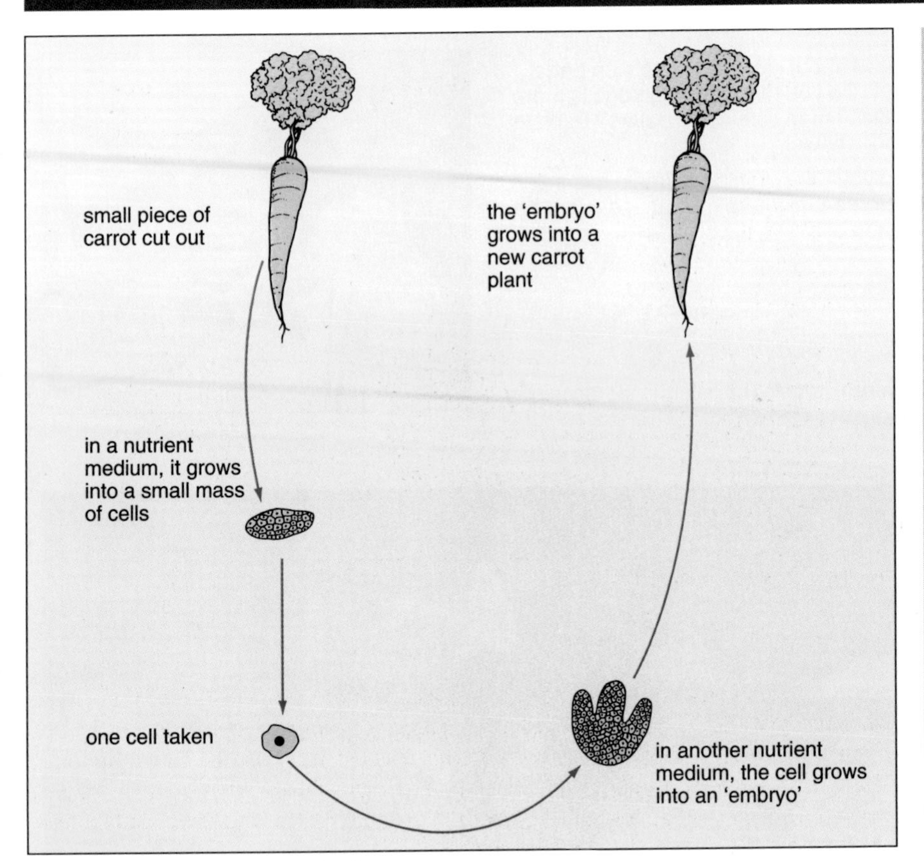

Picture 11 How to clone a carrot plant.

Choosing the right parent

Suppose you are a market gardener and you want to produce large numbers of identical plants by, for example, taking cuttings, grafting or cloning. Whatever method you use, one thing is very important: *you must choose the right parent plant from which to propagate the new individuals.*

What should you look for in choosing the right parent plant?

First and foremost, it must have the particular features you want – the right sort of flowers or fruits, for example.

Secondly, it should be genetically strong and healthy, and must not contain harmful genes. Otherwise the offspring may show defects of one sort or another. Ideally the parent should be the result of several generations of sexual reproduction with outbreeding (see page 295).

Thirdly, it should be free from disease and not contain harmful viruses or microbes such as fungi. Propagating plants from healthy parents is a good way of producing disease-free individuals

Cloning plants

The first experiment was done by an American scientist. He cut out a little piece of a carrot and put it in a dish containing various nutrients. The carrot cells divided, forming a small mass of unspecialised tissue. Growing tissues on their own like this is called **tissue culture**, and it can be done with plant or animal tissue.

Let's go back to the carrots. The scientist then separated the cells, and put one of them on its own in a nutrient fluid. A remarkable thing happened. The cell multiplied and developed into a tiny 'embryo', which grew into a new carrot plant (picture 11). It was as if the cell, away from its neighbours, thought it was an egg – so it developed all over again!

The great advantage of this technique is that lots of identical plants can be produced from one original plant. If the original plant has nice flowers, special chemicals in its cells, or grows well, so will all the new plants.

This kind of cloning opens up all sorts of possibilities. If carrots can be cloned, why not other species? Turning this experiment into a commercial process is not easy, but cloning plants is now done on a large scale (see page 252).

Cloning animals

Now for the second experiment. It was done by a British scientist, using frogs. The procedure is shown in picture 12. First he took an unfertilised frog's egg and destroyed its nucleus with ultraviolet radiation. Next he took a cell from a frog, and carefully sucked out its nucleus with a very fine pipette. He then injected the nucleus into the egg. So he now had an egg containing a nucleus from another individual. The egg developed into a tadpole, which in turn developed into a frog.

Our scientist went further. He took lots of nuclei out of the same frog, and injected them into eggs. In this way he produced lots of identical frogs. Can you see how this sort of thing might one day be used commercially?

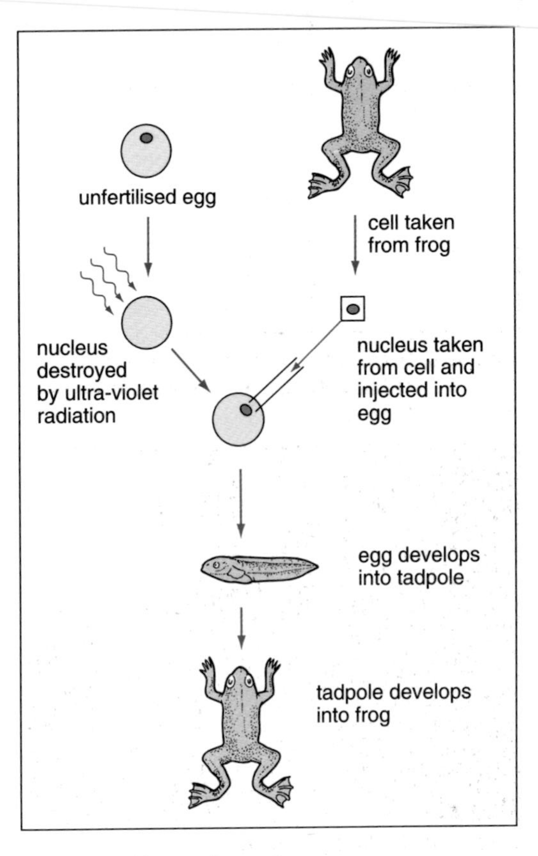

Picture 12 How to clone a frog.

Cloning for the benefit of humans: oil palms

The oil palm is a very important plant. The seed and the fruit both contain oil which is used for cooking and for making soap. When properly cultivated, the oil palm gives higher yields of oil per hectare than any other oil-producing crop.

The trees can be grown from seeds, but there is a lot of variation in the new plants. Some are good, others poor. This is the trouble with growing plants from seed – you cannot always be sure what you are going to get. So it was decided to try and clone the oil palms.

After ten years of painstaking research at Unilever's Colworth Laboratory in England, successful clones were made from small pieces of root. In 1976 the first fields of cloned oil palms were set up in Kluang, Malaysia. The parent trees were carefully selected for their good qualities. As expected, the new trees turned out to be just as good.

Today plantations of cloned oil palms can be seen in many tropical countries. It's a sure way of growing exactly the sort of plants you want.

1 What problems do you think the scientists at Unilever came up against in trying to clone the oil palm?

2 If you were cloning oil palms, what qualities would you want the parent plants to have?

3 What are the possible disadvantages of cloning oil palms on a large scale?

4 Does cloning mean that oil palms need never again be grown from seeds? Explain your answer.

5 Read about cloning animals on page 251. Do you think cloning frogs is more difficult than cloning oil palms? Explain your answer.

6 Do you think cloning animals has a commercial future? What use might be made of it?

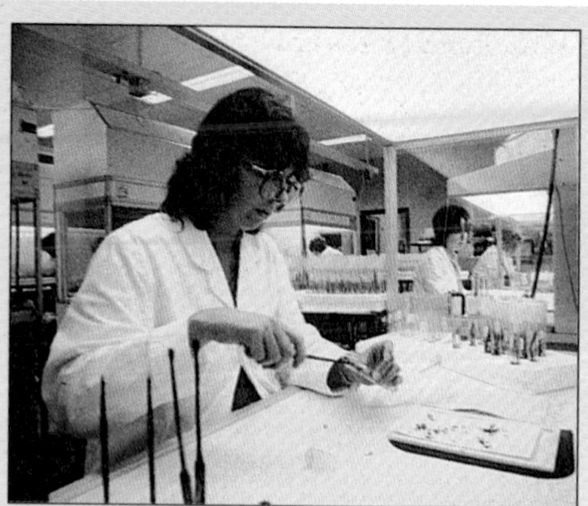

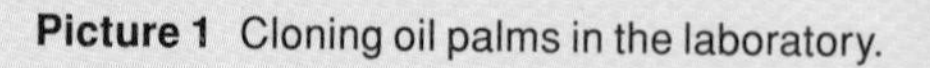

Picture 1 Cloning oil palms in the laboratory.

Picture 2 A plantation of young oil palms in Malaysia.

Sexual versus asexual reproduction

An examination candidate was asked to state one advantage of sexual reproduction over asexual reproduction. He wrote: 'It's more fun'. He may have been right, but there are other things that he might have said.

One of the great advantages of sexual reproduction is that the offspring differ from their parents and from each other (unless they are identical twins). They all differ because they contain different genes. So sexual reproduction gives rise to *variety*. Later we shall see that this enables the quality of the species to be improved.

On the other hand asexual reproduction produces offspring which are exactly alike, and like their parents. They all have the same genes. So asexual reproduction does not give rise to variety, and it cannot improve the species. Genetic defects and weaknesses are bound to be passed on to the offspring, and any infectious diseases are likely to be passed on too.

1 Market gardeners often produce new varieties of plants by sexual reproduction, and then propagate them by asexual reproduction. Why is this a good idea?

2 Explain very simply why sexual reproduction produces variable offspring, and why asexual reproduction produces identical offspring.

Activities

A Looking at yeast budding

1 With a pipette put a drop of yeast suspension on a slide.

2 Add a drop of a stain such as methylene blue or lactophenol.

3 Cover it with a coverslip.

4 Look at your slide under the microscope: low power first, then high power.

Can you see the yeast cells clearly?

Are any of them budding?

B Looking at mushroom spores

1 Obtain a mature mushroom and cut the cap off the stalk.

2 Place the cap, lower surface downwards, on a sheet of paper.

3 Cover it with an inverted dish.

4 After a day or two, remove the dish and lift up the mushroom cap.

What does the paper look like now?

Explain what you see.

5 Place a few spores on a slide and look at them under the microscope.

Why do you think they are so small?

C Looking at potatoes

1 Examine a potato tuber.

It is a swollen underground stem.

The 'eyes' are small buds.

2 Look at a second potato tuber which has been left for some weeks in a warm place and is 'sprouting'.

What structures are the new shoots growing from?

How does the tuber differ in appearance from the first one?

Where are the new shoots getting their food from?

3 Look at a complete potato plant which has been carefully dug up.

Can you see the old tuber from which it grew?

How many new tubers has it formed?

D Looking at starch in a potato

1 Slice open a potato to expose the white pulp.

2 Scrape off a little of the pulp and place it on a slide.

3 Put a drop of dilute iodine solution onto the tissue.

4 Cover the tissue with a coverslip.

5 Examine it under the microscope. Can you see starch grains? (They should have stained blue-black with the iodine solution.)

What do you think happens to the starch grains when the potato gives rise to a new plant?

How could you test your suggestion?

E Taking cuttings

1 Fill a test tube with water.

2 Cut off a side-branch from a busy lizzie plant. Make the cut just below a node.

3 Remove some of the leaves. This is to prevent it losing too much water.

4 Stick the cut end of the branch in the test tube of water.

5 Place the test tube in a warm, well-lit place, and observe your cutting at intervals during the next few weeks.

Do any roots grow out of it?

6 Plan an experiment to find out what conditions are needed for roots to grow out. Carry out the experiment after having it approved by your teacher.

F Examining storage organs

Your teacher will give you one of the plant storage organs illustrated in picture 3. Examine it carefully, cutting it open so as to see what's inside. Find out as much about it as you can. Try to answer these questions:

1 What sort of food does it contain, and how did it get there? (Tests for food substances are described on page 113).

2 How does the organ help the plant to survive the winter?

3 Does the organ enable the plant to reproduce, and if so how?

4 What use do humans make of this particular plant organ?

Questions

1 In good conditions a bacterial cell splits every twenty minutes. How many would be formed from one original cell after ten hours?

2 One mushroom can produce ten thousand million spores in a few days. Why does it need to produce so many?

3 A storage organ such as a potato tuber is also called a perennating organ. Why?

4 What are the advantages to a gardener of propagating a plant by vegetative means?

5 When taking a cutting, it is advisable:

a not to take a branch which has a flower on it,
b to cut the branch just below a node,
c to remove some of the leaves from the branch.

Suggest a reason for each of the above.

6 In choosing an individual for propagation, what features should the gardener look for and what precautions should be taken?

7 In theory it would be possible to produce dozens of genetically identical humans by cloning.

a How might this be done?
b What would be the main difficulties?
c Do you think it would be a good idea? Give reasons.
d Would you expect cloned humans to behave the same? Explain your answer.

8 The real value of cloning frogs is that it tells us important things about the way animals develop. In particular, it tells us that the nucleus decides what sort of individual an egg develops into. Explain this statement.

E10 How living things grow

How does a baby weighing 4 kg become an adult weighing 80 kg? This topic is all about growth.

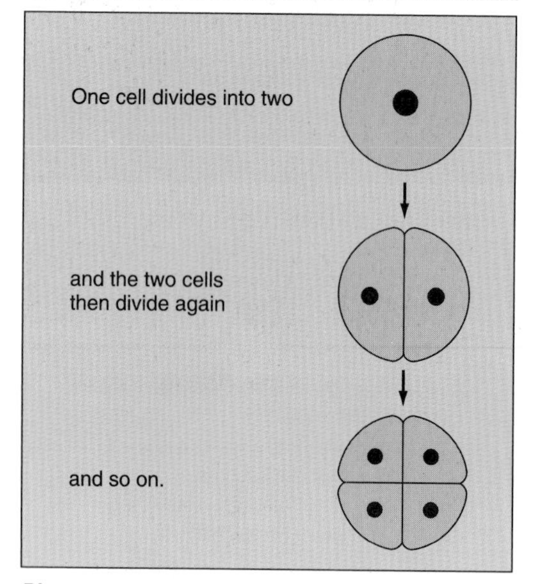

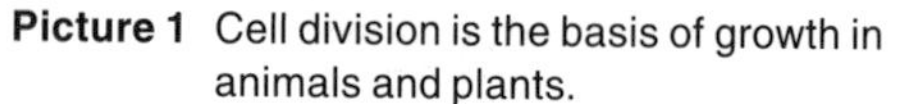

Picture 1 Cell division is the basis of growth in animals and plants.

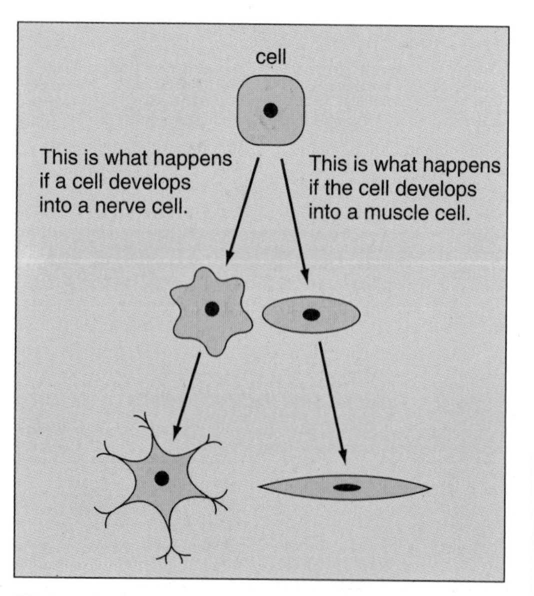

Picture 2 A cell may change its shape and form and develop into a particular type of cell with a specific function.

What is growth?

Growth is the permanent increase in size which takes place as an organism develops. It ensures that the adult organism is the right size to survive in its environment.

How do we measure growth?

There are many ways of measuring growth, depending on the organism in question. Usually we measure some parameter such as height, length or mass at regular intervals throughout the growth period.

When measuring growth we have to watch out for possible causes of error. For example, suppose you choose to study growth by measuring changes in an organism's mass. This sounds fine, but if the organism has a long drink just before you weigh it, the results can be misleading. Some of the questions and activities on page 257 will give you an idea of how we get round these sorts of difficulties.

How does growth take place?

Animals and plants start off as a single cell, the fertilised egg. This divides into two cells, which divide into two more – and so on. We call this **cell division**, and it is the basis of growth.

Picture 1 shows what happens when a cell divides. First the nucleus splits in two, then the rest of the cell divides across the middle. At first the new cells are smaller than the original cell, but they soon grow to full size. For this to happen, they must take in food substances to provide the necessary energy and materials. This is why a growing organism, such as a human baby, needs plenty of food.

Before long the cells start changing their shape and form, depending on where they are situated in the growing organism. In this way different types of cell arise, each in the right place. In the human body, for example, a cell might develop into a muscle cell if it happens to be in the wall of the stomach, or into a brain cell if it is in the head. In this way the cells become specialised to do particular jobs (picture 2). This process is called **differentiation**, and it plays a key part in the making of the full-grown adult organism.

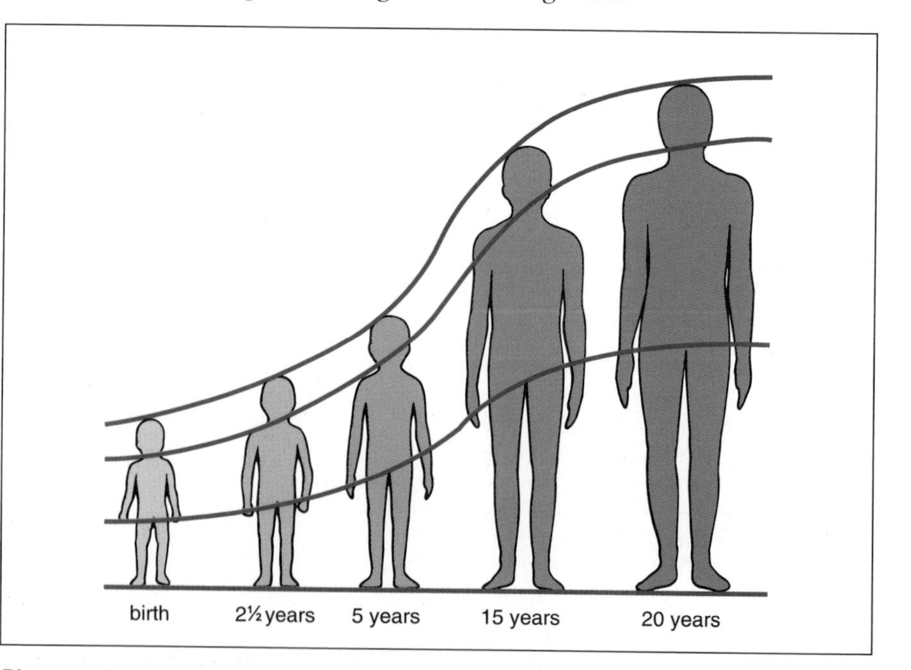

Picture 3 The growth pattern of a human. Notice that growth is slow at first, then speeds up, and then slows down again. This is typical of most animals.

Growth in humans

In a growing child, cell division takes place throughout the body. As a result, the child gets steadily larger. However, different parts of the body grow at different rates. This is because cell division occurs more quickly in some places than in others. For example, after birth the legs and arms grow much more quickly than the head (picture 3). The head has done most of its growing before birth. You can see this in picture 4.

Eventually no more new cells are added, so the person stops growing. This happens at about the age of eighteen in the human. However, certain cells need constantly to be replaced – red blood cells, for example. Blood cells are formed in the bone marrow, and here cell division goes on throughout life. Can you think of anywhere else in your body where cell divison goes on throughout life?

If the body is cut or damaged, cells which have stopped dividing will start dividing again. In this way wounds are patched up and damaged tissues replaced. Sometimes, cells start dividing when we don't want them to. This is what happens in cancer. There's more about that on page 261.

Picture 4 Father and baby son. Compare the sizes of the baby's head and hands with his father's.

Growth in plants

In a growing child cell division takes place throughout the body. This is true of most animals. However, in a young plant cell division only occurs in certain regions (picture 5). The two main regions are the tips of the shoots and the tips of the roots.

You can learn more about how a root grows by looking at it under the microscope (picture 6). The cells at the tip are constantly dividing. As new ones are formed, the older ones slightly further back expand. The expansion takes place mainly longways. This lengthens the root, helping it to push down into the soil.

The same sort of thing happens in the shoot. Picture 7 shows what a shoot looks like under the microscope. At the tip the cells divide and expand, and this makes the plant grow upwards. Similar activity at the sides enables leaves to develop, and branches to grow out.

While the cells expand, they differentiate into specialised tissues depending on where they are in the plant and the job they have to do. For example, the cells towards the centre of the root develop into transport tissues, whereas the cells further out develop into packing tissue.

Picture 6 A root grows by cells dividing at the tip and expanding further back. In the picture the root has been sectioned lengthways and is seen under the microscope.

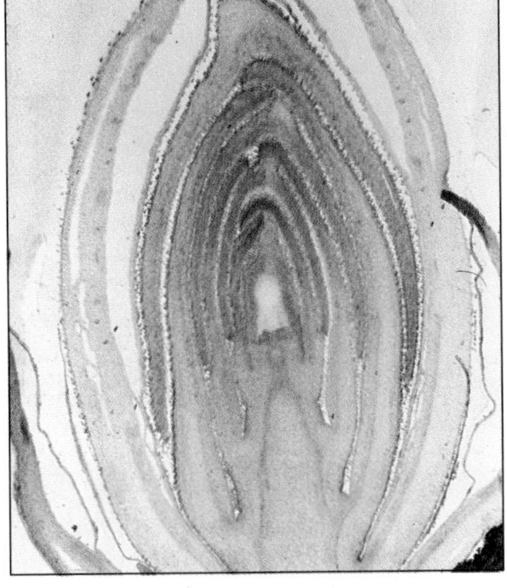

Picture 7 Here you see a shoot sectioned lengthways and looked at under the microscope.

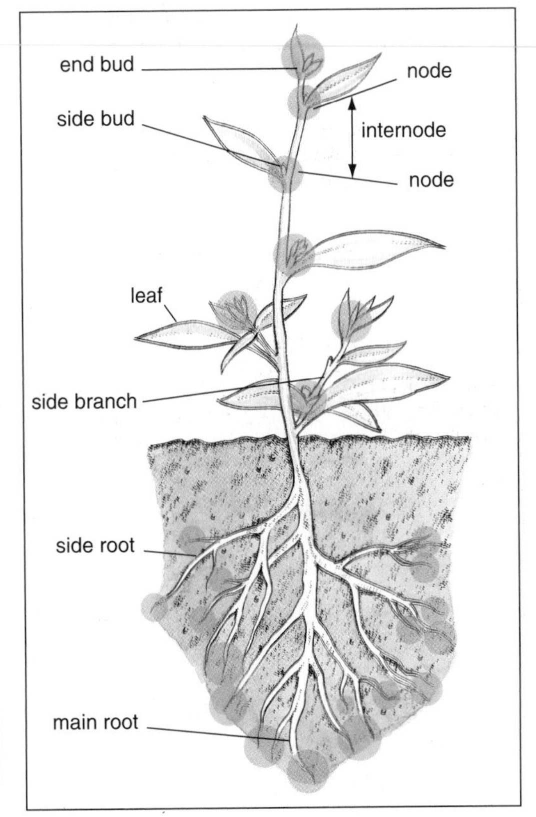

Picture 5 In this picture of a flowering plant the growing regions are indicated by the pink blobs.

Picture 8 Annual rings in the cut end of a felled elm tree.

Picture 9 Microscopic section through the stem of a tree showing parts of three annual rings. The smaller cells in each ring are the autumn wood.

How do stems get thicker?

The kind of growth just described makes the stem longer. However, stems get thicker too. Think of a tree, for instance, whose trunk and branches get thicker year by year. This occurs after the stem has increased in length, so we call it **secondary growth.**

How does secondary growth occur? As the stem lengthens, a circular layer of cells develops inside it. This layer of cells is called the **cambium**. Now the cambium cells are able to divide long after the other cells have stopped doing so. They divide to form new xylem cells towards the inside, and new phloem cells towards the outside. In fact more xylem is formed than phloem, and it forms the **wood** in a tree trunk and branches.

Secondary growth takes place during the warmer months of the year. In winter it slows down or stops. If you look at the cut end of a felled tree, you can see rings of wood corresponding to each year's secondary growth. These are called **annual rings** (picture 8). By counting the rings, you can tell the age of the tree.

The rings show up because the wood formed in the autumn is denser than the wood formed in the spring. The cells of the autumn wood are smaller and have thicker walls (picture 9).

Now imagine a tree trunk getting wider and wider. If the surface layer remained unchanged, it would be split wide open by the growing tissues underneath. However, a layer of **cork** develops on the surface, and more is added every year to keep pace with the ever-widening trunk. The cork is the hard part of the **bark** and it protects the living tissues underneath.

Two patterns of growth

If we plot a growing animal's height, length or mass against time, we get a **growth curve**. The graphs show growth curves for a human and an insect.

Notice how the two curves differ. The human has a *smooth* growth curve: growth takes place steadily and continuously. In contrast, the insect has a *stepped* growth curve: growth takes place in a series of spurts or steps.

The reason why the insect grows in spurts is that it has a hard cuticle which prevents it increasing in size. For the insect to grow, it has to shed its cuticle. This process is called **moulting**. A typical insect may moult four or five times before it reaches full size. A new cuticle is formed under the old one every time moulting occurs. The new cuticle is soft at first. The insect grows before the new cuticle has time to harden (see page 107).

1 In the insect growth curve, at what points would moulting have taken place?

2 Suppose mass, rather than length, had been used to construct the insect's growth curve. Would the growth curve have looked the same? Explain your answer.

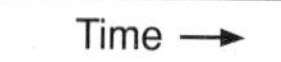

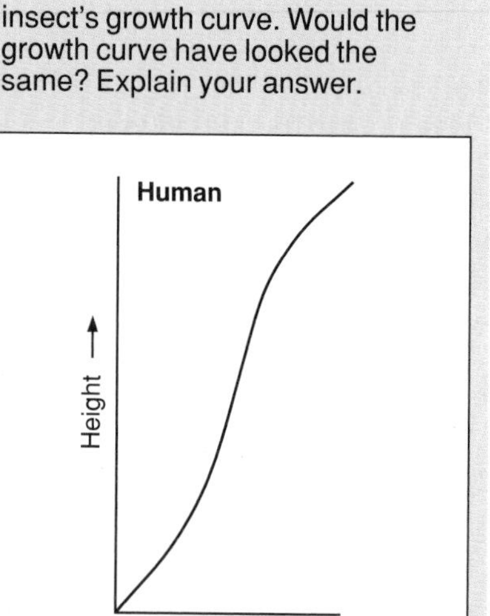

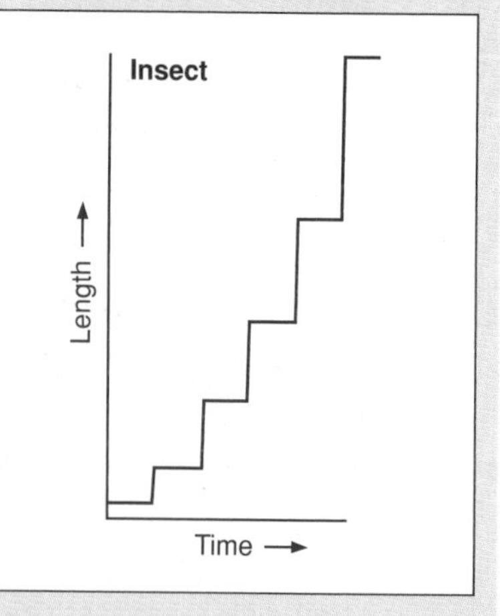

Activities

A Measuring the growth of an animal

If you have a young pet, such as a kitten or puppy you can carry out this investigation at home. Alternatively you can do it in the laboratory, using a small mammal such as a mouse or gerbil.

1. With a ruler measure the animal's length from the tip of its nose to the point where the tail is attached to the body.
2. Weigh the animal and find its mass.
3. Repeat steps 1 and 2 at regular intervals (at least once a week) until growth appears to stop.
4. Plot your results on a piece of graph paper, putting length and mass on the vertical axis, and time on the horizontal axis.

   Did the length and mass stop increasing at the same time? If not, can you explain the reason?
5. Work out the percentage increase in size between your first and last readings.

$$\text{percentage increase} = \frac{\text{final size} - \text{initial size}}{\text{initial size}} \times 100$$

B Measuring the growth of a plant

For this experiment use maize, wheat or oats.

1. Your teacher will give you some seeds which are about to germinate.
2. When the shoots appear measure their length and work out the average.
3. Repeat this at regular intervals (every day if possible) for at least a week.
4. Plot your measurements on graph paper, putting length of shoot on the vertical axis and time on the horizontal axis.

   How does the growth of the plant compare with that of a mammal.

C Where does growth take place in a root?

1. Obtain a bean seedling with a root at least 2 cm long.
2. With Indian ink, make a series of marks along the length of the root one mm apart. Use the special 'pen' shown in the picture below.
3. Pin the seedling to a piece of cork with the root pointing downwards.
4. Put the cork in a jar with a little water in the bottom to keep it moist, and cover it with a sheet of glass.
5. After several days examine the seedling. Where has growth taken place?

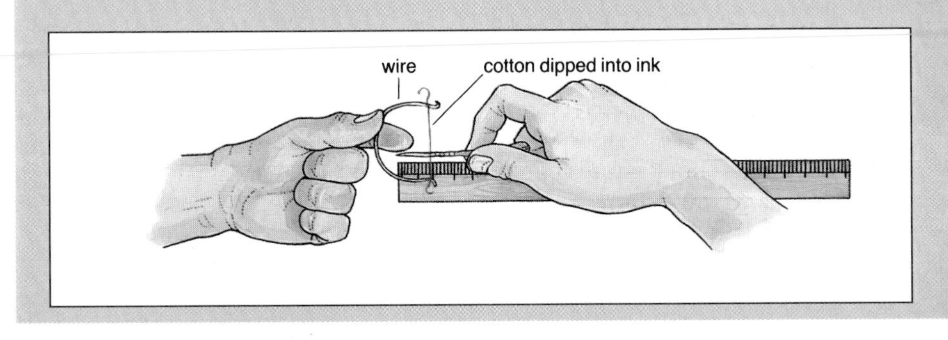

D Examining the inside of a young root

1. Look at a prepared longitudinal section of a young root under the microscope.
2. Observe the cells just behind the tip.

   Draw one of the cells in outline to show its shape.

   What were these cells doing when the root was alive?

   How do you know?
3. Now look at the cells further back.

   Draw one of them to show its shape.

   How did the cells come to be this shape?

   How does the change in shape help the root to grow?

   What happens to the various cells after they have changed their shape?

   How does the kind of growth seen in this plant differ from human growth?

E Secondary growth in a plant

1. Your teacher will give you prepared cross sections of a series of twigs of different ages.
2. Put the sections against a light background and if necessary look at them through a hand lens.

   How do they differ in appearance? Can you tell how old each one is?

   Look at the cut end of a series of older stems or branches.

   How do they differ in appearance?

   Can you tell how old they are?

Questions

1 Give *three* ways in which growth in humans differs from growth in flowering plants.

2 With a ruler measure in millimetres the width of the head and the length of the legs in the diagrams in picture 3. Plot the results on a sheet of graph paper so the curves can be compared.

a. Which grows more quickly between the ages of five and fifteen years, the head or the legs?
b. By how many times does one grow faster than the other?

3 A scientist sows a large number of seeds all at the same time, and she wants to measure the rate of growth of the seedlings. Here are three methods which she might use:

a. She measures the heights of the shoots of 50 plants every day and takes the average.
b. She digs up five plants every day, removes the soil from their roots, and estimates their **fresh mass** by weighing them.
c. She digs up five plants every day and dries them by heating them in a hot oven until all traces of water have been driven off. She then weighs them, thereby obtaining their **dry mass**.

Write down the advantages and disadvantages of each method.

In method (c) how could the scientist be certain that all the water had been removed from the plants before she found their dry mass?

Which is the most accurate method of measuring growth, and why?

4 If you look at the annual rings in a felled tree, you sometimes see that some rings are much wider than others. How would you explain this?

Tropical trees such as mahogany don't have annual rings. Why not?

E11
How growth is controlled

What makes an organism grow at a certain rate and to a particular size?

Picture 1 How the pituitary growth hormone can affect a person's size.

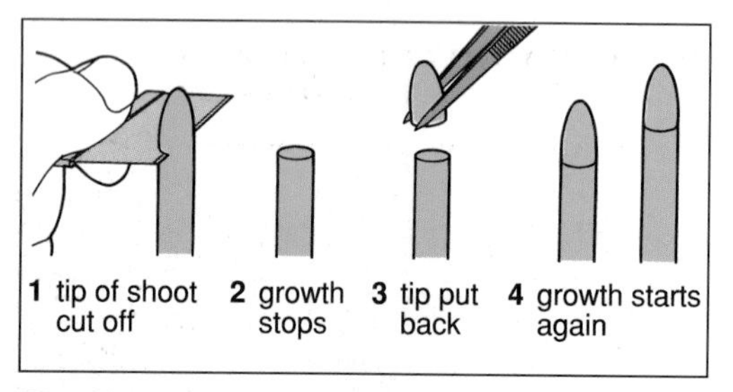

Picture 2 Experiment to find out if the tip of a shoot is needed for growth to occur.

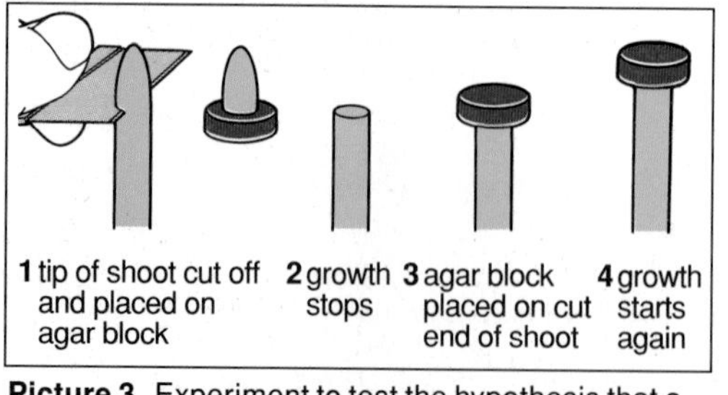

Picture 3 Experiment to test the hypothesis that a substance produced by the tip of a shoot makes the shoot grow.

The role of hormones

Look at the two people in picture 1. Both are adults. The person on the left is nearly three metres tall, whereas the one on the right is hardly a metre.

Why the difference? The answer is that the tall person produced too much **growth hormone** while she was growing. On the other hand, the short person produced too little growth hormone.

The growth hormone is secreted by the **pituitary gland** at the base of the brain (see page 200). This hormone plays a very important part in controlling the rate of growth and deciding how tall we will be.

Another hormone which helps us to grow is **thyroxine** from the thyroid gland. A growing child who does not produce enough thyroxine will become stunted in appearance and mentally retarded. This used to be common in places where there was insufficient iodine in the water, but it is rare now (page 114).

The sex hormones are also important in growth. For example, the male hormone **testosterone** increases the rate of growth during adolescence. It also affects the muscles, increasing their strength (see page 261). Equivalent hormones in the female have the same effects, but to a lesser extent.

Growth hormones in plants

If you cut off the tip of a shoot, the shoot will stop growing. However, if you put the tip back, the shoot will start growing again (picture 2). There seems to be something in the tip which stimulates growth.

What is it? Well, it might be a chemical substance which passes down the shoot, making it grow. Picture 3 shows an experiment which can be done to test this idea. You cut off the tip of a shoot and place the tip, cut end downwards, on a small block of agar jelly. Deprived of its tip, the shoot stops growing. After a few hours you place the agar block on the cut end of the shoot. Result? The shoot starts growing again! This result supports the idea that a substance produced by the tip makes the shoot grow.

These experiments were done some years ago. Since then scientists have found out what the substance is. It is called **auxin**. It is produced at the tip of the shoot, and then it slowly diffuses down the plant. As it passes down the plant, it brings about various effects. It therefore functions as a hormone.

Auxin does not always *stimulate* growth – sometimes it stops it. For example, as it passes down the stem, it prevents side branches growing out. So it makes the plant tall and straight. If you cut the top off a plant, the flow of auxin stops and side branches will grow out. Sometimes gardeners cut the tops off plants to make them more bushy. This is the secret behind making a thick hedge.

Growth substances

Since auxin was discovered, scientists have found other hormones which are important in plant growth. Today these substances, or very similar ones, are manufactured in chemical factories. They are known as **growth substances** and are used a lot in gardening and horticulture. For example, a substance similar to auxin is used for making cuttings 'take' (page 250). The cut end of the stem is dipped into the substance, and this stimulates roots to grow out of it.

A similar substance is used in **hormone weedkillers**. If applied in the right concentration, the substance makes the weeds grow so fast that they die. These are just two examples of how growth substances may be used. Growth substances are also used to control the growth of crops, bring about flowering, make fruits develop without fertilisation, and hasten the ripening of fruit.

How does light affect plant growth?

Look at picture 4. This shows the effect of lighting some seedlings from one side. The shoots have bent over towards the light. Most plants respond to light in this way. It ensures that the leaves get plenty of light for photosynthesis.

What makes the shoot bend? It is because the shoot grows faster on the dark side than on the light side.

If you cover the tip of a shoot with a little tinfoil cap, the shoot will not bend towards light. It seems that the tip senses the light stimulus, but the bending itself takes place *behind* the tip. This suggests that a message is sent from the tip to the part of the shoot a little further back.

We now know that the message is auxin. When you light a shoot from one side, more auxin gathers on the dark side than on the light side. This makes the dark side grow faster, with the result that the shoot bends towards the light (picture 5). Many experiments have been done which support this idea.

So plants respond to stimuli by growing in a particular direction. Such responses are called **tropisms**. A growth response to light is called **phototropism**.

Picture 4 These cress seedlings were lit for several days from the right hand side.

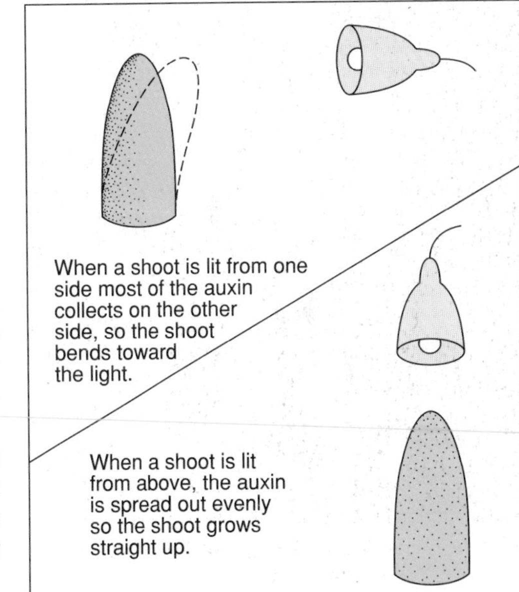

Picture 5 How light is thought to affect the distribution of auxin in a shoot.

How does gravity affect plant growth?

Look at picture 6. This shows what happens if you place a broad bean seedling in a horizontal position. The shoot bends upwards, and the root downwards. This is a growth response to gravity, and it is called **geotropism**. As you can see, the shoot grows away from gravity, and the root towards gravity. Why is this important? Well, it means that whatever way up a seed is, the shoot will always grow upwards, and the root downwards.

As with the response to light, the bending is brought about by growth. In the case of the shoot, growth is faster on the lower side than the upper side, so the shoot bends upwards. In the case of the root, growth is faster on the upper side than the lower side, so the root bends downwards. The red arrows in picture 6 show where growth is fastest.

How can we explain the gravity response? An explanation is given in picture 7. Auxin moves to the lower side of the shoot, speeding up growth on that side. The root is dealt with by a different hormone which is produced by the **root cap**. This hormone moves to the lower side of the root where it *slows down* growth.

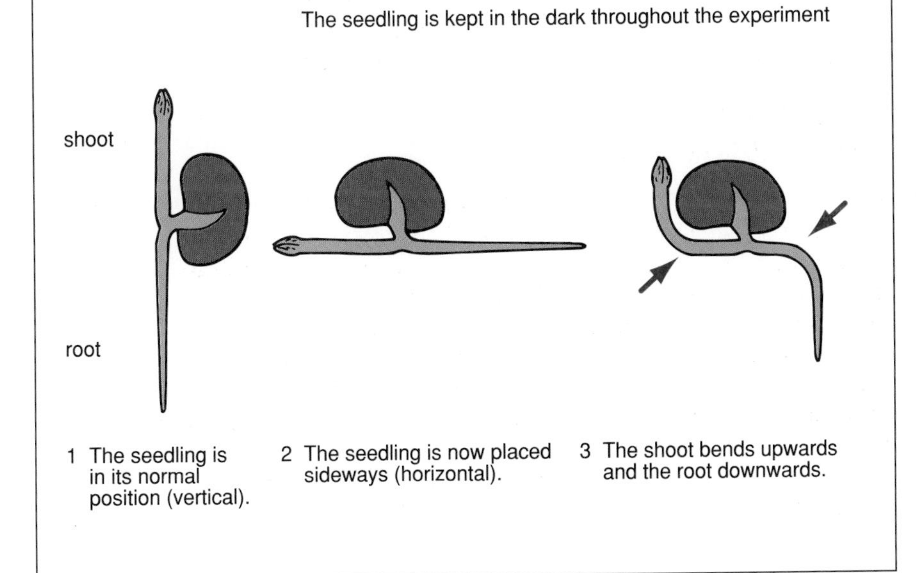

Picture 6 An experiment to show how a bean seedling responds to gravity.

shoot
root
root cap
1
auxin produced at tip moves to lower side and speeds up growth...
hormone produced by root cap moves to lower side and slows down growth...
2
...so shoot bends up
...so root bends down

Picture 7 The mechanism by which the shoot and root are thought to respond to gravity.

Questions

1 Why might children show poor growth in an area that lacks iodine? Poor growth because of lack of iodine is now very rare in Britain. Why?

2 The growth of a stem is speeded up by darkness. In what way is this useful to the plant?

3 Explain the reason behind each of the following:

a A gardener cuts the tops off a row of shrubs so as to make a thick hedge.
b Cuttings are dipped in 'rooting powder' before being stuck in the soil.

4 Mr Lewis spends the morning in the garden. He plants some seeds and puts many of them in the soil upside down.

Does this matter? Explain your answer.

5 How do shoots respond to light and gravity? how does the behaviour of the shoot help the plant to survive?

6 The diagram below shows an experiment which a scientist carried out on the shoots of three growing seedlings. In each case a thin piece of metal was placed between the tip and the rest of the shoot.

a What hypothesis was the scientist trying to test?
b What do you think the effect would be in each case, and why?

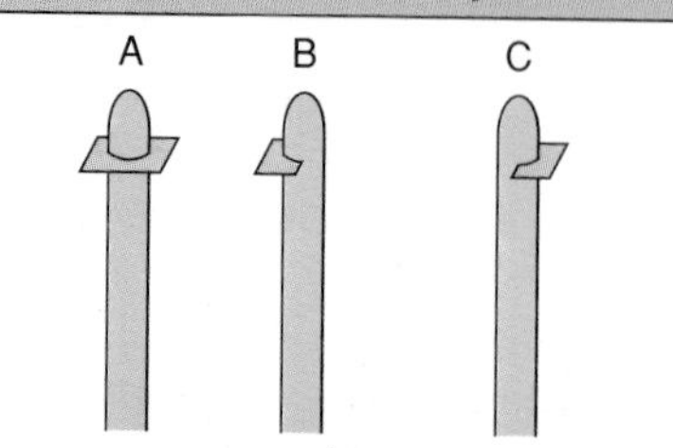

7 Which of the experiments in the pictures on the right best supports the hypothesis that growth of a shoot is stimulated by a hormone produced at the tip? Give reasons for your choice.

8 Describe an experiment which you would do to find out if a young root responds to light coming from one side.

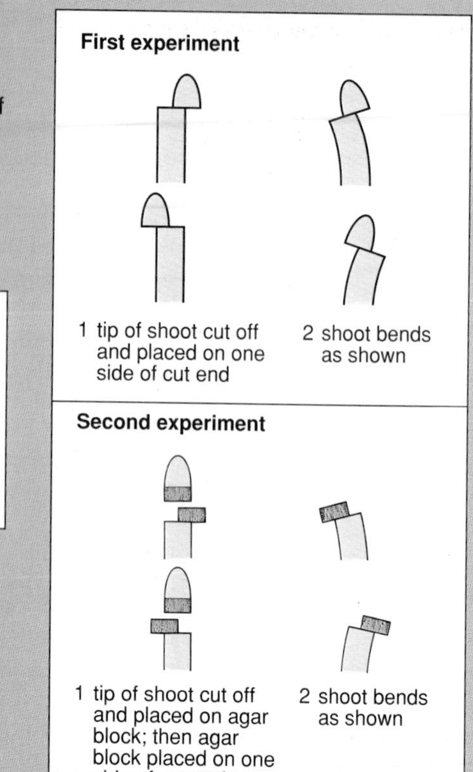

Activities

A Finding the effect of removing the top of a plant

1 Obtain two potted plants which do not have any side-branches.

2 Cut off the top of the stem from one of the plants, but not from the other one.

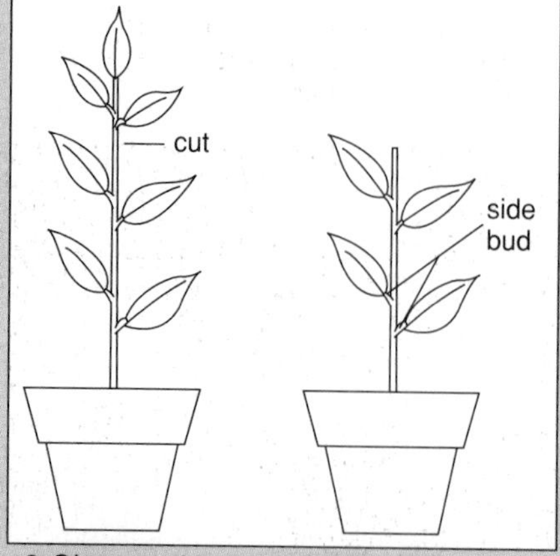

3 Observe the two plants at intervals during the next few weeks.

How do they differ in appearance?

What effect has been produced by removing the top of the plant?

B Does using a growth substance help cuttings to form roots?

On page 253 there are some instructions on how to take cuttings. Plan an experiment to find out if treating cuttings with a growth hormone helps them to form roots. Growth hormone preparations for this purpose are available from garden shops.

Get your plan approved by your teacher, then carry out the experiment. What conclusions do you draw from the results?

What advice would you give to someone who wants to take cuttings?

C Do shoots grow up and roots down?

Your teacher will give you the following: six bean seeds which are just beginning to germinate, a small tank with a glass top, a sheet of cork, some pins and a large cardboard box.

Using the above, set up an experiment to test the following hypothesis: *whatever way up a seed is, the shoot will grow upwards and the root downwards.* Bear in mind that it will take several days to complete the experiment, and the bean seedlings must be kept moist all the time.

Write an account of your experiment, describing your method and summarising your results. Do your results support the hypothesis?

D How effective is a hormone weedkiller?

Your teacher will give you a tray of soil in which is growing a mixture of grass and broad-leaved weeds. You will treat this with the selective weedkiller 2,4D which is toxic. Wear eye protection and avoid getting the substance on your skin.

2,4D is also a suspected carcinogen. It is very important to avoid skin contact.

Using a small watering can, water the plants with the weedkiller according to the manufacturer's instructions. Wash your hands carefully with soap and water afterwards.

Observe the result during the next week or so. Describe in detail how the grass and the broad-leaved weeds respond to the weedkiller.Explain your observations.

Selective weedkillers work by killing the broad-leaved plants but not the narrow-leaved grass. How important is it to get the concentration right, so as *not* to kill the grass?

Plan an experiment to answer this question. Have your plan approved by your teacher, then carry out the experiment.

Using hormones to make muscles grow

If you take a lot of exercise day after day, your muscles get stronger and stronger. The reason is that the muscles make extra protein, so they grow larger. This is what athletes are trying to achieve when they are training.

Certain hormones speed up this process of muscle growth. One such hormone is the male sex hormone **testosterone**. Artificial substances which have the same effect as testosterone have been made in laboratories. They were tried out on animals, and the muscles got larger. These substances are called **anabolic steroids** – 'anabolic' because they cause the body to make something, 'steroids' because this is the group of chemical compounds to which they belong. Over the last few years, certain athletes have tried to improve their performance by taking anabolic steroids during their training. This is not allowed, and athletes who do it are disqualified – as indeed they are if they take any kind of drug to improve their performance. Several Olympic athletes have had to go home in disgrace for breaking this rule.

Hormones have also been given to cattle. The idea is to make their muscles grow bigger, so that more meat can be obtained from them. The two sex hormones, testosterone and oestrogen, have both been used, and they had the effect of improving the growth rate. However, in 1986 the use of these growth promoters was banned in the European Community.

1. Do you think athletes should be disqualified for taking anabolic steroids?
2. Giving growth promoters to livestock was banned in the European Community, not because of possible health hazards but because it was felt to be wrong. Do you agree with the decision?
3. It's illegal to give growth promoters to cattle, but legal to give them to crop plants. What do you feel about this?

When growth goes out of control

We have seen that growth takes place by cells dividing. In animals there comes a time when this stops. Some kind of control process prevents the cells dividing any more.

On occasions this control process breaks down in some part of the body, and the cells start dividing again. The result is a disorganised mass of cells which don't do a useful job. This is called a **growth** or **tumour**.

There are two kinds of tumour: **benign** and **malignant**. A benign tumour stays in one place, and does not harm the surrounding tissues except by pressing on them. In contrast, a malignant tumour spreads (picture 1). Cells break away from it and are carried by the blood or lymph to other parts of the body. Here they invade and destroy the tissues and grow into new tumours. Tumours which spread like this are called **cancer** or carcinoma. The new tumours which start up are called **secondaries**.

Cancer is second only to heart disease as a cause of death, and people are very frightened of it. But there are many cases of people being cured. The surest remedy is for a surgeon to remove the tumour before it starts spreading, but cancer can also be treated with drugs (**chemotherapy**) and radiation (**radiotherapy**) (picture 2). Such treatment kills the tumour cells or at least stops them dividing.

For the treatment to be successful, it is important to discover the tumour as early as possible. A person who has a complaint that will not go away should see the doctor. A quick test will often show if it's serious. For example, a chest X-ray will show up cancer of the lung (page 134).

Women are particularly prone to cancer of the uterus. A doctor can find out if a patient has this by taking a smear from the neck of the uterus (cervix). This is called a **cervical smear**. The smear is examined under the microscope to see if any abnormal cells are present. If there are, it may be necessary to remove the uterus, or part of it, in an operation. This operation is called a **hysterectomy**.

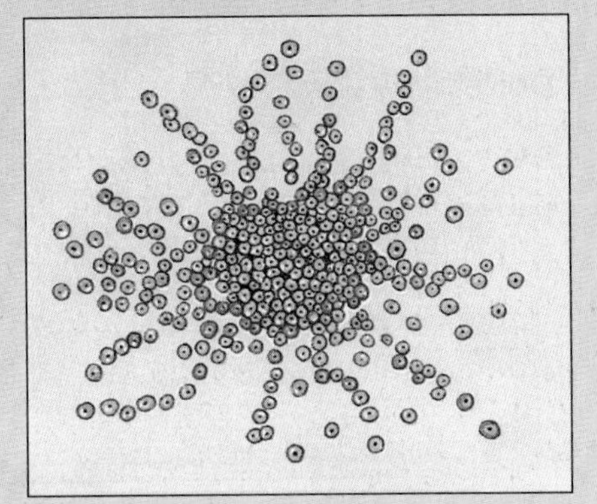

Picture 1 A malignant tumour. The arm-like extensions make the tumour look rather like a crab. This is how cancer got its name – it is the Latin word for crab.

What causes cancer? Basically, cancer is a type of mutation that occurs in the body cells. (Mutation is explained on page 284.) There are many possible causes of such mutations and a great deal of research is going on into this question. You cannot catch cancer from other people, but it can be brought on by environmental hazards such as radiation, asbestos dust and of course smoking. Certain kinds of virus may also be involved.

1. Cancer is more common now than it was eighty years ago. Suggest reasons.
2. In some countries people are tested regularly for the more common kinds of cancer.
   a. What benefits does this bring?
   b. What would be the problems of doing it in Britain?

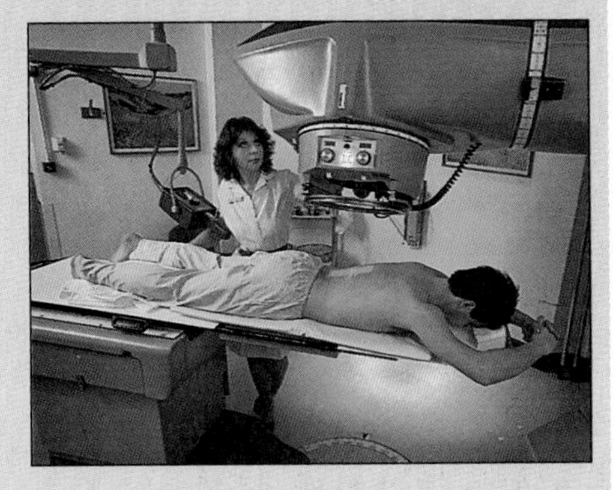

Picture 2 A patient being positioned on the treatment couch before receiving radiotherapy from a computer-controlled machine.

F1 Introducing heredity

Why are we like our parents, and why do we differ from them? This is the science of heredity.

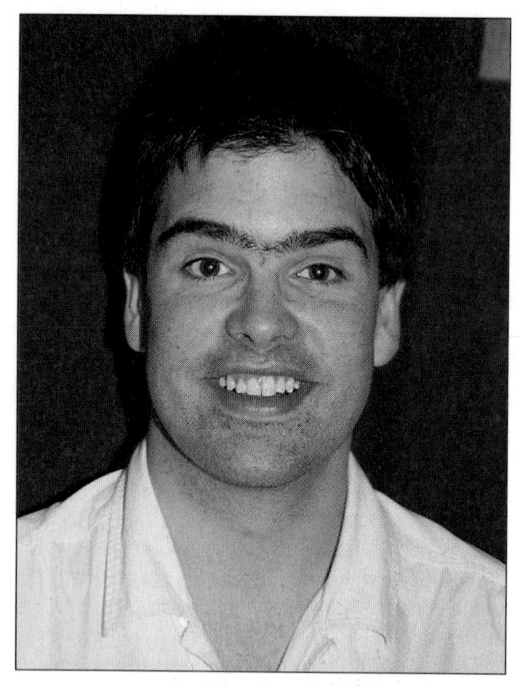

Picture 1 How did John get his brown eyes?

How John got his brown eyes

John has brown eyes. His mother has brown eyes too, but his father has blue eyes. How can we explain how John got brown eyes?

Let's suppose that people are born with **instruction cards** telling them what sort of eyes to have. John's mother has two eye-colour instruction cards which make her have brown eyes. John's father has two eye-colour instruction cards which make him have blue eyes. For convenience, we will call mother's cards 'brown cards', and father's cards 'blue cards'. John's parents and their cards are shown at the top of picture 2.

First of all, notice the eggs and sperm. Mother's eggs each contain one brown card, and father's sperm each contain one blue card. This is one of the most important things about heredity: *a person has two cards for controlling a particular feature, whereas the sperm and eggs have only one*.

John was conceived when one of his father's sperm fertilised one of his mother's eggs. When this happened, one of father's blue cards was combined with one of mother's brown cards. So John has two cards, a brown one from his mother and a blue one from his father.

John has a blue card as well as a brown card. Why, then, does he have brown eyes? The reason is that the brown card *overrules* the blue card. Putting it another way, the brown card is **dominant** to the blue card. So although John carries the blue card, it has no effect on his eyes.

Picture 2 How John got his brown eyes.

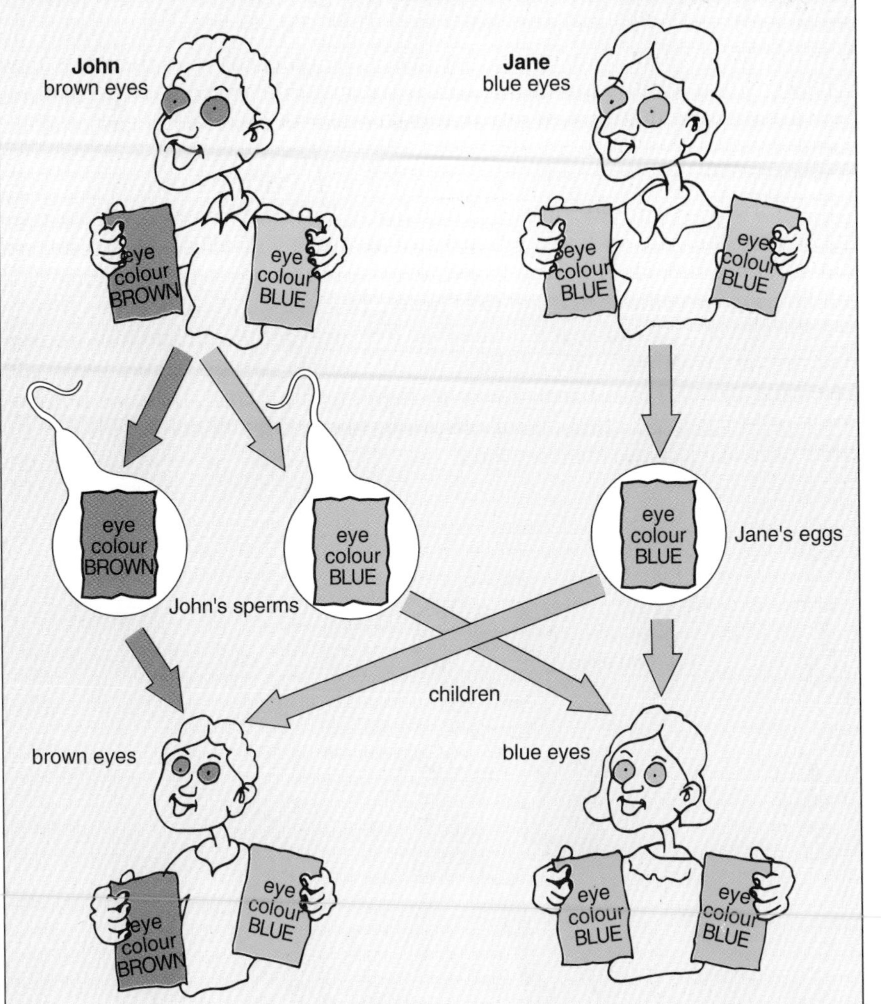

Picture 3 How John and Jane pass their eye colours on to their children.

John marries Jane

John marries Jane who has blue eyes. They have two children: one has brown eyes and the other has blue eyes. How can we explain this?

The explanation is given in picture 3. John has a brown card and a blue card, as we've already seen. Now half John's sperm contain one brown card, and the other half contain one blue card. This is because the two types of card are equally distributed amongst the sperm rather like dealing cards. All Jane's eggs contain a blue card.

When fertilisation takes place, Jane's egg may receive one of John's brown cards or one of his blue cards. Fertilisation is random – it's pure chance as to which kind of sperm fertilises the egg. So there's an equal chance of a brown card or a blue card combining with the egg's blue card. This means that there's a 50:50 chance (one in two) of any of the children being brown-eyed or blue-eyed. It happens that one of the children has brown eyes, and the other blue eyes. But both *could* have had brown eyes, or both blue eyes – it's just like tossing coins.

John has a sister

John has a sister called Sharon. Sharon has brown eyes, like John's. Sharon marries Kevin who also has brown eyes. They have lots of children: most of them have brown eyes, but to their surprise one has blue eyes, as shown on the right. How can we explain this? Think about it before you turn over the page.

Activity

Heredity in action

To study heredity you need animals or plants that reproduce quickly and have clear-cut features which you can observe easily. Suitable animals are the fruit fly *Drosophila* and the flour beetle *Tribolium*. Suitable plants are tobacco, tomato, pea and maize.

Whatever organism you use, you must first decide which feature or features you wish to study the inheritance of. In the fruit fly it might be the colour of the eyes (some have white eyes, others red). In maize it might be the height of the plants (some are tall, others dwarf).

The next step is to choose the parent organisms, and get them to reproduce. Male and female fruit flies will reproduce if you put them together in a container with the right food. Once the offspring have developed, you can anaesthetise them and count the different types.

With plants the parents have to be cross-pollinated by hand (see page 295). You then collect the seeds and sow them. Later, when the seedlings have grown sufficiently, you count the different types.

Your teacher will give you a batch of seeds which were obtained by crossing two particular plants. Sprinkle the seeds on fine soil, or agar jelly, in a tray or dish, as instructed by your teacher. Leave in a warm, well lit place. Look after the seedlings as they grow, and water them frequently.

When the seedlings have grown sufficiently, observe them carefully. Are they all of one type, or do they differ? If they differ, count each type. What is the ratio between them? Explain your results.

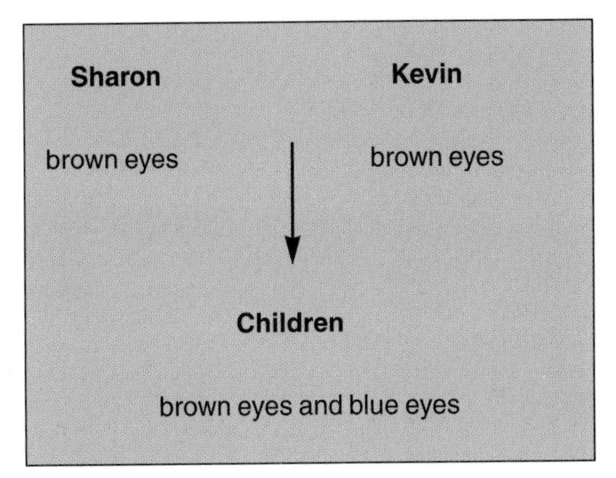

The explanation is given in picture 4. Half of Sharon's eggs contain the brown card, and half contain the blue card. Half of Kevin's sperms contain the brown card, and half contain the blue card. The different ways the cards may combine when fertilisation takes place is shown at the bottom of the picture. Fertilisation is random, so there's an equal chance of each combination taking place.

What colours are the children's eyes? Remember that the brown card is dominant to the blue card: as long as a child has at least one brown card, the eyes will be brown. If you look at the bottom of picture 4 you will see that there is a 3 in 4 chance of a child having brown eyes. And there is a 1 in 4 chance of a child having two blue cards, and thus having blue eyes.

Another way of looking at it is like this. If Sharon and Kevin had hundreds of children, approximately 3/4 of them would have brown eyes, and 1/4 would have blue eyes. This is hardly likely to happen in practice, because humans don't produce large numbers of offspring. But it does happen with certain other organisms, as we shall see later.

Picture 4 How Sharon and Kevin pass their eye colours to their children.

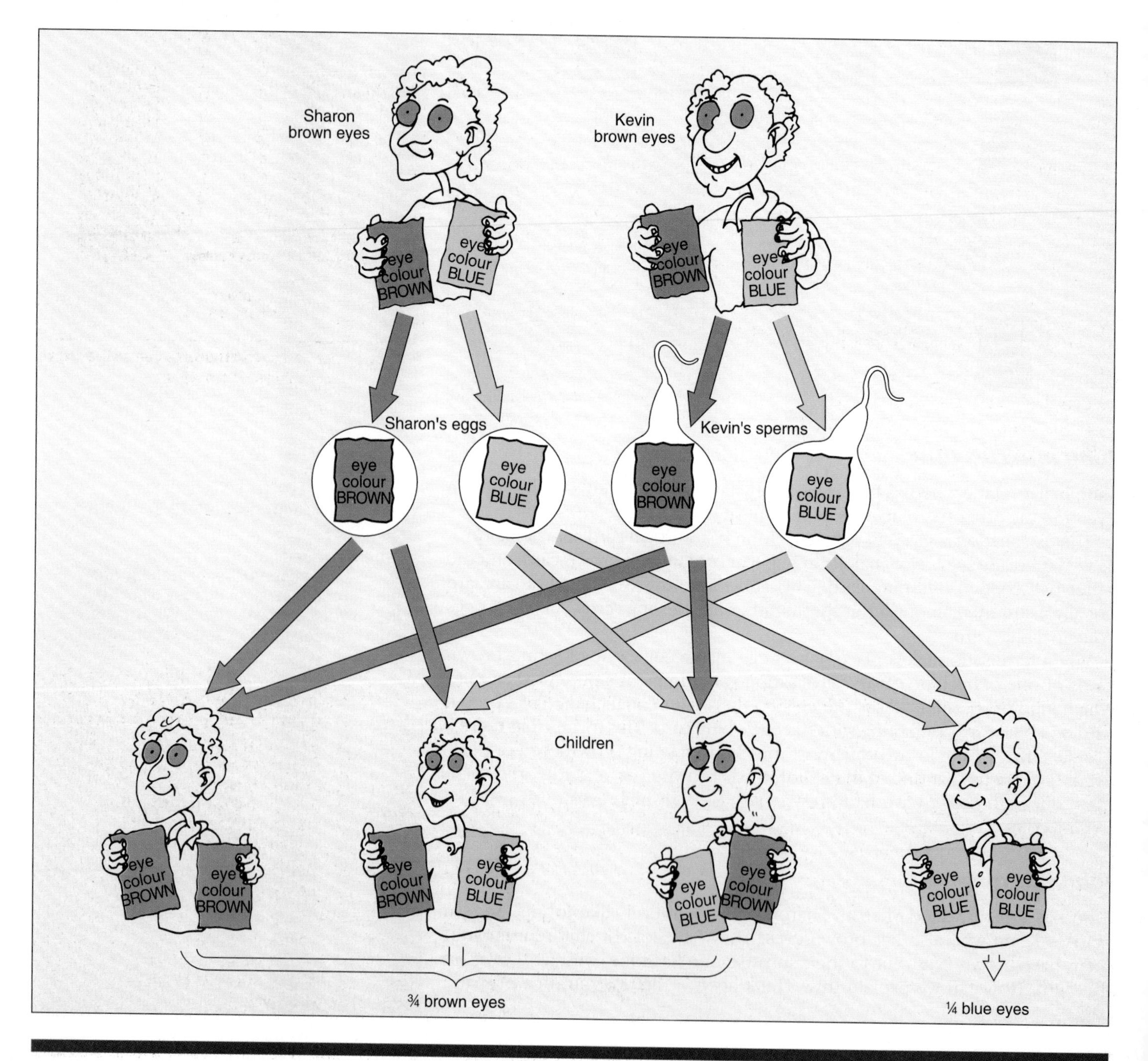

What are the instruction cards?

What we have been calling instruction cards are really our **genes**. Genes are found in all our cells and they control the way we develop, causing us to have certain features. Later we shall see what genes are made of. For the moment let's use the idea of instruction cards to illustrate some important things about genes.

There are different versions of the instruction card for eye-colour. One version says 'have brown eyes', another version says 'have blue eyes'. In the same way the gene that controls eye colour may exist in different forms, each one telling the person to develop a particular eye colour. These different forms of a gene are called **alleles**.

The cards are in packs

We don't just have instruction cards for eye-colour. We have cards for hundreds of other features as well. These cards are arranged in 'packs'.

In the same way, our genes are grouped together into **chromosomes**. Chromosomes occur in the nuclei of all living cells (see page 13). They are like pieces of thread. The genes are strung out along the chromosomes, like strings of beads. Each gene controls a particular feature such as eye-colour, hair-colour, the length of the nose and so on (picture 5).

The card that says 'have brown eyes' is in a separate pack from the card that says 'have blue eyes'. But these two cards are in exactly the same position within each pack. In the same way, the alleles that call for brown or blue eyes are in the same positions within two separate chromosomes. These two chromosomes look exactly alike: they belong to a pair. And just as the chromosomes are in pairs, so too are the alleles which they carry.

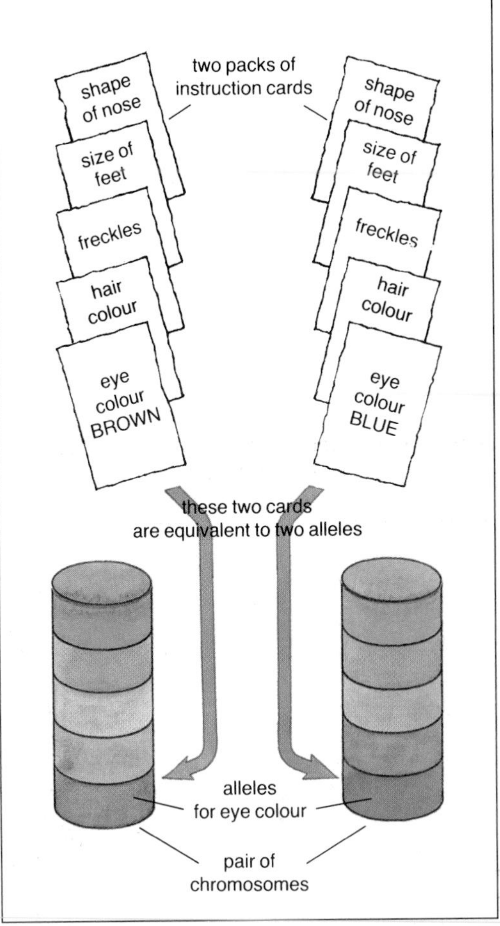

Picture 5 Genes in a chromosome can be likened to instruction cards in a pack.

Life is full of surprises!

From this topic you may have got the idea that blue-eyed parents can only produce blue-eyed children. This isn't always the case. Sometimes one or more of the children have brown eyes! The reason is that the inheritance of eye colour is complicated – in fact it's not fully understood. Eye colour is controlled not just by a single gene, but by a *group* of genes acting together. Sometimes the group of genes produces an effect which we may not be expecting. And of course there are other eye colours besides blue and brown – green and hazel, for example. The author of this book breaks all the simple rules by having eyes of two different colours – one green and one brown.

Questions

1 The blue-eyed child in picture 3 thinks she got her blue eyes from her mother. Is this true? Explain your answer.

2 John, in picture 3, has a sister with brown eyes. Could she have had blue eyes? Explain your answer.

3 Suppose Kevin in picture 4 had two brown cards instead of a brown card and a blue card. What difference would this make to the colour of the children's eyes? Explain your answer.

4 Human features sometimes 'skip a generation'. What does this mean, and why does it happen? Use the pictures in this topic to illustrate your answer.

5 In picture 3 both children might have had brown eyes, or both might have had blue eyes. Explain this. (Hint: when you toss a coin, what decides whether you get heads or tails?)

6 A black mouse mates with a brown mouse, and all the offspring are black.

a Why are no brown offspring produced? Use instruction cards to illustrate your answer.

b If two of the black offspring mate with each other, what kind of offspring would you expect and in what proportions? Draw a diagram to show what happens.

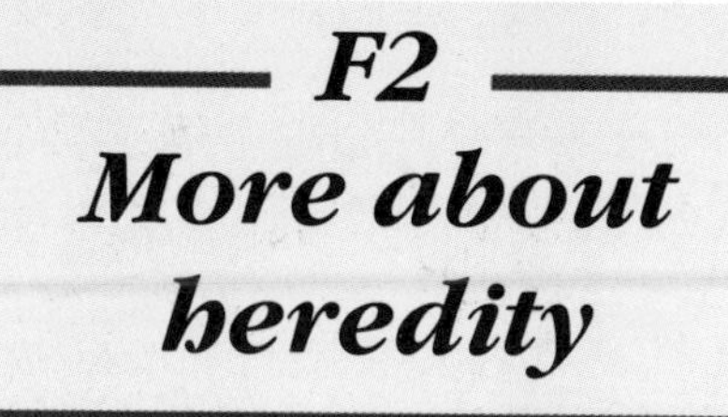

F2
More about heredity

In this topic we explain heredity properly in terms of genes and their alleles.

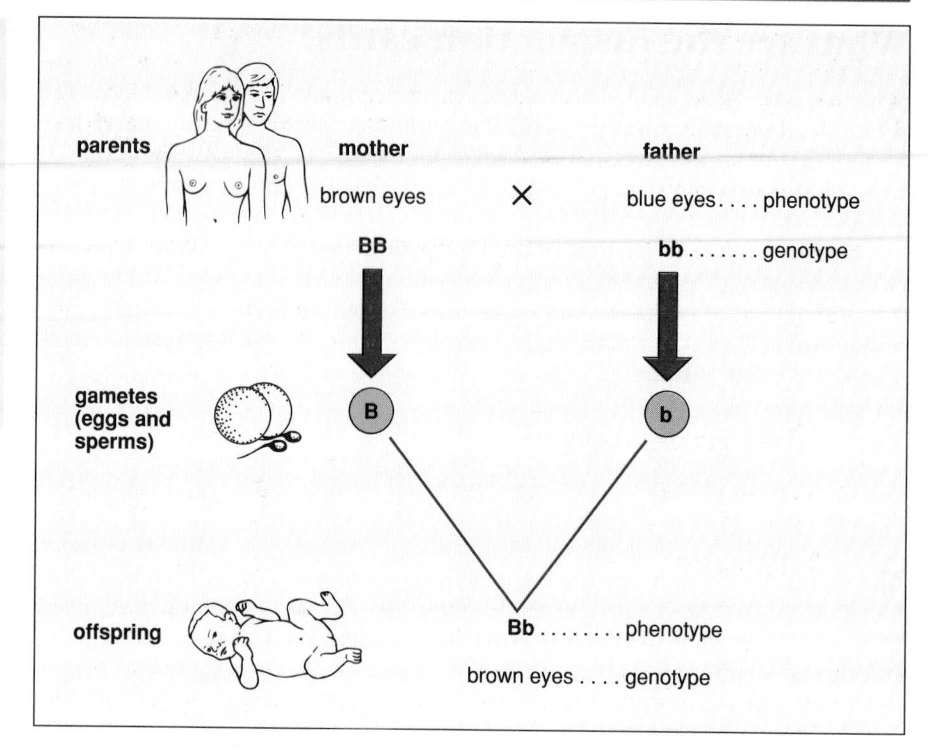

Picture 1 Diagram showing how John inherited his brown eyes. The alleles are indicated by letters: **B** is the allele for brown eyes, **b** is the allele for blue eyes. In diagrams of this sort it is usual to represent the alleles by the same letter, a capital letter for the dominant allele and a small letter for the recessive allele.

How are genes passed from parents to offspring?

Let's go back to John with his brown eyes (see page 262). You will remember that his mother has brown eyes, but his father has blue eyes. On page 262 we explained this, using the idea of instruction cards. Now we shall explain it in terms of genes.

John and his parents have in all their body cells a gene which controls eye-colour. This gene has two alleles which we will call **B** and **b**. **B** is the allele for brown eyes, and **b** is the allele for blue eyes. The **B** allele is **dominant** to the **b** allele – that is why it is written with a capital letter. The **b** allele is described as **recessive**. So brown eyes are dominant to blue eyes.

John's mother contains in her cells two **B** alleles which give her brown eyes: we can call her **BB**. John's father contains in his cells two **b** alleles which give him blue eyes: we can call him **bb**. The way these alleles are passed to John is shown in picture 1.

The gametes (eggs and sperm) contain only one allele. This is because of the way the gametes are formed (see page 274). Each of mother's eggs contains a **B** allele, and each of father's sperm contains a **b** allele.

When a sperm fertilises the egg, the **B** and **b** alleles are brought together. John develops by the fertilised egg dividing repeatedly in such a way that each of his body cells contains a **B** allele and a **b** allele. Because **B** is dominant to **b**, he has brown eyes. However, he is a carrier of the allele for blue eyes and can pass it on to his children.

Picture 1 on this page is really the same as picture 2 on page 262, but it explains how John got his brown eyes in terms of alleles rather than instruction cards. Try re-drawing pictures 3 and 4 on pages 263 and 264 in the same way.

Some technical terms

There are two ways of describing John and his parents. We can describe their outward appearance, e.g. brown eyes or blue eyes. Alternatively, we can describe their alleles, e.g. **BB**, **Bb** or **bb**.

The outward appearance of a person is called the **phenotype**. The person's alleles make up the **genotype**. So John's phenotype is 'brown eyes' and his gentoype is **Bb**.

Picture 2 Red and white flowered Busy Lizzie plants. As in many other kinds of plants, the colour of the flowers is controlled by genes.

When the genotype for a particular feature consists of two identical alleles, e.g. **BB** or **bb**, we say that the person is **homozygous**. If the two alleles are both dominant, e.g. **BB**, the person is **homozygous dominant**. If the two alleles are both recessive, e.g. **bb**, the person is **homozygous recessive**. John's mother is homozygous dominant for eye colour, whereas his father is homozygous recessive.

When the genotype for a particular feature consists of two different alleles, e.g. **Bb**, we say that the person is **heterozygous**. So John is heterozygous for eye colour.

These technical terms are applied not just to humans but to other organisms as well. They may seem long-winded and difficult, but they are a useful shorthand. They help us to describe examples of heredity without using too many words.

We shall now apply the principles of heredity which we've learned about in the human, to a completely different kind of organism – plants. This is important for market gardeners and others who grow plants to sell.

Crossing plants

Suppose we have a bed of plants like the ones in picture 2. Some have red flowers, others white. We take some pollen from a red flower and place it on the stigma of a white flower: in this way we cross the two plants. When the seeds develop, we sow them in the soil. In time new plants grow up from the seeds. They all have red flowers.

How can we explain this? Look at picture 3. Each of these plants has in its cells a gene which controls the colour of its flowers. The alleles of this gene include two which we shall call **R** and **r**. **R** is the allele for red flowers, and **r** is the allele for white flowers. **R** is dominant to **r**. The red-flowered parent is homozygous dominant (**RR**), whereas the white-flowered parent is homozygous recessive (**rr**).

Now the gametes (pollen grains and egg cells) contain only one of these alleles. Each gamete produced by the red-flowered plant contains an **R** allele, and each gamete produced by the white-flowered plant contains an **r** allele.

When fertilisation takes place, the **R** and **r** alleles are brought together, so the offspring are heterozygous (**Rr**). As **R** is the dominant allele, the offspring have red flowers.

These plants get their red flowers from their parents in exactly the same way as John got his brown eyes from his parents. So the principles of heredity apply just as much to plants as they do to humans.

Another plant cross

The red-flowered offspring from the previous cross belong to the **first filial generation**, or **F1** for short.

Now suppose we take two of these red-flowered plants and cross them. Or alternatively we could self-pollinate one of them. The resulting seeds are planted, and the new plants grow up and bear flowers. They belong to the **second filial generation** or **F2**.

This time we get a mixture of red-flowered and white-flowered plants. On counting each type, we find that roughly three-quarters are red, and one quarter white. In other words, they are in a ratio of 3 to 1.

How can we explain this? Look at picture 4. The two parent plants are both heterozygous (**Rr**), as we have already seen. Now the gametes produced by these plants contain either an **R** allele or an **r** allele. In fact there should be equal numbers of each.

Fertilisation is random, and it is sheer chance as to which kind of pollen fertilises which kind of egg cell. Picture 4 shows the possible combinations, and the offspring resulting from each. Can you see why we get a 3 to 1 ratio between the red and white-flowered offspring?

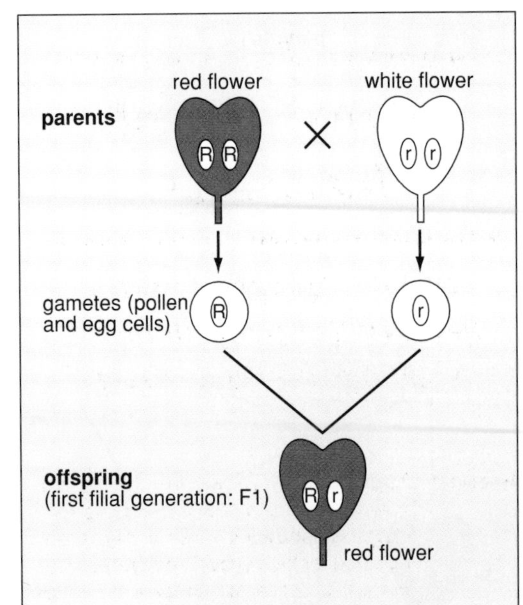

Picture 3 The result of crossing two plants with red and white flowers. The alleles are indicated by letters: **R**, red; **r**, white.

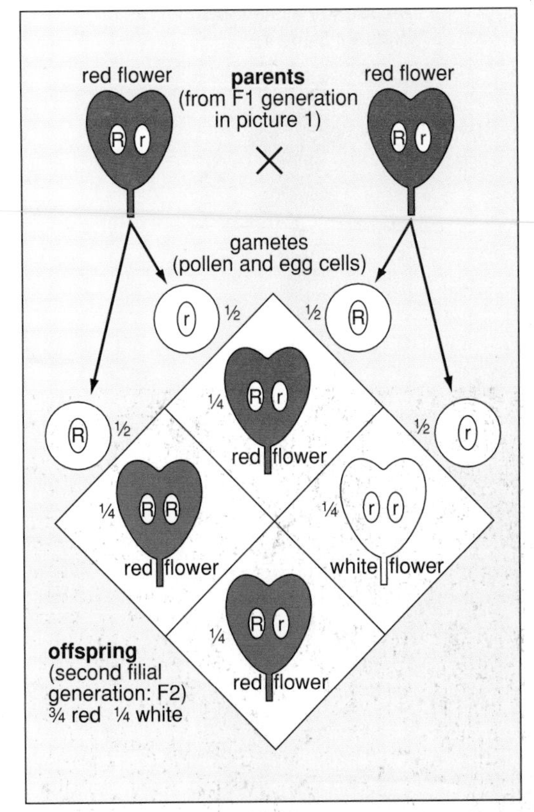

Picture 4 The result of crossing two red-flowered offspring from picture 3. The same result would be obtained by self-pollinating one of them. Symbols as in picture 3. Instead of using lines to show which gamete joins up with which, here the same thing is shown by means of a grid. A grid has the advantage of not having lots of lines crossing each other, like there are in picture 4 on page 264.

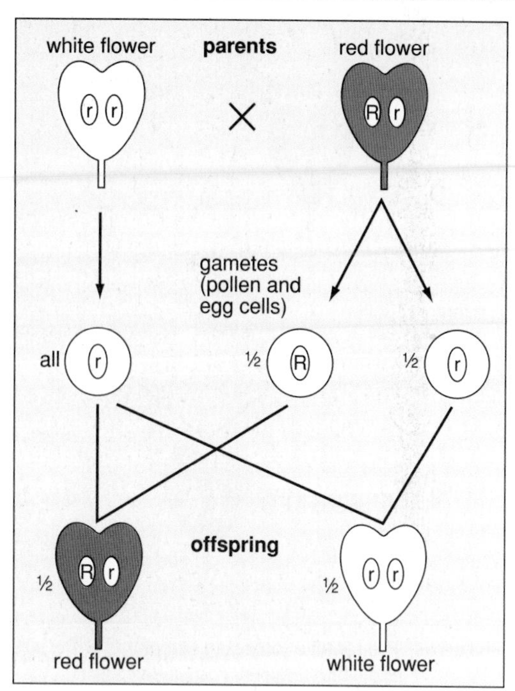

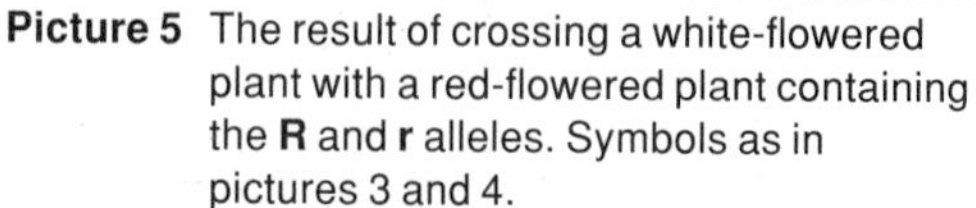
Picture 5 The result of crossing a white-flowered plant with a red-flowered plant containing the **R** and **r** alleles. Symbols as in pictures 3 and 4.

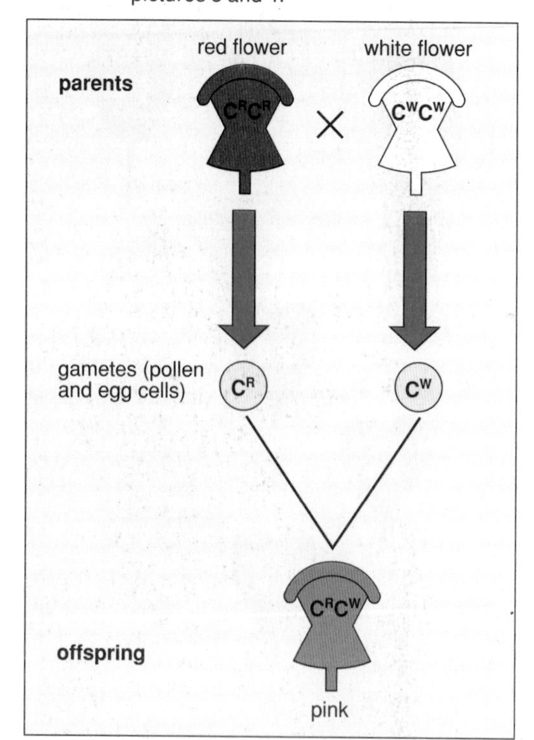

Picture 6 The result of crossing a red- and white-flowered plant which shows codominance. **R** and **W** stand for the alleles: **R**, red; **W**, white. They are represented by different letters because neither is dominant over the other. **C** stands for the flower-colour gene, of which **R** and **W** are the two alleles. This is the accepted way of writing the genotypes when codominance is involved.

Doing a test cross

The red-flowered offspring from the previous cross have two genotypes: some are homozygous (**RR**), and others are heterozygous (**Rr**). Both look exactly alike, so you can't tell which is which just by looking at them. How, then, could you find out if a particular red-flowered plant is homozygous or heterozygous?

One way would be to cross it with a white-flowered plant. If the red-flowered plant is homozygous (**RR**), the offspring will all be red-flowered (as in picture 3). On the other hand, if the red-flowered plant is heterozygous (**Rr**), we would expect to get a mixture of red-flowered and white-flowered plants in roughly equal proportions (picture 5).

A cross which is done to find the genotype of an organism is called a **test cross**.

Producing plants with the same flower colour

Suppose you are a market gardener and you find that your customers want mainly white-flowered plants. How could you produce nothing but white-flowered plants?

The answer would be to cross two white-flowered plants, or self-pollinate one of them. You know that these plants must be homozygous recessive (**rr**), so the offspring are bound to be homozygous recessive too – in other words they will be white, just like the parents. And if you cross two of these offspring, or self-pollinate one of them, their offspring will also be white, and so on down the generations. This is called **breeding true**, and it results in a **pure line**. In a pure line all individuals have the same genotype for a particular feature.

Suppose your customers want only red-flowered plants. In that case you would have to cross two homozygous red-flowered plants (**RR**), or self-pollinate one of them. You would need to make sure that these plants were homozygous and not heterozygous (**Rr**). If both plants were heterozygous you'd get some white-flowered plants amongst the offspring – and that might put you out of business!

Do genes always show dominance?

Look back at picture 3. Here a cross between a red-flowered and white-flowered plant produced nothing but red-flowered offspring. However, with some types of plant you get a different result. Instead of the offspring having red flowers, the flowers are pink.

The allele for red flowers and the allele for white flowers are present together in these pink-flowered offspring. In other words, the offspring are heterozygous. But in this case both alleles have an effect on the outward appearance of the plants. Because of this, these alleles are described as **codominant**. Picture 6 shows how they pass from the parents to the offspring.

Codominance occurs in both plants and animals. For example, flower colour is inherited in this way in snapdragons (picture 7). An animal example is seen in American shorthorn cattle: if a red-haired bull mates with a white-haired cow, the calves may have a mixture of red and white hairs. This is called the roan condition and gives them a sort of light red colour. There are examples of codominance in humans too, as we shall see in the next topic.

Picture 7 Snapdragons, an example of a plant that shows co-dominance.

Questions

1 In pea plants, the allele for long stem is dominant to the allele for short stem. A long-stemmed pea plant was crossed with a short-stemmed pea plant. 123 offspring were produced. Of these, 68 had long stems and 55 had short stems.

Explain this result, using the following terms: phenotype, genotype, homozygous and heterozygous. You may draw a diagram if you like.

2 Re-read the section entitled *Doing a test cross* on page 268.

Suppose you had no white-flowered plants available with which to do a test cross. Suggest another way of finding out if the genotype of a red-flowered plant is **RR** or **Rr**? Would the way you suggest be suitable for an animal? Explain your answer.

3 In a certain type of plant, white fruit colour is dominant to yellow. A fruit-grower crosses a white-fruited plant with a yellow-fruited one. About half the offspring have white fruit, and half have yellow fruit.

a What are the genotypes of the parent plants? How do you know?
b If you were to self-pollinate one of the white-fruited offspring, what phenotypes would you expect to obain, and in what proportions?

4 A farmer has a bull with a dark red coat, and a cow with a white coat. He allows them to mate, and the cow gives birth to a calf with a light red coat.

a Explain this result with a diagram.
b The calf grows into a fine cow and is mated with a white-coated bull. What sort of calves might she produce? Explain your answer.

Gregor Mendel, the father of genetics

Gregor Mendel was a monk. He belonged to an Augustinian monastery in a town called Brunn in Austria (now Brno in the Czech Republic). He was a teacher in the local school and everyone liked him. One of his pupils remembered him as a cheerful clergyman, kind to everyone while contemplating the world through his gold-rimmed spectacles.

Picture 1 Gregor Mendel, the Austrian monk who discovered the principles of genetics.

Mendel had always been interested in heredity, so he decided to do some experiments on pea plants. He chose pea plants because they had a number of clear differences which were easy to tell apart. For example, some plants had red flowers and others white flowers, and some plants had short stems and others long stems.

Starting about 1856, Mendel carried out a series of experiments in the garden of his monastery. He carefully isolated certain plants and transferred pollen from one to another. He then collected and sowed the seeds, and when the offspring grew up he counted the different types. In the next ten years or so, he set up hundreds of crosses and produced thousands of offspring.

By counting the offspring, Mendel discovered the ratios which are described on page 267. But he did more than that. He also drew the right conclusions. This is remarkable when you think that he knew nothing about genes – they hadn't been discovered. But he realised that such things must exist – he called them 'factors'. He described how they must be passed from parents to offspring, and we now know that he was right.

In 1866 Mendel published his results in the journal of the local scientific society, but no one took any notice. He even sent a copy to a famous Swiss professor, but he ignored it too. He said that Mendel's experiments were incomplete and that he should plant more peas. In fact Mendel's data were based on more than 21 000 plants.

In 1868 Mendel was made Abbot of his monastery, and he became so busy that he had no more time for research. He died in 1884, unrecognised as a scientist.

Sixteen years later a Dutch biologist called Hugo de Vries was looking through scientific journals in a library when he came across Mendel's paper. He immediately realised its importance and told his colleagues about it. Only then was Mendel's work recognised.

1 What difficulties do you think Mendel might have come up against in his research?

2 From his experiments Mendel was able to make certain predictions about living organisms which were confirmed later. Suggest two such predictions.

Picture 2 Mendel in the garden of the monastery where he carried out his famous experiments on pea plants.

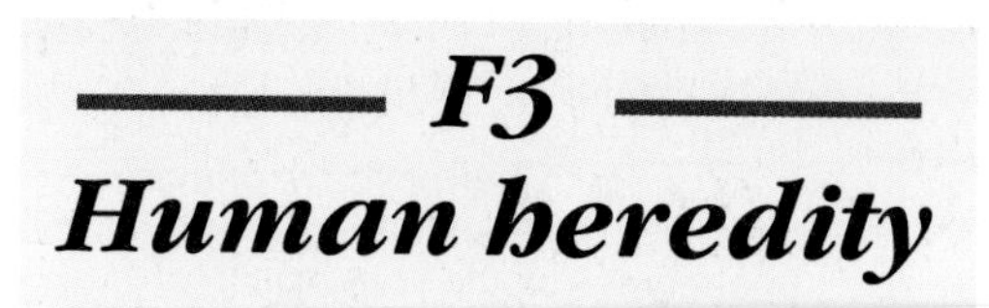

F3
Human heredity

Here we look at some examples of human heredity and the way it can affect our lives.

Picture 1 Cystic fibrosis is a disease of the pancreas, lungs and certain other organs. It is an inherited condition occurring about once in every 2500 births. Scientists have been working hard to find a way of preventing this disease which is known to be caused by a particular gene. Much of the research is funded by voluntary donations.

Inherited diseases

Certain genes can cause diseases which may be passed from parents to their children. In some cases the dominant allele is harmful, but more often it is the recessive allele.

An example of such an inherited disease is **cystic fibrosis** (picture 1). It is caused by a recessive allele which is inherited in a straightforward way. Only people who are homozygous for the harmful allele get the disease. Heterozygous people do not get the disease. However, they are **carriers** of the allele and may pass it to their children, as you can see in picture 2.

If a couple give birth to a child with a disease such as cystic fibrosis, their doctor can arrange for them to see a **genetic counsellor**. The genetic counsellor will try to work out the chance of their next child being born with the disease. Knowing the risks, the parents can then decide whether or not to have any more children.

Sometimes couples who have not yet had any children know that there is a history of a particular disease in one or other of their families. For example, the wife may have had a grandparent or an uncle with the disease. Here too a genetic counsellor can be helpful.

In order to advise people, the genetic counsellor must know as much as possible about the parents' **pedigrees**.

What is a pedigree?

You come from your parents, and they came from *their* parents – and so on. This is your pedigree.

Building up a person's pedigree involves tracing back his or her history through the parents, grandparents, great-grandparents and so on. A pedigree can be built up for any kind of organism whose ancestors are known. You can then use it to show how certain features or **traits** are inherited. This is done in the form of a **family tree.**

A family tree is shown in picture 3. It shows the occurrence of **night-blindness** in the males and females of a family over three generations. Having built up a chart like this, we can work out the genotypes, or possible genotypes, of the various individuals. From this it may be possible to work out the chance of the problem arising in the next generation.

It is remarkable how some features persist in a family. A famous example is the drooping lower lip of the Habsburg family (picture 4). By looking at family portraits, this feature can be traced back through several centuries. It is thought to be caused by a dominant allele.

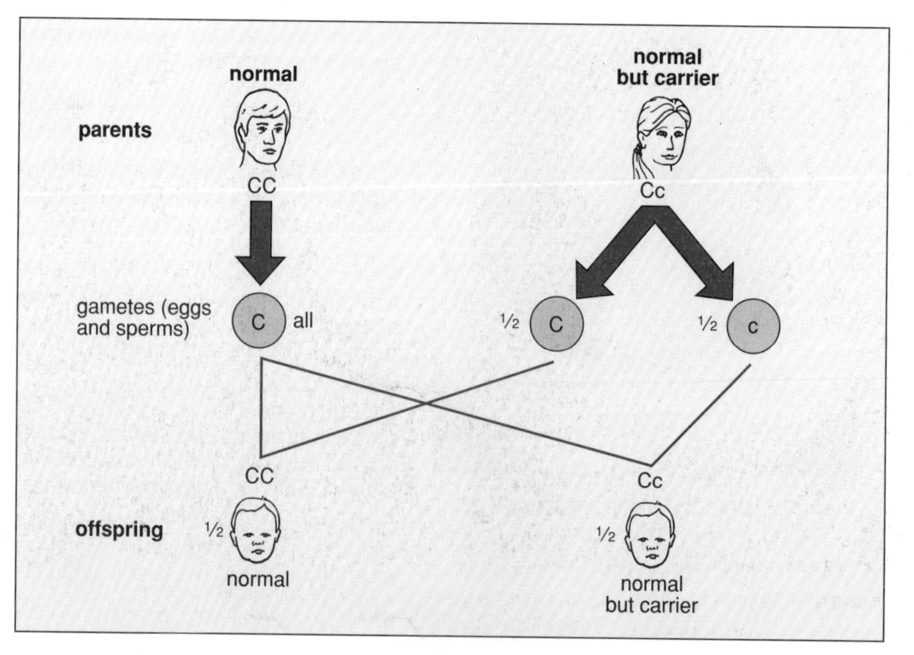

Picture 2 How cystic fibrosis is inherited. The diagram shows what happens if a normal person (genotype **CC**) mates with a carrier (genotype **Cc**). None of their children will have the disease, but there is a one in two chance that a child will be a carrier. If two carriers mate there is a one in four chance that they will produce a child with the disease. Can you see why?

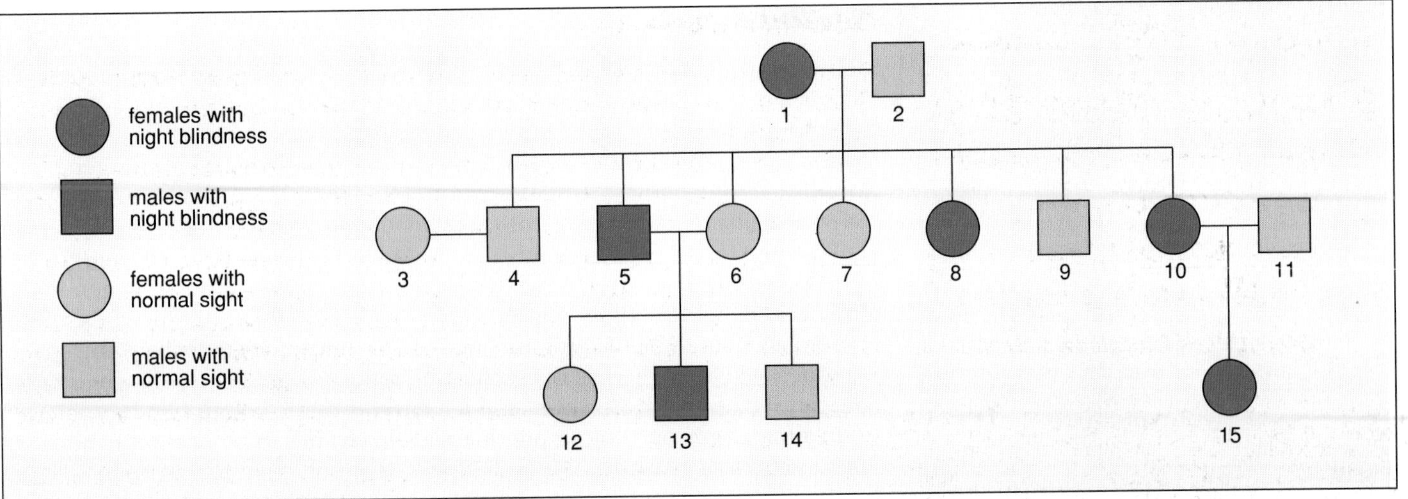

Picture 3 Pedigree showing the inheritance of night-blindness in a family. In this condition it is difficult to see in dim light. This type of night-blindness is controlled by a gene which has two alleles; the allele for night-blindness is dominant to the allele for normal vision.

A boy or a girl?

What decides whether the fertilised egg develops into a boy or a girl? Well, it depends on what kind of chromosomes the fertilised egg contains.

We have in our cells a pair of **sex chromosomes**: they decide the person's sex. One is longer than the other. The long one is called the **X chromosome**, and the short one is called the **Y chromosome**. Males possess an **X** and a **Y** chromosome in their cells, whereas females possess two **X** chromosomes.

Now the sperm which the male makes in his testes contain only one of these two chromosomes, either **X** or **Y**. In fact, half the sperm should contain an **X** chromosome and half should contain a **Y** chromosome. (For shortness we will call them **X** and **Y** sperm.) On the other hand, the eggs which the female produces in her ovaries will all contain an **X** chromosome. This is shown in the top part of picture 5.

When fertilisation occurs, the egg may be fertilised by either an **X** sperm or a **Y** sperm. In fact, fertilisation is random so there is an equal chance of either happening. If an **X** sperm fertilises the egg, the fertilised egg will contain two **X** chromosomes, and this will develop into a female. On the other hand, if a **Y** sperm fertilises the egg, the fertilised egg will contain an **X** and **Y** chromosome and will develop into a male. This is shown in the bottom part of picture 5.

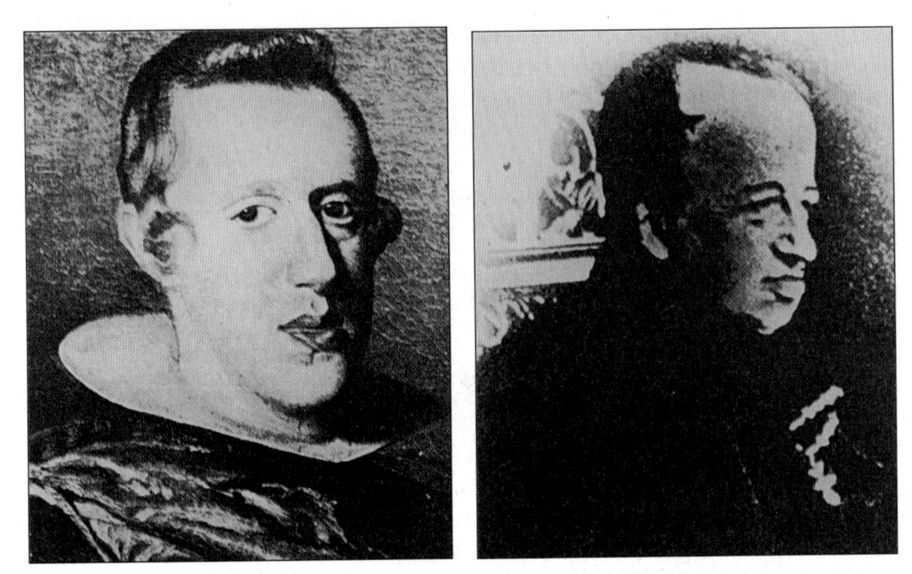

Picture 4 Two members of the Habsburg family, showing the famous 'Habsburg lip'. Left: Philip IV of Spain, 1605–1665. Right: Ferdinand I of Austria, 1793–1875.

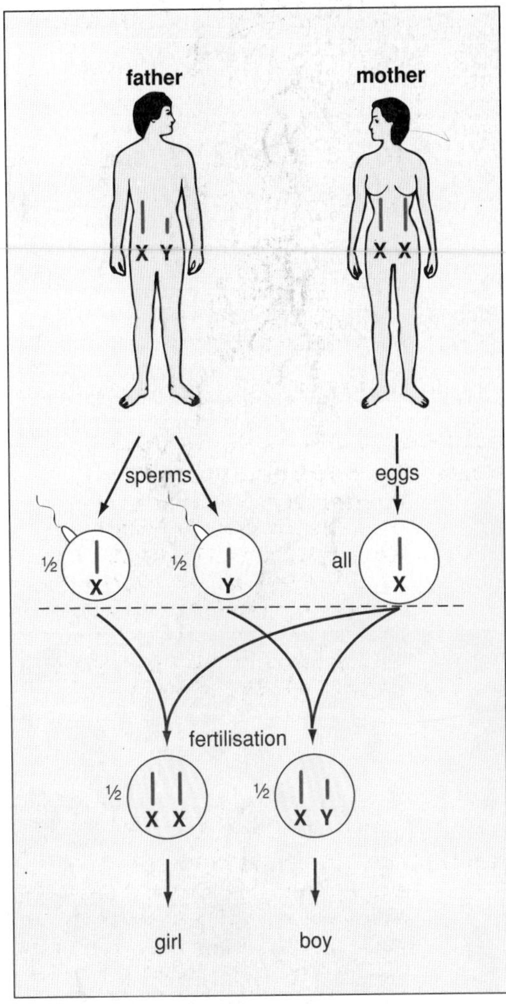

Picture 5 A boy or a girl? It depends on the sex chromosomes. The chromosomes are shown in red. There is a pair of sex chromosomes in every body cell.

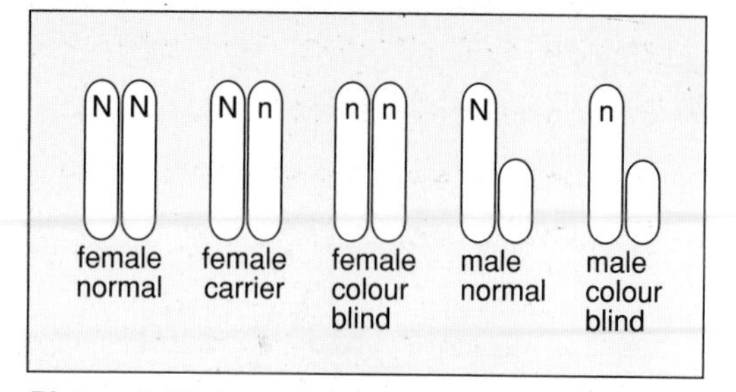

Picture 6 Various possibilities for a sex-linked disease such as colour blindness. The long chromosomes are **X** chromosomes, and the short ones are **Y** chromosomes. The normal allele (**N**) is dominant to the defective allele (**n**).

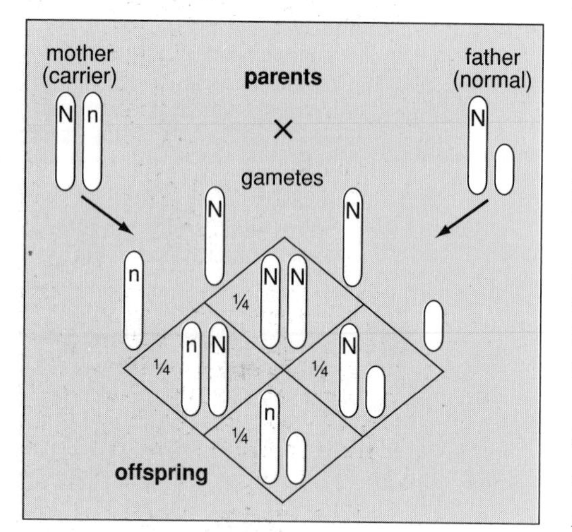

Picture 7 Possible offspring of a mother who is a carrier for colour blindness, and a normal father. Use picture 6 to help you describe the phenotype of each of the offspring. The symbols are the same as in that picture.

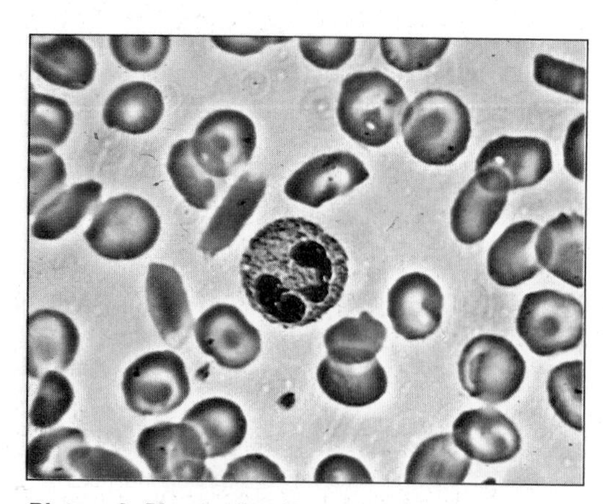

Picture 8 Blood cells of a person with sickle cell disease as seen under the microscope. notice the sickle-shaped red blood cells.

Sex linkage

Some people can't tell the difference between red and green. Both look grey. This is a type of **colour blindness** (see page 205).

Colour blindness is caused by a gene which is situated on the sex chromosomes. We say the gene is **sex-linked**. There are two alleles of this gene, one for normal sight and the other for colour blindness. The allele for normal sight (**N**) is dominant, and that for colour blindness (**n**) is recessive.

The gene for colour blindness occurs only on the **X** chromosome. It does not occur on the **Y** chromosome. picture 6 shows all the possible genotypes and phenotypes for colour blindness.

For a woman to be colour blind, both **X** chromosomes must have the colour blindness allele. If she has one colour blindness allele and one normal allele, she can see normally. However, she is a **carrier** of the colour blindness allele.

In a man, only one colour blindness allele (**n**) has to be present to show its effect. For this reason colour blindness is more common in men than in women.

Picture 7 shows what can happen if a woman who is a carrier for colour blindness mates with a normal man. The genotypes of the children are given. Can you fill in their phenotypes?

Work out the offspring that might be produced by matings between a normal mother and colour blind father, and between a carrier mother and colour blind father.

Another example of sex linkage is **haemophilia**. In this disease the blood takes a very long time to clot (see page 159). A haemophiliac who gets a bad cut may bleed to death. Like colour blindness, haemophilia is caused by a recessive allele carried on the **X** chromosome. It is inherited in the same way as colour blindness.

Sickle cell disease

This is an inherited disease of the blood which is particularly common in Africa. A person with the disease has sickle-shaped red blood cells which contain an abnormal kind of haemoglobin (picture 8). This makes it difficult for the blood to carry oxygen, resulting in severe anaemia.

The disease is caused by a gene which has two alleles, one normal and the other harmful. If you are homozygous for the normal allele you don't get the disease, but if you are homozygous for the harmful allele you do get the disease. This is what you would expect.

The interesting thing is what happens if you are heterozygous. In this case you get a very mild form of the disease called **sickle cell trait**. So the harmful allele can have an effect even when the normal gene is present. This is an example of **codominance** which we met in the last topic. We shall meet sickle cell disease again in connection with evolution (page 292).

How blood groups are inherited

Everyone belongs to one of four blood groups called **A**, **B**, **AB** and **O** (see page 161). This blood group system is controlled by a gene which has three different alleles. We shall call these alleles $\mathbf{I^A}$, $\mathbf{I^B}$ and $\mathbf{I^o}$. Alleles $\mathbf{I^A}$ and $\mathbf{I^B}$ show no dominance over each other (i.e. they are codominant). However, both are dominant to the $\mathbf{I^o}$ allele.

Although there are three alleles, any one individual can only have two of them. Your blood group depends on which two alleles you possess:

- To belong to **group A**, your genotype must be either $\mathbf{I^AI^A}$ or $\mathbf{I^AI^o}$.
- To belong to **group B**, your genotype must be either $\mathbf{I^BI^B}$ or $\mathbf{I^BI^o}$.
- To belong to **group AB**, your genotype must be $\mathbf{I^AI^B}$.
- To belong to **group O**, your genotype must be $\mathbf{I^oI^o}$.

Now suppose a man belonging to group **A** marries a woman belonging to group **O**, and they have a child. What blood group will the child belong to?

The answer depends on whether the husband is I^AI^A or I^AI^o. If he is I^AI^A then the child must belong to group **A**. On the other hand, if the husband is I^AI^o, there's an equal chance of the child belonging to group **A** or group **O**. If you are uncertain about this, look at picture 9.

Blood groups are sometimes used in court cases. For example, Mrs Green claims that Mr White is the father of her child. Their blood is tested, and it turns out that Mrs Green belongs to group **B**, Mr White to group **O** and the child to group **AB**. This shows that Mr White could not be the child's father. Not all cases are as clear-cut as this.

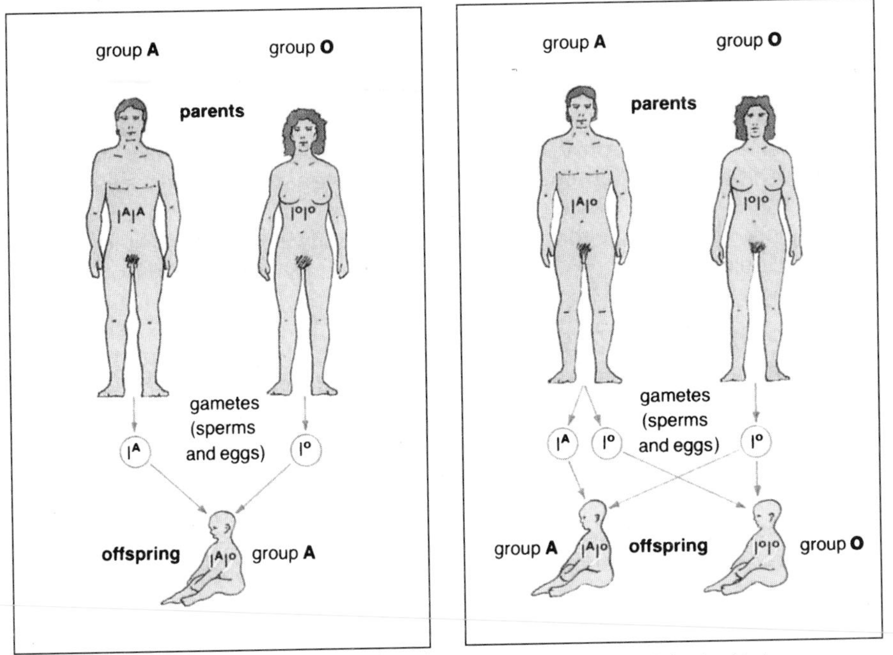

Picture 9 These diagrams show the way the ABO blood groups are inherited in humans. **I** represents the gene that controls the blood groups. **A**, **B** and **o** are its three alleles.

Activity

Haemophilia in the royal families of Europe

One of the most famous examples of haemophilia was in the royal families of Europe. The story starts with Queen Victoria, seen in the portrait below.

From books find out as much as you can about the way haemophilia spread through the royal families of Europe in the 19th and early 20th centuries. How did the allele arise and to whom was it passed? In particular how did it affect the Russian royal family?

Haemophilia did not spread through the British royal family. Why not?

Does haemophilia still occur in the royal families of Europe and their descendents?

Questions

1 Suppose that in humans the allele for short fingers is dominant to the allele for long fingers. Short-fingered Sue marries long-fingered Larry, and they have five children. Three of the children have short fingers, and two have long fingers. Using **S** as the allele for short fingers, and **s** as the allele for long fingers, give the genotypes of Sue, Larry and their five children.

2 Albinos have no pigment in the skin, so they are very pale. This condition is caused by a recessive allele. Dave marries Joan. Dave and Joan are both normal, but each of them has a parent who is an albino.

a How likely is it that Dave and Joan's first child will be an albino? Give your reasons in full.

b Certain individuals in this family are 'carriers'. Which ones are carriers, and what does this word mean?

3 Look at the family tree in picture 3 on page 271. In answering the following questions, use **B** as the symbol for the night-blindness allele (dominant), and **b** for the normal allele (recessive).

a Write down the possible genotypes of all the people in the chart.

b Explain how you know the genotype of person 1.

c How do you know the genotypes of persons 13 and 15?

d If persons 13 and 15 should marry, what is the chance that any of their children will be night-blind? Explain your answer.

e If persons 14 and 15 should marry, what is the chance that any of their children will be night-blind? Explain your answer.

4 Imagine you are a genetic counsellor. You are visited by Mr and Mrs Flap. They are worried because Mrs Flap and her father both have huge ears, and they don't want to bring a child into the world with ears like Mrs Flap's. There is no history of huge ears in her husband's family. What advice would you give them? Explain your reasoning, and state any assumptions that you need to make.

5 Mr Smith is dismayed to find that his son has a green nose, just like Mr Tree next door. So he accuses Mr Tree of being the boy's father. He takes the matter to court. Blood tests are carried out. Mr Smith's blood group is A, his wife's is B and the son's is O. Mr Tree's is AB. What advice would you give the court?

6 Mr and Mrs Cross have three children, all boys. They are sure that their next child will be a girl. Do you agree? Explain your answer.

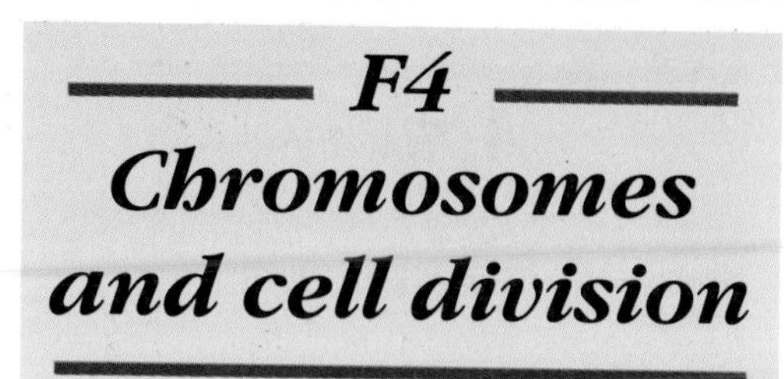

F4

Chromosomes and cell division

Here we shall look more closely at chromosomes and what they do.

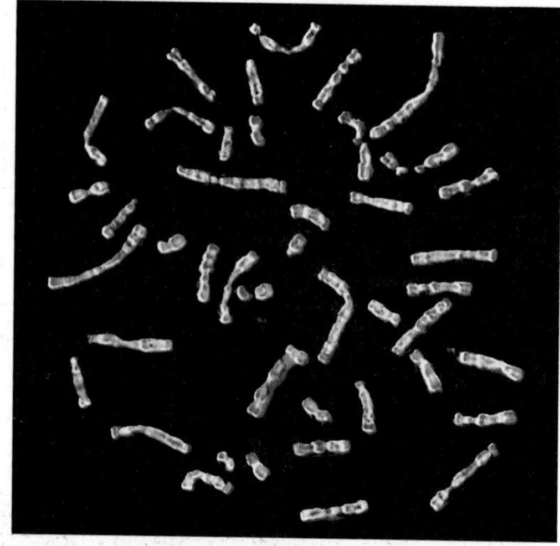

Picture 1 The full set of chromosomes from a cell of a human. This picture was obtained by breaking the cell open and staining the chromosomes with a special dye.

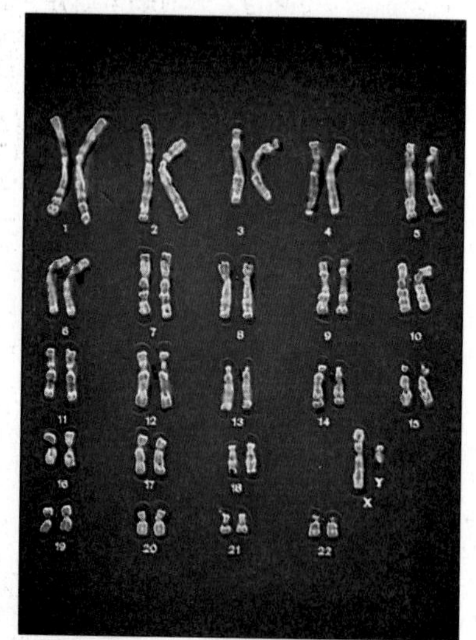

Picture 2 Human chromosomes cut out and arranged in pairs. The chromosomes differ in size and in various features. Each chromosome is given a number for identification purposes. The **X** and **Y** chromosomes are sex chromosomes which make this person a male. A female would have two **X** chromosomes.

Looking at chromosomes

Picture 1 shows the chromosomes of a human. Every cell of the body contains a set of chromosomes like this. Now look at picture 2. Here the chromosomes have been sorted out and arranged according to their sizes and various others features.

In picture 2 notice that the chromosomes are in pairs. The matching chromosomes in each pair are called **homologous chromosomes**.

As you can see in picture 2, the total number of chromosomes in a human cell is 46 (23 pairs). Other species have different numbers of chromosomes. For example, a chimpanzee has 48 (24 pairs), a dog 78 (39 pairs), a Siamese fighting fish 42 (21 pairs) and a cabbage 18 (9 pairs).

The chromosomes in pictures 1 and 2 come from a cell which was about to divide. Chromosomes can be seen clearly only when a cell is dividing or about to divide. At other times the chromosomes are very long and thin and cannot be seen properly. Notice that each chromosome consists of two strands joined together at some point along their length. They look like this because each chromosome has produced a copy of itself. This always happens before a cell divides so that the new cells have the right number of chromosomes.

What happens when cells divide?

When a cell divides, it splits into two **daughter cells**. 'Daughter' does not mean that they are female – it simply means that they are new cells formed as a result of the division. To begin with, the daughter cells are smaller than the parent cell. however, they quickly grow to full size. After that, they too may divide. In this way a single cell may multiply into a mass of cells.

Cells divide in two different ways according to how the chromosomes are shared out between the daughter cells. These two types of cell division are called **mitosis** and **meiosis**.

In **mitosis** the daughter cells finish up with *exactly the same* number of chromosomes as the parent cell. This is called the **diploid number** as there are two of each type of chromosome. Mitosis occurs during growth and asexual reproduction. It means that all the body cells have a full set of chromosomes, and a full set of genes. Similarly, all the offspring resulting from asexual reproduction have a full set of chromosomes, and a full set of genes.

In **meiosis** the daughter cells finish up with *half* the original number of chromosomes – one member of each pair. This is called the **haploid number**. Meiosis occurs during the formation of eggs and sperm.

Why is meiosis important?

Eggs and sperm *must* be formed by meiosis. You can see why if you look at picture 3. This shows the human **life cycle**, that is the events which take place from a person's conception (when fertilisation takes place) to when he or she reproduces. Meiosis results in the egg and sperm having half the full number of chromosomes, i.e. 23. When fertilisation takes place, we get the full number of chromosomes back again. So the number of chromosomes in the cells of an adult human is always 46.

Think what would happen if eggs and sperm were formed by mitosis instead of meiosis. They would have 46 chromosomes.This number would double on fertilisation, so the offspring would have 92. And when the offspring reproduced, the number would double again – and it would go on doubling every generation!

Think back to heredity. Do you remember that the adult has two alleles controlling each characteristic, but the eggs and sperm have only one? The reason is that the eggs and sperm are formed by meiosis. The chromosomes that carry the two alleles get separated from each other and each goes into a different egg or sperm.

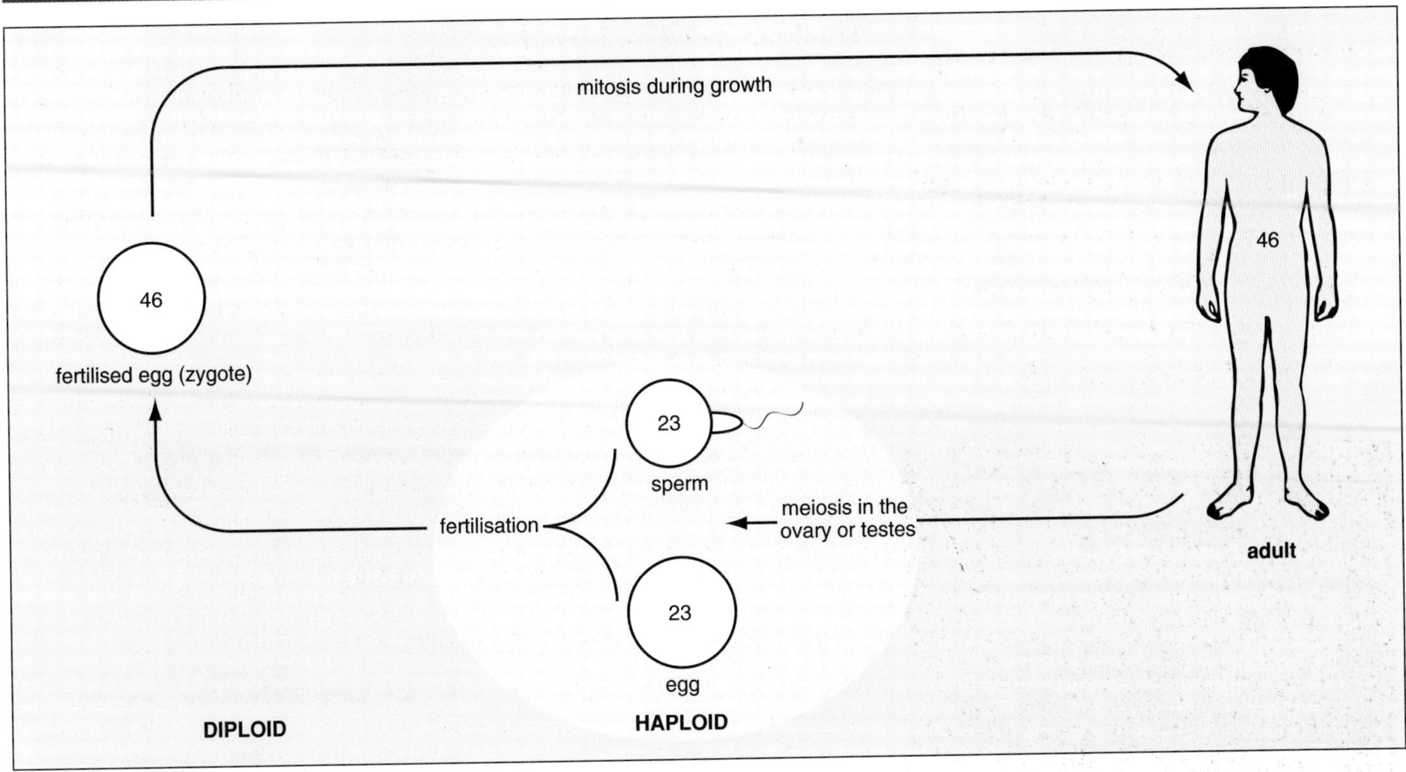

Picture 3 The human life cycle. The figures refer to the number of chromosomes in each of the cells.

When chromosomes go wrong

Occasionally a baby is born with too few, or too many, chromosomes. Or a bit of a chromosome may be missing, or an extra bit is present.

In some cases this does not matter, but in other cases it does. For example, there is a condition known as **Down's syndrome**. A person with Down's syndrome has an extra chromosome in his or her cells – number 21 to be exact. The presence of this extra chromosome causes physical and mental problems.

How does the person get this extra chromosome? It results from an abnormal type of meiosis which usually occurs in the formation of the egg within the mother's ovary. This happens more often in older women. It is a type of **mutation** (see page 284).

Nowadays doctors can find out if a baby has any abnormal chromosomes before it is born. A few cells are obtained from the fluid surrounding the baby in the uterus. The cells are stained to show up the chromosomes and examined under the microscope. If any abnormalities are seen, the mother and father are told. In serious cases an abortion may then be considered.

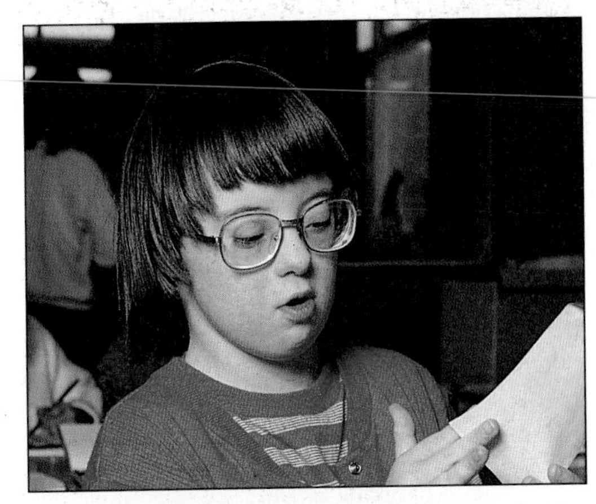

Picture 4 This little girl has Down's syndrome but is able to lead a fairly normal life.

Questions

1 Look at the cell shown on the right, then answer the following questions.

a How many chromosomes are there altogether?
b How many *pairs* of chromosomes are there?
c If this cell divided by mitosis, how many chromosomes would there be in each daughter cell?
d If the cell divided by meiosis, how many chromosomes would each daughter cell contain?

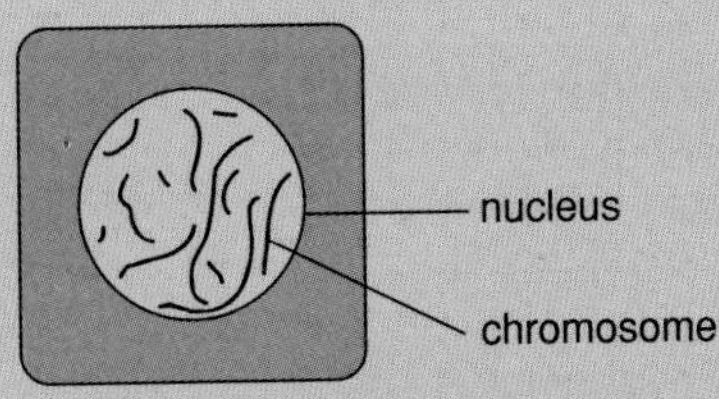

2 In a human, how many chromosomes are present in:

a a brain cell,
b a sperm cell in the testis,
c an egg which has just been produced by the ovary,
d a skin cell,
e a fertilised egg?

3 The number of boys and girls born in the world is approximately equal. Explain the reason for this.

Activity

Looking at chromosomes

Try doing this activity *before* you read about mitosis below.

You can only see chromosomes clearly in cells that are dividing or about to divide.

The cells in the tip of a root are constantly dividing, so this is a good place to look for chromosomes.

Your teacher will give you a prepared slide of a root tip.

Can you see chromosomes in some of the cells? What do they look like? The cells which you are looking at are in various stages of division: some are just about to divide, others are in the process of dividing, and others have just finished dividing. Some are not dividing at all.

From your observations, can you get any idea of what the chromosomes do when a cell divides?

Now read about mitosis below and see if you are right.

Mitosis and meiosis

Imagine a cell with two pairs of chromosomes, two long ones and two short ones. Now suppose this cell divides by **mitosis**. What sort of chromosomes will the daughter cells have? They will have the same number and types of chromosomes as the parent cell, namely two long and two short.

Now look at picture 1 on the opposite page. This shows what happens to the chromosomes during mitosis. Study the diagrams carefully, starting at the top and working your way down. Notice that the chromosomes behave in such a way that the daughter cells are bound to have the *same* number and types of chromosomes as the parent cell.

Now suppose this same cell divides by **meiosis**. What sort of chromosomes will the daughter cells have this time? Well, each daughter cell will have only *two* chromosomes: one long and one short. In other words, the cells will contain only *half* the original number of chromosomes.

Picture 2 shows how this comes about. Study the diagrams carefully, as you did for mitosis. Notice that the chromosomes behave in such a way that the daughter cells are bound to contain only one of each type of chromosome.

There is something about meiosis that makes this halving of the chromosome number inevitable. What is it? It is the way the chromosomes line up in the middle of the cell. In mitosis matching chromosomes line up separately. But in meiosis matching chromosomes line up side by side. They then separate to opposite ends of the cell which subsequently splits in two. This is followed by a *second* cell division in which the chromatids part company and finish up in separate cells.

It is a basic feature of meiosis that there are always two cell divisions, one after the other. So meiosis always results in the formation of four daughter cells.

1 First read about *diagrams* on page 6. The picture below shows what the chromosomes *really* look like during mitosis. This picture was made by taking a photograph of a root tip down a microscope. How does the photograph differ from the diagrams in picture 1 on the opposite page? Does it matter that they are different?

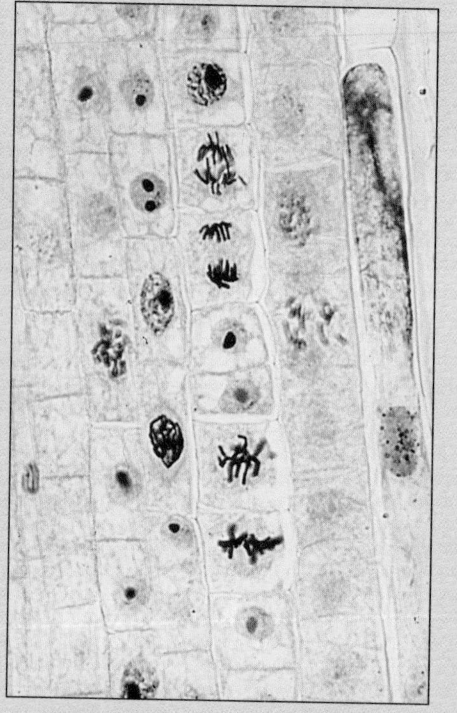

2 If you look very carefully at pictures 1 and 2 on page 274, you will see that each chromosome consists of two strands, side by side. What are these strands, and what is their significance?

Meiosis and variation

Gametes (eggs and sperm) are normally produced by meiosis. Now the gametes produced by an individual – the sperm produced by a man for example – all contain different sets of alleles. In fact *no two gametes ever contain exactly the same set of alleles.*

Two features of meiosis ensure that this is so. They depend on the fact that the two chromosomes of a matching (homologous) pair may carry different alleles. For example, one may carry an allele for brown eyes whereas the other carries an allele for blue eyes (see page 265).

- When matching chromosomes line up in the middle of the cell, and then separate, they do so independently of all the other pairs of matching chromosomes. It is sheer chance as to which chromosomes of each matching pair finish up together in the daughter cells, so all sorts of combinations are possible in the gametes. This is called **independent assortment**.
- When the matching chromosomes line up, they don't just lie side by side as shown very simply in picture 2. What really happens is that they wrap round each other in an intimate manner. At certain points where the chromosomes touch each other, they may break and change places, taking their alleles with them. This is called **crossing over** and it leads to new combinations of alleles in the gametes.

Independent assortment and crossing over are entirely random. The way they take place in one meiosis differs from the way they occur in another. The result is that *every gamete receives its own unique set of alleles.*

When fertilisation takes place and gametes unite, a completely new combination of genes is established in the fertilised egg (zygote). This is one of the main reasons why human beings, even close relatives, differ from each other.

Being different from each other is called **variation**. Variation is very important in biology and we shall look at it again later (page 284).

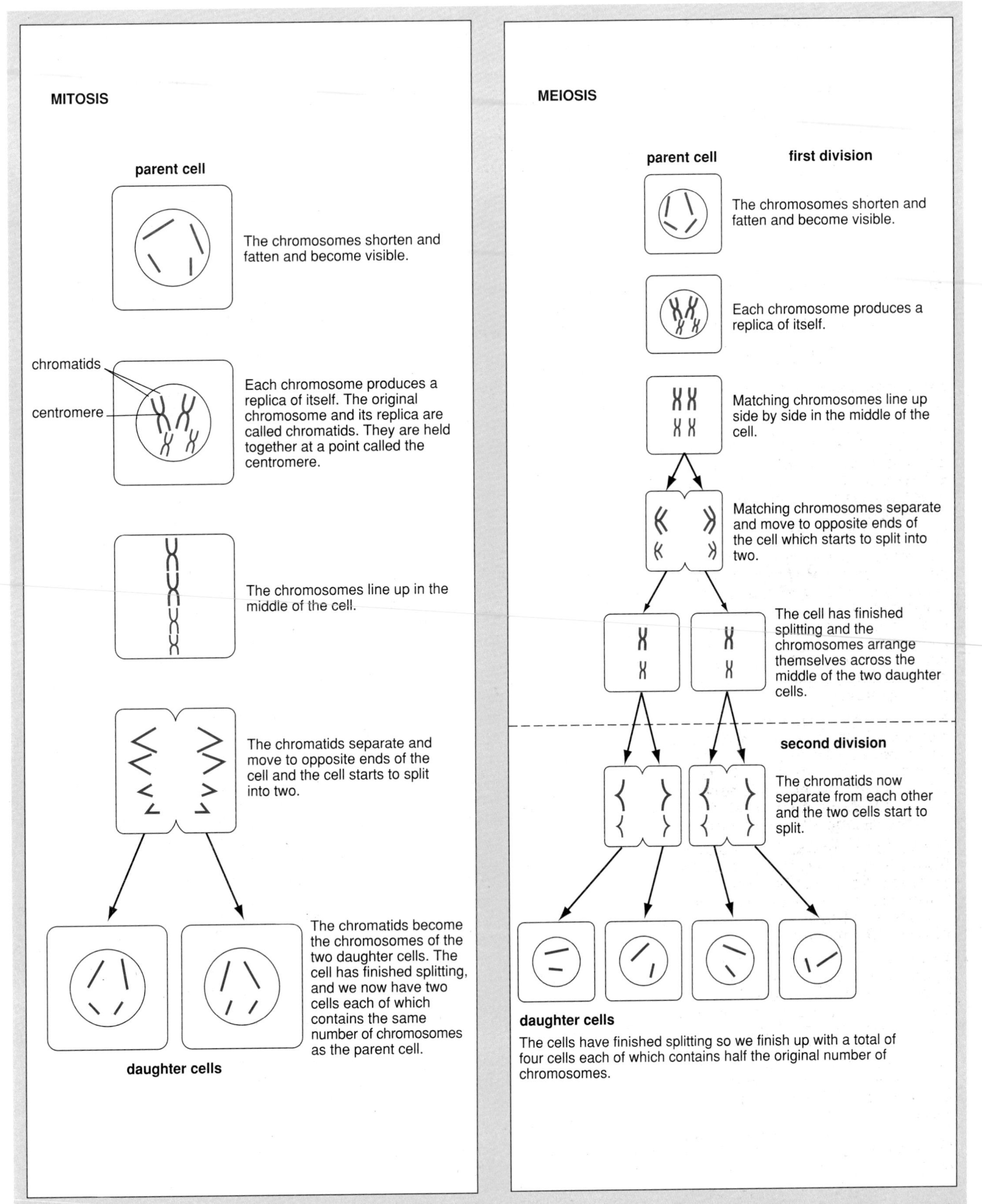

Picture 1 What the chromosomes do during mitosis.

Picture 2 What the chromosomes do during meiosis.

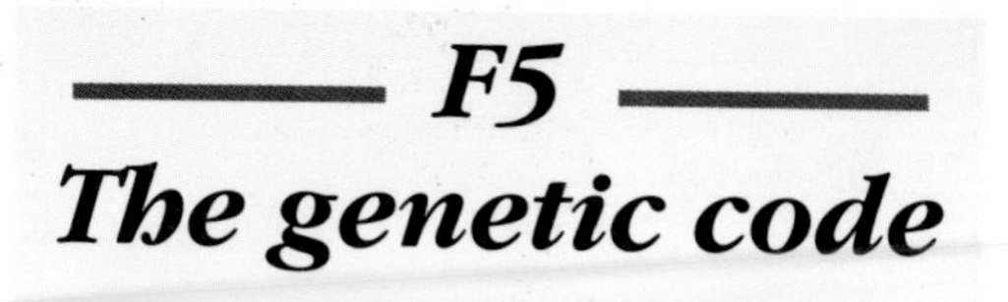

F5 The genetic code

What are genes, and how do they work? Modern research has given us the answer.

Inside a chromosome

Suppose we could enlarge a chromosome so much that we could see exactly what it consists of. Unfortunately we can't do this even with the most powerful microscope – but if we could, what would we see?

Picture 1 shows a chromosome greatly enlarged. it consists of a very long thread, tightly folded. The thread runs from one end of the chromosome to the other.

At the bottom of the picture the thread has been pulled out and enlarged even more. The thread consists of a chemical called **DNA**. DNA stands for **deoxyribonucleic acid**.

DNA is a very important chemical because it is what our genes are made of. The structure of DNA was discovered in the early 1950s by two scientists at Cambridge, James Watson and Francis Crick. Their story is told on page 281.

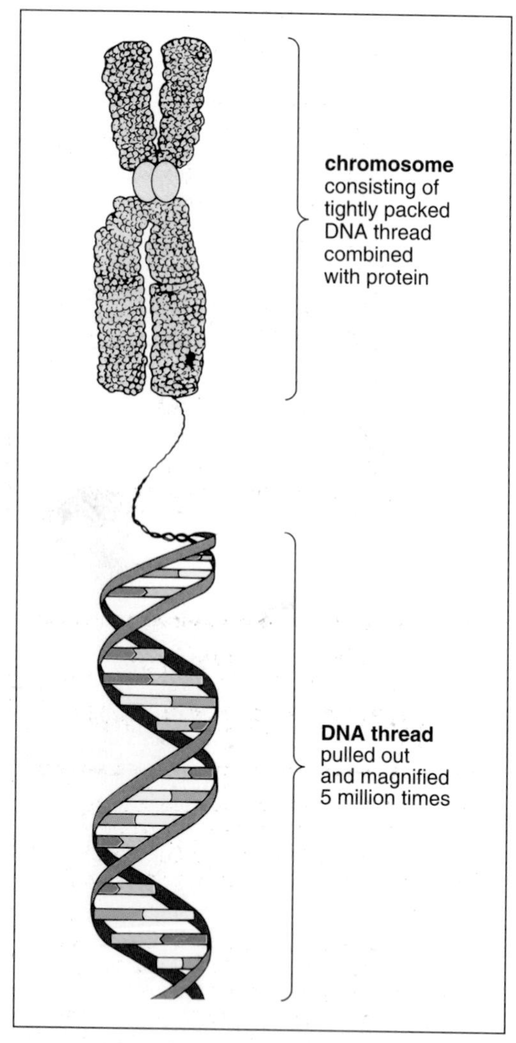

Picture 1 A chromosome consists of a long, tightly packed thread of DNA. Inside the chromosome itself the DNA is combined with protein.

The structure of DNA

Picture 2 shows part of a DNA molecule. It is a polymer consisting of two parallel strands linked together by cross-pieces. It's rather like a ladder, with the cross-pieces as the rungs. However, it's a ladder with a difference because it's twisted into a spiral. The whole thing is called a **double helix**. (A helix is any spiral-shaped object.)

From the biological point of view, the most important parts of the molecule are the cross-pieces. Each one is made up of a pair of organic bases: we shall refer to them as **base-pairs**.

There are four different bases. Don't worry about their names. We'll call them by their initial letters: **A**, **C**, **T** and **G**. The bases fit together as shown at the bottom of picture 2. Because of their chemical structure, **A** always pairs with **T**, and **C** with **G**.

What is a gene?

A **gene** consists of a short length of the DNA molecule. Several thousand base-pairs may make up one gene. Now the order in which the base-pairs are arranged within a gene varies, and this is how genes produce their effects. For example, one order of bases may produce brown eyes, while another order produces blue eyes – and so on.

What the bases are really doing is to provide *instructions*, telling the organism what features to develop. The instructions are contained in the order of the base-pairs. However, there are only four different bases. So the bases are like letters in a four-letter language. We call this language the **genetic code**.

How does DNA cause a particular feature to develop?

Here's a very simple example to illustrate what happens. Brown eyes are caused by the presence of a certain pigment in the iris of the eye. The iris makes this pigment because the person's DNA tells it to. How does the DNA 'tell' the iris what to do? It does so by making the iris cells produce a particular enzyme. The enzyme then causes the pigment to be made.

Enzymes are proteins, so the DNA is really telling the cells what proteins to make. Putting it another way, *DNA controls protein synthesis*.

How does DNA control protein synthesis?

Proteins consist of long chains of amino acids (see page 22). There are about 20 different amino acids in nature. Each type of protein is made of a certain number of specific amino acids linked together in a particular order.

DNA tells the cell how to join its amino acids together so as to make a particular protein. DNA consists of a chain of base-pairs, and protein consists of a chain of amino acids. *The order of base-pairs in the DNA decides the order of amino acids in the protein.*

Scientists have shown that three consecutive bases (i.e. three bases in a row) stand for one amino acid. The genetic code is therefore known as a **triplet code**. When the cell makes a protein, the amino acids join up in an order which is decided by the order of base triplets in the DNA.

DNA can make copies of itself

Before a cell divides, each chromosome makes a copy or *replica* of itself (see page 277). In reality a chromosome consists of a very long thread of DNA, and it is the DNA that produces replicas. We call this process **replication**.

Picture 3 shows what happens. We start with a single molecule of DNA, like a ladder. The two strands of the DNA separate, starting at one end. It's rather like opening a zip fastener. Meanwhile, new bases which are lying about in the cell come along and join up with each of the two strands. So we finish up with two DNA molecules.

Now you remember that the bases fit together in a particular way: **A** with **T** and **C** with **G**. This means that the two new molecules of DNA *must* have the same order of bases as the original DNA. So, when a cell divides by mitosis, the daughter cells have exactly the same genetic instructions as the parent cell.

A new look at the gene

Now that we know about DNA, we can say exactly what a gene is. *A gene is a stretch of DNA which contains instructions for making a particular protein, or part of a protein.*

Our genes play a crucial part in making us what we are. DNA is truly the molecule of life.

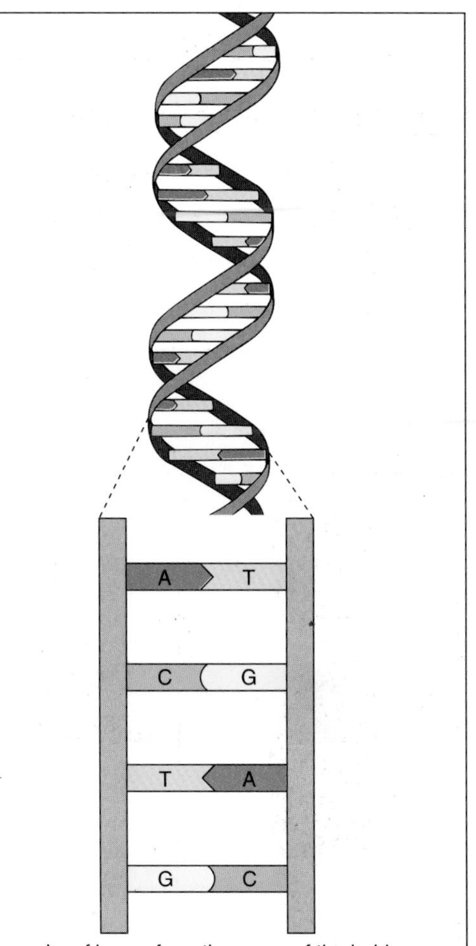

Picture 2 The DNA molecule is like a twisted ladder – a double helix.

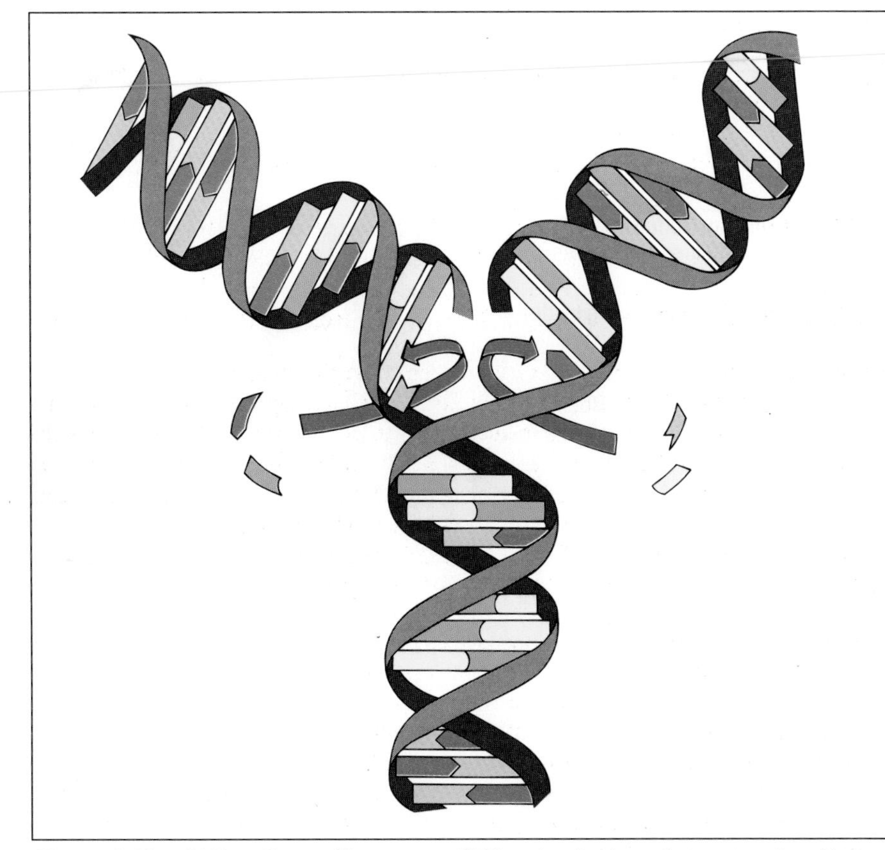

Picture 3 How DNA replicates. The two new DNA molecules have the same order of bases as the parent DNA.

The human genome project

An organism's **genome** is its full set of genes. It is not known for certain how many genes the human has, but it could be as many as 100 000 consisting of a total of 3000 million base pairs.

Techniques are now available which make it possible to find out not only how many genes we have, but whereabouts they occur on the individual chromosomes and which features they control. We already know about some of our genes, and scientists the world over are working hard to discover the rest. It is called the **human genome project.**

It will take years to complete the human genome project, but when it *is* complete it will pave the way to finding remedies for some of the diseases caused by faulty genes. At present these diseases afflict about 60 million people worldwide.

Activities

A Making a model of DNA

Watson and Crick discovered the structure of DNA by making models of the molecule. You can follow in their footsteps.

Using whatever materials you have available, make an upright model of a short length of a DNA molecule. The model should show the shape of the double helix in three dimensions.

Use picture 2 on page 279 to help you build your model. You may find it useful to use more advanced books as well. Be sure that you pair up the bases in the right way: **A** must pair with **T**, and **C** must pair with **G**.

B Watching DNA replicate

You can't really watch DNA replicate, but you can do the next best thing which is to make a model of it.

Using plasticine or pieces of card, make a model of a short length of DNA which lies flat on the table. Don't try to make an upright three-dimensional model. Also make some spare bases, the same ones as are in your DNA.

Now make your model replicate. Separate the two strands of the DNA, then bring in new bases so as to make new DNAs, each exactly like the original one.

Questions

1 Explain the meaning of the term genetic code.

2 Why is it necessary for DNA to be able to produce exact copies of itself?

3 Except for identical twins, everybody's DNA is unique. In what way is a person's DNA unique? Why are identical twins an exception?

4 Read about Gregor Mendel on page 269, then write a short letter to him bringing him up to date on genetics.

5 DNA is mainly in the nucleus of a cell, but the enzymes and other proteins are made in the cytoplasm. So the message in the DNA has to get from the nucleus to the cytoplasm. Suggest how this might happen.

Genetic fingerprinting

With the exception of identical twins, everybody's DNA is different. In other words, we all have our own unique genetic code. In recent years this has been used by the police to investigate crimes.

Suppose Mr X is suspected of murder. The handle of the murder weapon has a few fragments of clotted blood and broken skin on it, possibly from a cut on the hand of the murderer. The fragments are sent to a laboratory. Here forensic scientists extract the DNA from the sample, Then, by means of enzymes, they cut the DNA into small pieces. The cuts occur wherever there is a particular sequence of bases.

A solution of the DNA is then put in an electric field. This causes the pieces of DNA to separate from each other. After this treatment, they become visible on film as a series of bands (see picture).

The same procedure is then carried out on a sample of blood taken from Mr X. If the banding patterns of the two blood samples are identical, it's bad news for Mr X. The chances of them being the same by coincidence is about one in a million million.

Although people call this procedure 'genetic fingerprinting', forensic scientists prefer to call it **genetic profiling**. Otherwise it might be confused with ordinary fingerprinting which has been carried out by the police for years.

Genetic profiling can be done on any tissue that contains DNA – blood, semen and even cells from the roots of hairs. This makes it useful for investigating all sorts of crimes including murder and rape. It can also be used to find out if people are related, and if so how closely related they are.

1 Why is the term genetic profiling better than genetic fingerprinting?

2 Do you think suspects should be convicted on the basis of their genetic profiles?

3 In what circumstances might it be necessary to find out if two people are related?

4 Some people feel that genetic profiling is an invasion of people's privacy. What do you think? In what ways might the technique be misused?

5 The picture shows the DNA banding patterns of seven suspects and a bloodstain found on the handle of a murder weapon. Which suspect is most likely to be guilty? Give a reason for your answer.

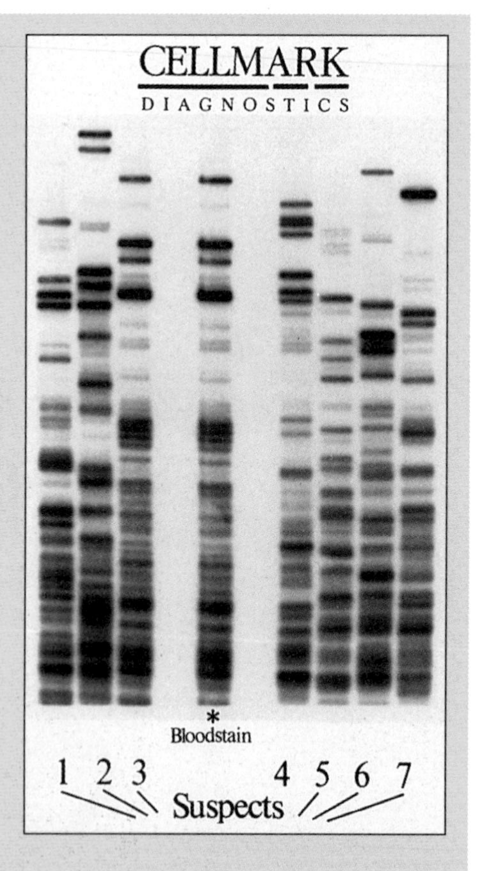

Discovering the structure of DNA: two different approaches

Consider the problem. We have a molecule. It's a very complex molecule, with millions of atoms linked together in an elaborate three-dimensional structure. By molecular standards it's large, but still too small to be seen in any detail, even with the most powerful microscope. How can you work out the structure of such a molecule? This was the problem facing the scientists who were working on DNA in the early 1950s.

There are two possible approaches. One is to build a **model** of the molecule, putting all the atoms where you think they should be. To make your model accurate, you need to know as much as possible about the properties of DNA: its density, how much water it contains, and so on.

In building your model you bear in mind what you believe the molecule has to do. In the case of DNA, you suspect that it is the molecule of heredity. This means that it must contain some kind of code, and must be able to produce copies of itself, that is replicate. Any model you build must fit in with this idea.

To some extent model-building is guesswork, but it's guesswork based on a sound knowledge of chemistry. For example, you need to know a lot about the way atoms fit together and how far apart they should be. If your model doesn't work, or if it breaks any rules of chemistry, you must take it to pieces and start again.

Picture 1 James Watson and Francis Crick with one of their successful models of DNA.

This was the approach of **James Watson** and **Francis Crick**, working in the Cavendish Laboratory at Cambridge. You can see Watson and Crick with one of their successful models of DNA in picture 1.

The other approach is to use a special technique that gives you a kind of 'picture' of the molecule. The technique is called **X-ray diffraction**. It depends on the discovery that DNA forms crystals. First you obtain some purified DNA from a cell, and you make a perfect crystal of it. Then you fire a beam of X-rays at the crystal. The X-rays are scattered by the atoms in the DNA within the crystal. The way they are scattered is recorded on a photographic plate placed behind the crystal. Between each firing, the crystal is rotated slightly so that the X-rays hit all sides of it in turn. This enables the whole molecule to be analysed.

From the X-ray diffraction pattern, as it's called, you can work out the exact positions of the various atoms in the DNA molecule. This requires some complicated mathematics, but it can be done with the aid of a computer. It enables you to build up a three-dimensional picture of the molecule. It's the closest you can get to 'seeing' DNA.

This was the approach of **Rosalind Franklin**, working at King's College, London. She produced some superb X-ray diffraction photographs of DNA. One of them is shown in picture 2.

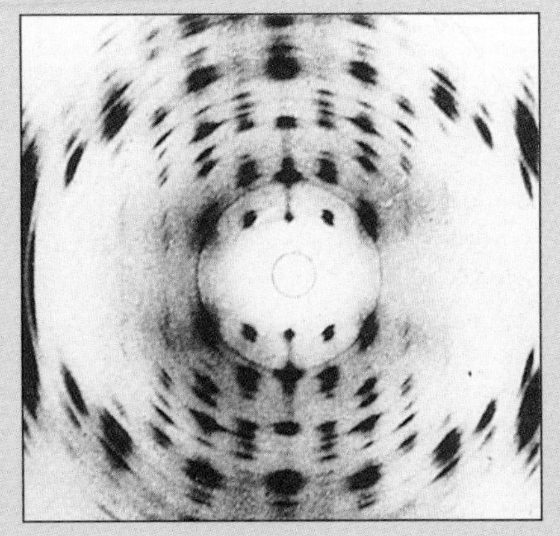

Picture 2 One of Rosalind Franklin's X-ray diffraction photographs of DNA.

Which approach is better, model-building or X-ray diffraction? It's impossible to say. Both are important, and both should go together. As one scientist has put it, models are no use unless they are accurate, and the only way of making them accurate is to base them on X-ray diffraction. In fact, Watson and Crick used Rosalind Franklin's X-ray diffraction photographs to help them get their models right.

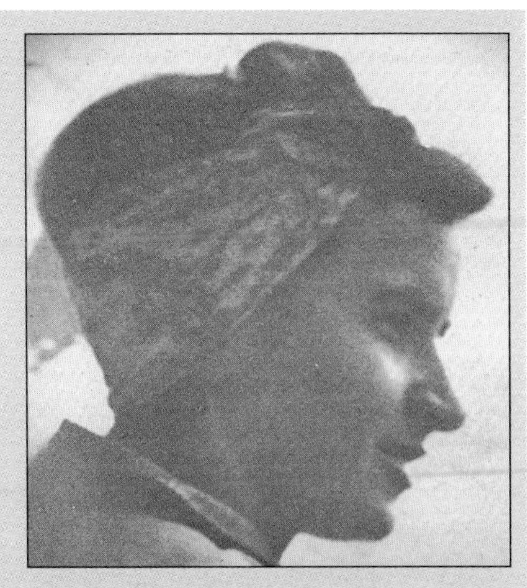

Picture 3 Rosalind Franklin, whose X-ray diffraction photographs paved the way to discovering the structure of DNA.

In the end it was Watson and Crick who first described the structure of DNA and pointed out its genetic significance. In 1962 they were awarded the Nobel Prize, the highest honour that can be bestowed on a scientist. Also honoured was Maurice Wilkins, Rosalind Franklin's head of department at King's. Rosalind Franklin herself had died of cancer four years before, at the age of 37. Nobel Prizes are never given posthumously. Undoubtedly she too would have been given one, had she lived.

1 One of the properties of DNA which Watson and Crick took into account in their model-building concerns the bases **A, T, C** and **G** (see page 278). In any sample of DNA, the amount of **A** is always the same as the amount of **T**, and the amount of **C** is always the same as the amount of **G**. What do you think this suggested to them about the structure of DNA?

2 Some people say that Watson and Crick's approach was inferior to Rosalind Franklin's because it involved guessing the structure of DNA. Is this a fair comment? Do you think guesswork has a part to play in scientific discovery? Which approach would you prefer to follow if you were doing research on DNA, Watson and Crick's or Rosalind Franklin's?

3 Some people have said that the structure of DNA would have been discovered sooner if Watson, Crick and Franklin had all been working together in the same university. Do you think scientific research should be organised so that people who are working on the same problem are all in the same place? How might it be achieved?

Scientists have learned how to alter an organism's genes so that they do useful things for us.

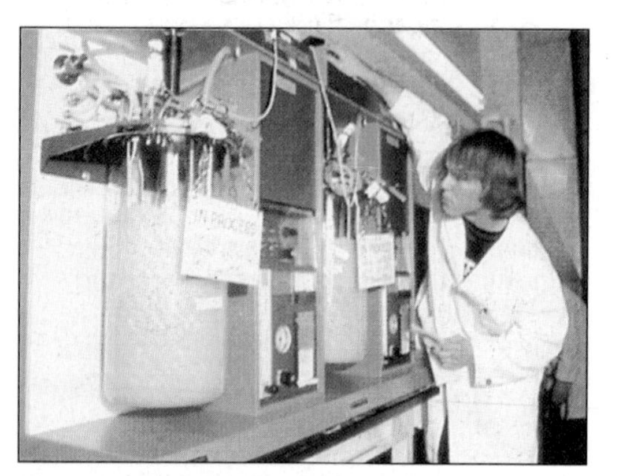

Picture 1 A hormone being produced by genetic engineering.

Making insulin

Insulin is a hormone which lowers the concentration of sugar in the blood. It is produced by the pancreas. Some people are unable to produce insulin, or at least not enough of it. As a result, they suffer from **diabetes** (see page 193).

People with diabetes have to inject themselves with insulin every day. They therefore need a constant supply of it. For many years insulin has been manufactured by extracting it from the pancreas of cattle and pigs after they have been slaughtered. The insulin is then purified and made suitable for human use.

The trouble with this way of manufacturing insulin is that it is costly, and there are so many people with diabetes that it is difficult to produce enough to go round. Also animal insulin is not quite the same as human insulin, and it may not be as good at combating diabetes.

What makes the human body produce insulin? The answer is that one of our genes tells the pancreas cells to make it. Now suppose we could take this gene out of a human cell and put it in a bacterium. Might the bacterium then make insulin for us?

Not long ago this kind of thing was science fiction, but it has now been done. Scientists have identified the gene that makes human cells produce insulin. They have removed this gene from human cells and transferred it to bacteria. The bacteria multiply rapidly, and the human gene replicates along with the bacteria's own genes. Once inside the bacterial cells, the human gene causes the bacteria to produce insulin.

The procedure just described is illustrated in picture 2. It's an example of **genetic engineering**, a branch of **biotechnology**. Because bacteria multiply so quickly and can be grown in such large numbers, genetic engineering provides a way of producing insulin on a large scale.

Genetically engineered insulin was first tried out on a group of volunteers in a London hospital in 1980. The trial was successful, and this kind of insulin is now widely used, though animal insulin continues to be used as well.

Other products

Genetic engineering is particularly useful when there is no other satisfactory way of making what you want. Such is the case with **growth hormone**.

Lack of growth hormone in a child causes a type of **dwarfism** (see page 258). It can be put right by giving the child injections of the hormone, but in this case only human growth hormone will do. Animal growth hormone is not suitable.

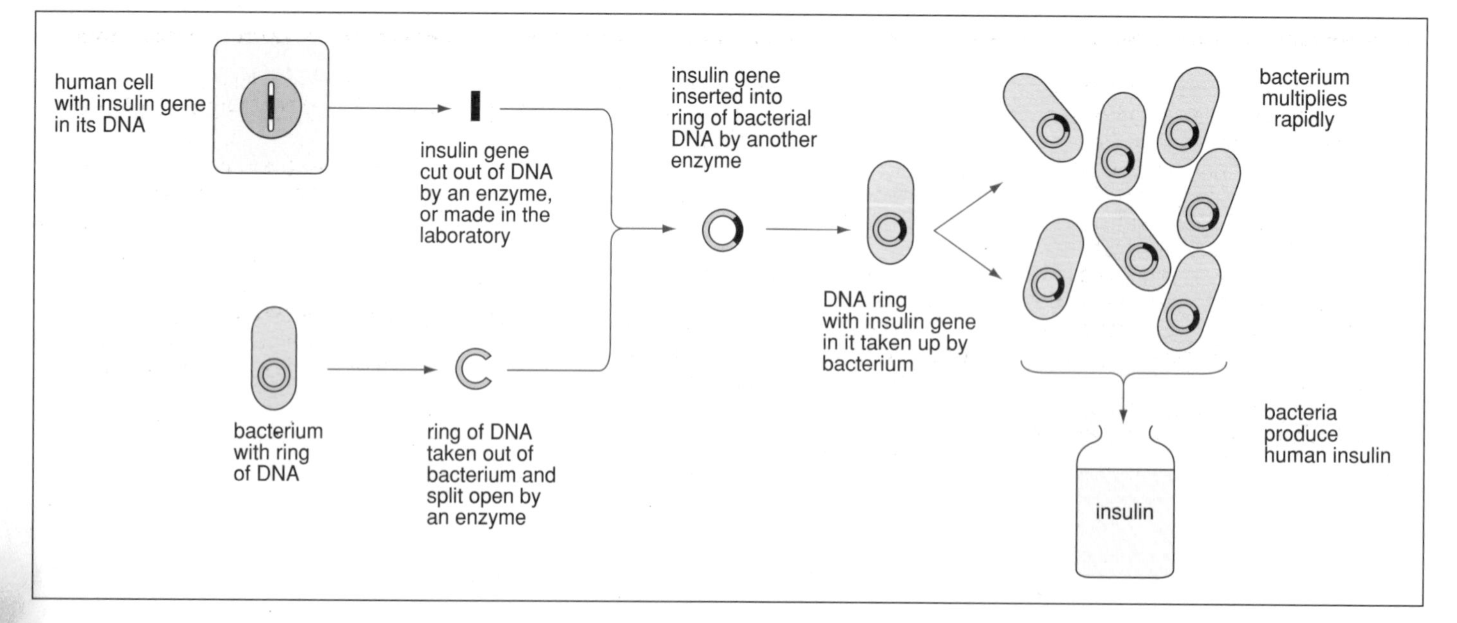

Picture 2 The principle of genetic engineering is illustrated here by insulin production.

Growth hormone is produced by the pituitary gland at the base of the brain. Until recently the only way we could get the hormone was by extracting it from the pituitary glands taken from human corpses. Dwarfism is rare – much more so than diabetes. Even so, it was impossible to get enough hormone to go round. Now genetic engineering has come to the rescue. The growth hormone gene has been transferred to bacteria. These bacteria produce the hormone in factories such as the one shown in picture 1.

Genetic engineering is a new, booming industry, rather like micro-electronics was a few years ago. As time goes on, we shall see genetically engineered bacteria mass-producing all sorts of useful things for us, including hormones, antibiotics, vaccines, enzymes and even cheap fuels.

Genetic engineering and the future

You can grow plants in the laboratory from small groups of cells (see page 251). Now suppose you were to take a gene from bacteria and put it into such cells. The new plants would possess the bacterial gene. Why should we want to do this? You may remember that certain bacteria in the soil can make use of nitrogen from the air (see page 37). This is called **nitrogen-fixation** and it is something that only certain microbes can do. Plants cannot do it, unless they happen to have nitrogen-fixing bacteria inside them.

Now think how useful it would be if we could transfer the genes responsible for nitrogen-fixation into a crop plant such as wheat. The wheat could then use nitrogen from the air. Less fertiliser would be needed, saving money and reducing nitrate pollution.

Scientists have been trying to produce nitrogen-fixing wheat for many years, so far without success. However, other useful genes have been transferred from bacteria to plants. An example is given below.

What about humans? Eggs can be taken out of a woman's body, fertilised and then put back again (see page 235). While the egg is out of the body, it might be possible to alter its genes in various ways. For example, harmful genes might be taken out of it, and useful ones put in.

In this way it might be possible to prevent children being born with inherited diseases such as cystic fibrosis and sickle cell disease. This kind of thing is still a long way off, but one day it might become possible.

Questions

1 Suggest *three* advantages of producing human growth hormone by genetic engineering rather than by extracting it from human corpses.

2 Insulin was the first product of genetic engineering to be made available to the public. Why do you think insulin was chosen? Think of as many possible reasons as you can.

3 There has been some debate as to which is better for treating diabetes, human insulin or animal insulin. If you were comparing them, what sort of things would you look for?

4 Look up tissue culture on page 27. Tissue culture is of great use in genetic engineering. Why?

5 Why do you think scientists have found it so difficult to produce nitrogen-fixing wheat by genetic engineering?

6 Suggest how genetic engineering might be used to prevent inherited diseases such as cystic fibrosis.

7 Some scientists are worried that bacteria produced by genetic engineering might escape from the laboratory into the environment. What could be the problems if this happened?

8 Suppose it was possible to use genetic engineering to make people more intelligent. Do you think this should be allowed?

Tomato toxin spells end to caterpillar's salad days

A gene responsible for a toxin from a harmless bacterium has been inserted into tomatoes by the Monsanto company so that the engineered fruit can ward off caterpillar-type insects without resorting to artificial pesticides.

'When the insect eats the plant, it ingests that bacterial toxin and dies,' said Dr David Hulst, director of Hulst Research Farm Services in Hughson, California.

The toxin of the bacterium, *Bacillus thuringiensis*, is dangerous only to the caterpillar family, said Dr Hulst. With genetic engineering, this attribute can now be given to a range of crops.

Researchers predict widespread use of this biological control because of growing concern about environmental effects from pesticides that do not always keep insects at or below acceptable levels. Monsanto estimate such genetically engineered tomatoes will be in use in the mid-1990s.

By Roger Highfield

An example of genetic engineering, reproduced from *The Daily Telegraph*, 12 July 1989.

1 How do you think the toxin gene might have been inserted into the tomato plants?

2 What are the possible dangers of inserting this kind of gene into tomato plants?

3 This research was announced in 1989, but it won't be till the mid-1990s that such genetically engineered tomatoes will be in use. Suggest reasons for the long delay.

F7 Variation

The people in the picture are all different. This topic is about variation and its causes.

Picture 1 People differ from each other in all sorts of ways.

Continuous variation

Suppose you measure the heights of a whole lot of people. You will find that there is a smooth gradation from very short to very tall. The same applies to body mass, shape of the face and many other features. This sort of variation is called **continuous variation**.

You can show continuous variation as a histogram, like the one in picture 2. If we join the tops of the bars, we get a bell-shaped curve. This is called a **normal distribution curve**. Before reading on, describe in words what the curve in picture 2 tells us about the way people's heights vary.

What causes continuous variation?

The main cause is that every individual possesses a unique combination of genes which is different from everyone else's. This is because of sexual reproduction. Sexual reproduction involves the formation of gametes, and no two gametes contain exactly the same alleles. The reasons for this are explained on page 276.

Still further variation is created when fertilisation occurs. This is because fertilisation is random: it is sheer chance as to which particular sperm fertilises an egg.

So in an organism which reproduces sexually, there is always variation in the offspring. This explains why brothers or sisters differ from one another, and why children differ from their parents. Only identical twins are truly alike (see page 229).

Discontinuous variation

Do you remember cystic fibrosis? This inherited disease is explained on page 270. Now you either have the disease or you don't; there are no 'in-betweens'. This is an example of **discontinuous variation**.

Another example in humans is the ABO blood group system. Every person belongs to a particular blood group and there is no smooth gradation between them. Examples in other organisms include the different coloured flowers and different kinds of fruit flies mentioned in the topic on heredity.

What causes discontinuous variation?

Discontinuous variation is caused by a process called **mutation**. A mutation is a sudden change in the genetic make-up of an organism. It sometimes leads to people being born with a defect such as a missing arm or an extra toe (picture 3). In Britain, about two per cent of babies are born with a physical or mental defect of some kind.

Mutation occurs when eggs and sperm are being formed. It will therefore be present in the fertilised egg, and will finish up in all the cells of the body. And, since it affects the reproductive cells, it may be passed on to the next generation.

There are two kinds of mutation. In **chromosome mutations** a major change occurs in one or more of the chromosomes. For example, a person may lack a particular chromosome, or part of one – or an extra one may be present, as in Down's syndrome (see page 275).

In **gene mutations**, a chemical change occurs in an individual gene. Scientists have shown that there is a change in the order of bases in the DNA molecule. The change may be very small indeed, but it may have a great effect on the organism. For example, sickle cell disease (page 272) is caused by a change in just two of the bases in DNA.

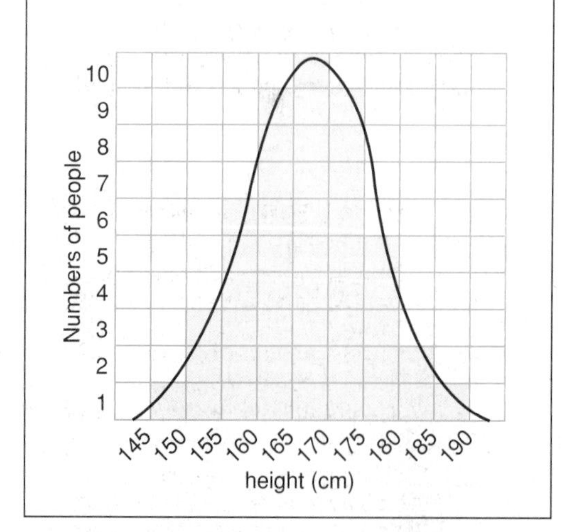

Picture 2 Histogram showing how height varies in an adult human population. Each bar represents the number of people who fall within a particular height group, e.g. 145–150 cm, etc.

What causes mutation?

Mutations happen from time to time for no apparent reason. However, the rate at which they occur can be greatly increased if the organism is exposed to ionising radiation or to certain chemical substances. This is why these things are so dangerous. Many of the people who survived the two atom bombs which were dropped on Japan at the end of the Second World War received massive doses of atomic radiation. As a result they developed lots of mutations and many of their children were born with defects, far more than in a normal population.

Mutations are not *always* harmful. The man in picture 3 does not suffer because he has an extra finger and toe. Indeed, on rare occasions a mutation may be helpful. We shall return to this later.

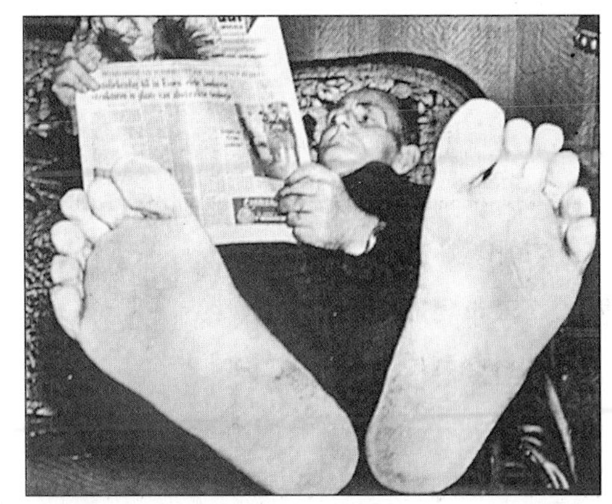

Picture 3 What is unusual about this man? This is one way a mutation can affect a person.

Differences caused by the environment

Many people grow hydrangeas in their gardens. This plant has large clumps of flowers which may be white, pink or blue. What decides the colour of the flowers? In this case it's not the genes. Instead it's the type of soil the plant is growing in. If the soil is acidic, blue flowers develop; if the soil is alkaline, white or pink flowers develop (picture 4).

This is an example of variation being caused by the environment. Many differences between people, particularly in their behaviour and attitudes, can be explained by the fact that they have been brought up in different environments. Environment here includes our diet, surroundings, home, school and the people with whom we live and work.

We still don't know how important the environment is, compared with our genes, in making us different from each other. This particularly applies to features like intelligence and artistic ability.

In trying to find an answer to this question, studies on identical twins can be useful. Identical twins have the same genes, so any differences between them must be due to the environment. Such studies suggest that our environment helps us to develop motivation and plays an important part in overall achievement.

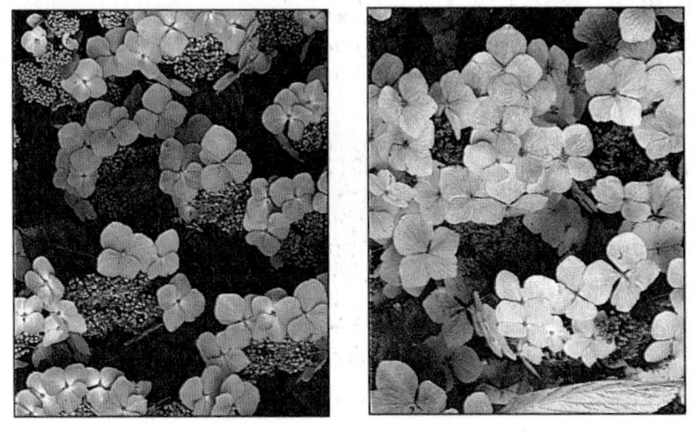

Picture 4 The colour of hydrangea flowers depends on the environment.

Activity

Looking at an example of variation

1 Measure the height of each person in your class.

2 Divide the heights into 5 cm groups, starting with 120 cm and finishing up with 180 cm (i.e. 120–25, 125–130, 130–135, etc.) and write down the groups in a list.

3 Work out how many people in your class fall into each group. Write the numbers alongside the groups in your list.

4 Construct a histogram (like the one in picture 2) showing how height varies in your class.

Which group contains (a) the largest number of people, and (b) the fewest people?

What is the **height range** (that is, the difference in height between the shortest and tallest pupils)?

What is the average height in your class?

Does it correspond to the tallest bar?

5 Join the tops of the bars in your histogram with a smooth curve.

What does the shape of the curve tell us about the way height varies?

Suggest reasons why the members of your class should vary in height.

What other variable features in humans, or in animals or plants, could you measure and present in this way?

Questions

1 Make a list of your own features which you think you have inherited, and those which you think you have got from your environment. Which aspects of your environment have affected you most, and in what ways?

2 The man in picture 3 has an extra finger and toe. Can you think of any circumstances in which it might be useful to have an extra finger?

3 Cyclamates are artificial sweeteners, about thirty times as sweet as sugar. They used to be added to drinks by the manufacturers. It was then discovered that they could cause 'chromosome aberrations', so they were banned.

a What do you understand by the term 'chromosome aberration'?

b What precautions should be taken to prevent potentially dangerous chemicals being added to our food?

4 You have been asked to plan investigations on identical twins to find out the extent to which intelligence is inherited from parents or acquired from the environment. Let us see your plan.

5 The chairman of a local Conservative club saw a blue hydrangea plant in the Lake District, and liked its colour. So she took a cutting of it and planted it in her garden. To her annoyance, the cutting developed pink flowers instead of blue flowers.

a Explain why the cutting developed pink flowers instead of blue flowers.

b Describe an experiment which you could do to test your explanation.

c What could the lady do to make the hydrangea develop blue flowers?

F8
The history of life

In this topic we take a step back in time.

Fossils

Normally when an animal dies, the soft parts of its body decay, leaving only the hard parts, such as the bones and teeth. Given enough time, even bone will decay completely, but occasionally something else happens. As the organic matter decays, the bone becomes full of tiny holes and spaces. If the bone is buried in mud, the mineral particles work their way into it and gradually replace it. As a result, the bone becomes rock-hard and eventually turns into a **fossil**.

Meanwhile, the surrounding mud hardens to form **sedimentary rock**. This is usually softer than the fossil. Millions of years later a scientist may come along with a hammer and chisel and chip away the surrounding rock, exposing the fossil (picture 1).

Bones, teeth and shells become fossilised this way. So, in certain circumstances, does wood. In Arizona there is an area of desert where complete tree trunks have been turned into rock (picture 2).

Picture 1 This man is a palaeontologist who studies fossils in order to find out about animals which lived in the past.

Other kinds of fossils

Occasionally footprints are left by an animal. They then become covered by mud which hardens into rock. If later the overlying rock is removed, you can see these fossil **impressions** (picture 3). Many surface structures have been shown up this way, including leaves, feathers and scales.

Sometimes a dead organism gets buried in mud which in time hardens into rock. Meanwhile, the organism decays, leaving a space or **mould**. If later the rock is split open, the mould can be exposed for everyone to see. Moulds can tell us a lot about the shapes of organisms, even soft-bodied ones like jellyfish.

Sometimes a mould fills up with tiny particles which seep into it from the surrounding rock. The particles then harden, forming a natural **caste** of the mould which later may be separated from the surrounding rock.

On occasions animals have become covered by a natural **preservative** which stops them decaying. For example, insects have been preserved in amber, sabre-toothed tigers in tar, and mammoths in ice. Not long ago the preserved body of a man, estimated to be over a thousand years old, was found in a marsh in Cheshire, buried in peat. He was nicknamed Pete Marsh!

Picture 2 A fossilised tree trunk, about 160 million years old. Although it has been replaced by rock, you can see the annual rings clearly.

Telling the age of a fossil

Very rough estimates can be made from knowing which layer of sedimentary rock the fossil comes from. The oldest fossils are found in the deepest layers, the youngest ones in the topmost layers. This enables us to say that one fossil is older than another, but of course it does not tell you their exact ages.

More accurate estimates are made by measuring the amount of radioactivity in the rock that the fossil is found in. The older the rock, the less radioactive it is.

Reconstructing the past

Over the years, thousands of fossils have been discovered. They range from scattered bones or fragments to almost complete skeletons. By carefully piecing the bones together, scientists have been able to reconstruct the animals and get some idea of their appearance.

Picture 4 shows some of the animals and plants of the past. If we go back 400 million years, we find little sign of life on land. But the oceans and lakes were teeming with backboneless animals (invertebrates) and strange armoured fishes.

Between 200 and 300 million years ago there was lots of life on land: insects, amphibians and reptiles. Between 100 and 200 million years ago the world was dominated by the ruling reptiles, the great dinosaurs and their allies. Two of

Picture 3 This fossilised dinosaur footprint was found in Arizona, USA.

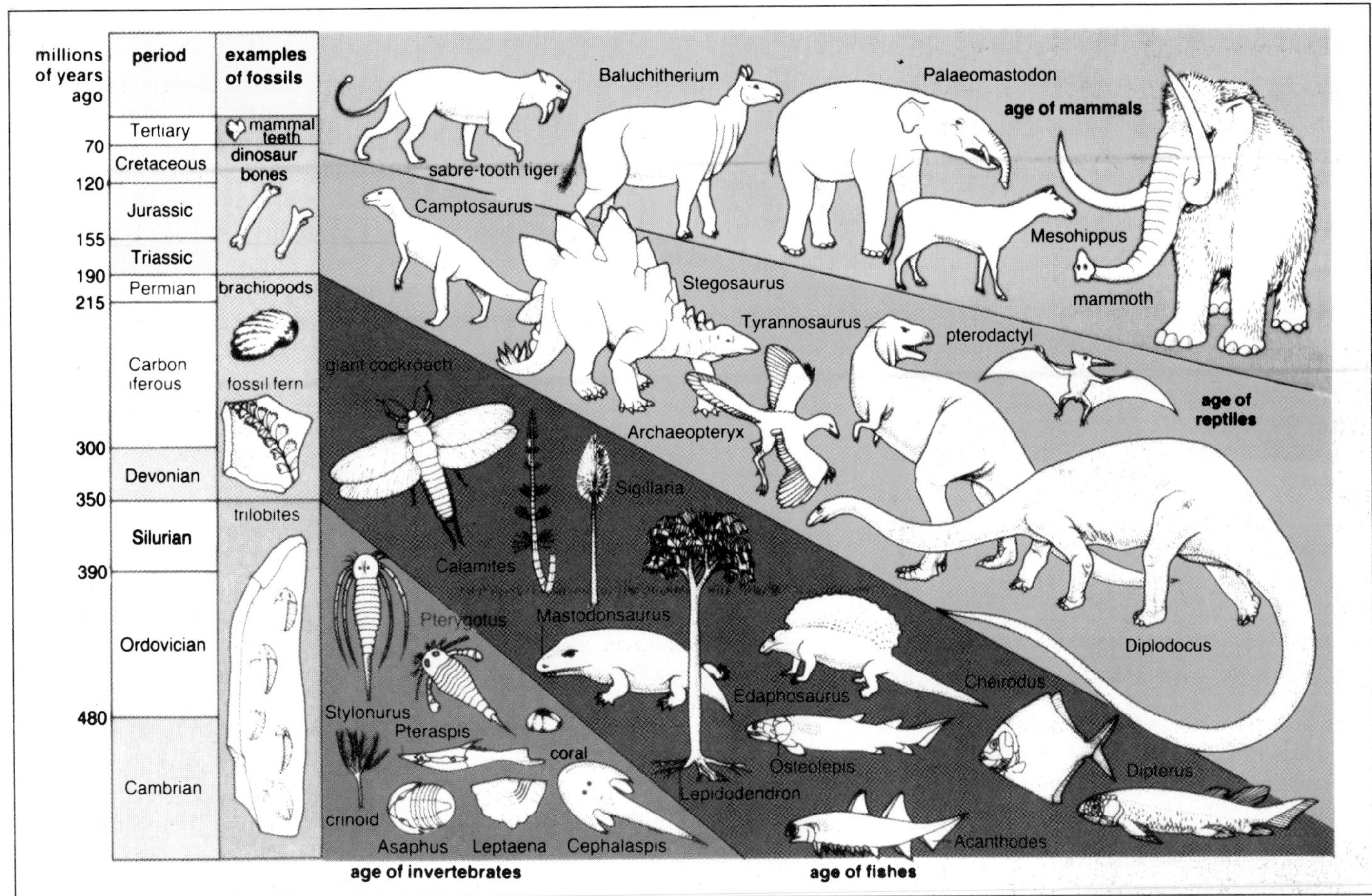

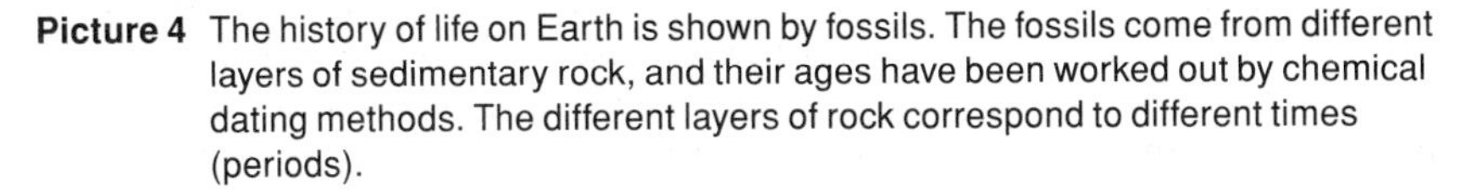
Picture 4 The history of life on Earth is shown by fossils. The fossils come from different layers of sedimentary rock, and their ages have been worked out by chemical dating methods. The different layers of rock correspond to different times (periods).

the more notorious ones are shown in picture 5. By about 60 million years ago these reptiles had been replaced by mammals, and about 3 million years ago the first ape-like humans appeared.

While the animals were changing, the plants also changed. During the age of dinosaurs the larger plants were almost all ferns and conifers. Later on the flowering plants developed, and spread across the world.

And throughout this time the environment, too, was changing. For example, mountain ranges were formed, the climate altered and there were tremendous variations in the amount of water available to organisms. The organisms which survived were those that were adapted to cope with the conditions at the time.

The fossils themselves give us a clue about the environment. For example, nowadays corals are only found in warm topical seas. So if a rock contains corals, we may conclude that the rock was laid down in a warm tropical sea.

Extinction

Many types of animals and plants that inhabited our planet in the past have died out and become **extinct**. Probably the most famous case was the dinosaurs. Nobody knows what happened to them, though there are lots of theories. All we can say is that, after many millions of years of success, they became unable to survive in the environment as it then was. On the other hand, the mammals did survive, and eventually they took over.

Extinction still happens. Sometimes an animal or plant becomes extinct for perfectly natural reasons. It might be because the environment changes, or because another more successful species competes with it, or because it is

Picture 5 *Tyrannosaurus rex* (top), a savage meat-eater, was over six metres tall and fifteen metres long. Its skull alone was over a metre long.
Diplodocus (bottom), one of the largest land animals which has ever lived. It was about thirty metres long, and the main part of its body was the size of a two storey house. It was a plant-eater.

Picture 6 African elephant, under threat.

preyed on too much by other animals. But all too often the cause is human interference. Over the centuries we have killed, eaten and exploited wild animals and plants with the result that many have died out. We have also taken our own favourite animals to other lands and let them loose there. This has upset the balance of nature, and has led to the native species becoming extinct. Two stories of extinction are told on the opposite page.

Today we show concern about conservation, yet many animals and plants face the risk of extinction. In Britain these include fifty species of wildflower, ten species of butterfly, seven species of birds together with otters, bats, water voles, grass snakes and newts. The gorilla, Nile crocodile and African elephant face complete extinction from the world unless we make sure that they are protected.

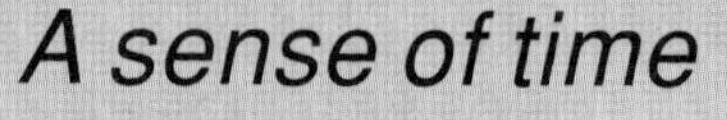

A sense of time

Think how long a year is. Now try to imagine two thousand million years. Life is believed to have existed on our planet for at least as long as that.

The first true humans (*Homo sapiens*) appeared less than one million years ago. This may seem a long time, but it's very short compared with the length of time that life has existed on this planet.

It may help you to understand the relative lengths of these times if you imagine the whole history of life taking place in 24 hours, from one midnight to the next (see picture). On this time scale, modern humans have been around for less than one minute, and the whole of recorded history has taken place in the last quarter of a second!

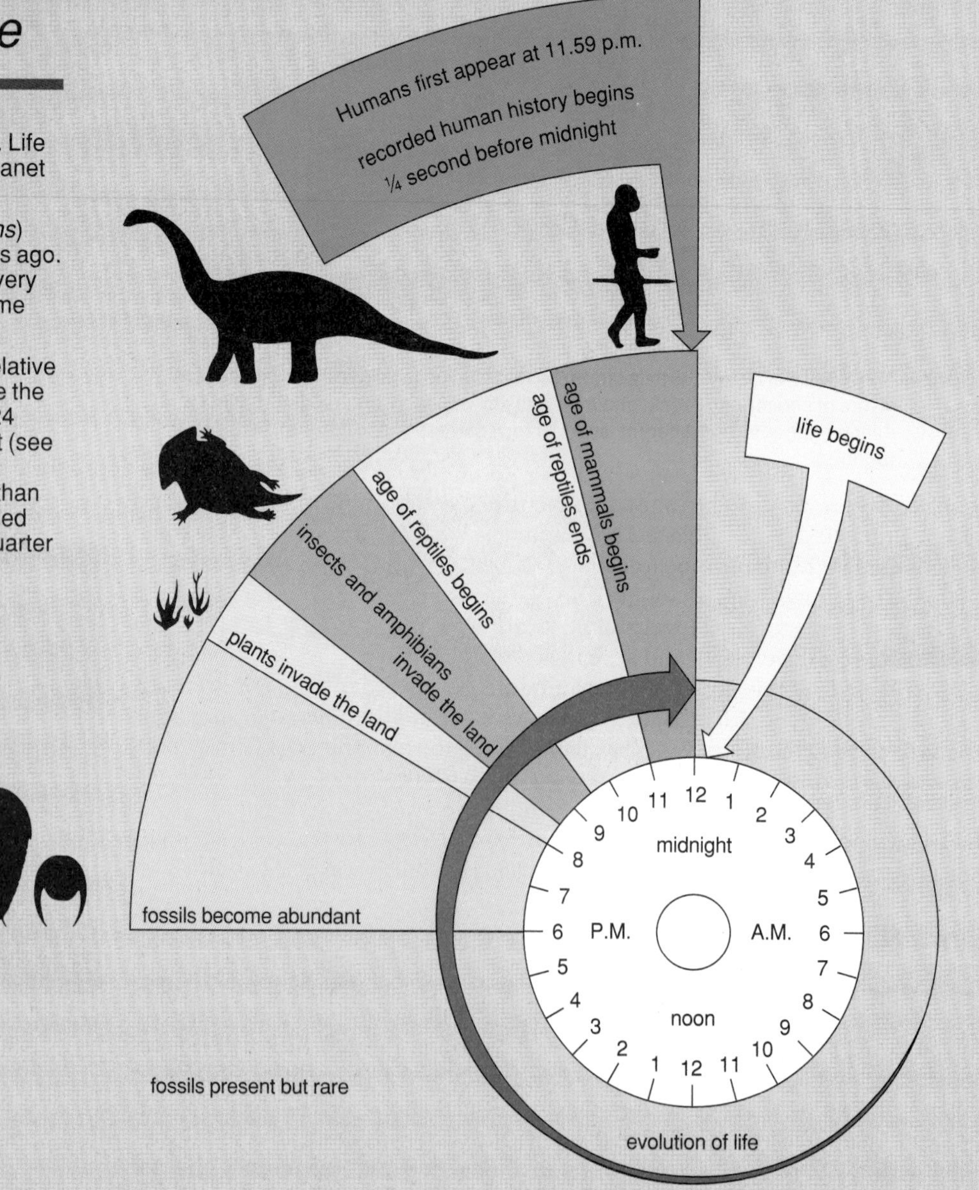

Activity

Stepping back into the past

Choose an animal or plant which is well known from fossils. You might choose an armoured fish, dinosaur, mammoth, or sabre-toothed tiger, though there are many other possibilities.

Use books, and if possible a museum, to find out as much as you can about your animal or plant. In particular try to answer these questions:

1 How long ago did your animal or plant live?

2 How much do we know about its structure from its fossil remains?

3 Whereabouts in the world did it live?

4 What was its environment like?

5 What did it feed on, and what (if anything) fed on it?

In finding the answers to these questions, try to distinguish between scientific fact and imagination. For example, a fossil bone in a museum is a fact, but an artist's painting of a dinosaur is imagination, although it may be based on facts.

Prepare a short talk on your animal or plant. Give your talk to other members of your class.

Questions

1 A fossil bone is harder than the surrounding sedimentary rock. Why do you think this is?

2 Bones fossilise more readily than tree trunks. Suggest a reason.

3 Suppose a mould gets filled with silicon oxide, and the surrounding rock is limestone. Suggest a chemical method for getting the silicon oxide out of the limestone. Why would we want to do this?

4 Only a tiny fraction of the organisms that have died have become fossilised. Suggest reasons for this.

5 Why didn't Pete Marsh decay?

6 Suggest as many reasons as you can think of why the dinosaurs became extinct.

7 Do you think humans will eventually become extinct? If so, why? Which species might take over as the dominant form of life? Argue your case.

Two case histories of extinction

The Dodo

The Dodo inhabited the island of Mauritius in the Indian Ocean. It was large and heavily-built, and could not fly.

It became extinct. Why? It seems that during its evolution it lost any means of defending itself against predators. It was hunted by humans for its meat. Moreoever, the animals that humans introduced – pigs, goats, cats and monkeys – probably preyed on its eggs and chicks. In addition, the early settlers felled large areas of forest, destroying the Dodo's habitat. By the early 1690s it was 'as dead as a Dodo'.

Picture 1 The Dodo.

Writers have described the Dodo as degenerate, over-specialised and badly adapted. This is totally wrong. There is only one way of judging evolutionary success and that is survival, and the Dodo certainly survived and flourished for many millions of years. What the Dodo could not do was to cope with the rapid changes that occurred to its environment with the arrival of the human.

Based on *Vanishing Birds* by Tim Halliday (Sidgwick and Jackson)

The Great Auk

The Great Auk, a native of the Atlantic, was the original penguin.

The numbers of Great Auks were already falling rapidly by the beginning of the 19th century as the result of reckless persecution by humans. Continued hunting and natural disasters finally wiped them out completely.

On 3 June 1844, on the Island of Eldey, off the coast of Iceland, three fishermen discovered the last two living Great Auks. They were a breeding pair with a single egg. Jon Brandsson and Sigourer Isleffson killed the two adult birds with clubs while Ketil Ketilsson smashed the egg with his boot.

David Bintley, based on the *Doomesday Book of Animals* by David Day (Ebury Press)

Picture 2 The Great Auk.

1 Suppose a present-day species of bird met the same fate as the Great Auk. What should society's attitude be towards the people responsible?

2 Suppose it was you who discovered the last breeding pair of Great Auks. What steps would you have taken to help the species to survive?

3 Overheard in a cinema foyer after a performance of 'Gorillas in the Mist': 'I'd do anything to save those gorillas but I don't care what happens to horrible crocodiles.' What should a biologist's attitude be?

(There is more about threatened species on pages 51 and 90.)

F9 Evolution

How have all the species which inhabit the world today come into being?

Picture 1 Charles Darwin (1809–1882).

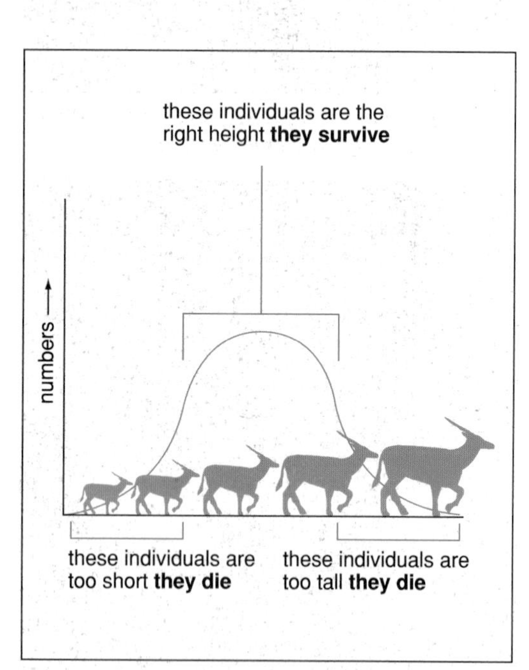

Picture 2 This simple diagram shows how natural selection ensures the survival of the fittest. The graph shows the numbers of different-sized antelopes in a large population on the African plains. The fittest individuals are those best adapted to the environment, and they are the ones that survive.

The theory of evolution

Most scientists consider that the organisms present in the world today are descended from other species that inhabited the world in a previous age. The changing of one type of organism into another is called **evolution**.

This theory was first put foward in the middle of the 19th century by Charles Darwin (picture 1). Darwin was a naturalist. In the 1830s he sailed round the world in a ship called The Beagle. He visited many countries and islands, and he studied the animals and plants there. Gradually he became convinced that the various species which he observed on his travels had come into being by a process of slow and gradual evolution.

In 1859 Darwin published his famous book *The Origin of Species* in which he put forward evidence to support his idea. He also put forward a theory to explain how evolution may have taken place.

Independently of Darwin, another naturalist called Alfred Wallace came to the same conclusions, though with less evidence. Darwin and Wallace wrote a short paper together a year before *The Origin of Species* was published.

Creation or evolution?

Evolution is based on the idea that similar organisms are descended from the same type of ancestor. For example, humans may share the same ancestor as the chimpanzee, gibbon, orangutan and gorilla. We have a lot in common with these apes, particularly chimps.

Darwin put forward his theory at a time when most people believed that all living creatures were created by God. They found it hard to accept his theory, particularly the idea that humans might be descended from apes.

Today many people feel that the theory of evolution can fit in with creation if you interpret the meaning of creation flexibly. For example, it is possible that God's way of creating living things is through a process of evolution. However, this is a very difficult question to which we can never *know* the answer.

How does evolution take place?

Darwin put forward an explanation of how evolution may take place. Most biologists today still believe that Darwin's explanation is correct. The explanation goes like this:

In any population of animals or plants there is usually competition for food and other scarce resources, and the ever-present threat of being attacked by enemies. This creates a **struggle for existence** in which every individual is striving to survive. Darwin saw this idea in an essay on human populations written in 1798 by an English clergyman, Thomas Malthus. Malthus argued that the human population always grows faster than the food supply, resulting in 'famine, pestilence and war'.

Now within a species there is **variation** between individuals. The reason for this was not known in Darwin's time, but we know the reason now. It is because every individual possesses a different combination of genes (see page 284). As a result, some individuals are better adapted to the environment than others. For example, some may be particularly strong or good at running. These individuals are more likely to survive than the others. This idea is called the **survival of the fittest.**

We are not using the word 'fittest' in the same sense that your PE teacher would use it. In evolution the fittest individuals are the ones most likely to reproduce successfully. If they do, they will hand on their good qualities to their offspring. On the other hand, the less fit individuals are more likely to die before they have a chance to reproduce.

This process is called **natural selection.** Nature, as it were, selects the fittest individuals and rejects the less fit ones. Picture 2 shows how we can apply this to a herd of antelopes on the African plains.

How does natural selection bring about evolution?

Look again at picture 2. According to this diagram the antelopes that survive are the medium-sized ones. The very short and very tall ones die. But suppose the environment changes so that it becomes an advantage to be very tall. In these circumstances the very tall antelopes survive best and reproduce successfully. They pass their tallness on to their offspring. As a result, a taller race of antelopes emerges. Evolution has taken place from a medium-sized antelope to a taller one.

What sort of variation is important in natural selection?

Natural selection can only take place if organisms show variation. The kind of variation that is needed is *genetic* variation, particularly the sort that results from mutation in the reproductive cells (see page 284). Mutations are usually harmful, but occasionally a **mutation** may occur which makes an individual *better* adapted to its environment. A famous example of this is seen in the peppered moth.

Natural selection in action: the peppered moth

The peppered moth *Biston betularia* is found throughout Britain. The moths rest on tree trunks. They are eaten by thrushes which peck them off the trees.

Two hundred years ago most peppered moths were speckled white. Tree trunks were mainly covered in lichens which were light in colour. Against this background the moths were well camouflaged, and this helped to protect them from the thrushes. Occasionally a mutation would take place and a black moth would appear. However, it stood out so much on the tree trunks that it was quickly pecked off by the birds (picture 3).

Then something happened which changed the environment: the Industrial Revolution. Smoke from factory chimneys blackened the tree trunks. In industrial areas the black form of the moth was now at an advantage. Against the blackened tree trunks it was perfectly camouflaged. On the other hand, the white form stood out and was quickly pecked off by the birds (picture 4).

This was the situation in the 1950s when a scientist called Bernard Kettlewell decided to investigate the peppered moth. He studied its distribution in Britain (picture 5). He found that in industrial areas such as Manchester and Birmingham the black form was the more common, whereas in non-industrial areas such as Cornwall and the north of Scotland the white form was the more common.

More recent studies have shown that in industrial areas the white form is becoming more common again. This is because we now have much cleaner air, and the trees are less black than they used to be.

Other examples of natural selection

We see natural selection in the way new kinds of germs arise. For example, certain bacteria have mutated into new types (strains) which are resistant to penicillin. In the same way we now have new types of mosquito which are resistant to DDT, new types of malarial parasite which are resistant to certain anti-malarial drugs, new types of rat which are resistant to the rat poison warfarin, and new varieties of grass that can grow on the toxic tip from mines.

In all these examples a change in the environment gives the new organisms an advantage. In the case of the peppered moth the change in the environment was the blackening of the trees. In the case of bacteria, it was the arrival of penicillin. Being at an advantage, the new organisms survived and reproduced; their numbers increased until they had replaced the original organisms.

Mutation and penicillin

Sometimes mutations can be useful to humans. Take the antibiotic penicillin, for example. Penicillin was discovered by Fleming in the late 1920s. He extracted it from the fungus *Penicillium notatum.* As a result of mutation, we now have strains of this fungus that produce at least 300 times more penicillin than Fleming's original strain.

Picture 3 White and black peppered moths resting on a light-coloured tree trunk.

Picture 4 White and black peppered moths resting on a dark-coloured tree trunk.

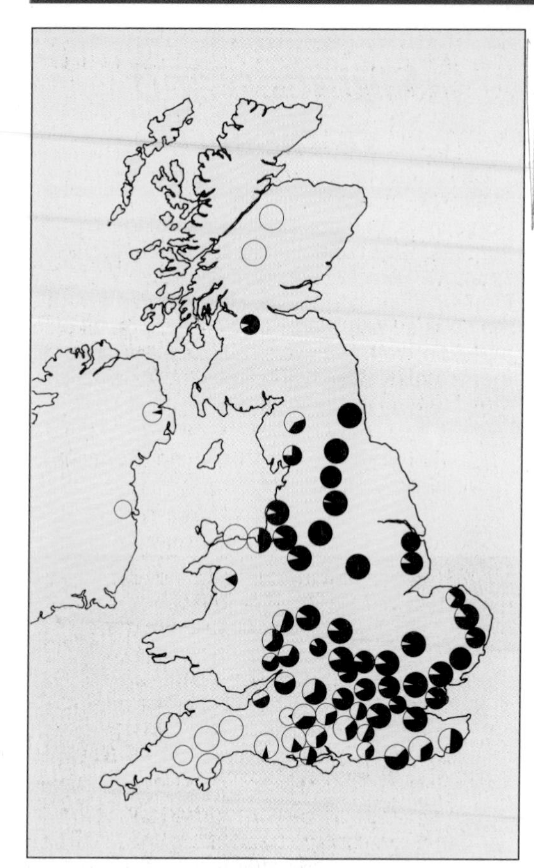

Picture 5 Map of Great Britain showing the distribution of the peppered moth in the 1950s. Within each pie chart, the clear area represents the number of white moths in the population, and the dark area represents the number of black moths. Large numbers of black moths were found in areas where the trees were darkened by industrial smoke.

It's really alleles that are selected

An organism possesses lots of alleles. Some of them are 'good' alleles which help the organism to survive. Others may be 'bad' alleles which work against the organism's survival.

Suppose an individual possesses a good allele which gives it an advantage over other individuals. This individual will reproduce and pass its allele to its offspring, and they will pass it to their offspring – and so on. In this way good alleles spread through a population. It becomes part of the **gene pool**.

Now suppose an individual possesses a bad allele which puts it at a disadvantage. In this case the individual dies, or at least fails to reproduce, so its bad allele does not get passed to its offspring. It gets lost from the population for ever and ceases to be part of the gene pool.

Sickle cell disease: the allele that won't get lost

First read about **sickle cell disease** on page 272. The allele that causes this disease is very harmful. Few people who have the disease survive to reproduce, so you might have expected the allele to disappear. But it hasn't. In fact it is surprisingly common. Why?

For a long time the answer was not known. Then someone noticed that the much less serious **sickle cell trait** seemed to be particularly common in those parts of the world where malaria occurred. Was there a connection between the two? One way of finding out would be to study the occurrence of people with sickle cell trait and the occurrence of malaria. If there are lots of people with sickle cell trait in places where there are also lots of cases of malaria, this would suggest that the two are connected.

Such a study was carried out, and the results are shown in picture 6. You can see at once that there is a close similarity between the distribution of malaria and the sickle cell trait.

Why should they be connected? The answer is that people with the sickle cell trait have some protection against malaria. Many people die of malaria, particularly in areas where anti-malarial drugs are not used. However, people with sickle cell trait have a better chance of surviving, and they pass their sickle cell alleles to their children. So a large number of sickle cell alleles remain in the population.

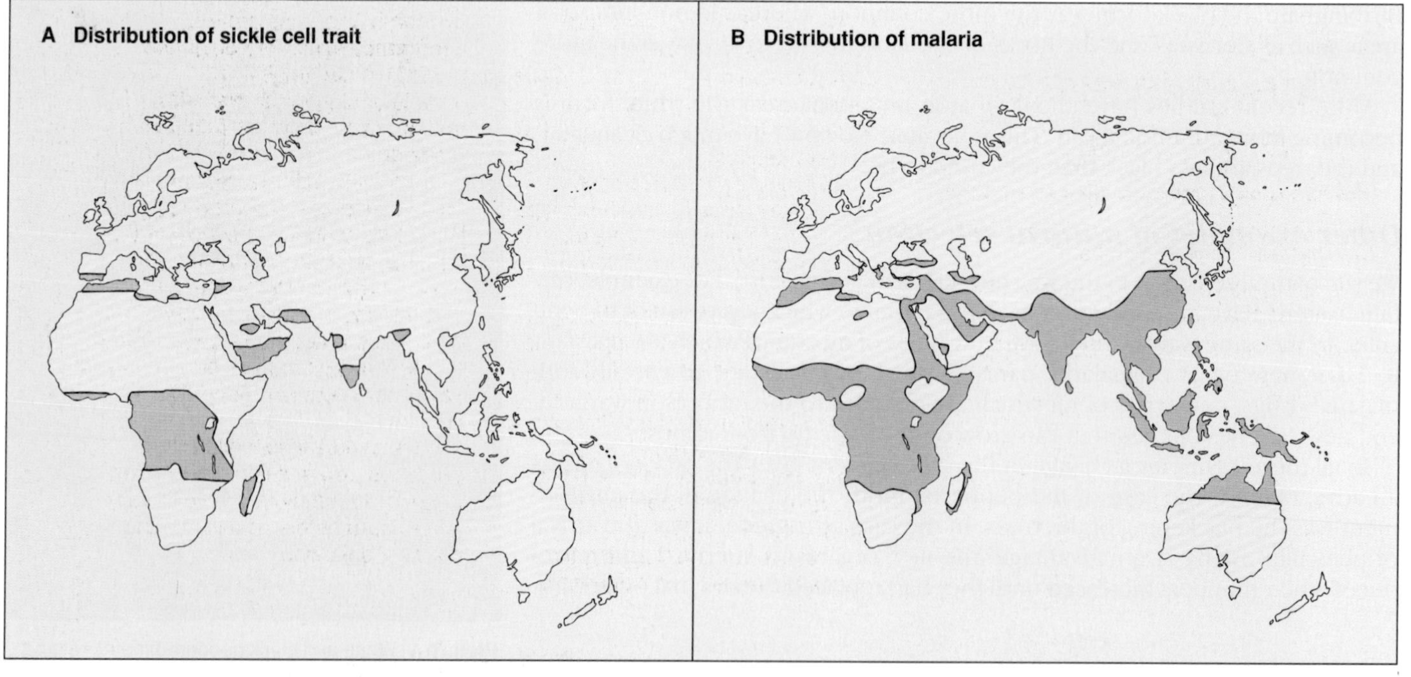

Picture 6 These two maps compare the distribution of sickle cell trait and malaria.

Activities

A Looking for common ancestors

Your teacher will give you a group of animals or plants, or photographs of them. Examine them carefully, and see if you can find any evidence that they are closely related in evolution. Do you think they all came from the same type of ancestor? If so, what might the ancestor have been like? Try drawing it.

If you think that one or more of the specimens in the group did not share the common ancestor, then say so with reasons.

Test your ideas by getting additional information from more advanced books.

Do you think the idea of common ancestors provides good evidence for evolution? Present your ideas to other members of your class.

There is more about common ancestors on page 221.

B Making sense of the distribution of the black peppered moth

Look at picture 5 on the opposite page. On a copy of this map, mark the positions of the major industrial towns which used to produce a lot of smoke.

On another copy of the map, draw arrows to show the direction of the prevailing winds.

On a third copy of the map, shade in those areas of the country which were likely to be heavily polluted by industrial smoke.

Does this exercise help you to explain the distribution of the white and black forms of the peppered moth?

Questions

1. Explain in no more than one page how evolution is thought to have taken place. Divide your account into three parts with the following headings: Variation, Natural selection, Reproductive success.
2. Warfarin is a rat poison. However, some rats are not affected by it, and their numbers are increasing. How would you explain this?
3. Doctors don't like giving a patient an antibiotic such as penicillin unless it's really necessary. Why?
4. Look at picture 2 on page 290.
   a. Suggest reasons why it is a disadvantage for an antelope to be (i) very short, and (ii) very tall.
   b. What sort of changes in the environment might make it an advantage to be very tall?
   c. If being very tall *did* become an advantage, what effect would this have on the shape and position of the curve in the graph?
5. Look at picture 6. What further studies could be carried out to confirm the close connection between the distribution of sickle cell trait and malaria?
6. A well known lady, on hearing about Darwin's theory, exclaimed: 'Descended from the apes! My dear, we will hope that it is not true. But if it is, let us pray that it may not become generally known.' This lady lived in the 1860s. How would you explain her attitude? Do similar attitudes towards scientific theories occur today?
7. Read how limbs are adapted for different jobs on page 221.
   a. In what way do these limbs provide evidence for evolution from a common ancestor?
   b. Use the theory of evolution to illustrate the difference between *evidence* and *proof*.

Engineers copy Darwin to evolve best design

Read the newspaper article on the right, then answer these questions:

1. What do you think 'fittest' means in the context of a gas turbine?
2. Why is a computer needed to design a machine by evolutionary methods?

DARWIN'S theory of evolution, which shows how plants and animals developed over successive generations, is being used by engineers to come up with more efficient designs.

Employing a computerised version of the evolutionary principle "survival of the fittest", engineers have improved the efficiency of gas turbines.

The technique, using computer software called genetic algorithms, "evolved" a gas turbine blade design, achieving twice the improvement in efficiency managed with conventional methods, said Miss Anne Finnigan, of the design and development agency Cambridge Consultants.

The agency is employing this computerised evolutionary design method to make an artificial intelligence system to diagnose faults in diesel engines.

"It is analogous to evolution, though much simpler," said Miss Finnigan.

A problem with developing the best version of a complex design like a turbine blade is choosing the best value for length, height, material and overall shape.

Changes to one factor can affect others and often the only way to get the best design is to fix one or two key parameters and keep the rest constant.

This makes life simple but produces poorer results. However, genetic algorithms are able to vary all the parameters at once to find the best solutions.

F10 Selective breeding

The dogs in the pictures were all produced by selective breeding.

Picture 1 Three breeds of dog. There are over 100 breeds of dog altogether. They have been produced by selective breeding from a wild wolf-like ancestor.

Picture 3 Sperm can be frozen and stored in 'sperm banks' like the one shown here. Embryos can also be frozen: the embryo is taken from the uterus of a selected female before implantation (or obtained by *in vitro* fertilisation as described on page 235) for future transfer to the uterus of another female.

What is selective breeding?

In the last topic we saw how nature selects the fittest individuals and rejects the less fit ones, and we saw how this can result in evolution.

The same thing can be done by humans. Suppose we have a population of animals or plants. We can select those with good qualities and let them breed, whereas those with poor qualities can be prevented from breeding. In other words, we *choose* the ones we want to breed – that is why it is called *selective* breeding.

Selective breeding is like natural selection but instead of nature doing the selecting, humans do it. To distinguish it from natural selection, we call it **artificial selection**.

How is selective breeding carried out?

If you are planning a breeding programme, you have to remember certain things. For example, it's no good trying to cross two different species. Different species can't normally interbreed – it is biologically impossible.[1] However, you can cross two different varieties of the same species – different breeds of dog, for example. In this way you may produce a new variety which combines the best qualities of the two original ones.

Another thing you can do is to stick to one variety and always choose the best individuals for breeding. You might choose the fastest Greyhounds or the biggest Great Danes. In this way you may gradually improve the quality of that variety.

Now for some details, animals first, then plants.

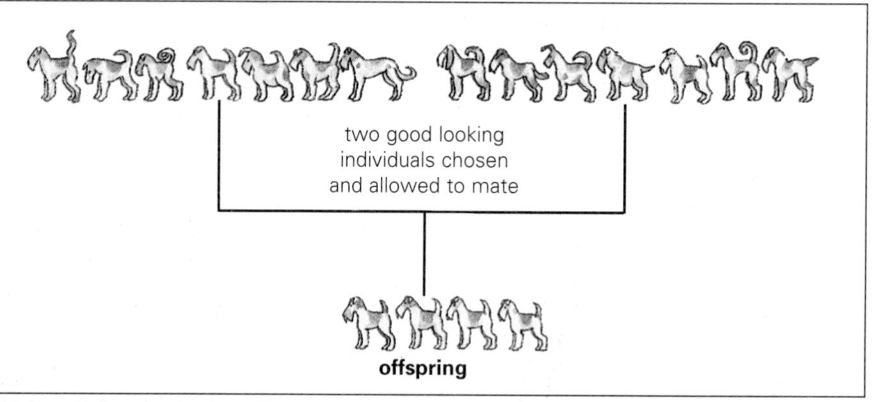

Picture 2 In selective breeding, particular individuals are chosen and allowed to breed. The others are prevented from breeding.

Breeding animals

First you choose a male and a female with the features you want. You then put them in a pen together and let them mate. This is the sort of thing farmers do with sheep.

Another way is to obtain some semen from the male and place it in the vagina of the female. This is called **artificial insemination**.

Artificial insemination is often used for breeding cattle and other farm animals. In the case of cattle, bulls are kept at special centres. Semen is taken from them by a vet, and either frozen or mixed with a preservative for storage. Frozen bull semen can be kept for a very long time and used when needed. Embryos can also be frozen for future use (picture 3).

[1] There are exceptions to this general rule. For example, the mule is a cross between a donkey and a horse. However, the mule is sterile and cannot breed.

Breeding plants

First you choose two plants with desirable features. Suppose they are the kind of plants which, left on their own, might pollinate themselves. How can you make sure that one pollinates the other?

The best way is to remove the anthers from one of the plants before it starts producing pollen (see page 240). This is quite easy to do with a needle and forceps. When the carpels of this plant are ripe, you pollinate it with pollen from the other plant, using a small paintbrush or probe (picture 4).

You must make sure that pollen from other plants does not get to your plant. So you cover the plant with a transparent bag, except of course when you are pollinating it.

Picture 4 A wheat flower being pollinated by hand. The pollen is transferred by means of a metal probe.

What has selective breeding done for us?

All our familiar breeds of farm animals, and domestic animals such as cats and dogs, have been produced by selective breeding (picture 5). So have the various varieties of garden plants, such as roses (picture 6).

In the same way plant breeders have produced new varieties of crop plants which are better than the older ones. For example, we now have varieties of wheat and rice which grow more quickly, give more grain and are more resistant to disease.

All this is good for humans. Improved crop plants, in particular, have led to enormous advances in agriculture. This has been particularly helpful in developing countries.

Outbreeding and its advantages

When you breed animals or plants, it is best to choose individuals that are not closely related. The offspring produced by crossing unrelated parents are called **hybrids**, and the process by which they are produced is called **hybridisation**. They are usually strong and healthy, and are most likely to reproduce successfully. Breeding from unrelated individuals is called **outbreeding**.

Picture 5 The main breeds of cattle seen in Britain: Friesians (top left), our most common breed, are used mainly for milk; Herefords (top right) are used for beef; Jerseys and Guernseys (bottom left and right) produce very rich milk.

Picture 6 Three varieties of roses. Selective breeding carried out over centuries has produced many different varieties of rose and more are created every year.

Picture 7 The poodle then and now. The top picture is a portrait of a working poodle painted by the artist George Stubbs around 1780. The lower picture is a photograph of a modern miniature poodle.

Inbreeding and its dangers

Sometimes breeders cross close relatives: a brother and sister perhaps, or father and daughter. This is called **inbreeding**.

Inbreeding is all right if it is only carried out every now and again. But if you go on doing it generation after generation, the offspring start to decline and show various defects. For example, they may be smaller than usual or less resistant to disease, or they may have certain physical abnormalities. Eventually they may fail to reproduce, and the line dies out.

The effects of inbreeding are more noticeable with animals than with plants. So if you want to breed animals, it is particularly important to cross unrelated individuals. This is what the best dog breeders do. Often several different breeders exchange dogs for mating, and in this way the bad effects of inbreeding are avoided. A highly inbred dog may have physical defects such as weak hips or a bad heart. It may have mental defects too. For example, it may suddenly turn savage and attack people.

An example of a highly inbred type of dog – the poodle – is shown in picture 7. The top picture is a portrait of a poodle, painted around 1780. It shows what poodles used to look like. In those days poodles were quite large. It was a working dog and was used for hunting ducks because it was a strong swimmer. Later it was used as a retriever. In the last hundred years or so, breeders have turned this breed into a pet – the miniature poodle shown in the lower picture. This dog is much smaller than its ancestor. It is soft and pretty, easy to keep in a small house or flat, and it fits snugly under its owner's arm. Unfortunately breeding for such a small size has resulted in a number of problems such as kneecaps which slip out of place, and hips which collapse.

It is sad to think that a dog which wins first prize in a show may have serious health problems as a result of the way it has been bred. Fortunately the British Kennel Club has introduced rules and regulations which should reduce the amount of inbreeding in the future. Hopefully the health and fitness of pedigree dogs will then improve.

What about humans? Inbreeding in humans is called **incest**. This applies to matings between brothers and sisters, and between parents and their children. In most countries incest is illegal. Laws against it go back thousands of years, and you can see that there is a good biological reason for them.

Activity

Investigating the different breeds of cattle

The various breeds of cattle in picture 5 were all developed by selective breeding for particular purposes. Find out as much as you can about one or more of these breeds. In particular how the breed was formed, and what its main job is. Don't just use books; if possible find a friendly farmer to talk to!

Write an illustrated account of your discoveries.

(Note: If you are particularly interested in some other domestic animal such as horses, you may prefer to investigate that instead.)

Questions

1 Explain the difference between natural selection and artificial selection.

2 What is artificial insemination? Think of as many reasons as you can why artificial insemination is better for cattle production than letting the animals mate in the normal way.

3 The various breeds of dog have been developed for particular purposes. Suggest five different uses to which different breeds of dogs are put. What particular features will each breed need to possess to carry out its job?

4 Thanks to artificial selection, we now have cattle with good milk yields.
   a What does the word *good* mean in this context?
   b Suppose your great ambition was to produce a new breed of cow with an even better milk yield. Outline your programme, step by step.

5 Inbreeding of livestock such as cattle is sometimes carried out in order to bring out a particularly good feature. However, it should only be done by very skilled breeders. What special skills do you think inbreeding requires?

6 Suggest a reason, to do with the genes, why repeated inbreeding should result in a decline of the offspring.

Producing a new variety of wheat

Britain used to import lots of high-quality wheat from Canada and still imports some. Here is an account of how the Canadians developed a new variety of wheat which was particularly well suited to their climate.

In the early 1900s, most Canadian farmers were growing a variety of wheat called **Red Fife**. This produced a lot of grain and the flour was good for bread-making. But Canada often has frosts in late summer which may damage the ripening crops, and Red Fife did not grow fast enough to be sure of avoiding the frosts.

So scientists decided to try to create a fast-growing variety of wheat. They did this by cross-breeding Red Fife with an early-ripening wheat from India. The result was a new variety of wheat which ripened about six to ten days earlier than Red Fife – in good time to beat the frosts. It was called **Marquis**. It was so successful that in a few years more than half the wheat grown in Canada was of this new type. Although no longer grown there it is the standard against which quality Canadian wheats are judged.

In Britain much of our wheat is **winter wheat**: the seeds are sown in the autumn, and the shoots come up before the winter sets in. Growth stops during the winter, and then starts up again the following spring. The advantage of winter wheat is that the young plants are already established by the time spring comes, so the wheat can be harvested earlier.

However, in Canada they have very hard frosts in the winter which would kill the young wheat plants. So most Canadian farmers plant their wheat in the spring. Marquis grows fast enough for the farmer to be sure of harvesting it before the late summer frosts occur.

As well as growing quickly, Marquis gives a high-quality grain. The flour produced from the grain contains a lot of gluten and makes good, well-risen loaves (see page 246).

1 Name one feature of Marquis which makes it particularly suitable for growing in countries with long winters.
2 Name one feature of Marquis which makes its grain particularly useful in bread-making.
3 Describe in detail how the cross between Red Fife and the Indian wheat would have been carried out.
4 What is gluten and why does it help to make 'good, well-risen loaves'?
5 It can take 12 years from when a new variety of wheat is first developed by scientists to when it becomes available to farmers. Why do you think it takes so long?

Capsticks' fleeces are just champion

Cumbrian hill farmers Edmund and Frank Capstick are leading the way in exploiting the full earning potential of wool. Their 450 Rough Fell ewes (females) are not only producing heavier fleeces but wool quality is in the top grade and yielding premium prices.

This year's wool clip was worth £1500 and won them the new reserve champion award in the CIBA Wool Producer of the Year competition. The whole flock is hand-clipped.

"Sheep have to be clipped no matter what the value of the fleece so you might as well have a good fleece to sell as have a bad one," say the Capsticks of Birkhaw, Sedbergh.

The brothers' dedication to producing top-quality wool spans 30 years of stock selection, i.e. choosing the right individuals to breed. Rejecting unsuitable individuals from the ewe flock and meticulous evaluation of fleece quality in stock tups (males) has paid dividends.

But fleece weight has also been addressed by breeding bigger "framier" Rough Fell ewes which are now yielding an average clip of 2.5 kg.

Tups and ewes are selected for white wool with long fibres; lambs showing dark fibres in the neck or body wool are not kept for breeding and all ewes are removed from the flock and sold after three crops.

"This maintains our high standard by selling ewes before coarse hair starts to appear in the fleece," said Edmund Capstick.

This year's clip fell into the British Wool Marketing Board's 714 grade and earned up to 108p/kg for hog fleeces which averaged 3 kg. Most Rough Fell wool this year was earning 102p/kg.

Ewes are hand clipped in late June to early July leaving on more of the wool that has grown since the spring than would be left if the flock was machine-clipped.

"We can often get a cold night, even at shearing time. By leaving a little more fleece on we help to avoid the risk of ewes becoming chilled which can affect milk yield," says Frank Capstick.

Liz Ambler of the British Wool Marketing Board, which organised a flock inspection, said: "The Capsticks are now producing a heavy, top-quality fleece which will always provide them with a clip capable for earning premiums despite market fluctuations."

Improving fleece quality while still retaining the sheep's shape and form has been a constant challenge for the Capsticks. Their swing towards a bigger type of Rough Fell has not only produced more wool but has helped to maintain the size and shape of prime lambs.

"Our mature ewes weigh 120–140 lbs. We achieve 150% lambing and want all our prime lambs away by the end of autumn. We averaged 40.77 kg for all lambs sold this year and had individuals up to 48 kg," said Edmund Capstick.

by Jeremy Hunt

Adapted from *Farmers Weekly* 29 December 1995

Picture 1 Edmund Capstick, champion wool producer of 1995, with one of his Rough Fell sheep.

Index

[Page numbers followed by 'a' refer to activities, those followed by 'c' refer to case studies]